86-000101 44

Marine Minerals

Advances in Research and Resource Assessment

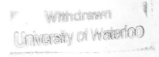

NATO ASI Series

Advanced Science Institutes Series

A series presenting the results of activities sponsored by the NATO Science Committee, which aims at the dissemination of advanced scientific and technological knowledge, with a view to strengthening links between scientific communities.

The series is published by an international board of publishers in conjunction with the NATO Scientific Affairs Division

A	Life Sciences	Plenum Publishing Corporation
B	Physics	London and New York
C	Mathematical and Physical Sciences	D. Reidel Publishing Company Dordrecht, Boston, Lancaster and Tokyo
D	Behavioural and Social Sciences	Martinus Nijhoff Publishers
E	Engineering and Materials Sciences	Dordrecht, Boston and Lancaster
F	Computer and Systems Sciences	Springer-Verlag
G	Ecological Sciences	Berlin, Heidelberg, New York, London, Paris, and Tokyo

Series C: Mathematical and Physical Sciences Vol. 194

Marine Minerals

Advances in Research and Resource Assessment

edited by

P. G. Teleki

U.S. Geological Survey, Reston, Virginia, U.S.A.

M. R. Dobson

University of Wales, Aberystwyth, Wales, U.K.

J. R. Moore

University of Texas, Austin, Texas, U.S.A.

and

U. von Stackelberg

Bundesanstalt für Geowissenschaften und Rohstoffe, Hannover, F.R.G.

D. Reidel Publishing Company

Dordrecht / Boston / Lancaster / Tokyo

Published in cooperation with NATO Scientific Affairs Division

Proceedings of the NATO Advanced Research Workshop on
Marine Minerals: Resource Assessment Strategies
Gregynog, Wales
June 10-16, 1985

Library of Congress Cataloging in Publication Data

NATO Advanced Research Workshop on Marine Minerals: Resource Assessment
    Strategies (1985 : University of Wales)
    Marine minerals.

    (NATO ASI series. Series C, Mathematical and physical sciences; vol. 194)
    "Published in cooperation with NATO Scientific Affairs Division."
    Includes index.
    1. Marine mineral resources—Congresses. I. Teleki, P. G. (Paul G.), 1937–   .
II.   North Atlantic Treaty Organization. Scientific Affairs Division. III. Title. IV. Series:
NATO ASI series. Series C, Mathematical and physical sciences; vol. 194)
TN264.N38   1985   553'.09162   86–31503
ISBN 90–277–2436–9

Published by D. Reidel Publishing Company
P.O. Box 17, 3300 AA Dordrecht, Holland

Sold and distributed in the U.S.A. and Canada
by Kluwer Academic Publishers,
101 Philip Drive, Assinippi Park, Norwell, MA 02061, U.S.A.

In all other countries, sold and distributed
by Kluwer Academic Publishers Group,
P.O. Box 322, 3300 AH Dordrecht, Holland

D. Reidel Publishing Company is a member of the Kluwer Academic Publishers Group

PREFACE

     Discoveries of new types of marine mineral occurrences during
the last decade, and specifically the massive sulfide deposits at
spreading ridges on the ocean floor, have significantly advanced
geologic concepts about the origin of ore deposits in a very short
period of time. These discoveries also renewed interest in all
marine mineral occurrences including the well-known manganese
nodules, and led to more wide-ranging and thorough examination of
cobalt-rich manganese crusts, expanded mapping of phosphorites of
continental shelves, and the initiation of several new surveys for
placer minerals in shallow waters. The result of these activities
is already noticeable in an increasingly broader variety of minerals
being found on and below the ocean floor.
     This upsurge of scientific interest and research in marine
minerals provided the impetus to organize an Advanced Research
Workshop under auspices of the NATO Science Council and its Special
Program Panel on Marine Sciences. The workshop was held in the
United Kingdom at Gregynog Hall of the University of Wales, June
10-16, 1985, under the theme "Marine Minerals--Resource Assessment
Strategies." The timing of this workshop was propitious in many
ways. First, marine surveys and expeditions to chart the mineral
resources of the world's oceans had increased in number in recent
years, involving a growing number of nations interested in obtaining
firsthand information. Secondly, legal and jurisdictional
developments, such as claims made by the United Nations under the
Law-of-the-Sea statutes for marine minerals in the deep oceans and
claims made by maritime nations for minerals within 200 nautical
miles of their shores had accentuated interest in the potential
value of deposits and in the extractability of the valuable
constituents. Recent, well-publicized discoveries have fueled
efforts to establish national claims and rights to future resources
because considerable economic benefit is anticipated from their
eventual exploitation. However, the depth of scientific knowledge
about the genesis, location and extent, constituent minerals, and
engineering know-how about mining and beneficiation is not yet
commensurate with desires to extract and exploit these deposits.
The practical aspects of raising capital for exploitation depends on
accurate economic evaluations. The value, or worth of a marine
mineral occurrence at any time, is determined by means of resource
assessment, availability and cost of mining technology, and
information about market conditions of the commodities to be
extracted. Costs and market conditions are constantly fluctuating.
     Adequate resource assessments of marine mineral deposits have

been few; examples to date have involved manganese nodules of the equatorial Pacific Ocean and the metalliferous brines of the Red Sea. While the economic evaluation of surficial marine-mineral deposits, such as crusts and nodules, should be relatively easy as these deposits are practically two-dimensional, a thorough assessment of the crusts has yet to be made. Buried deposits, such as placers, phosphorites, and massive sulfides are much more difficult to distinguish from their host marine sediments or rocks and because knowledge about their vertical distribution is missing, their resources are not easily quantified. The uncertainty of resource estimates presents mineral economists with great problems in performing economic evaluations of these deposits.

Conventional seismic surveying, grab sampling, and coring methods, that are used to determine the extent, composition, quantities, and grade variations of placer or phosphorite deposits are slow and costly; the need to invent methods for their rapid and less ambivalent detection is clearly urgent. Resource assessments of placers and phosphorites also depend on much better understanding of their genesis than exists today. Both types of deposits however, could be exploited in the near future as their locations commonly are in shallow water, not too distant from shore, and technology to mine the deposits exists.

Renewed interest in manganese crusts, on account of their cobalt content, has been demonstrated by the frequency of recent surveys in the Pacific Ocean. When the extent and valuable metal content of these crusts and the appropriate technology to mine them become known, their economic worth will be more easily determined. For now, technology to extract samples or to measure crust thickness is sought; this may lead to the design of suitable mining equipment and methods.

Metalliferous sulfide occurrences are being noted with increasing frequency; their status is just passing from the stage of serendipitous discoveries toward more systematic surveys and mapping of likely occurrences along spreading centers of mid-ocean ridges and back-arc basins. Attempts to estimate the resources of these deposits are premature, as information about the composition, extent, and volume of polymetallic sulfide occurrences is still rudimentary. As lateral or vertical variations in mineralogy are virtually unknown so far, little can be said about metal grades within these deposits.

Although the economics of marine polymetallic sulfide deposits may preclude near-term development, investigations of these deposits yield benefits by improving the understanding of and exploration strategies for mineral deposits on land that were formed in a similar manner.

Consequently, with the exception of manganese nodules, surveys designed to find and map marine mineral deposits are largely at the reconnaissance stage. In research, emphasis continues to be placed on deciphering the origin of deposits, on understanding their relation to the geologic framework and tectonic settings of their location, and on gaining knowledge about the physical, chemical, hydrological, and biological processes governing their origin, deposition, and alteration. The ocean-going tools of the marine scientific community are growing in number and sophistication; nonetheless, they are still not sufficient in scope, number, and sophistication to obtain all the data needed now to characterize these resources. Surveying and measurement techniques need to be improved, sampling and drilling methods suitable to explore these deposits need to be invented, and long-term monitoring of hydrothermal tectonic and volcanic processes at spreading centers and back-arc basins, using the full suite of scientific disciplines, should be implemented.

The future mining of marine mineral deposits was certainly the rationale behind the jurisdictional claims made for them by the United Nations and by individual nations. Improvements and advances in technology are needed before the marine mining industry can fulfill aspirations to extract mineral resources. Surveying techniques and methods of stationary or towed measurements exist, by and large, for scientific "exploration," but to suit the need of the mining industry for closely spaced, three-dimensional, site-specific information many of these techniques will have to be improved or refined. Data analyses will have to become more sophisticated, and more analyses will have to be performed at sea, on site. What tools, methods, and approaches are appropriate for either surveys or analyses remain largely defined because their choice depends heavily on better and broader knowledge about the characteristics of deposits, their location, and extent.

The bridge between the science of marine minerals and the engineering of future commercial exploration and exploitation is built in part by resource assessments. Assessments bring forth specific information on the value of mineral deposits and promote them from the esoteric to the practical. Methodology exists for assessing mineral resources; whether such may be suitable to apply to marine mineral resources is an important topic in this book. It is safe to say that substantial improvements are needed in this field too.

Most of the aforementioned issues were discussed in the NATO workshop: knowledge about the occurrence and origin of four classes of deposits, the status of surveying and measurement techniques, the needs for improvements in these techniques, the status of resource-assessment methodologies, and the needs of those who may exploit the mineral wealth of the oceans eventually. Participants in the workshop, which included geologists, geochemists, geophysicists, mineral economists, oceanographers, and mining experts, reviewed the current state of knowledge and probed future needs for information. The disparate level of information existing about each mineral commodity group was noted. An even more arduous task was to suggest the proper approaches to resource assessment for each commodity or type of mineral occurrence. Prior experience in this field is slim and data needed for such assessments are insufficient. The results of these deliberations are conveyed by the reports of the working groups. Thirty two background papers

summarize much of the state of knowledge about marine minerals and
approaches to resource assessment. These reports and contributed
papers are eloquent testimony of the participants' dedicated,
intensive interest in the subject of marine minerals, and their
findings and recommendations ought to guide research and development
during the next few years.

   We, the editors, also hope that the contents of this book will
expand professional interest in the subject of marine minerals. We
are well aware, however, that in this fast-moving era of discoveries
and developments in the oceans, this volume could soon be superseded
by results of exciting new findings.

P.G. Teleki
M.R. Dobson
J.R. Moore
U.v. Stackelberg

ACKNOWLEDGMENTS

This volume represents the sincere interest and collaboration of many individuals. Special thanks are due the participants in the workshop who had labored hard to provide new and significant information about their research. Particularly gratifying was the interchange of concepts, information, and suggestions between the geologists, geophysicists, geochemists, oceanographers on one hand with the resource assessment experts, mineral economists, and mining experts on the other.

Special appreciation is due the NATO Science Council and its Special Program Panel on Marine Sciences, the U.S. Geological Survey, and the Bundesanstalt fur Geowissenschaften und Rohstoffe (BGR) of the Federal Republic of Germany, all of whom cosponsored the NATO Advanced Research Workshop. Individuals meriting special acknowledgments are Dr. Ferris Webster on behalf of the NATO Panel on Marine Sciences, Dr. Terry Offield of the U.S. Geological Survey, and Dr. Friedrich Bender of the BGR.

Thanks are due the University of Wales who had provided the excellent facilities for the meeting, the Department of Geology at the University helping with logistics, Mrs. Sonia Dobson for her untiring efforts to organize and administer the activities, Mrs. Connie Conte and Mrs. Helana Cichon in organizing travel, and Mrs. Sue Pitts for her exemplary efforts in typing the summary reports and most of the contributions contained in this book.

CONTENTS

Phosphorites

Manganese nodules and crusts

Metalliferous sulfides

Resource assessment and geostatistics

REPORT OF THE WORKING GROUP ON PLACER MINERALS

Robert M. Owen, Chairman
H. Kudrass, Rapporteur
Dennis Ardus
John DeYoung
David B. Duane
J. Robert Moore
Angela Frisa Morandini
C. Perissoratis

INTRODUCTION

Placer deposits are detrital sediments which contain potentially economic concentrations of valuable heavy minerals concentrated by hydraulic processes. Among the four types of marine mineral deposits that are the subject of this book, only placers are currently mined. Assessment of the resource potential of these deposits presents the same problems that are inherent to land-based placer deposits, namely, the diversity of placer types and the mineral commodities contained therein, the difficulty in defining a deposit, and the effects of the vagaries of mineral market conditions. Exploration and assessment in the marine realm, compared to onshore deposits, also presents a different set of problems because offshore deposits (1) are concealed by water as well as by solid material; (2) may not fit the genetic descriptions that are based on knowledge of onshore deposits; (3) are subjected to submarine processes that effect their alteration and preservation; and, (4) are in an environment that makes sampling more difficult and less precise than that on land.

ORIGIN AND DESCRIPTION

Placer deposits can be grouped into three genetic categories: fluvial, estuarine, and beach deposits. Fluvial and beach deposits are found not only in modern depositional environments, but also in ancient raised beaches of emerging coastlines, and in submerged beaches and drowned stream channels and estuaries extending onto the continental shelf. Deposits of each category are composed of one or more mineral species in various concentrations.

Placers that formed in the fluvial environment typically consist of minerals with a density greater than 6.5 g/cm^3, such as platinum, gold, cassiterite, and wolframite, and by virtue of their high density

1

P. G. Teleki et al. (eds.), Marine Minerals, 1–8.
© *1987 by D. Reidel Publishing Company.*

are concentrated along the deepest parts of river beds. In addition to
those representing bedload transport, deposits also form by elutriation
where detritus from weathered source rocks is removed by rain, wind, or
marine currents, leaving a concentrate of chemically stable heavy
minerals on the erosional surface.

In the coastal environment, the beach and the neighboring dunes
are usual sites for the deposition and subsequent enrichment of those
minerals, whose density ranges from 3.0 to 5.0 g/cm^3. Mineral suites
are usually composed of rutile, leucoxene, ilmenite, monazite,
magnetite, zircon and garnet; at some locations chromite and diamonds
are found in beach sands, as well as fine-grained gold and platinum.
The process known as "hydraulic sorting" is responsible for aggregating
grains of heavy minerals into distinct size and/or density fractions.
This concentration process eventually leads to the formation of beach
placers.

Although Ti-group heavy minerals are the most commonly occurring
placers, enrichments of ultra-fine-grained (5-50 microns) gold and
platinum group metals have been found during the past decade in
quiescent coastal embayments in Alaska and the Southwest Pacific Ocean.
Deposition of ultra-fine grained placer particles is not governed by
the same processes as those that control high-energy (lag) placer
deposits; instead, the settling behavior is dominated by physical and
chemical phenomena of the grain surface, chiefly high surface tension,
electrical charges, selective coating or adsorption, catalytic activity
and surface geometry. Because of these characteristics, ultra-fine
particles of ore minerals, including native gold and platinum, are
deposited with gangue minerals of the same size. Little is known about
the geochemistry or about the mineralogy of the source rocks of these
precious-metal deposits. It has been suggested these deposits may form
in estuaries where fine-grained fluvial particles agglomerate upon
encountering the relatively high electrolyte concentrations in saline
water. Hence, these deposits could guide reconnaissance surveys toward
related source rocks and fluvial placers.

MINERAL DEPOSIT POTENTIAL OF SHELF AREAS

All types of placer deposits could have been formed on the
Continental Shelf, when the shelf was exposed to subaerial processes
during much of the Pleistocene epoch. The probability that any of
these deposits could have been preserved varies from place to place
and, within each placer category,depends on the present and
paleo-geologic and -oceanographic setting of that area on the shelf.
The placer deposits are, thus, closely related to the geological
history of shelf sediments.

The first step in assessing the mineral prospectivity of an area
on the shelf is to review existing data from publications and files on
field measurements concerning the heavy-mineral assemblages and their
distribution along the coast and within the rivers. If the source
rocks for fluvial placer deposits occur near the coast or on the shelf,

it is necessary to probe the lowermost sedimentary fill of river beds and/or the lag sediments on the floor of tidally scoured channels. Traces of mineralization could instigate investigations further offshore, using, e.g., seismic profiling and coring, as these fluvial deposits are very likely to be preserved on the shelf also. However, the distal portions of most shelves consist of thick sedimentary wedges, and if source rocks and fluvial placer deposits exist there, these may be deeply buried.

When placer deposits are found along a modern beach or in nearby Pleistocene beach ridges, the probability that similar deposits also exist on the shelf depends mainly on the shelf's sedimentary history during the Holocene and Pleistocene interglacial transgressions. On narrow shelves, beach barriers with their placer deposits were probably transported shoreward during the transgression. On shelves of this type investigations should, therefore, concentrate on locating trap structures, among which are former scour channels, barrier islands, headlands, and reefs, all of which might contain residual segments of the migrating barriers.

On broad shelves, and especially those subjected to high sedimentation rates of terrigenous detritus, former beach placer deposits could still be present in disseminated form. As long stillstands of sea level, by and large, provided the necessary time to form beaches and other environments amenable to placer formation, a very precise sea-level curve (and calibration of the oxygen-isotope curve) for the last glacial periods would be a useful exploration tool.

In exploring for placers, particularly in the shallow, coastal-zone waters, knowledge of the adjacent uplands geology is important, and its investigation should precede offshore surveys. The basic requirements for marine placer formation--provenance/source rocks, weathering and erosion, transport and processing conduits, and a depositional site--all relate to the adjoining geologic framework. In the case of exploring for seaward extensions of terrestrial placer deposits (formed during the Pleistocene under lowered sea-level), the knowledge of local terrestrial geology does, indeed, provide for directly extending the survey seaward.

EXPLORATION METHODS

Positioning and Bathymetry

Navigational control must provide adequate positioning accuracy in order to locate and relocate ships and equipment, to enable correlating geophysical, sampling and coring operations and data, at a precision sufficient for resource appraisal. The accuracy of existing systems is adequate for these tasks; the major limitations are the awareness about alternative positioning methods and about technical developments in other fields that could be adapted to placer exploration. The progressive development of satellite-based global positioning systems along with their increased accuracy will improve locating ships or equipment at sea.

    Information on bathymetry is an integral part of surveying or
mining. Most bathymetric data exists in the form of hydrographic
charts, but echo sounding, as an integral part of geophysical
traversing, or preferably swath bathymetric surveying with its rapid
areal coverage, should supplement this data base locally.

Geophysics

    The location and three-dimensional extent of deposits can be
discerned by using one or more acoustic profiling systems, the
selection of which depends on the thickness and lithology of the strata
being investigated. The highest resolution compatible with adequate
penetration should be employed if necessary by using several systems
simultaneously. Using side-scan sonars in addition to acoustic
profiling is also desirable to determine seabed morphology and perhaps
sediment texture. It may be advantageous to tow a magnetometer to
gather information about magnetite concentrations and buried rock types.
Gamma-ray spectrometers or other radiometric equipment towed along the
seabed are new tools to detect the presence of certain types of
surficial deposits.
    Down-hole geophysical sensors, such as natural-gamma probes, seem
to have limited use in placer studies, because the thickness of these
deposits is considerably less than the sensors' ability to detect them.

Sampling and coring

    In coring, a principal objective should be the recovery of
continuous undisturbed cores; such cores are needed to fully calibrate
the geophysical data, to determine stratigraphic sequences, lithology
of sedimentation layers and modes of deposition. Currently, this is
done by vibracoring or to a lesser degree by rotary drilling.
    As vibracoring in sands and in coarse sediments will not always
achieve the desired depth of penetration, and insofar that large
volumes of material may occasionally be needed to be retrieved, systems
employing air lift, jet lift or counterflush, at times in conjunction
with vibratory methods, can be used. However, these must be monitored
carefully to ensure that representative bulk samples are recovered.
Incremental penetration/recovery techniques are means of achieving some
control in the recovery of representative samples. Conventional
techniques employed to obtain surficial sediment samples rarely produce
an acceptably representative sample either, because there are
imperfections in extricating material from the seabed in a
non-selective manner and because the fine material in cores, en route
to the surface during core recovery, becomes winnowed.
Significantly, there is a strong need for establishing better
communication between designers and users of the various sampling and
coring techniques in the pursuit of optimal performance.

Geochemistry and Mineralogy

Geochemical and mineralogical laboratory methods currently in use are considered to be adequate for analyzing different types of placer minerals. However, improvements are needed in current methods of chemical analysis and in separating aboard ship heavy minerals from very fine-grained particulates. Recent and proposed advances in the technology of geochemical-analysis systems towed near or along the seafloor are expected to influence marine placer exploration significantly.

Particulate matter trapped in the surface microlayer of stream waters (the upper 200 microns of the water column) may be representative of drainage basin lithology because it is less subject to chemical or physical segregation during transport than suspended matter in the water column. Chemical and mineralogical analyses of samples of microlayer particulates should be especially helpful to studies of regional geology, especially in areas where the lithology of possible source rocks is not well known, or where rough terrain limits the mobility of exploration teams.

Geostatistics

Compared to other types of marine mineral deposits, marine placer bodies are relatively small and highly localized, and geostatistical analysis, as an exploration tool in the search for marine placers, appears to hold much promise in detecting and delineating these deposits. Applied statistical techniques can be useful to analyzing geochemical data from reconnaissance-type surveys. Clearly, subtle trends of both pathfinder elements and target (commodity) elements can be enhanced by applying such techniques. Incorporating search theory in the design of exploration survey cruise plans should help to minimize the duration and costs of exploring shelf areas. Simple correlation and regression analysis techniques have been applied to the relationship between geochemical signals used in exploration and the lithology of the source rocks from which the placer minerals are derived. However, these methods provide little insight into the possible cause-and-effect relationships that may underlie an observed statistical association. Multivariate statistical techniques, such as comparative principal-component analysis and end-member composition-factor analysis, could be useful in addressing these problems.

Various algorithms exist already for modelling regional sediment dispersal patterns, and these appear to be useful to improving the understanding of marine placer genesis. Probably the most significant advance in this regard has been the introduction of linear programming methods, with which source areas and sediment dispersal patterns can be identified. Preliminary studies have shown that the sources and dispersal pathways of placer minerals in a relatively small coastal depositional regime can be determined and mapped by statistical treatment of the several types of geochemical data. While such an

approach has been applied with good success to the noble metals, such
as platinum and gold, the lower-density placer minerals, such as
rutile, chromite, cassiterite and scheelite, for example, have not yet
been so tested. Hence, future research on placer exploration should
include application of geostatistical techniques to the study of
closely-sampled placers containing ore minerals of varying specific
gravities at several sites with different geologic settings and
hydraulic regimes.

LABORATORY PROCESSING

     The current techniques for the beneficiation of placer minerals
are hydrogravimetric, magnetic, and electrostatic separation. These
techniques are suitable for processing coarse grained particles;
however, particles smaller than about 40 microns are very difficult to
separate. For those very fine particles that are commonly present in
placers associated with low (hydrodynamic) energy environments, it must
be ascertained whether the results of chemical analysis actually
represent heavy mineral constituents or non-crystalline material. In
case of fine-grained minerals, flotation or processing using
superconducting magnets might be suitable ways for enrichment. Further
research is needed in this area.
     Ultra-fine particles are not amenable to gravity-based separation,
such as by means of sluices, jigs, spirals, etc.; therefore, other
methods, mainly chemical recovery, must be used. This is one of the
thorniest problems encountered in mining muddy marine placers.

ENVIRONMENTAL CONCERNS

     Marine placer mining takes place in open, relatively shallow
waters, and the substrate is unconsolidated sediment. Biota of this
region have adapted to an environment characterized by episodes of
sediment movement of varying intensity, including occasional high
levels of turbidity caused by storm-driven waves and currents.
     Most placer deposits are associated with sand and gravel and
recovery of placer deposits can be considered analogous to the
extraction of these two sediment types. Where marine sand and gravel
have been dredged for use in building aggregates and for beach
nourishment, adverse environmental consequences have been largely
limited to benthic biota in the dredged area. Research has shown that
recolonization of the seafloor in and adjacent to mined areas occurs
within a short time.
     It is judged that there need be no long-term detrimental
environmental damage on account of marine placer mineral extraction, if
mining is carefully planned and monitored. Furthermore, it is probable
that any short-term effect could be lessened by carefully scheduling
dredging operations. Toward that end, it would be wise to involve
marine ecologists in the environmental assessment and particularly in
the process of planning mining operations.

MINERAL - RESOURCE ASSESSMENT

Three recommendations concerning the assessment of offshore placer resources, either as individual deposits or as groups of deposits present in a large region, address some of the problems inherent in the appraisal of any poorly known resource, especially those beneath the ocean. The suggestions below should lead to assessments that convey useful information to its users.

First, good communication between the prospective users of resource assessments, namely, those who prepare the assessment report and those who collect and analyze field and laboratory data, is essential in order to make efficient use of data collection opportunities and to focus assessment methods on a desired data.

The geoscientist needs to know what kinds of data are needed for resource assessments, what types of samples or measurements are desirable, and what density of observations or estimates are required (and in how many dimensions). Sampling programs that enable geologists to achieve a multitude of scientific goals and also satisfy the data requirements of resource assessment, necessitate an early and continuing dialogue between the earth scientist and the resource analyst/mineral economist. Neither all offshore areas nor all placer mineral occurrences discovered should be subjected to resource-assessment studies routinely. The most appropriate time for a review, at which field sampling plans as well as data needs are evaluated, is after the reconnaissance phase, but before detailed data collection or detailed exploration.

A second recommendation concerns the methods used for placer resource assessment. A comparative evaluation of resource-assessment methods currently applied to onshore placer deposits should provide insight into their suitability of use with offshore placers. However, offshore deposits that were formed by a succession of fluvial, coastal, and submarine processes and became preserved, may be particularly difficult to evaluate with standard deposit-assessment approaches.

The third recommendation is that the limits of error resulting from all previous steps must be accounted for in the resource appraisal. The appearance of several significant digits in analytical data or in resource estimates may give a false impression that the geologic source data were highly precise and accurate. Resource appraisals should reflect uncertainties of estimates by using error bars or confidence intervals.

SUMMARY OF RECOMMENDATIONS

A. Origin of placer deposits

1. Refinement of the Pleistocene sea-level curve, to permit determining former stillstands of sea level better on the present day shelf, and at sites where placers could have formed, is strongly needed.

2. Conceptual models of concentration of placer minerals and
    preservation of placer deposits need to be confirmed.

3. Ultrafine materials, that may be viewed either as a resource or
    an indicator of other deposits, still require development of
    genetic models, assessment techniques, and beneficiation
    processes.

4. The hypothesis, that the particulate composition of the
    surface microlayer of the seabed will indicate provenance
    of placer minerals, should be validated.

B. Exploration methods

1. Better communication is needed between designers and users of
    the various sampling and coring techniques available for
    exploration and assessment.

2. Planning of data acquisition in reconnaissance and evaluation
    programs should include a valid statistical approach; the data
    obtained should have confidence limits clearly indicated.

C. Resource assessment

1. The analog provided by resource appraisals on onshore deposits
    should be applied in the calculation of offshore resource
    estimates.

2. Timely communication between those who collect and analyze data
    and the preparers and users of resource assessments is
    essential.

3. Limits of error from previous steps in data collection and
    analysis should be built into the assessment.

REPORT OF THE WORKING GROUP ON MARINE PHOSPHORITES

Stanley Riggs, Chairman
Robert Whittington, Rapporteur
Jean-Paul Barusseau
William Burnett
Ulrich von Rad
Richard Sinding-Larsen
Paul Teleki

OCCURRENCE OF PHOSPHATE

Various types of phosphate deposits occur within the marine environment. The location, geometry, physical and chemical characteristics of each type of deposit varies as a function of extreme differences in their genesis. The exploration, development and resource potential of each type of deposit will consequently also differ. Thus, any resource assessment strategy for phosphate must take first into account its origin. The following brief summary of major types of deposits, accompanied by Table I, outlines five categories of marine phosphate occurrences.

Much of the knowledge about marine phosphorites has been derived from studying fossil deposits occurring on land. Comparison of this data base with the rapidly accumulating information about phosphorites in the modern marine environment has led to a good understanding of the genesis of various types of phosphorites, however, the level of this understanding is unequal (Table I). Most available information is for continental margin type deposits, which have many examples exposed on land and consequently, have been studied in considerable detail for many decades.

A tremendous increase in research on models of phosphorite formation has taken place during the past decade. This has been inspired partly by the increasing world-wide need for phosphate fertilizers necessary for improving and expanding world food production and partly by the inception of an International Geological Correlation Program (IGCP) on phosphorites. IGCP Project 156 on Phosphorites, sponsored by UNESCO and the International Union for Geological Sciences (IUGS), has stimulated research in all scientific areas of phosphate investigation on land and under the sea. A four volume series, containing the results of this research, are being published during the next few years by Cambridge University Press and consist of the following: 1) Proterozoic and Cambrian Phosphorites, edited by Cook and Shergold, 1986; 2) World Phosphate Resources, edited by Notholt,

9

P. G. Teleki et al. (eds.), Marine Minerals, 9–19.
© *1987 by D. Reidel Publishing Company.*

TABLE I. OCCURRENCES OF MAJOR MARINE PHOSPHORITE DEPOSITS*

TYPE OF DEPOSIT	EXAMPLES	CHARACTERISTICS OF DEPOSITS			RESOURCE POTENTIAL
		Geometry	Mineral Associations	Texture	
CONTINENTAL MARGIN a. Inner Shelf & Coastal Area b. Outer Shelf c. Upper Slope	Congo & Baja California SE United States Peru/Chile	3D**, Wedge-shaped, Regular, 10^3 to 10^4 km^2	Dolomite, Silica, Organic matter, Mg-rich clays, Glauconite, Manganese	Sands	Very Large
CONDENSED SECTIONS a. Plateau (Pavements) b. Rise (Nodules)	Blake Plateau & Agulhas Bank Chatham Rise	2D**, Highly irregular, 10 to 100 km^2	Manganese, Glauconite, Calcite	Slabs & Nodules Nodules	Moderate Moderate
INSULAR	Christmas Island Matiava Atoll Nahru Island	3D**, Irregular 10's km^2	Dolomite, Calcite	Sand & Nodules	Local to Moderate
SEAMOUNT	Mid-Pacific Mountains	2D**, Unknown, 10's km^2	Calcite, Fe-Mn, Volcanic material	Slabs & Nodules	Unknown
EPICONTINENTAL SEA	No contemporary seafloor examples known; a fossil example is the Aptian-Albian of Saxony Basin	2D**, Irregular 10 to 100 km^2	Calcite, Silica, Siderite, Organic Matter	Sand	Local

* This is a generalized table--individual deposits may vary in detail.
** 2D means essentially two-dimensional bodies with bed thicknesses of $<$1 meter.
    3D means essentially three-dimensional bodies with bed thicknesses of $>$1 meter.

Sheldon, and Davidson, in press; 3) Genesis of Neogene and Recent
Phosphorites, edited by Burnett and Riggs, in preparation; and 4)
Cretaceous and Paleogene Phosphorites, edited by Lucas and Al-Bassam,
presently being formulated.

Continental Margin Deposits

Continental margin phosphorites presently supply 70 to 80 percent
of the world's need for phosphate and include major deposits throughout
the world. These deposits, whose ages range from the Precambrian
through the Holocene, are products of major global episodes of
phosphogenesis. Although many aspects of their genesis are still
unclear, such as the mechanism of phosphate precipitation, rates of
pellet formation, role of bacteria, etc., it is widely accepted that
these deposits formed in areas of enchanced biological activity
associated with zones of upwelling. It is recognized that mechanical
concentration of phosphate grains is occasionally important in
producing an economic deposit, although the origin of many deposits is
clearly primary.
    The probability is high that continental margin deposits of
Tertiary age also occur beneath the coastal zone, shelves, and upper
slope environments throughout many of the world's present continental
margins between 45° north and south latitudes. Most of these deposits
occur buried in the subsurface; however, the deposits may locally occur
either in outcrop or in the shallow subsurface below thin and variable
shelf sediments which contain reworked phosphate in varying
concentrations. The phosphate model for southeastern United States
(probably the best understood continental margin phosphate system at
this time) is applicable to many other Tertiary continental shelves as
described by Riggs (this volume) and should be tested at selected sites
on other margins.

Condensed Sequence Deposits

Condensed sequences form on major plateaus and rises at
intermediate oceanic depths (100's to 1,000's of meters).
Occasionally, elevated levels of cobalt, platinum, and manganese are
associated with these deposits. Condensed sequences form by layered
accumulation or by replacement of carbonate on surfaces subjected to
long periods of non-deposition. The exact circumstances which lead to
this type of phosphatization are unknown. The Chatham Rise
phosphorites off New Zealand are a well known example of a condensed
sequence; even though these deposits have been studied in considerable
detail, many aspects of their genesis are still not understood
completely.

Insular Deposits

Insular deposits form when avian guano accumulated on oceanic
islands and interacted chemically with underlying bedrock, particularly
limestone. If these islands subsequently become submerged, then the

associated deposits become potential shallow marine resources. To
improve our understanding of these deposits, more precise age-dating
techniques are essential, especially for relating periods of
phosphogenesis to paleoceanographic and paleoclimatic conditions.

Seamount Deposits

Seamount phosphates form contemporaneously with manganese on
oceanic seamounts at intermediate oceanic depths (100's to 1,000's of
meters). The distribution and characteristics of these deposits, as
well as their resource potential, are very poorly known due to few
studies and inadequate sampling techniques. Seamount deposits are
often associated with cobalt-rich ferromanganese crusts. They are
clearly the replacement type, but circumstances which led to the
phosphatization of limestone and other materials remains unknown. To
improve the understanding of these deposits, analytical procedures must
be developed for more precise absolute and/or biostratigraphical dating
for correlating periods of phosphatization with paleoceanographic
events.

Epicontinental Sea Deposits

Epicontinental sea phosphates are not known to form in modern
marine environments. However, because they formed repeatedly during
past geologic times, this type is included here as a potential source
that may yet be discovered in recent marine sediments.

RESOURCE ASSESSMENT

Hard-mineral resource assessment must be done separately for each
type of deposit and requires the application of specific geologic
information in order to define the phosphate resource potential. For
example, continental margin deposits require quantification of the
spatial relationships of large-scale temporal and geologic parameters,
especially those that indicate paleoceanographic conditions necessary
to delineate favorable areas of previously formed phosphate deposits.
Within these areas the accurate description of the tectonic setting,
and its relation to paleoceanographic and paleoclimatic conditions, are
important steps in defining sites where favorable sediment facies are
expected to occur. Endowment of a target area can then be estimated by
assessing the percentage of sediment volume which may be suitable to
contain recoverable quantities of phosphate.

A resource assessment strategy as described by Harris (this
volume) can be used for deposits such as occur on seamounts. By
quantifying information on bathymetry, paleoceanography, and seamount
morphology, and by identifying favorable geologic settings together
with parameters of direct or indirect abundance, a regional assessment
of the potential phosphate and associated platinum resources can be
made. Before the full benefits of such a resource assessment can be
realized, additional detailed geologic information from these
environments is required.

EXPLORATION

Surveying and sampling schemes rarely consider the fundamentally different requirements of geologists and resource analysts. The first phase of exploration should consist of a reconnaissance survey with widely spaced grid lines, and an initial program of sampling based upon current theories of phosphogenesis. Layout of ship tracks, areal density and volume of samples, and navigational accuracy should be carefully considered to include requirements of both the geological evaluation and the future estimation of grade, tonnage and other characteristics of phosphate deposits. The second phase should delineate the deposit, and the third phase should provide data for the exploitation of the phosphate, which will require navigational accuracy better than 10 meters.

Reliable navigation is critical to determining the quality of marine data for use in resource evaluation and development. Satellite-aided navigation, integrated with doppler sonar, is the primary method of position fixing in open ocean areas. More detailed studies of small areas are dependent upon acoustic underwater transponders. A broad range of shore-based radio positioning systems, whose location-fixing accuracy improves shoreward, are currently employed in surveys conducted on continental margins. However, the future worldwide availability for obtaining position fixes by means of the Global Positioning System (GSP) will supercede most previous methods of marine navigation.

Methods of marine phosphate exploration include both remote detection and direct sampling techniques. The primary methods of remote detection include seismic reflection profiling, gamma-ray logging, and side-scan sonar imaging, all of which need close coordination with the sampling program. A number of high-resolution seismic techniques including echo sounder, pinger, boomer, sparker, and air guns, in combination with deeper multi-channel seismic data provide powerful tools for detailed stratigraphic and structural analysis of continental margins. In deeper water (>400 meters), the deep-tow boomer system is an important seismic tool. Enrichment of uranium and its daughter products in phosphates, relative to other marine sediments, make the gamma-logger an important exploration tool for surficial deposits or down subsurface drill holes. Measurement of acoustic reflectivity and the strength of back-scattered signals of the sea bottom from side-scan sonar may help in determining qualitative and quantitative information on seabed phosphorites. Sea-bottom photography and underwater television are also important methods of identifying and mapping seafloor phosphorites on a small scale. Simultaneous acquisition of biological information from remotely sensed data will provide valuable knowledge necessary for understanding environmental consequences of future resource extraction.

Methods of direct sampling of phosphates range from the relatively inexpensive dredge-hauls that do not obtain representative samples, to the expensive, but vital, rotary drill holes. Grab sampling combined with side-scan sonar and high-resolution seismic profiles are used to map textural and mineralogical variations over large areas of the

seabed relatively quickly. Sub-surface samples can be obtained by
gravity-, box-, piston-, and vibra-coring in order of increasing
reliability, expense, and complexity of operation, respectively. For
hard-rock and phosphate crusts, dredges and large, heavy,
power-assisted grabs are the only available sampling tools.

In the exploration stages, oceanographic data on currents and
waves, and biologic data on benthic and other organisms, should be
obtained and monitored for subsequent use in environmental evaluations.
The establishment of a more thorough research base will facilitate
subsequent evaluation and exploitation of the resources.

RESOURCE EVALUATION

Several factors govern the evaluation of phosphate resources in
identified deposits. Among these is availability, which includes the
fundamental information on size of the deposit, water depth, burial
depth, location, and distance to shore. Infrastructure and proximity
to markets are added costs necessary in order to deliver a sellable
product. Value of the ore is dependent on quality and quantity of the
recoverable resource. Quality is mainly a function of grade and nature
of the gangue materials. Reactivity of the phosphate grains is a
measure of the ore's value for agricultural use. Another increasingly
important aspect relates to by-products, whereby associated elements
contribute to the ore value while others are regarded as harmful.
Concentration of elements like cadmium, with its impact on human
health, influences the value of specific phosphorites used in
agriculture.

Quantification of resource parameters is becoming increasingly
important. Availability, infrastructure, market parameters, and the
parameters of ore value should be expressed as probability
distributions. In this manner the data are made more suitable for
entry into computations of a mineral supply system.

RESOURCE DEVELOPMENT

For the development and production phase of any offshore mining
activity, there are several key technical, logistical, environmental,
socio-political and economic considerations that must be taken into
account. The Working Group has identified a sampling of these,
realizing that it was beyond its resident expertise to either make a
comprehensive list, or to examine each in detail.

Offshore enterprises in the design stage must take into account
physical factors such as water depth, prevailing and extreme
oceanographic conditions, and distance from shore to the proposed
mining site. Water depth beyond the shallow portions of inner shelves
dramatically influences costs in development and maintenance of
electrical, mechanical and hydraulic systems. Hence, with current
technology, continental margin types of phosphate deposits are more
likely to be exploited in the near future than other types of deposits
in deeper waters (Table I). However, many continental shelf deposits
have narrow occurrences and face the open sea; these areas are subject

to high waves and wind speeds and strong currents. Areas under the
influence of long shoaling waves and tropical and extratropical storms
impose similar constraints. Such oceanographic conditions will impose
high engineering costs for mooring equipment, position keeping, and for
alleviating tension on bottom-to-surface pipelines and ocean cables.
Furthermore, as distance increases between the operational site and
port facilities on shore, costs associated with transportation of
material mined, equipment deployed and maintained, and movement of the
labor force increases substantially.

Design of offshore mining plants for phosphates must take into
account two significant engineering considerations. The first pertains
to the environment, the second pertains to the physical and
physiological safety factors. Environmental considerations must
include special biological provisions to minimize pollution of nearby
and down-current marine environments, i.e. ensure that benthic and
nektonic species are not suffocated by intolerable quantities of
discharged sediment, that prime benthic habitats are not covered by
mining waste, and that discharge sites are reinhabitable by marine
species. Safety considerations must include all operating equipment
and, from the physiological perspective, control and containment of
potentially toxic materials extracted during mining or refining
including such elements as fluorine, cadmium, and uranium.

Future development of offshore phosphate resources depends on
market conditions. Traditionally, the price of phosphate has
fluctuated on world markets with cycles averaging 6 to 7 years.
Long-term growth rates of fertilizer usage are linear and rising fairly
steeply. While the latter portends reasons for offshore mining,
fluctuating prices could seriously affect plans for this wholly untried
manner of extraction from the marine realm. However, changing patterns
of land use in many parts of the world are in favor of extracting
phosphates from the marine environment. Many resources and potential
mine sites are no longer available for use; this may eventually force
phosphate mining companies toward considering the exploitation of
deposits in the oceans. As a result of declarations of Exclusive
Economic Zones (EEZ) by many nations, greater assurance of protection
may now be available to offshore operators, provided that legal and
regulatory framework for concessions, operations, and fair return are
ensured and stable. In this regard, small phosphate deposits near the
shore may be economically viable to extract and also beneficial to
countries that want to be self-sufficient in fertilizer supply,
especially for direct application of unprocessed phosphate to
agricultural lands.

At present, technology useable in offshore mining is only an
extension of land-based techniques modified for and adapted to the
marine environment. Conventional equipment that is suitable for
offshore include many types of dredging and suction systems, scrapers,
cutters, and hydraulic slurry mining. Of these, hydraulic slurry
mining probably has the greatest potential for economically recovering
continental shelf phosphate deposits. New mining technology is being
developed and includes: 1) subsurface borehole excavation utilizing
water jets to slurry weakly consolidated phosphate sediments in

caverns, followed by backfilling with waste materials, and 2) a
remotely controlled mining device for recovery of surficial deposits
such as those occurring on the Chatham Rise.

RESOURCE UTILIZATION

Phosphate deposits are the major source of phosphorus, a critical
nutrient element essential for agricultural food production. This
element is economically important to both developed and developing
countries, however, in totally different economic contexts.
Consequently, it is necessary to consider phosphate resource
utilization from two quite different points of view.
Developed countries grow food-stuffs on large agricultural
production scales and are actively involved in the world export-import
markets for both fertilizers and agricultural products. Consequently,
demands for and utilization of phosphate resources is also large scale,
dependent upon sophisticated chemical processing, and entirely subject
to world-wide market conditions of supply and demand. For chemical
processing, it is essential that ores be high grade and high quality.
However, large-scale manufacturing and chemical processing does permit
extraction of by-product materials (ie., fluorine, uranium, etc.) from
the ore.
Presently, the main sources of phosphate are from on-land,
continental margin type deposits of various geologic ages, and from a
few insular deposits. It is likely that future development of
phosphate resources from the marine environment will necessitate
utilizing the same two types of deposits, as well as deposits of
condensed sections where special economic conditions would so dictate
(i.e., the Chatham Rise deposits off New Zealand).
Utilization of phosphate resources from the marine environment may
encounter certain impediments, such as the preference for traditional
mining techniques, sources, and existing infrastructure. However,
utilization of marine resources can be encouraged if appropriate
incentives emerge, such as the need to develop self-sufficiency,
changes in land-use patterns, environmental considerations, taxation,
etc.
Developing countries are increasingly experiencing severe problems
in their food supply. The problems result, in part, from high
population growth rates, exacerbated by the fact that this growth is
taking place in areas with tropical and subtropical soils which are
highly leached. Economically unable to purchase needed processed
fertilizers on the world market, these countries are compelled to
develop their own, internal fertilizer resources. Such resources can
be derived from very small-scale mines employing unsophisticated mining
techniques. Where soils are acidic, phosphate ore can be applied
directly to the soils, eliminating the need for finding and using high
quality phosphate ores and for establishing costly chemical treatment
plants. In fact, components such as shell, dolomite, glauconite, etc.,
often deemed detrimental for processed ores, are actually beneficial to
soils through direct application techniques. For those countries not
having land-based deposits, exploitation of small, low-grade deposits

from their EEZ may be feasible with simple and available equipment. Consequently, almost any type of phosphate deposit close enough to both the seafloor and the coast could be utilized.

RECOMMENDATIONS

Recommendations of the Working Group are keyed to the topics previously discussed.

A. Occurrence of Phosphate

1. Test various models of phosphogenesis including:
   a. Western boundary current non-upwelling versus geochemical recycling from older to younger sediments,
   b. Eastern and western boundary current upwelling, and
   c. Overflow versus upwelling for seamounts, plateaus, and rises.

2. Develop integrated mass balance and geochemical models of associated authigenic minerals (ie., ferro-manganese, silica, glauconite, dolomite, etc.).

3. Improve age determinations of all kinds of phosphorites by biostratigraphic and radiometric methods (ie., U-He, Nd-Sm, $Sr^{81}-Sr^{86}$, Be^{10}, etc.).

4. Define the relationship of organic materials to and the role of micro-organisms in the formation of phosphorites.

5. Study the diagenetic alteration, reworking, and transport of previously formed phosphate.

6. Study the processes of formation of phosphate grain types and their relationship to diagenetic reactions.

7. Investigate modern analogs of types of phosphate deposits that are known to occur only in the fossil record (ie., epicontinental seas, condensed sequences, seamounts).

B. Resource Assessment

1. Develop a methodology to relate phosphate deposits to geological, physical, and chemical parameters that control phosphogenesis on a regional basis.

2. Study specific areas in which Tertiary and Modern phosphate deposits occur, and that are considered to be important in understanding and testing various models of phosphogenesis.

a. New Areas   b. Type Areas
   Somalia - Oman - Pakistan   Peru - Chile
   E. Australia   Namibia
   Brazil - Argentina   Baja California
   Western India   S.E. United States
   Mid-Pacific Seamounts
   Senegal - Mauritania - Morocco
   Congo - Cabinda
   Caribbean
   Portugal - Spain

3. Integrate quantitatively, including using new image analysis techniques, the geological, geophysical, geochemical, and oceanographic data to improve recognition and analysis of patterns and trends on these data.

C. Exploration

1. Develop a passive, preferably remotely operated or tethered instrumentation system that incorporates acoustic (seismic and side-scan sonar), chemical, and radioactive detection devices together with TV and still photography.

2. As present sampling techniques are inadequately designed for use in certain seafloor morphologies, such as seamounts and plateaus, equipment such as a seabed rotary drill for vertical samples, a grab sampler for large areal sampling, and a large vibra-box-corer, should be developed which can be fitted with TV and be capable of remotely-controlled maneuvering. Explore the use of new techniques in equipment development including: a. patterned directional explosives to blast shaped samples, b. thermic lances to cut shaped samples, etc.

3. Examine the factors which control the total seismic response of phosphates to help direct locating and identification of such deposits.

D. Resource Evaluation

1. Analyze samples from known phosphate deposits for the presence of associated minerals or trace elements that may be potentially valuable (or detrimental) to extract.

2. Plan and conduct surveys and sampling programs in full recognition of the need for geo-statistically significant data needed for resource appraisal and for engineering concept development and mine design.

E. Resource Development

    1. Place increased emphasis on understanding associated ecosystems
and oceanographic conditions for environmental considerations
in light of possible mining of deposits and the variety of
mining techniques that may be used in the extraction of
resources.

F. Resource Utilization

    1. Study the interaction of phosphate supply and demand in
response to possible new technological developments, and as a
function of the availability and accessibility of phosphate
deposits.

    2. Carry out long-term field trials of direct application of
unprocessed fertilizer for a variety of soil types and climatic
conditions.

REPORT OF THE WORKING GROUP ON MANGANESE NODULES AND CRUSTS

David S. Cronan, Chairman
Helmar Kunzendorf, Rapporteur
Alan A. Archer
Mustafa Ergun
DeVerle P. Harris
Benjamin W. Haynes
Augusto Mangini
William D. Siapno
Ulrich von Stackelberg

MANGANESE NODULES AND CRUSTS: ORIGIN AND OCCURRENCE

Ferromanganese oxides, some containing potentially economic concentrations of Ni, Cu, Co and Zn, in addition to Fe and Mn, occur in the form of nodules and crusts on the ocean floor.

Manganese nodules, first reported in 1873, are found throughout the deep oceans. Broad regional variability in nodule occurrences and composition are reasonably well known, but much less is known publicly about the variability of the deposits on smaller and more detailed scales. Furthermore, our understanding of the deposits is heavily influenced by the preponderance of data from the Clarion-Clipperton Fracture Zone (CCFZ). Can we extrapolate the present knowledge about genetic and geological controls, largely based on this zone, to nodules elsewhere on the ocean floors? If so, can this assist us in developing exploration techniques for potentially economic varieties of the deposits?

The formation of manganese nodules is dependent on several factors. Growth nuclei (often termed "seeds") must be present, around which the iron and manganese oxides, usually derived from seawater, can be deposited layer by layer. The particles that serve as nuclei, vary in kind from place to place; most frequently though they are coarse rock particles or biogenic fragments. To start growth of the layers of potentially economic nodules, such as those in the CCFZ, high-intensity biological activity must be present in the surface waters to support diagenetic metal supply in the sedimentary surface, and the water depth in the area must be near or below the calcium-carbonate compensation depth (CCD). Notably, high-grade nodule deposits have not been found where these conditions could not be met. In the Peru Basin and the S.W. Pacific and Central Indian Oceans, conditions similar to those in the CCFZ appear to have

21

P. G. Teleki et al. (eds.), Marine Minerals, 21–27.
© *1987 by D. Reidel Publishing Company.*

generated nodules of above-average grade. Thus, genetic models
developed to account for the composition of CCFZ nodules could
possibly be applied to other areas. This possibility needs to be more
fully evaluated, although not without full recognition that the
geological history of these areas will differ, and modify the original
model.

Conditions which have led to the formation of potentially
economic nodules, and resulted in certain nodule grades and abundances
are not uniform; instead they vary considerably over short distances.
These variabilities are caused among other factors, by small-scale
lateral variations in the depositional environment and by
oceanographic, chemical, depositional, and other factors that change
during the period of nodule growth. The variations in the
depositional environment are reflected in the composition of sediments
associated with the nodules. The sediments should, therefore, also be
studied, to permit reconstructing the nature of the paleoenvironments
that favored periods of nodule growth. Another uncertainty is how
nodule growth is initiated. Substrate variability and types of growth
nuclei are known to influence local nodule variability significantly.
All variables interact in the genesis of a deposit.

A thorough understanding of what controls variability in nodules
can lead to genetically-based exploration models, whereby areas that
appear to have limited economic value can be excluded, ab ovo, from
initial consideration, and efforts can be focused on the more
promising areas with a concomitant saving in time and survey costs.
However, it should be realized that mining a manganese nodule area
selectively, wherein only the highest grade deposits are mined, will
likely never take place, and instead, judicious blending of nodules
with varying grades will occur, at best, because the selective
retrieval of nodules is technologically limited. Such blending is
best done during the collection process. Furthermore, metal market
conditions, i.e. supply, demand and price, are likely to change during
the anticipated 20-year lifetime of a mine site. For example,
compared to other metals, the relative importance of Mn and Co have
increased in international metal markets in recent years. For this
reason alone, too severe constraints should not be applied to the
grade and composition of nodules retrieved, as other metals could be
recovered from them at a later time. Similar considerations are valid
for ores from terrestrial deposits.

Manganese crusts are widespread on the seabed, and are commonly
found on flanks of seamounts in zones where water depths range from
300 to 2000 m. Crusts precipitate from the water column, therefore,
their growth is most likely controlled by hydrogenous processes.
Thus, understanding the chemistry of the water column is critical to
deciphering their genesis and further studies on crust/seawater
interactions are much needed. Adsorption and other surface chemical
reactions are also important to determining their composition.

Should the chemical and physical controls on modern crust
formation become elucidated, the paleoceanographic conditions
attendant on their formation in the past could possibly be determined
by studies of the geochemical history of the crusts, layer by layer,

as these layers often represent a relatively long depositional history. One advantage that crusts have over nodules in this regard, is that they are firmly fixed to the basement substrate, and thus, probably provide a better continuous record of the paleoceanographic conditions that existed at a given site than nodules do, which may have been moved around by benthic processes. Various existing, thick, ancient manganese crusts in the CCFZ area could be used for such paleoceanographic studies.

It is too early to evaluate the potential economic importance of cobalt-rich manganese crusts. Presently, only those deposits that contain Co in excess of 1% are considered to have any possible economic value. Co-rich manganese crusts are thin, on the average about 2 cm. Their thickness varies over short distances, and this in turn influences calculations of their economic potential. Thickness variations, therefore, need to be investigated further. Investigations have been so few that it can be easily concluded that substantially more surveys are needed, that the physical properties of crusts should be investigated, and that critical evaluation of different possible mining systems, with deference to these physical properties should be carried out.

Phosphate is often associated with ferromanganese crusts on seamounts, as substrate, admixtures and thin horizons separating different generations of metallic crust (see the report of the Working Group on phosphorites). The genetic/paleoceanographic significance of this ferromanganese-phosphorite association should be probed, as it could be a possible avenue to understand the origin of both these minerals. From the economic point of view, a method has to be developed to separate phosphate from Co-rich manganese crusts, because phosphorus is deleterious in crust processing and in the end-products.

NODULE AND CRUST EXPLORATION

Given the importance of manganese nodules and crusts as potential metal resources, one must ask what visual and sampling techniques could be appropriately applied in their exploration and how these may be improved. To date, the principal visual tool in nodule exploration has been the video-camera. Advanced photographic cameras of the charge-couple device type have also been used. More sophisticated imaging exploration techniques, e.g., the use of the Argo-Jason system, are yet to be tested over nodule fields and encrusted seafloor areas, hence, their utility, no doubt tied to resolution capability, is still unknown. Attention should be paid also to the development of acoustic image processing techniques, such as multiple- and variable-frequency interferometry, because, if deployed in deep-tow systems, these can survey larger areas more rapidly than cameras lowered from ships can. Such techniques should enable estimating nodule abundance during the course of surveys, even when nodules are thinly covered with sediments, and would, therefore, be of considerable help in future mine-site evaluations. The development of deep-towed geophysical multipurpose platforms, having operational

capability in water depths as much as 6 km, would also be desirable,
because mapping and reconnaissance of potential manganese-nodule sites
could be facilitated at higher speeds and with concurrent time
savings.

It is now possible from satellite altimetry data to deduce some
of the major topographic features of the seafloor. Refinement in the
spatial resolution of such techniques would enable examination of the
seafloor topography and morphology in greater detail. Improved
satellite navigation and communication systems is also desirable, and
could be met by the Global Positioning System (GPS) being implemented
at present. SEABEAM, a multibeam bathymetric profiling system, is
useful in initial bathymetric mapping of an area, unfortunately, at
present it seems to be too expensive for routine exploration. Most
importantly, subareas of a prospective area must be mapped in detail,
and to a level that allows results to be extrapolated to adjacent
areas.

In addition to remote (e.g., acoustic) detection, nodule and
crust evaluation will rely on sampling and analysis of recovered
material well into the future. Improvements in sampling techniques,
such as those based on box cores and free-fall grabs, are anticipated,
and simultaneous deployment of various types of sampling tools could
assist nodule/sediment interrelationship studies. For sampling of
Co-rich crusts, a spot-sampling device with improved capability over
existing techniques (mainly dredges) is needed.

In addition to improved sampling techniques, statistically-based
samplings, as employed in terrestrial exploration, should also be
implemented, identifying minimum spacing of samples for gathering
optimal information about the deposit.

Improvements in shipboard and laboratory chemical analyses would
be useful to accelerate the data supplied for statistical evaluations.
In-situ chemical analyses of nodules and crusts on the seafloor,
however, is not yet considered to be a viable alternative to analyses
of recovered samples.

RESOURCE ASSESSMENT

In principle, the approach to estimating the mineral endowment
for marine minerals has much in common with estimating the endowment
of terrestrial mineral deposits. For manganese nodules, however, such
estimation in practice may differ from that of crusts, partly on
account of their differing physical and chemical attributes, and
partly on account of the more limited base-line (statistical) data
available for the latter.

For manganese nodules it may be desirable to replace tonnage per
deposit by the number of nodules per unit area, nodule size
distribution, the variation in nodule density and the correlation
between metal content and nodule size. Similarly, average grade per
deposit may be replaced by element concentrations and size. Another
feature of nodules and crusts that may indicate differences in details
of endowment estimation is the definitions of deposit-occurrence.

Suppose the presence of one or more nodules per unit area indicates a deposit. Such a nodule deposit, per se, may exhibit an extremely large volume, one that is much larger than appropriate to a single mining unit. By imposing minimum nodule abundance as a definition of a deposit, the large-volume deposit may transform into several smaller deposits of higher nodule abundance than the original single deposit had. In general, number of deposits, deposit size, average nodule abundance, and average metal concentration become functions of the "cut-off" abundance. Such a state contrasts sharply with those of certain terrestrial deposits, such as porphyry-copper deposits, and requires appropriate modelling of mineral endowment estimation. It seems likely that endowment estimation for nodules and crusts will initially rely more heavily upon multivariate geostatistical techniques and less heavily upon computer-based expert systems. At this time, as the availability of needed data is limited, endowment estimation is more difficult and is attended by greater uncertainties.

For manganese nodules, it is necessary to define the number of mine-site descriptors that can be related to variations in the environmental settings of nodules. Descriptors are the number of nodules per unit of area, the frequency distribution of nodule size per unit area, the frequency distribution of metallic element concentration, the number and sizes of deposits as a function of endowment, cutoff values for nodule density and metallic element concentration, and correlations between the nodule abundance, grades, and size.

Estimation of potential mineral supply requires the probabilistic description of the mineral endowment, other components of which are subroutines for endowment characterization, exploration, mining, processing and economic analysis. If it is desired that potential supply estimates be made for selected regions, high priority must be given to the construction of this system. When completed, the geologists' description of endowment would be input to the system, which would produce probability distributions for potential supply for each relevant combination of economic parameters. Initial work at Massachusetts Institute of Technology on a cost model for manganese nodule mining, the recent updating of this work at Texas A&M University, and the pending modelling of endowment characterization could be integrated into a potential supply system. Such studies could possibly provide a priori endowment values for other regions, where information is even scarcer. Ultimately, variations in these descriptors could be related to variations in the environment in which the nodules are found.

RECOMMENDATIONS

A. Manganese nodules and crusts: origin and occurrence.

  1. Knowledge about manganese nodules in the CCFZ area need to be more fully evaluated especially by taking the geological evolution of the deposits into account.

2. Local variability in nodule grade and abundance needs to be investigated in more detail, and an attempt should be made to obtain results that can be transferred from one area to another, in order to lessen the need for routine surveys.

3. Studies on crust/seawater interaction, including laboratory experiments are desirable to establish what processes and conditions control crust growth.

4. Paleoceanographic conditions might be elucidated by the study of older crusts, whose growth history represents relatively long periods of time.

B. Nodule and crust exploration.

1. Development of acoustic imaging techniques appropriate to nodule exploration would be desirable; specific requirements are the determination of nodule densities and crust thicknesses.

2. Satellite imaging of the sea surface to provide derivative data that would improve the resolution of seafloor topography, is desirable.

3. Selective evaluation of subareas of a prospective area containing nodules or crusts, to a level where results can be extrapolated to other areas, should be pursued.

4. Advances in the designs of box-corers and free-fall grab samplers are required to improve the success rate of sample recovery.

5. The development of simultaneous deployment mechanisms should be investigated in order that samples and measurements all come from the same positions on the seafloor.

6. Improving and standardizing techniques for statistical evaluation of a deposit would be desirable in order to minimize the frequency and/or areal density of sample retrievals.

7. Concepts for systems design for mining crusts should be developed soon, and in regard to the physical properties of the crust and the substrate below.

C. Resource assessment

1. From the point of view of resource estimation, more data are needed on the global distribution of nodules and crusts, their physical and chemical properties, and their areal and internal

variability. Phosphorite-crust interrelationship should also be investigated in more detail in this context.

2. Special studies should be designed to obtain data on mine-site descriptors; i.e., the number of nodules per area or thickness of crust, the frequency distribution of nodule or crust variability per unit area, etc.

3. The integration of an existing model for manganese-nodule mining and modelling of endowment characterization into a potential supply system, should be pursued.

4. In response to any future change in the market demand and prices for metals, elements other than Mn, Co, Ni and Cu associated with crusts and nodules, should be evaluated more carefully from both the scientific/technical and resource potential prespectives.

REPORT OF THE WORKING GROUP ON MARINE SULFIDES

John M. Edmond, Chairman
F. P. Agterberg, Rapporteur
Harald Bäcker
John R. Delaney
Peter Diehl
M. R. Dobson
T. J. G. Francis
Randolph A. Koski
José H. Monteiro
S. A. Moorby
Elizabeth Oudin
Steven D. Scott
F. N. Spiess

INTRODUCTION

The significance of the discovery of submarine hydrothermal processes has been felt throughout the Earth Sciences, and the interest it has generated is evident in the many parallel investigations and in continued discoveries. Studies of the thermal budget of oceanic spreading centers, of the composition of marine sediments and seawater and of the environment and mechanisms of formation of ore deposits have led to the realization that a large portion of the massive copper, lead, zinc sulfide deposits, presently exploited on land, were in fact formed on or beneath the seafloor. Concommitantly, deposits or ore minerals presently forming in the ocean can be viewed both as "working models" of the fossil land deposits and as potential resources in themselves. These developments are very recent and have involved a very diverse group of investigators and a full range of experimental and theoretical approaches.

Hot springs were first discovered in 1977 on the Galapagos Spreading Center. Two years later, high temperature activity was found on the East Pacific Rise where small mineral deposits were observed to form from metal-transporting hydrothermal solutions, whose temperatures were approximately 350°C. Since then, similar occurrences have been discovered at numerous locations in various spreading centers of the Eastern Pacific. All sulfide deposits, discovered to date and related to these hydrothermal systems, have been small relative to those currently exploited on land, and relative to the anticipated size necessary for feasible submarine exploitation.

P. G. Teleki et al. (eds.), Marine Minerals, 29–37.
© 1987 by D. Reidel Publishing Company.

What has been established beyond doubt is that hydrothermal activity
is intimately associated with submarine volcanism at spreading
centers, seamounts and island arcs, and that volcanism is as
characteristic of this phenomenon as are pillow lavas and transform
faults.

Many problems in understanding the phenomenon remain to be solved,
however. There is very little information on the frequency, duration
and scale of the activity at any particular location. Insofar that a
large, actively forming deposit has not yet been found, it is difficult
to assess to what degree are the small systems true analogs of larger
ones. If they continue to be active, will they eventually grow to
contain millions of tons of sulfides, or are there additional factors
that govern genesis? Only a small sub-set of the known types of
submarine volcanic activity has been explored and that at the
reconnaissance level. Although ore deposits are known to form in a
diversity of environments, these settings have not yet been found in
the contemporary ocean. It is not known whether there are settings
that are particularly advantageous to ore genesis, that is where, for
tectonic or environmental reasons, the formation of large deposits is
favored.

For the reasons cited above, we recommend the initiation of a
long-range, coordinated research and exploration program, with at least
a ten-year time-scale, international in participation and global in
scope. It would involve earth scientists with expertise in both
oceanographic and terrestrial problems. Its objectives would be to
produce an explicit model for the submarine formation of volcanogenic
massive sulfide ore of all types, as an aid to exploration on land and
to be used in detailed assessment of the resource potential of the
contemporary sea-floor.

COMPARATIVE STUDIES

Close interaction between marine scientists and continental earth
scientists will be crucial to the success of any program of research
and resource evaluation directed at contemporary submarine massive
sulfide deposits. It is unlikely, given the great differences in the
expertise and in the exploration techniques used by the two groups,
that more than a few individuals will feel equally at home in the two
domains; it must be recognized that all present criteria used in
determining what constitutes a massive sulfide resource, and most of
the information on the tectonic and geologic environment of their
formation, come from studies of "fossil" systems on land. The original
submarine discoveries on the East Pacific Rise were made, almost
entirely, on the basis of oceanographic considerations. However, the
present impetus to extend exploration to island arcs and especially to
seamount calderas and rifted margins has been advanced, in large part,
by land geologists.

Geologists concerned about resource appraisal recognize four
essentials for the formation of massive sulfide deposits: (1) high
heat flow, i.e., a local heat source; (2) a permeable substrate of

significant base-metal content; (3) ground and/or standing waters of
relatively high ionic strength with chloride as the dominant anion; and
(4) a mechanism of ore localization, i.e., containment of the ore
metals transported by the convecting fluids resulting from the
coincidence of the first three requirements.

Heat flow is determined by the plate-tectonic framework, namely,
by the phenomena occurring at accreting and consuming margins in all
their manifestations. The substrate is tholeiitic basalt alone or in
combination with partially lithified sediments of marine and/or
continental origin. The fluid can be seawater or river and lake waters
concentrated by evaporation. The localization mechanisms are diverse
but usually involve either an increase of pH by reaction with an
external fluid or solid phase or a reduction in temperature.

Several tectonic environments are recognized. The most obvious
are the sediment-starved "mature" accreting margins, either the
open-ocean or the back-arc type. While there is some controversy
regarding the distinction between these types in the geologic record,
there is also consensus that the classical Cyprus- or ophiolite-type
massive sulfides are closely analogous to the much smaller deposits
recently found on the East Pacific Rise.

Compared to the previous types, a much more complex collection
(perhaps continuum) of environments is formed during the evolution of
unrifted continental (or island-arc) crust to mature (in the sense of
sediment-starved) ocean basin. Contemporary crustal thinning appears
to be rare, of which the Salton Trough, Northern Red Sea, Afar Triangle
and East African Rifts are examples. Often, in such settings
sedimentary basins form and host sediment accumulations several
kilometers thick. The basins host petroleum and occasionally very
large ore bodies. The ores are known to have formed in the presence of
very high geothermal heat flow, but corollary volcanism is at most,
minor. The Sullivan deposit of British Columbia is an example. As
spreading becomes organized the sedimentary fill in the trough
gradually changes from predominantly fluvial to marine sediments. This
unconsolidated material is subsequently intruded by tholeiitic dykes
and sills. The accompanying hydrothermal circulation metamorphoses the
sediments and volcanics to greenschist facies and results in forming
ores that are closely syngenetic with sediment deposition. These are
the so-called Besshi-type deposits, that are similar to those forming
today in the Guaymas Basin of the Gulf of California and, apart from an
apparently unique evaporitic association, also in the Red Sea. The
common distinction between marginal and intracratonic rifts probably
needs refinement, because it is not clear how deposits in the Red Sea,
East African Rift and Gulf of California could be distinguished, should
activity cease and be followed by an episode of compression, uplift and
denudation. Presumably, the likelihood that ore deposits formed in
failed rifts are preserved in higher than that in deep-water bodies.
Certainly, sediment-hosted bodies, formed at various stages of early
seafloor opening, belong to a major ore deposit category. The
significance of these bodies can be underlined further in that they
derive supplementary metals from the metamorphosed sediment fill in
addition to those derived from the relatively barren tholeiites.

Seamounts, that are discrete volcanic centers, are observed in all rifting environments and appear especially prominent at slow and intermediate spreading rates. Many ore bodies, e.g. the kuroko-type, have been noted to form in submarine calderas. Recent exploration has found these to be features common to seamounts. As caldera formation requires that a shallow magma chamber exist and as the fracturing associated with its collapse produces permeability, seamounts are prime sites for the next "wave" of oceanic exploration.

EXPLORATION PROGRAM

Our basic assumption is that the deep-sea sulfide deposits, that may initially compete with terrestrial metal deposits, will be those found exposed on the seafloor or only superficially covered. To evaluate the resource potential of seafloor sulfide deposits, we recommend a major thrust in establishing a method of efficient inventory of exposed rather than subsurface sulfide deposits. Two likely types of exposed sulfide masses are actively venting systems and dormant or inactive deposits. Exploration strategy for each deposit type will differ.

We outline below a hierarchical approach to exploration and sampling of these deposits that is applicable to exposed deposits regardless of the type of volcano-tectonic association. Further, we recommend that real-time observations and experiments be performed on active systems to define the nature of the dynamic processes, such as variations in temperature, flow rate, composition, biological (particularly bacterial) activity, rates and nature of chimney building, etc., in which time scales are expected to range from hours to years.

As our understanding increases, exploration activities will logically expand to include buried deposits. Specific geophysical techniques will have to be developed in order to detect such deposits and provide "first-cut" evaluations of their form.

Focused exploration rationale

Broad-scale exploration.

Ideally, exploration for polymetallic sulfides has to involve a logical hierarchy of investigation. At the first level, exploration of major volcano-tectonic features of the order of 100 km areal extent should be undertaken. It is important that navigational techniques, as accurate as +50 m, be developed, for geographic positioning of surface vessels, and positioning as good as +1 m be achieved within local areas on the seafloor, particularly for drilling or for long-term instrumented observations and experiments at a particular site.

Multibeam swath charting and side-looking sonar mapping should be undertaken on 100-km size segments of ridges. The produced maps should allow the tectonic and volcanic features of these areas to be evaluated.

In addition, seismic profiling with airguns, and magnetic and gravity surveys should be undertaken at this scale. This is a conventional stage of surface exploration, conducted at conventional ship speeds with interspersed measurements at hydrographic stations (salinity, temperature, conductivity, suspended matter and chemical composition of sea water).

Intermediate-scale exploration.

Results of these broad-scale surveys permit setting priorities for intermediate-scale examinations using instrument packages towed near the bottom. Instrument packages should consist of a side-looking sonar, a precision sounder and a sediment profiler; equipment to measure water temperature, conductivity, magnetics and electrical potential, supplemented by photography and television, as well as in-situ real-time chemical (e.g. for CH_4, Mn) detectors.

Technology for performing most of these observations is developing rapidly. Special emphasis should be given, however, to further development of real-time chemical sensors and optical systems that could image from heights greater than 10 m above the bottom. With the combined systems one should be able to determine physical and chemical patterns, indicative of a mineral deposit, simultaneously.

Fine-scale exploration

At this scale, observations and sampling should refine the findings based on the near-bottom instrumented measurements. This phase, that involves seeking information at considerably greater detail, and at sites smaller than in previous surveys, would include on-bottom T.V. and photographic and submersible operations. Submersible operations are particularly helpful from the viewpoints of direct contact with the seafloor, to note the presence of features and processes that might not otherwise be detected and for detailed sampling.

Evaluations and Experiments.

Based on the exploratory phases, further activities could branch in two directions. One of these is to establish the detailed nature of the deposits, the other to design and carry out in-situ experiments necessary for the understanding of processes and development of models.

Evaluation of the nature of sulfide deposits beyond superficial collection of small pieces of samples requires development of new sampling techniques following two approaches. One is the development of drilling equipment capable of operating on the seafloor and able to drill many holes. The other approach is to fracture and dislodge rock material, then retrieve these fragments using dredge-hauls whose precise position is known and can be controlled.

Using of geophysical techniques, such as in-situ resistivity measurements and down-hole or inter-hole logging will provide additional information for the description of these bodies.

In the area of experimentation there is particular concern for establishing the nature of hydrothermal dynamic processes, such as variations in hydrothermal discharge rates and composition, biological (particularly bacterial) activity, and rates and nature of chimney building, in which time scales are expected to range from hours to years.

ANALYTICAL PROGRAM

An analytical program for the study of sulfides has two components, one relating to fluids, the other to solids.

Regarding fluids, solutions recovered from active systems have compositions which reflect processes presently inaccessible to direct study. It is important, therefore, to exploit the known chemistry of the various elements, their compounds and isotopes to model all recorded reactions as a means to understand the principal controlling mechanisms. This will require extensive and accurate measurements encompassing about one hundred chemical entities present in the fluids.

Concerning solids, research should be distinct from routine analyses that are singularly related to determining resource potential. Scientific analyses include mineral chemistry, mineralogy paragenesis and textures, chemical and physical properties of fluid inclusions, age determinations and organic geochemistry. To the extent possible the chemical measurements made on the solids should be complementary to those made on the fluids. A parallel program must be carried out on fossil analogues.

Routine analyses are aimed at the resource assessment of the deposits and should be statistically representative. Chemical analyses include those of the valuable metals (Zn, Cu, Pb, Ag, etc.) and of the components which may affect metallurgical tests and processing (Fe, Si, etc.). Measurements of the physical and geotechnical properties of deposits and samples are required.

Physical modeling

Continued refinement of genetic models for ridge-crest hydrothermal systems and for the formation of massive-sulfide deposits must incorporate the volcano-tectonic evolution of the systems involved. The relative timing of magmatic, volcanic, tectonic and hydrothermal events is crucial to the development of an efficient and accurate predictive capability incorporating the distribution, abundance and size of submarine deposits associated with spreading centers, volcanic areas and major fault systems. Particular emphasis must be devoted to identification of structural, or stratigraphic controls that contribute to the formation of larger metal deposits.

Chemical modeling

Computer experiments and simulation studies depend on accurate thermodynamic data for reactions occurring in the pressure-temperature range of interest (150-750 bars, 150-450°C) and access to large fast computers for the numerical solution of coupled chemical and hydrodynamic models. Thermodynamic data are sparse or non-existent especially in the region close to the critical point.

RESOURCE EVALUATION

In comparison to costs associated with mining of massive sulfides on land, ocean mining is attractive in several ways. Because the principal exploratory facilities, namely ships, are mobile, many relatively small deposits could be mined sequentially using the same vessel. Infrastructure costs, therefore, are not captive to a single mined unit. On land, these costs are considerable including construction and maintenance of towns, communications and power lines, water supplies, shaft sinking and subsurface development or stripping. Not much of this investment can be recovered when the mine is exhausted. Hence, the criteria for successful exploitation, in terms of deposit size and grade, could be considerably relaxed for submarine deposits compared to those on land.

Estimation of the frequency of occurrence of submarine deposits of exploitable size and grade, based on statistical methods, is difficult at present. Only about 100 kilometers of ridge crest (less than 1% to the total) has been explored at the required level of detail. While a range of sizes of deposits have been observed, none of them are remotely comparable to the sizes of exploitable land deposits, whose economic statistics are based on the very small ratio of commercially viable deposits to uneconomic deposits of the same type. Thus, the terrestrial experience cannot be "patched" to the oceanic data, though, two exceptions sustain hopes for commercial exploitation of submarine sulfide deposits; first, that exploitable deposits with sizes and grades acceptable for mining will require infrastructures lower in cost than on land; secondly, that continued exploration beyond the miniscule attempt to date will discover progressively larger deposits. The suggested research directions should contribute substantially toward evaluating both these expectations.

RECOMMENDATIONS

A. Comparative studies

1. Joint participation of marine scientists and continental earth scientists in all aspects of field programs should be taken as the norm. Ore geochemists and petrologists should work on the precipitates formed in both active and fossil ore systems.

Aqueous geochemists and hydrologists should make every effort
to relate the composition and mechanics of the reacted
substrates exposed on land to the chemistry and physics of the
advected products sampled in active oceanic systems.

2. The oceanic exploration program should direct its efforts at
the complete spectrum of tectonic environments that have been
identified to host fossil ore systems. In addition to
open-ocean ridges, these include back-arc and intra-arc rifts,
both marginal and cratonic continental rifts, areas of
incipient rifting and young seamounts.

3. The three-dimensional nature of ore deposits is common
knowledge, and data on the vertical extent (depth) of deposits
are needed in oceanic settings, as much as on land where they
are more accessible. By analogy, information on land-based
deposits should be incorporated into the results of oceanic
studies and exploration.

B. Exploration Technology

1. Development and validation of sensing systems for determining
the chemical properties of the water.

2. Evaluation of how electrical methods could be used in
determining of the existence and nature of seafloor sulfide
deposits.

3. Development of methods for optical imaging of the seafloor
from distances of 10-100 m off the sea bottom.

C. Sampling Technology

1. Development of techniques for breaking up large sulfide rocks
and recovering the resulting fragmented samples.

2. Development of reliable capability for drilling sulfide
deposits using equipment located on the seafloor, and that can
recover cores as long as 50 m.

D. Sampling Operations

1. Drilling deep into sulfide deposits on ridge crests or
seamounts, by the Ocean Drilling Program's vessel, to begin
developing an understanding of their underlying structure.

2. Application of techniques for using drill holes to log
subsurface rock-type distribution down-hole.

E. Experimental Studies

   1. Developing instrumental techniques/procedures for monitoring
       real-time changes in geologically active environments; to
       measure various physical-chemical and biological processes the
       results of which would provide needed constraints on models of
       formation of sulfide systems, and the evaluation of their
       mineral potential.

F. Physical modeling

   1. Fundamental research in this subject area should include
       experimental verification of the failure criteria for in-situ
       rocks.

   2. Coupled thermal stress models are necessary to interface with
       chemical/hydrodynamic models to examine the evolution of rock
       permeability.

   3. New techniques should be developed to evaluate integrated heat
       loss from an area populated by several vents, including losses
       associated with low-temperature diffused flow.

G. Chemical Modeling

   1. Implement a rigorous experimental program on the physical
       chemistry of the secondary alteration minerals and sulfides
       found in hydrothermal systems.

   2. Conduct theoretical (and if possible, experimental) studies on
       solid-solution equilibria.

   3. Using field data, develop coupled chemical and hydrodynamic
       models that simulate processes deduced to occur beneath the
       seafloor.

SEDIMENTARY MODELS TO ESTIMATE THE HEAVY-MINERAL POTENTIAL OF SHELF
SEDIMENTS

H.R. Kudrass
Bundesanstalt für Geowissenschaften und Rohstoffe
Stilleweg
3000 Hannover 51
Federal Republic of Germany

ABSTRACT. Three models, based on data from several prospecting
campaigns are presented, which should aid in estimating the heavy-
mineral potential of specific shelf areas. High-density minerals, such
as gold and cassiterite are preferentially enriched in river beds not
far from their source rocks. Shelf areas with outcropping or shallow
buried ore-bearing bedrock have a high potential (Indonesian model).
Vertical differentiation of valuable and other heavy minerals in beach
sand, in combination with transgressive beach-barrier migration, determine the distribution of ilmenite, rutile, magnetite, monazite, and
zircon. Transgressions tend to move these placer minerals landward,
especially on narrow shelves that have a moderate terrigenous sediment
influx (Australian model). On broad shelves that experience high
terrigenous influx, large bodies of disseminated placers may be present
(Zambezi model).

INTRODUCTION

Marine placer deposits are an important source of various minerals.
Beach placer deposits are mined for rutile, ilmenite, magnetite,
zircon, monazite, garnet, gold and diamonds (see map in the report of
the working group on heavy minerals). Shelf placer-deposits are dredged
for cassiterite and scheelite. About 80% of the total rutile production,
about 50% of the total ilmenite and about 30% of the total cassiterite
is provided by these marine deposits.

The selective enrichment of heavy minerals to form a beach placer
can be studied at many beaches in statu nascendi and the processes
involved are well understood (Komar & Wang, 1984). The various
enrichment processes, including those which produce fluviatile placer
deposits, are described briefly below. Three models of the origin of
placer deposits in shelf sediments are illustrated by case histories of
heavy-mineral prospecting expeditions. These models should aid in the
estimation of the heavy-mineral potential of shelf areas.

P. G. Teleki et al. (eds.), Marine Minerals, 39–56.
© *1987 by D. Reidel Publishing Company.*

HYDRAULIC PROCESSES OF ENRICHMENT

The differences in density of heavy minerals (>2.85 g/cm^3) and light minerals (<2.85 g/cm^3) cause them to have different hydraulic properties. Grains of heavy minerals are less easily entrained and transported by currents than grains of light minerals of the same size and shape. Therefore, in any bulk sample of sandy sediment, the heavy-mineral grains tend to be smaller than grains of light minerals. This difference is explained by the concept of hydraulic equivalence (Rubey, 1935), which states that all groups of grains that are simultaneously deposited by a current have the same average settling velocity. A closer examination shows that systematic deviations from this principle occur (Rittenhouse, 1943), especially in sediments rich in heavy minerals.

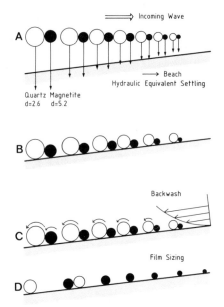

Fig. 1. Sorting of minerals with different densities in the surf zone (modified from Seibold and Berger, 1982). The incoming wave transports various grain-sizes of light and heavy minerals (A) which settle corresponding to their hydraulic equivalent diameter (B). In the thin water-layer of the backwash, current velocities fall off steeply near the water-sand interface and the larger light-mineral grains are more easily removed than the smaller heavy-mineral grains (C). This film sizing produces a lag deposit rich in heavy minerals (D).

Consequently, other processes must be effective in sorting grains according to their density, grain size and shape. One of the most efficient processes operates in the swash zone of the beach (Engelhardt, 1937). The highly turbulent flow of the nearshore surf zone transports a broad spectrum of grain sizes towards the shore (Fig. 1A). When the current velocity of the wave running up the beach looses its power, the suspended grains are deposited according to their settling velocities (Fig. 1B). In the thin water layer of the backwash, current velocities fall off steeply near the water-sand interface. The larger light-mineral grains are more easily entrained and rolled away as bedload than the smaller heavy-mineral grains (Fig. 1C). This so-called film-sizing produces a fine-grained lag deposit of heavy minerals (Fig. 1D). Additional factors may improve the sorting, such as the higher roll-ability of larger light minerals (Veenstra and Winkelmolen, 1976) or "shear sorting" by which larger grains are pushed upwards in a stirred-up sand flow (Sallenger, 1979). It is apparent that only one layer of heavy minerals can be deposited at a time, but when new unsorted sediment is continuously supplied by longshore currents, thick layers of heavy minerals may accumulate. A possible way of permanently separating light and heavy minerals is to divert the heavy-mineral-enriched upper shore sediments into other sinks; the light minerals are most commonly washed back into the foreshore zone. For example, landward blowing winds may remove the surface heavy-mineral-rich layer of the upper shore and store it in adjacent coastal dunes. Mechanically stable minerals, which were not easily worn down by grain abrasion, such as ilmenite, rutile, zircon, magnetite, and monazite are concentrated by these processes. Occasionally fine-grained gold and cassiterite are also enriched as accessory minerals in beach placer deposits.

The huge panning system of the surf zone has no analogue in the fluviatile environment. Economically valuable placer deposits in fluviatile sand contain gold, platinum, cassiterite and scheelite, i.e. minerals which have a much higher density than those enriched along the beaches. In contrast to the lighter heavy-minerals (<5.5 g/cm^3) which are transported by saltation or in suspension, the heavier heavy-minerals (>5.5 g/cm^3) are predominantly moved as bedload. During normal flow conditions, the latter fluviatile placer minerals are not transpor-ted at all, because, being small grains, they are usually "hidden" among gravel-sized sediment particles from where they cannot be entrained (Slingerland, 1984). During flood conditions, when the bulk of the sedimentary fill of the river bed becomes mobilized, these minerals tend to sink to the bottom (shear sorting: Sallenger, 1979), where they are easily trapped in potholes, behind boulders (kinetic sieving: Middleton, 1970), at divergent river sections, at tributary confluences and other places where the current velocity along the river bottom is reduced. Consequently, economically important deposits of the heavier heavy-minerals are usually found within a few kilometers of the source rocks.

PROSPECTING THE SHELF FOR HEAVY MINERALS - CASE HISTORIES

Knowledge of the processes of enrichment of the lighter heavy-
minerals (2.85 - 5.5 g/cm^3) at the beaches and of the heavier heavy-
minerals (>5.5 g/cm^3) at the bottom of riverbeds was used as the basis
for a model for prospecting shelf areas. As the present sea level was
reached about 6000 years ago and the last comparable high sea level
occurred during the last interglacial about 100,000 years ago, the
long glacial period when the sea level was considerably lowered is
likely to have permitted extensive transport of fluviatile sand to the
present shelf area. There it could have been sorted, resulting in large
beach placer deposits. In particular, the shelf seaward of areas where
beach placer deposits and fluviatile placers are mined can be regarded
as highly prospective.

Sunda Shelf

The onshore prospecting concept for cassiterite, one of the heavier
heavy minerals, originally developed for the Indonesian tin islands, was
successfully applied also on the Sunda Shelf (Sujitno, 1977, Bon, 1979).
Seismic profiles were used to trace buried tin-bearing valleys into the
offshore area. Drilling in these valleys usually proved that the high
cassiterite concentrations continued from the land into the offshore
area. As in the onshore deposits, the most promising sediments are sandy
and gravelly sediments directly overlying the bedrock. Beside this
terrestrial type of tin placer, there are other typically marine types
of cassiterite enrichment (Aleva, 1973), e.g. generated by the process
of elutriation on wave-cut platforms. Another type consists of
cassiterite-rich lag sediments at the bottom of tidal-scour gullies in
the Malacca Strait (Kudrass and Schlüter, 1983). The typically marine
cassiterite-rich sediments on the Sunda Shelf are usually limited in
extent, contain fine-grained cassiterite in low concentrations and are,
therefore, of marginal economic interest. The Andaman Shelf off southern
Thailand is more exposed to waves and currents and consequently the
proportion of marine stanniferous sand is considerably greater than that
on the Sunda Shelf (Rasrikriengkrai, 1983). To date, however, no studies
have been published about the genesis of these deposits, although they
are currently mined.

Northern Mozambique Continental Shelf

The 80-km-wide north Mozambique Continental Shelf off the Zambezi
Delta is broken by a 3 m to 10 m high scarp which approximately coincides with the 50-55 m isobaths. Seaward of this and landward of the
shelf break at 120 m water depth, an underwater dune field provides
evidence that velocities of the south-flowing Mozambique Current at such
depths are high. On the middle shelf, small ripples oriented parallel
to the coast are formed by wave action and tidal currents. On the inner
shelf a countercurrent flows northeastwards and entrains sand and silt
from the Zambezi River. Fluvial sand thus remains within the nearshore

zone and is eventually added to the zone of Holocene barrier beach, that is locally as wide as 10 km (Jaritz et al., 1977). A large placer deposit of ilmenite, rutile and zircon exists in Holocene beach ridges and in adjacent dunes on a retreating part of the coast near Micaune. Immediately offshore, silt accumulates in a belt that extends parallel to the coast at water depths of 10 to 20 m. This belt extends as far as 600 km to the northeast of the delta.

Terrigenous, medium grain-sized sand covers most of the Shelf except for a 100-km-wide zone of fine sand crosscutting the shelf off the Zambezi Delta (Beiersdorf et al.,1980). This fine-sand zone marks the retreat path of the river during the Holocene transgression. The average heavy-mineral concentration in the shelf sand based on analyses of 456 surface samples is 3.58 wt. %. There are three areas where the average heavy-mineral concentration exceeds 5%: in a dune field at a water depth of between 60 and 100 m, and in two areas off the Zambezi Delta at water depths of between 50 and 60 m and between 50 and 25 m (Fig. 2). The valuable minerals are the same as in the onshore deposit: mainly ilmenite, some zircon and lesser amounts of rutile.

The vertical dimension of the heavy-mineral-rich sand was investigated by vibracoring (cores as long as 6 m taken) and counterflush drilling (holes to 10 m subsurface depth, Beiersdorf et al., 1980). The sand in the dune field on the outer shelf is thin and is moving over gravelly lag sediments which overlie a stiff Pleistocene clay. The sand thickness is considerably greater in the area off the main mouth of the Zambezi where more than 9 m of Holocene sand were penetrated (Fig. 3). Areas with high surficial heavy-mineral concentrations were noted commonly to have high concentrations in cores also. The highest concentrations were found in a Holocene sand lens occurring between the 50 m and 60 m isobaths (Fig. 3). Heavy minerals are disseminated throughout the cores but do not contain the heavy-mineral-rich layers typical of beach placer deposits.

A preliminary assessment of the heavy-mineral-rich areas indicates, that a resource potential of 50 million tons of ilmenite, 4 million tons of zircon, and one million tons of rutile exists. The heavy-mineral suite contains more ilmenite, rutile and zircon than fluviatile sand from the Zambezi, indicating that the process of enrichment occurred in a beach environment. During the Holocene transgression (indicated by reworked peat with a ^{14}C age of 9800 years), the beach placer deposits were reworked and their sand partially spread over the directly adjoining hinterland (Figs. 2 and 3) producing an enlarged, but disseminated placer deposit.

Southeast Australian Continental Shelf

The southeast Australian Shelf has an oceanographic setting similar to that of the Mozambique Continental Shelf. The East Australian Current flows southwards and transports sandy sediment along exposed parts of the middle and outer shelf. It induces a countercurrent flowing northwards, which is accelerated by wind and waves from the south and southeast, resulting in a northward littoral drift (Roy and Thom, 1981).

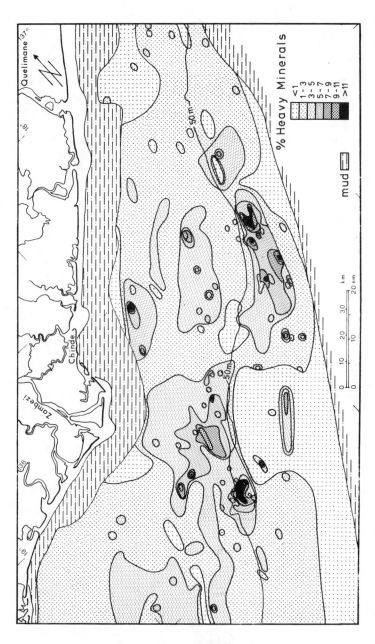

Fig. 2. Heavy-mineral distribution of the sandy surface sediments (456 samples) of the Mozambique Shelf between Zambezi Delta and Quelimane River (from Beiersdorf et al., 1980)

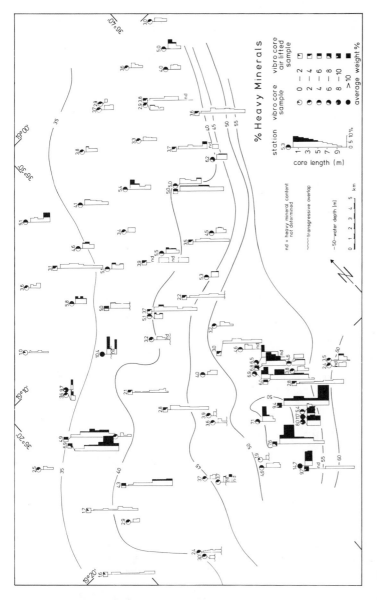

Fig. 3. Heavy-mineral concentrations in the core samples offshore of the Zambezi Delta. Symbols and numbers above the diagrams show the average heavy-mineral concentration for each core (Beiersdorf et al., 1980)

    The geological setting is different from that of Mozambique. The
shelf is narrow (10-20 km) with a shelf break located at 70-150 m.
Modern rivers do not supply significant quantities of sand to the
coast (Jones et al., 1982) as the Zambezi does on the Mozambique coast.
    Holocene and Pleistocene (Last Interglacial) beach placer deposits,
which are mined for rutile, zircon, monazite, and ilmenite, are spread
along the Australian coast. Many of the rich beach placer deposits are
exhausted and mining activities have been extended to the large low-
grade accumulations in coastal eolian dunes (Jones and Davies, 1979).

Fig. 4.   Areas of the East Australian Shelf surveyed and sampled for
heavy minerals (from Jones et al., 1982).

    Four shelf areas (Fig. 4) were investigated in detail (Jones et
al., 1982). This involved taking closely-spaced air-gun seismic-reflec-
tion profiles and taking 640 grab samples and 650 m of vibracores. In
most areas, the Holocene sand of the outer and middle shelf is only a
few meters thick. In a few nearshore areas south of headlands, thick
Holocene sand lenses have accumulated, because at these locations the
northward longshore drift interferes with the East Australian Current
and current velocities drop.

The Pleistocene sediments on the shelf are confined by two major unconformities. Shelf areas where large rivers enter the sea show a thicker Pleistocene sedimentary sequence, especially in the middle shelf region. Near the mouth of the large Clarence River, elongated sand lenses (Fig. 5) were found parallel to the coast and in water depths of 50 m, 57 m and 64 m (Schlüter, 1982). These lenses consist of gravelly sand that incorporate thick well-rounded shell fragments partly cemented together in a carbonate matrix. These sand bodies are interpreted to be drowned Pleistocene beach barriers (Kudrass, 1982). Landward of them, Pleistocene mud, silt, very fine sand, and sand-mud intercalations indicate a back-barrier or lagoonal environment. The barrier systems are interrupted off the mouths of the main rivers, gaps which represent the outlets for the ancient drainage system (Fig. 5). Further to the north, the barriers are drastically reduced in height and width, indicating that the fluviatile sand source is more distant.

These barrier systems could be potential sites for Pleistocene beach placer deposits as rich as the deposits along the modern coast. But the cores revealed great differences between the heavy-mineral assemblages of the Holocene and Pleistocene beach ridges (Riech et al., 1982). The modern beach sand contains ten times more heavy minerals than the Pleistocene sand on the shelf. Furthermore the heavy-mineral suites in Holocene beach sand are relatively rich in valuable minerals, such as rutile, ilmenite, zircon and tourmaline. In contrast, the heavy-mineral suites in Pleistocene beach sand contain predominantly such minerals as amphibole, epidote and pyroxene. The Pleistocene beach sands can easily be related to their fluviatile source by comparing their heavy-mineral suites. The low concentration of heavy minerals in the total sediment, and the low concentration of valuable minerals in the heavy-mineral suite, greatly reduce the prospective value of Pleistocene offshore sand, which is almost unaltered fluviatile sand deposited after travelling only a short distance.

The origin of the significant differences between Holocene beach - sand and Pleistocene shelf sand can be explained by studying the transgressive Holocene shelf sand, which is the link between both sediment - types. Within this sediment, the proportion of the mechanically stable and relatively dense heavy minerals (rutile, ilmenite, zircon and tourmaline) in the heavy-mineral assemblage shows a positive correlation with water depth (Fig. 6). Mid-shelf sand contains only 50-60% of this group of heavy minerals and modern beach sand contains less than 15%. This correlation was observed for all four shelf areas of Fig. 4. Regional differences are caused by differences in the provenance of heavy minerals from the hinterland.

Correlation of the composition of heavy-mineral suites with water depth results from a combination of two processes: vertical differentiation of heavy minerals in beach sand caused by wave action, and barrier beach migration caused by the Holocene transgression.

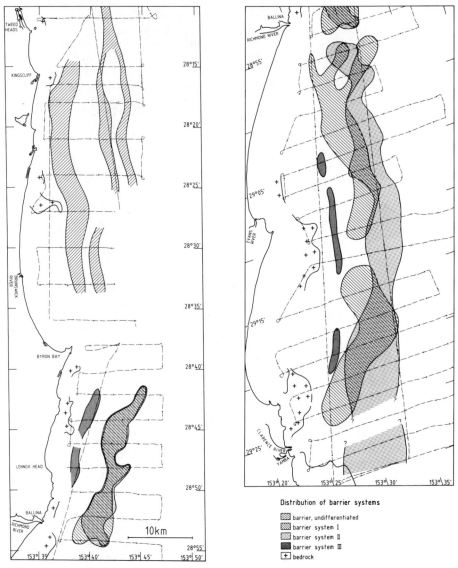

Distribution of barrier systems

barrier, undifferentiated
barrier system I
barrier system II
barrier system III
bedrock

Fig. 5. Regional distribution of barrier-beach systems within the
Pleistocene shelf sediments as recorded by a seismic reflection system
(from Schlüter, 1982). The left map is located at the northern prolon-
gation of the right map. Water depths for the uppermost parts of scarps
at the sea floor are given in meters.

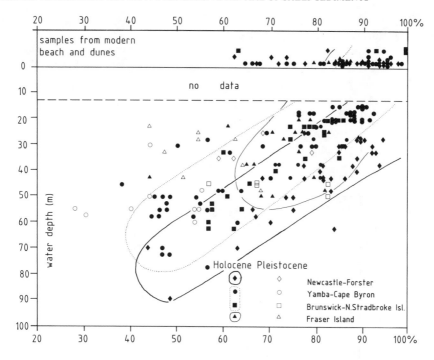

Fig. 6. Proportion of mechanically stable and dense heavy-minerals (rutile + ilmenite + zircon + magnetite + tourmaline) of the total heavy-mineral fraction is plotted as a function of water depth. Heavy, light, and dotted lines indicate the different shelf areas shown in Fig. 4.

Sorting processes in the surf zone, described earlier, produce a vertical differentiation of heavy minerals in beach sand based on density differences. Rutile, ilmenite and zircon become enriched in the upper part of the barrier beach (Fig. 8). Eolian landward transport further accentuates this vertical differentiation. The less dense minerals such as garnet, epidote, pyroxene, and amphibole are enriched in the heavy-mineral suites of the lower offshore parts of the barrier beach. In Eastern Australian beach and dune sands tourmaline is much more frequent than in the nearshore and shelf sediments. The reason for this unexpectedly high proportion in onshore sediments is not known, although it might possibly be a result of beach mining.

   Gow (1967) describes a comparable vertical differentiation in
samples from the Taranaki coast, New Zealand (Fig.7). Komar and Wang
(1984) analyzed the relative sorting and transport processes, and found
that they produce a similar effect in beach sands on the Oregon coast.

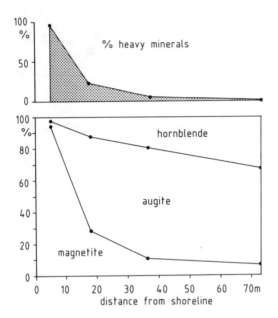

Fig. 7. Concentration of heavy minerals and composition of the heavy-
mineral suites in 4 samples from the Taranaki coast, New Zealand.
Concentration and composition of the samples are plotted as a function
of their distance from the shoreline (recalculated data from Gow,
1967).

   On the Mozambique coast, by comparison, the nearshore zone is a
sink for amphibole which is removed from the beach sand (Jaritz et al.,
1977, Beiersdorf et al., 1980), and on one of the barrier islands off
Holland, garnet, which is has the highest specific gravity in the
heavy-mineral assemblage, is significantly enriched in dunes and sand
of the upper beach compared to sand of the lower beach (Veenstra and
Winkelmolen, 1976). Density and grain-size sorting presumably produces
this vertical differentiation in beach sands of most high-energy coasts.
   The second process involved in the genesis of Holocene shelf sand
and beach sand is the transgressive migration of barrier beaches
(Swift, 1975). The early Holocene rise of sea level moved the barrier
beach system by storm-washover processes like a bulldozer across the
shelf. On the seaward side of the beach, the estuarine sediments of the
migrating system were subjected to erosion, as is shown by pebbles
of beach rock, wood and reworked Pleistocene estuarine shells directly
overlying older Pleistocene sediments (Kudrass, 1982). On the transgressed shelf, a trailing sheet of sand was left behind with reworked

Pleistocene sediment at its base (Fig. 8). This sand, which was originally deposited as nearshore sand, has been partly reworked and today
represents the Holocene cover of the middle shelf.

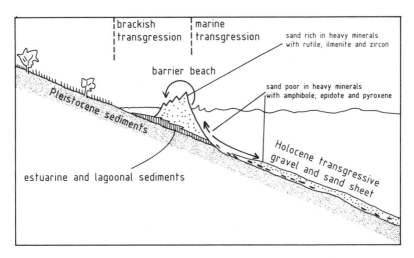

Fig. 8.   Early Holocene transgressive migration of a barrier beach and
vertical differentiation of heavy-mineral suites (Australian model).

During the transgression the lighter heavy-minerals, first pyroxene
and later epidote, were gradually removed from the heavy-mineral suites
of the barrier sand by density sorting and preferentially deposited in
the lower part of the barrier sand. From this the Holocene transgressive
sand sheet originated as the sea level rose. Thus, the economically
important minerals rutile, zircon and ilmenite were preferentially
transported landwards and their concentration in the barrier sand
continuously increased. The sand arriving at the modern coastline was
only a fraction of the total volume of sand initially moved, but it was
an ideal substratum for further differentiation processes which in turn,
led to the formation of the economically important beach placer deposits
along the east coast of Australia.

Pilkey and Lincoln (1984) described a similar distribution of heavy
minerals from the Puerto Rico Shelf. The inner shelf sand is rich in
heavy minerals and the predominant mineral is magnetite, while sand poor
in heavy minerals on the middle and outer shelf contains mainly pyroxene
and amphiboles.

MODELS FOR HEAVY-MINERAL PROSPECTING OF SHELF AREAS

The different processes by which heavy minerals are transported and concentrated give rise to different types of deposits. To aid the prospecting for these deposits, three genetic models are presented, one for the heavier heavy-minerals (Indonesian model) and two for the lighter heavy-minerals (Australian and Zambezi models).

Indonesian model

The heavier heavy-minerals, such as gold, cassiterite and wolf-ramite, are concentrated in the immediate neighbourhood of their source rocks. Their high density (> 5.5 g/cm^3) and primary occurrence as sand--sized minerals prevent transport over long distances, so that they usually become trapped at the elutriated surface of the source rock, or after minimal transport by bed load at the bottom of river valleys. Onshore prospecting for cassiterite on the Indonesian tin islands of Bangka and Belitung was guided by this model, which was later success-fully applied to the surrounding Sunda Shelf (Sujitno, 1977, Bon, 1979). Most of the cassiterite recovered in the offshore area comes from fluviatile deposits directly overlying the basement rocks. Prospecting in the other productive offshore area, the shelf of the Andaman Sea off Thailand, probably also relies on this Indonesian model.

Australian model

The Australian model concerns placer deposits of the lighter heavy-minerals along high-energy coasts with narrow shelves and a low to moderate supply of terrigenous material. The model is based on the vertical differentiation of heavy minerals within the sediments of the beach profile: economically valuable minerals, such as rutile, ilmenite, leucoxene, magnetite, zircon, monazite, and garnet, are enriched in the upper part of the barrier beach, while the less dense heavy-minerals, such as amphibole, pyroxene, epidote, and sillimanite, are more frequent in the lower section of the barrier. During the Holocene or the Last Interglacial transgression, the sediments of the barrier beach migrated across the shelf. During this migration, the upper layers of the barrier sediments were preferentially transported landward, and thus the concentration of economically valuable minerals was continuously upgraded. The uneconomic minerals were predominantly deposited in the trailing sand sheet left behind on the shelf. Conse-quently, Pleistocene barrier beaches are usually not preserved on the present shelf. Sometimes, the truncated, and possibly carbonate cemented, lower parts of the barriers are preserved in which heavy-mineral suites strongly resemble the original fluviatile source material.

The Holocene or Last Interglacial placer deposits found along the present beach originated by heavy-mineral sorting and selective trans-gressive migration and by erosional sorting during the subsequent period of steady sea-level. These placer deposits are no indication that similar deposits exist on the shelf, but rather that most of the valuable minerals have been removed from the shelf. Disseminated placer deposits may be preserved on narrow shelves wherever the migrating barrier beach has been trapped in channels or behind reefs or head-lands.

There are several thoroughly investigated shelf areas adjacent to extended beach-placer deposits which have a sedimentary history corre-sponding to the Australian model. On the west coast of North Island, New Zealand, large magnetite placer deposits are exploited, and on the adjacent shelf magnetite, as shown by coring, occurs only as a minor component (Carter, 1980). The bedding typical of beach placers is not present in the shelf sediments and it seems that most of the magnetite currently mined onshore was deposited by the Holocene transgression. A similar situation is found in the offshore parts of the Japanese magnetite placer deposits (Maruyama, 1969).

Meyer's (1983) investigation of the seaward extension of the ilmenite-rutile-zircon beach deposit in Pulmoddai, Sri Lanka, showed that the heavy-mineral-rich sediments are confined to the beach and nearshore zone, and that the proportion of rutile and zircon in the heavy-mineral fraction decreases strongly seaward. This deposit has no offshore equivalent either. Similarly, seismic profiling and coring showed that the exhausted ilmenite deposits on the beaches of southern Senegal also have no equivalent on the shelf (Tixeront, 1978).

Zambezi model

This model refers to placer deposits of the lighter heavy-minerals along high-energy coasts, which have a broad shelf and a moderate to high supply of terrigenous sediments. The model is based on the concept that during the long periods of low sea level in the Pleistocene, placer deposits formed along the barrier beach of the delta of a large river. During the Holocene transgression, sediments of the barrier beach containing placer minerals were reworked, homogenized and depos-ited near the Pleistocene coastline. It is assumed that the large volume of beach sand, deposited during the pre-transgressive phases, could not be totally entrained by the transgressive sea, and that the broadness of the shelf, providing a formidable expanse for the trans-gression, prohibited barrier-beach migration comparable to that of the Australian model. This led to the formation of a disseminated placer deposit on the middle Zambezi Shelf. There is no analogous deposit presently known from other shelf areas.

CONCLUSION

Overall, the heavy-mineral potential of shelf areas is low in comparison to the potential of the present coastal plains. As gold, cassiterite and wolframite are usually concentrated near their respective source rocks (Indonesian model), which are predominantly magmatic rocks, the shelves are not prospective for these minerals.

Concerning the more common heavy-minerals such as ilmenite, rutile, magnetite, zircon, and garnet, most narrow shelf areas with limited terrigenous supply may have been depleted in these minerals by barrier-beach migration during the Holocene or former transgressions (Australian model). On certain shelves, trap structures may have preserved disseminated placer deposits. The prospective value of narrow shelves with a copious terrigenous supply may also be low, because during long periods of low sea level in the Pleistocene, most of the terrigenous material was channeled directly into the deep sea and lost to beach-sorting processes on the shelf. The widespread occurrence of Pleistocene carbonate sediments on the tropical and subtropical outer shelf, either in the form of reefoidal sediments, aragonitic mud or oolitic sand, is compatible with the total lack of terrigenous sedimentation.

Broad shelves with a high supply of terrigenous sediments may have a good potential for the occurrence of disseminated placer deposits (Zambezi model), especially if beach placer deposits occur along the present coast.

REFERENCES

ALEVA, G.J.J., 1973, Aspects of the historical and physical geology of the Sunda Shelf essential to the exploration of submarine tin placers; Geol. Mijnbouw, v. 52, no. 2, pp. 79-91.

BEIERSDORF, H., KUDRASS, H.R., and STACKELBERG, U. von, 1980, Placer deposits of ilmenite and zircon on the Zambezi shelf; Geol. Jahrb., v. D-36, pp. 5-85.

BON, E.H., 1979, Exploration techniques employed in the Pulau Tujuh tin discovery; Transact. Inst. Mining & Metall., Sect. A, v. 88, pp. A13-A22.

CARTER, L., 1980, Ironsand in continental shelf sediments off western New Zealand - a synopsis; New Zealand Jour. Geol. Geophys., v. 23, pp. 455-468.

ENGELHARDT, W. von, 1937, Über die Schwermineralsande der Ostseeküste zwischen Warnemünde und Darßer Ort und ihre Bildung durch die Brandung; Zeitschrift angew. Min., v. 1, no. 1, pp. 30-59.

GOW, A.J., 1967, Petrographic studies of ironsands and associated sediments near Hawera, south Taranaki; New Zealand Jour. Geol. Geophy., v. 10, no. 3, pp. 675-696.

JARITZ, W., RUDER, J., and SCHLENKER, B., 1977, Das Quartär im Küstenge-
biet von Mocambique und seine Schwermineralführung; Geol. Jahrb.,
v. B-26, pp. 3-93.

JONES, H.A. and DAVIES, P.J., 1979, Preliminary studies of offshore
placer deposits, eastern Australia; Marine Geol., v. 30, pp.
243-268.

JONES, H.A., KUDRASS, H.R., SCHLÜTER, H.U., and STACKELBERG, U. von,
1982, Geological and geophysical work on the east Australian shelf
between Newcastle and Fraser Island- a summary of results from the
SONNE cruise SO-15; Geol. Jahrb., v. D-56, pp. 197-207.

KOMAR, P.D. and WANG, C., 1984, Processes of selective grain transport
and the formation of beach placers; Jour. Geol., v. 92, pp. 637-655.

KUDRASS, H.R., 1982, Cores of Holocene and Pleistocene sediments from
the east Australian Continental Shelf (SO-15 cruise, 1980); Geol.
Jahrb., v. D-56, pp. 137-163.

KUDRASS, H.R. and SCHLÜTER, H.U., 1983, Report on geophysical and
geological offshore surveys in the Malacca Strait (1976); In: Exploration for offshore placers and construction materials; Committee
Co-ord. Joint Prosp. Min. Res. in Asian Offshore Areas, CCOP/TP 12,
pp. 115-154.

MARUYAMA, S., 1969, Exploration of iron sand deposits in Volcano Bay,
Hokkaido Japan; Rept., 6th Session Committee Co-ord. Joint Prosp.
Min. Res. in Asian Offshore Areas, Bangkok 1969, pp. 93-102.

MIDDLETON, G.V., 1970, Experimental studies related to problems of
flysch sedimentation; In: Lajoie, J. (ed.), Flysch sedimentology in
North America; Geol. Assoc. Canada, Spec. Publ. 7, pp. 253-272.

MEYER, K., 1983, Titanium and zircon placer prospection off Pulmoddai,
Sri Lanka; Marine Mining, v. 4, pp. 139-166.

PILKEY, O.H. and LINCOLN, R., 1984, Insular shelf heavy mineral
partioning in Northern Puerto Rico; Marine Mining, v. 4, no. 4, pp.
403-414.

RASRIKRIENGKRAI, P., 1983, Geological - geophysical investigations for
detrital heavy minerals in the south eastern Andaman Sea, Southern
Thailand; In: Exploration for offshore placers and construction
materials; Committee Co-ord. Joint Prosp. Min. Res. in Asian
Offshore Areas Project Office, CCOP/TP 12, pp. 54-99.

RIECH, V., KUDRASS, H.R., and WIEDICKE, M., 1982, Heavy minerals of the
east Australian shelf sediments between Newcastle and Fraser Island;
Geol. Jahrb., v. D-56, pp. 179-195.

RITTENHOUSE, G., 1943, The transportation and deposition of heavy
minerals; Geol. Soc. America Bull., v. 54, pp. 1725-1780.

ROY, P.S. and THOM, B.G., 1981, Late Quaternary marine deposition in New
South Wales and southern Queensland - an evolutionary model; Jour.
Geol. Soc. Australia, v. 28, pp. 471-489.

RUBEY, W.W., 1933, The size distribution of heavy minerals within a
    water-laid sandstone; Jour. Sed. Petrol., v. 3, pp. 3-29.

SALLENGER, A.H., 1979, Inverse grading and hydraulic equivalence in
    grain-flow deposits; Jour. Sed. Petrol., v. 49, no. 2, pp. 553-562.

SCHLÜTER, H.U., 1982, Results of a reflection seismic survey in shallow
    water areas off east Australia, Yamba to Tweed Heads; Geol. Jahrb.,
    v. D 56, pp. 77-95.

SEIBOLD, E. and BERGER, W.H., 1982, The seafloor: an introduction to
    marine geology; Springer-Verlag, Berlin, 288 p.

SLINGERLAND, R., 1984, Role of hydraulic sorting in the origin of
    fluvial placers; Jour. Sed. Petrol., v. 54, no. 1, pp. 137-150.

SUJITNO, S., 1977, Some notes of offshore exploration for tin in
    Indonesia, 1966-1976; Committee Co-ord. Joint Prosp. Min. Res. in
    Asian Offshore Areas, Tech. Bull., no. 11, pp. 169-182.

SWIFT, D.J.P., 1975, Barrier-island genesis : evidence from the central
    Atlantic shelf, eastern USA; Sediment. Geol., v. 14, no. 1, pp.
    1-43.

TIXERONT, M., 1978, Ilmenite prospection on the continental shelf of
    Senegal : methods and results; Marine Mining, v. 1, no. 3, pp.
    171-187.

VEENSTRA, H.J. and WINKELMOLEN, A.M., 1976, Size, shape and density
    sorting around two barrier islands along the north coast of Holland;
    Geol. Mijnbouw, v. 55, pp. 87-104.

EXPLORING THE OFFSHORE AREA OF N.E. GREECE FOR PLACER DEPOSITS:
GEOLOGIC FRAMEWORK AND PRELIMINARY RESULTS

C. PERISSORATIS, I. ANGELOPOULOS AND D. MITROPOULOS
Institute of Geology and Mineral Exploration
70 Mesoyion str.,
Athens,
Greece

ABSTRACT. This paper presents the preliminary results of investigation
on the general geology and placer deposits in nearshore and offshore
areas of Eastern Macedonia and Thraki. The research, carried out from
1981 to 1984, included surface sampling, coring and bathymetric and
seismic profiling. The examined shelf area is covered by sands, silts
and clays with finer sediments occuring to the west, and coarser to
the east. Most of the coarse grained sediments are palimpsest or
relict. The microscopic examination showed that the principal types of
heavy minerals are amphiboles, pyroxenes, garnet, olivine, magnetite
and pyrite. The heavy mineral content in the 2 to 3 phi fraction ranges
from less than 5% to more than 20%. High concentrations of heavy
minerals occur in nearshore and the outer shelf areas, but, on the
basis of the data at hand, no significant placer deposits are expected
to occur in area investigated.

INTRODUCTION: REGIONAL GEOLOGY

The offshore area of Greece is twice that of the land. The seafloor typically exhibits highly variable bathymetry and a complex
morphology. In the largely unexplored shelf area, a sector where most
of the requirements for formation of placer deposits seem to be present
is the area offshore of Eastern Macedonia and Thraki (Fig. 1), which is
characterized by an extensive shelf (compared with other shelf areas)
and by relatively shallow water depths. Land areas adjacent to this
sector are largely composed of metamorphic and igneous formations some
of which are mineralized (mixed sulfides, manganese, copper, gold,
etc.). The land is drained by a number of rivers, the largest of which
are Strimon, Nestos and Evros.
    This area was surveyed by the marine geology department of the
Institute of Geology and Mineral Exploration of Greece between 1981 and
1984, except for vibracoring, which phase is scheduled for 1985. During
the last two years this project was sponsored by the Commission of
European Communities.
    The study area extends from Chalkidiki Peninsula to the Greek-

P. G. Teleki et al. (eds.), Marine Minerals, 57–70.

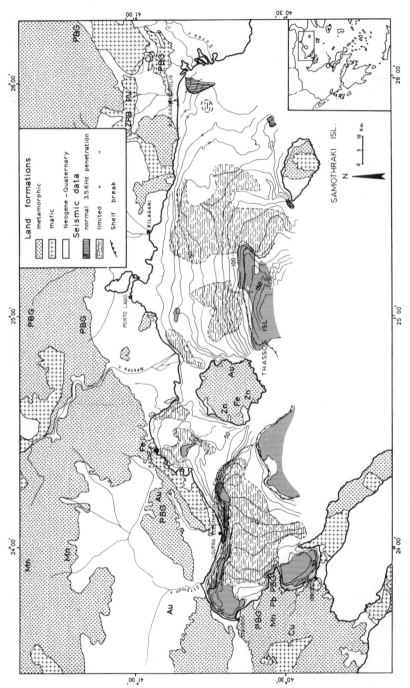

Figure 1. Geology of the surrounding area and bathymetric map. PBG, Mn, Au, etc., mineralizations on land. Depth in meters.

Turkish border at Alexandroupolis and to Samothraki Island. Morphologically it is divided at Thassos Island into two parts, Strymonikos Plateau to the west and Samothraki Plateau to the east (Fig. 2).

SURVEY METHODS

The surveys of these plateaus consisted of surface sampling and coring, and bathymetric, seismic and magnetic profiling. The research was carried out by using conventional sampling and gravity coring apparatus, two seismic profiling systems (a 3.5 KHz and a 300 joule Uniboom) and a Barringer proton magnetometer. All these instruments were installed on suitable fishing boats (trawlers). Navigation and positioning was controlled by means of a Motorola Miniranger system and by radar. A total of 500 surface samples, 80 gravity cores, and about 8,500 km of seismic, bathymetric and magnetic profiles were collected.

This offshore region was essentially unsurveyed until the commencement of this project except for a few local studies, such as in the Gulf of Strymonikos (Konispoliatis, 1984) and in the Gulf of Alexandroupolis (Marinos et al., 1978). The unconsolidated sediments (sands, gravels and their heavy mineral content) in the coastal and inshore ares were, however, investigated earlier (Marcoulis et al. 1978; Papadakis, 1975; Pe and Panagos, 1981).

BATHYMETRY AND SEISMIC FACIES

The bathymetric map constructed from the data gathered shows that Strymonikos Plateau, the area west of Thassos Island, is more complex morphologically than Samothraki Plateau, east of the island. The former extends into three minor bays (Gulfs of Ierissos, Strymonikos and Kavalla), while the latter and larger plateau is a low-lying submerged area. Several hills, ponds, terraces and channels, which occur mainly in the inner or landward parts of the gulfs, are of sedimentary or tectonic origin. The shelf break is usually easily distinguishable west of Thassos Island at a depth of 120 to 125 m. In contrast, the offshore area east of Thassos Island is nearly flat. There are two minor embayments, at Porto Lago and Xilagani and a few small submarine channels and hills at the easternmost extent of the plateau between Samothraki and Alexandroupolis. The shelf break is indistinct at about 130 m of water depth.

Neither the 3.5 KHz nor the Uniboom systems could achieve penetration of the sediments in the outer, more offshore, parts of the study area. In the inner part of the bays, near river mouths, and below the shelf break normal penetration was, however, obtained. Detailed examination of the profiles showed that the late Wurmian unconformity is, in several locations, well distinguishable and has the form of an erosional horizon. Above this horizon two distinct seismic units can be identified. The lower unit consists of alternating relatively thick (2 to 3 m) opaque and transparent horizons of variable configuration. This unit fills the channels and valleys of the erosional surface. Its

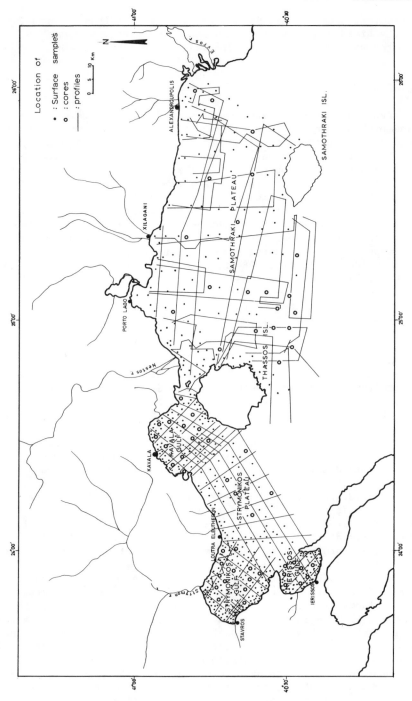

Fig. 2. Data map and physiographic provinces.

thickness ranges from one to more than 10 m, and represents coarse
grained sediments (sands, gravels) deposited apparently during the last
fluvial-transgressional phase. The upper unit consists of thin, horizontal to subhorizontal transparent and opaque reflectors with nearly
constant thickness locally, ranging in different areas from a few to as
much as ten meters. The sediments are finer than those comprising the
lower unit, principally fine sands, silty sands and silts which were
deposited after the Wurmian transgression when open marine conditions
were predominating. At several locations, where the upper unit attains
the greater thickness, masking effects were noted in the seismic profiles ascribable to the presence of organic gases within the sediments.

SEDIMENT TEXTURE AND COMPOSITION

     The sediment types together with the percentage of sand and the
principal types of sand are depicted in Figures 3, 4 and 5. The predominant types of surficial sediments on the Strymonikos and Samothraki
Plateaus differ from one another. On the Strymonikos Plateau, the
nearshore, as well as the outer shelf sediments are comprised of sands
and silty sands. The surficial sediments, however, which cover the
central part of the Gulfs of Ierissos, Strymon, and Kavalla, as well as
that of the inner shelf, are fine grained, namely silty clays and
clayey silts. On the Samothraki Plateau, on the other hand, the sand
and silty sand units cover a much more extensive area, and the finer
sediment types (sandy silts, clayey silts) occur only locally near the
river mouths and seaward of the shelf break (Fig. 3). All these sediments are generally poorly to very poorly sorted.
     Although the entire area is covered largely by sandy sediments
(Fig. 4, 5), the distribution of the principal modes of sand (Fig. 5)
shows that the sands are fine to very fine, with coarser sands being
present only in a narrow strip along the coastline of the mainland and
adjacent to the islands of Thassos and Samothraki.
     The examination of the sand fraction showed that it consists of
terrigenous and biogenic components, represented mainly by quartz,
rock fragments and feldspars, and by bioclasts and foraminiferal tests,
respectively. Locally, north of Thassos and east of Samothraki, glauconite occurrences were noted. Terrigenous components are abundant
in nearshore and mid-shelf areas. Bioclasts, which commonly have broken
and worn tests, occur in high concentrations at a few sites in the mid-
shelf sector.

HEAVY MINERALS

Previous studies on placer deposits in coastal sediments.

     The presence of heavy mineral concentrations in the coastal sedi-
ments offshore of Alexandroupolis was first reported by Marinos (1955)
who found magnetite in the beach sands, with a TiO_2 content of 6% to
8%. Based on a later study, Marinos et al. (1978) reported that the

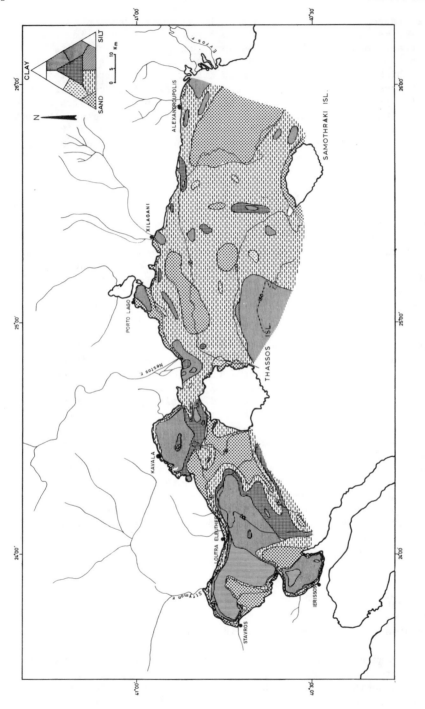

Figure 3.   Distribution of sediment types on the seafloor.

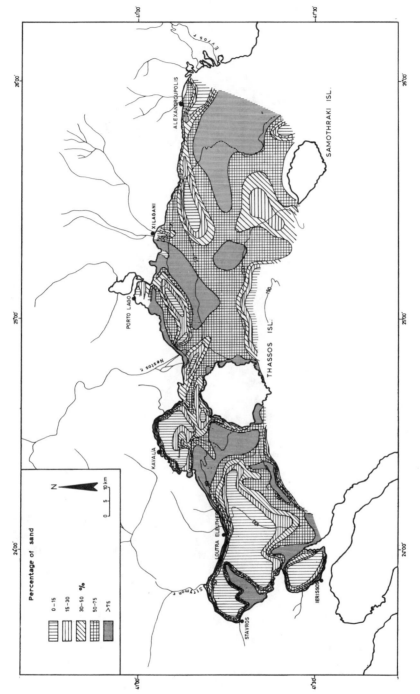

Figure 4. Distribution of sand (percentage in coarse fraction).

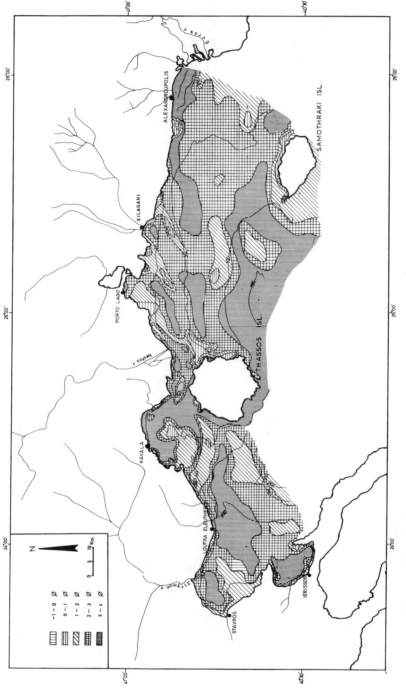

Figure 5. Principal size class distribution of sand.

Ti-magnetite content in the beach sands ranged between 0.2 and 2.7% with a Ti content of 2.7 to 4.8%. The same authors carried out a limited areal sampling of the seafloor off Alexandroupolis, where they found a Ti-magnetite content in the sand fraction of 1 to 2.8% with a Ti-content of 2 to 5%.

A study conducted at Loutra Eleutheron by Papadakis (1975) reported the presence of black sands in a coastal zone for a distance of about 3 kilometers. He identified, among others, the minerals magnetite, geothite, ilmenite, pyrolusite from the opaque fraction, and the minerals hornblende, garnet, epidote, zircon, and monazite from the transparent fraction. No concentrations were cited.

Subsequently Pe and Panagos (1979) sampled and studied the heavy minerals of the beach sands from the Nestos River eastward to Alexandroupolis. Their report states that the heavy mineral content of the 2.9 to 3.5 fraction ranges from 2.22% to 7.41% with a variable composition in the various heavy minerals (Table I).

Table I. Heavy minerals in the beach sands between the Nestor River and Alexandroupolis (based on 11 samples, after Pe and Panagos, 1979).

Heavy Fraction %	Composition of heavy fraction						
	Opaque % of total	Altered % of total	% of non-opaque, unaltered				
			Amp	Px	Gt	Ep+St	Res
2.22 to 7.41	3 to 60	3 to 83	5 to 83	10 to 58	2 to 58	1 to 7	up to 17

Amp: Amphibole   Ep: Epidote
Px: Pyroxene   St: Staurolite
Gt: Garnet   Res: Tourmaline+zircon+cassiterite+rutile+apatite

The most extensive exploration program for recent placer deposits at the Eastern Macedonia and Thraki beach sands was carried out during the years 1976 to 1978 by the Institute of Geology and Mineral Exploration of Greece (Markoulis et al., 1978). During this program the beach sands of this area were extensively sampled by trenching and drilling methods. The samples were taken at intervals of approximately one kilometer. A total of 361 trench and 48 drill samples, the latter providing 1188 m of cored material, were collected. This investigation aimed to locate concentrations of the minerals magnetite, ilmenite, rutile, zircon and garnet. The preliminary results were not encouraging for further research, as noted by McDonald (1979) when he reviewed the work of Markoulis. McDonald also cites concentration values of heavy minerals in the beach sands from selected areas between Ierissos and Alexandroupolis. The mean values of these concentrations are shown in Table II.

Table II. Concentration (mean values) of heavy minerals in selected
areas from Ierissos to Alexandroupolis (modified after McDonald, 1979).

| A R E A | Type of sampling | % heavy minerals in sand fraction | Types of heavy minerals in % | | |
			Combined magnetite ilmenite	Zircon	Rutile
E. of Alex/lis	core	4.16– 6.91	4.53– 5.36	0.12–0.16	0.72–1.2
Xilagani	core	5.9 – 7.77	3.41–17.76	0.01	0.39–0.88
E. of Nestos R.	core	3.85– 8.60	7.54–10.93	0.07–1.03	0.23–0.41
W. of Nestos R.	core	3.06– 5.09	5.21–10.68	0.06–0.07	0.04–0.06
Loutra Eleutheron	trench	4.75–11.77	20.21–73.25	0.13–0.15	0.06–0.46
Strymon R.	core	3.06– 5.09	5.21–10.23	0.06–0.32	0.04–0.07
Stavros	pit	7.98–12.71	2.99– 6.39	0.09–0.65	0.09–0.66

From the economic viewpoint these values were found to be very
low, hence, no further exploration was carried out on land. Later it
was suggested that additional investigations of areas further offshore
were warranted in order to examine whether Pleistocene sea level fluctuations could have concentrated heavy minerals in deeper waters.

Present Results

The research of the offshore area started in 1981 and was gradually extended to the shelf break, in order to construct first the overall
geological description of the area. As part of this survey, samples of
surficial sediments were collected, which led to an examination of their
heavy mineral content. Preliminary evaluation of the heavy mineral content of the 2φ to 3φ sand fraction is depicted in Fig. 6. As shown in
this figure, two subareas can be distinguished where high concentrations
of heavy minerals occur. One is the nearshore, the other is the outer
shelf. In nearshore areas high concentrations occur off the northern
and southern coasts of Ierissos Gulf, between Loutra Eleutheron and
Kavalla, around the mouth of the Nestos river, the east coast of
Thassos, and between Alexandroupolis and the Evros River. On the outer
shelf, areas of high concentrations occur west of Thassos Island and in
the outer and central parts of Samothraki Plateau. Heavy mineral concentrations are reported by Konispoliatis (1984) to occur also around
the coasts of Strymonikos Gulf, with concentrations of as much as 25%
in the sand of fraction. However, information about the percentage of
heavy minerals in the 2φ to 3φ fraction was not given, and as a result,
the relevant contents are not depicted in Figs. 6 and 7.
The heavy minerals recognized in this study are amphiboles, pyroxenes, epidote, garnet, olivine, zircon, rutile and tourmaline in the

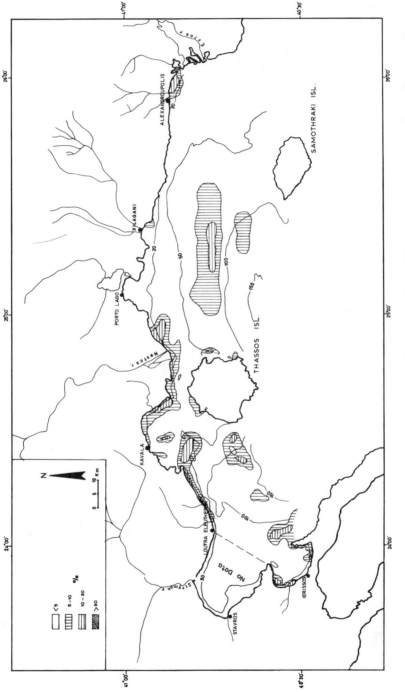

Figure 6. Distribution of the percentage of heavy minerals in the 2 Φ to 3 Φ size fraction.

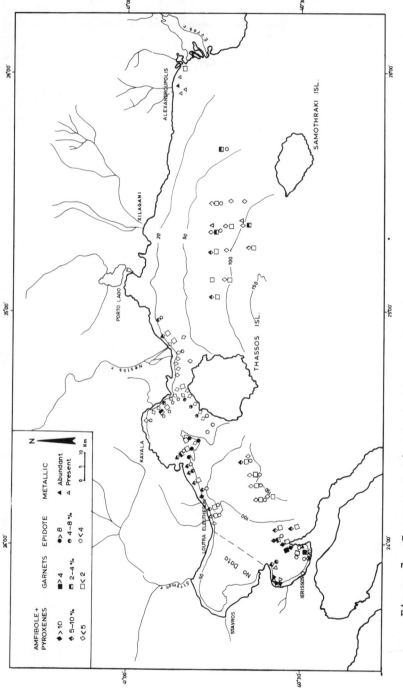

Figure 7. Percentage of selected heavy minerals in the 2Φ to 3Φ size fraction.

transparent category, and among the metallics, pyrite, Ti-magnetite
and magnetite. The distribution of the principal heavy minerals (amphiboles, pyroxenes, garnet, epidote, metallics) contained in the 2ϕ to 3ϕ
sand fraction shows that those most common are amphiboles and pyroxenes,
derived from the abundant metamorphic formations on land, or from
second cycle concentrations in the Neogene and Quaternary formations.
Among the remaining transparent heavy minerals, garnet shows occasionally significant concentrations along the southern coasts of Ierissos and
near Loutra Eleutheron, with its origin in local occurrences of ultramafic rocks and chromite deposits on the nearby land. Local concentrations of garnet also occur around the mouth of the Nestos River.
Epidote, on the other hand, is more abundant around the northern coast
of Thassos derived from the metamorphic land formations, which are rich
in this mineral.

Among the metallic minerals, pyrite is locally abundant near the
northern and southern coasts of Ierissos Gulf, a natural consequence of
the presence of numerous pyrite mines in the surrounding area. The
other metallic mineral present, magnetite, was identified offshore of
Alexandroupolis, as reported also by Marinos et al. (1978). The magnetite was probably deposited by the Evros River and other smaller rivers
draining the metamorphic rocks rich in Ti-minerals to the north.

On the outer shelf, landward of the shelf break, high concentrations of heavy minerals also occur, also consisting mainly of amphiboles, pyroxenes, garnets and epidote, which may have been deposited
during periods of earlier, lower sea-level stands (submerged shorelines).

CONCLUSIONS

The preliminary study of the area offshore of Eastern Macedonia and
Thraki indicates that:

Most of the surficial sediment cover is composed of fine sands,
silty sands and sandy silts. Although the adjacent land is drained by
many rivers, only a small part of the area investigated, mainly to the
west, is covered by recently deposited sediments. Thus, the greater
part of the area contains palimpsest of relict deposits. The principal
suite of heavy minerals in the sand fraction of these sediments consist
of amphiboles, pyroxenes, epidote, garnet, zircon, olivine, rutile and
tourmaline of the transparent type, and pyrite, magnetite and Ti-
magnetite of the opaque type. However, the heavy minerals represent
but a small portion of the sand fraction. Local concentrations (around
10% to 20% in the 2ϕ to 3ϕ size class) occur in a few nearshore areas
and on the outer shelf. The former is related to local mineralization
on nearby land, while the concentration of heavy minerals on the outer
shelf is connected with presence of old shorelines formed at earlier
low sea-level stands.

Based on the data available at this time, it appears that the
prospect for the occurrence of major placer deposits in the area offshore of Eastern Macedonia and Thraki is small. Possible reasons for
this include:

a) the sediments presently deposited on the seafloor by the larger
rivers contain only the fine fraction (fine sand and silt), the coarser
being deposited as alluvial fill prior to reaching the coast;
b) the relatively extensive occurrence of Quaternary and Neogene
sedimentary formations in the area close to the shore supply abundant
terrigenous sediments, which may mask possible placer concentrations;
c) the low energy waves and currents may not be sufficient to sort out
hydraulically the heavier sediment grains, a possible factor also noted
earlier by McDonald (1979).

ACKNOWLEDGEMENTS

    We thank Dr. K. Papavasiliou and Dr. G. Katsikatsos, Director
General and Director of IGME respectively, for their help in carrying
out the research. We also thank the Commission of European Communities
for partly funding the project. Dr. A. Moorby kindly reviewed the
manuscript.

REFERENCES

Konispoliatis, N., 1984, Study of the Strimonikos Gulf sediments; un-
    publ. doctorate thesis, National Technical Univ., Athens, Greece,
    109 p. (in Greek).

Marinos, G., 1955, Deposits of Ti-magnetite sand near Alexandroupolis;
    Inst. Geol. and Min. Expl., Athens, Bull. v. 1, p. 13 (in Greek).

Marinos, G., Mariolakos, H., Sabo, B., 1978, Ti-magnetite sand deposits
    of the beach of Alexandroupolis Bay (Thrace, Greece); Thalassogra-
    phica, v. 1, pp. 5-20, (in Greek).

Markoulis, M., Orfanos, B., Kaklamanis, N., and Takousis, D., 1978,
    Industrial minerals and rocks. Research for heavy minerals at
    sands of the coastal areas of Eastern Macedonia and Thraki; Unpubl.
    rept., Inst. Geol. and Min. Expl., Athens, 75 p. (in Greek).

McDonald, E.H., 1979, The mineral sand deposits of Thrace and Macedonia;
    Unpubl. rept., United Nations Project UN/TCD/GRE/77/007, 28 p.

Papadakis, 1975, The black sands from Loutra Eleutheron near Kavalla,
    Greece; Sci. Ann., Fac. Phys. and Math., Univ. of Thessaloniki, v.
    15, pp. 331-390.

Pe, G., and Panagos, A., 1981, Heavy mineralogy of river and beach
    sands, Continental Greece; N. Jahrb. Miner., Abt., v. 136, no. 3,
    pp. 254-261.

MARINE PLACERS: RECONNAISSANCE EXPLORATION TECHNOLOGY

David B. Duane
National Sea Grant College Program
National Oceanic and Atmospheric Administration
6010 Executive Boulevard
Rockville, Maryland 20852, USA

ABSTRACT. Active placer mines and known deposits provide the
stratigraphic clues and paleoenvironmental indictors useful to
offshore exploration programs. Existing marine mapping, surveying,
and sampling tools are judged capable of providing most of the kinds
of information needed to find prospective targets for detailed
exploration near coastlines. Beyond the nearshore area, and the
relatively facile extension of terrestrial knowledge, exploration
problems loom large. The problem becomes less the tools or how to
apply them, but rather where to apply them. A few new techniques in
the fields of bathymetric mapping, positioning systems, analytical
methods, and exploration models hold promise for improving the
accuracy and speed of exploration.

INTRODUCTION

     In the United States during the past decade, a series of govern-
ment sponsored studies have examined the topic of marine minerals as a
replacement for terrestrial sources (National Academy of Sciences,
1975 a, b; National Academy of Sciences 1977; National Advisory
Committee on Oceans and Atmosphere, 1983; U.S. Geological Survey,
1984). Groups empaneled have been comprised of representatives from
the academic, industrial, and federal government communities. Recommendations reached by them represented a consensus from the cross
section of groups to be involved in future marine mineral activities.
Each study group had specific purposes, but the overriding single
issue dealt with was the potential viability of exploiting various
marine mineral resources, with the secondary issues being technology
for exploration and recovery. These issues transcend the parochial
interest of a single national or local government and are concerns
common to economic and marine geologists or to any one who, for
whatever reason, feels the need or opportunity to seek marine sources
of minerals to substitute for traditional terrestrial sources.

71

P. G. Teleki et al. (eds.), Marine Minerals, 71–80.
© *1987 by D. Reidel Publishing Company.*

As extensions of continental land masses, continental shelves can be, and have been, explored by extending, and adopting, terrestrial exploration strategies and technologies to the marine environment. That this philosophy in the past has been (or is still) successful, at least near the shore, is attested to by placer recovery operations such as for gold off Nome, Alaska (Nelson, et al., 1972), tin in southeast Asia (Mitchell and Garson, 1981) and diamonds off southwest Africa, (Archer, 1973) and the recent discovery of heavy minerals in delta-like distributaries of the nearshore Gulf of Mexico (Woolsey, 1985, personal communication). These instances mark the successful discovery of marine mineral deposits, produced by terrestrial (mainly fluvial) processes, which were subsequently inundated by rising sealevel. In other instances, deposits formed by marine processes in earlier times are now exposed and mined on land, for example, the Trail Ridge ilmenite deposit in Florida (Pirkle, et al., 1977), the ilmenite deposits of the Kirkwood Formation of New Jersey (Markewicz, et al., 1978) and chromite sands in raised terraces of coastal Oregon (Griggs, 1945).

Interpretation of paleogeography and paleooceanography with a focus toward source area, environment of deposition, transport process and direction permitted development of conceptual models for formation of these deposits. Coupled to the application of terrestrial survey-ing, mapping, and sampling technologies to the nearshore marine environment, these models have facilitated discovery of deposits that were subsequently developed or at least served as good prospects for future development. In this regard there appears to be a parallel to the exploration methods of the offshore oil industry, wherein initial marine production of petroleum resulted from extension of terrestrial geology and of terrestrial exploration technologies into nearshore waters. Since those early days and successes, the petroleum industry has moved farther offshore and into ever deeper waters. Perhaps the placer mineral industry is poised to do so too; clearly the manganese nodule industry is (Siapno, 1986).

TECHNOLOGIES

If the hard mineral-placer industry is to develop offshore, an appraisal of the adequacy of the tools, technologies, and geologic and engineering information available is a very necessary first step. Adequacy of geologic information will be determined, in part, by examining whether the appropriate depositional models governing mineral accumulations exist, and whether information derived from the paleoenvironmental interpretation of deposit prone areas is accurate and sufficient to validate the models. Interaction between research on presently active processes producing accumulation should occur concomitantly with use (and development) of data collection technologies and the information they provide for interpreting paleoenvironments (Komar and Wang, 1984). In focusing on field methods (in contrast to laboratory methods) one needs to examine what is currently in use, what its capabilities are (accuracy, precision, sensitivity, sampling rates) and whether those methods produce the

requisite data for evaluating deposits. The following paragraphs
attempt to summarize the adequacy of present techniques.

Mapping and Positioning

A first necessity for prospecting in the marine environment is a
bathymetric (topographic) map of high quality: accurate, precise, and
large scale. Considering the high cost of at-sea operations, to map
in detail the large area of the seafloor yet unmapped requires a
bathymetric mapping system capable of combining a high speed of survey
with accurate determination of position as well as elevation (depth
below a datum). A number of new high-quality tools are available.
Within line of sight of the coastline, a variety of commercially
available range/range systems give high degrees of accuracy (Sikorski,
1985). Other state of the art positioning systems such as LORAN C and
Raydist are satisfactory for reconnaissance studies for those coastal
areas covered. By 1988, the satellite global positioning system (GPS)
will be in full operation. The GPS will provide worldwide, 24-hour
coverage to 100 m accuracy for ships at sea (Kalafus, 1985). A
differential GPS, in the planning stages, should consistently be
capable of producing 15 m accuracy, and perhaps as much as 5 m under
benign sea state conditions and along a straight survey traverse
(Kalafus, 1985).
Bathymetric mapping has experienced a quantum jump in survey
speed, precision, and accuracy since the recent advent of swath
mapping systems. Two systems, by General Instruments, Inc., are
presently operational: a deep water system, SeaBeam, and a shallow-
water system, BS^3 (Glenn, 1970; Perry, 1985). Both systems sonically
illuminate a swath of the seafloor at survey speeds as high as 7 m/sec
(14 knots). Vertical accuracy is a function of sound speed in water
and having a high quality timing system. The stated depth accuracy is
1% of depth for both systems. The BS^3 system is suited to operating
in water depths of 30 to 650 m, acoustically illuminating and recording a bottom swath 2.5 x water depth. However, best accuracy is
achieved if the swath is 1.5 x water depth. The deep water SeaBeam
system maps a swath 0.8 x water depth, and is suited to depths greater
than 300 m. With the aid of computers, contoured swaths are provided
routinely in near real-time. Subsequently, maps are made either by
mechanically merging hard copy versions, or by computer processing of
data. Oceanographic research- and survey-ships of several nations are
presently outfitted with these systems. The United States is now in
the initial stages of mapping its continental shelves using both
swath-mapping systems and a sidescan sonar system (Perry, 1985).
The swath-bathymetric systems and positioning systems provide a
satisfactory mapping capability. However, insofar as initial placer
mineral exploitation will likely be in shallow waters, the deep water
SeaBeam system is not now suited to placer exploration. For most
continental shelf depths, the BS^3 system is satisfactory, but inside
the 30 m contour-line, standard narrow-beam sounding is more workable.

Another type of acoustic imaging system useful to exploration is
the sidescan sonar. Because the sound-pulse return is a function of
seafloor roughness, it readily provides geomorphological and textural
information. A large number of near-surface towed shallow water
systems are available which have found wide application for geologic
and engineering surveys and investigations. Clearly, they are
applicable to heavy-mineral placer exploration. Deep towed sidescan
sonar systems are now available and the acoustic images obtained can
be coupled directly to bathymetric data and maps, or physically
overlain on swath bathymetry (Crane, et al., 1985). Either way
provides a very powerful small-scale exploration method. The near-
surface towed deep-water GLORIA (Geologic Long Range Inclined Asdic)
system illuminates much larger areas with a consequent reduction in
detail and accuracy. GLORIA performs best in water 150 m deep and
greater, illuminating a swath 8 x water depth. The scale makes it
valuable for rapid reconnaissance over large areas (Laughton, 1983).
Optimum water depths probably preclude its immediate utility to placer
exploration, though the regional scale image may provide prospective
targets.

Another use of acoustic energy is in sub-bottom seismic reflection profiling--the acoustic equivalent to a stratigraphic section
(Moore and Palmer, 1968). The offshore petroleum industry uses this
seismic technique searching for large, deeply buried structures. The
association of placers with fluvial and beach systems makes them also
amenable to exploration with seismic techniques. Channel cut and
fill, beach ridges, and channel courses are small- to medium-scale
features discernible in many seismic sections. However, heavy-
mineral-placer miners will seek near-surface deposits to minimize the
need to remove thick overburden. Accordingly, shallow-penetration and
high-resolution seismic systems that can detect small scale subsurface
structures and lithologic/stratigraphic units are needed. Unfortunately, high-frequency pulses (short wave length) which are required
for detecting and outlining small features, do not have sufficient
power to penetrate. Physical constraints thus seem to override the
desirability for obtaining better resolution of small, shallow
structures than is presently possible. A subbottom analog to the
swath bathymetric mapping systems, if it could be invented, would be
very helpful.

Sampling

Remote sensing systems described above are necessary precursors
to, and partners of, site-exploration. After prospective sites have
been selected, the next step is sampling sediments and analyzing their
contents for heavy mineral volume and species. Two basic methods are
available, in-situ analysis, and shipboard or laboratory analysis.
In-situ geochemical sampling usually results in obtaining data on
surficial data, or at best, data from subsurface depths of only a few
cm. These geochemical analytical systems make use of a variety of
techniques. Some are passive, such as those that detect naturally
emitted gamma radiation (Noakes, et al., 1974; Miller, 1977). Towing

speeds of the gamma radiation detection systems are on the order of
1 m/sec (2 knots). Heavy minerals such as monazite, epidote, sphene,
and zircon, which contain thorium, are natural emitters of radioactivity and subject to radiometric detection techniques. Valuable in
themselves, they are also valuable in so far as they are associated
with other sought after species.

Another technique also exists now for non-destructive continuous
underway sampling. In the towed continuous seafloor sediment sampler
(CS^3) system of Noakes, et al., (1985) the seafloor component of the
system agitates the bottom surficial sediments creating a cloud of
material which is pumped as a slurry to the tendering ship. Onboard
ship, samples are automatically affixed and bound to filter paper and
subjected to elemental analysis by x-ray fluorescence (XRF). The CS^3
system can analyze sediment samples for 25 elements at a rate of two
samples a minute (Noakes, et al., 1985). A prototype unit towed at a
speed of 1.5 m/sec (3 knots) was used in the Alaskan panhandle (Duncan
Canal) prospecting for a chromalloy barite lode (Noakes 1984, personal
communication).

A modification to the standard XRF system will offer, to field
systems, sensitivity to the level of a few ppm, which is in the range
of good laboratory instruments. A system under development would make
use of mercuric-iodide detectors which operate at room temperature but
have the resolution of cryogenic silica-lithium detectors (Dabrowski,
1982). Mercuric-iodide detectors are effective for a wide range of
x-ray energies making them suitable for detecting elements from sodium
and magnesium to high atomic number elements such as gold and
platinum.

Another potentially valuable exploration tool, new to the marine
environment, is induced polarization (IP), which takes advantage of
the magnetic susceptibility of certain iron-bearing minerals. Recent
field studies and laboratory experiments indicate that ilmenite in
deposits of the northeastern Florida placer belt have a strong IP
response (Wynn, and Grosz, 1983). The authors add that they could
correlate IP response and volumetric percent of ilmenite. While this
work suggests some promising progress has been made using IP as a land
based exploration tool, it has only been suggested as a possible tool
for marine exploration.

In the preceding discourse, discussion focused on systems for
data collection on the seafloor, essentially in the horizontal plane.
Assessment of a resource, of course, requires samples in the vertical
dimension. Cores, which can be obtained by a variety of methods, are
still necessary for obtaining samples of this dimension. Gravity and
free-fall corers, are not efficient penetrators of granular materials.
Vibratory type corers, introduced in the early 1960's continue to be
the workhorse coring device for unconsolidated marine sediments,
whether clays or coarse sand and fine gravels. These devices, which
use electrically, pneumatically, or hydraulically operated vibrating
heads, have been very successful in obtaining long (in excess of 10 m)
cores of coarse granular material. The apparatus consists of a
standard core barrel, liner, shoe, and core catcher with the driver

element fastened to the upper end of the barrel. These are enclosed
in a self-supporting frame which allows the assembly to rest on the
seafloor during coring, thus still permitting the support vessel to
move up and down in response to waves. Power is supplied to the
vibrator from the deck by means of a flexible conductor. Information
on this and numerous other seafloor sediment sampling techniques is
provided in a survey of capabilities by Ling (1972).

Two other generic systems require mention: sediment-lift and
cuttings-lift systems. Sediment-lift systems are those where water,
air, or a combination of water and air are used to raise material from
the seafloor to a surface vessel. The lifting fluid is forced into
the cutting barrel near the cutting surface, where it entrains
particles and carries them to the surface in a hose. This system can
penetrate to as deep as 20 m, which is greater than with vibra corers,
and can perform better in a wider range of grain sizes. However,
sample control in the vertical dimension is lost. Cuttings-lift
systems are a combination percussion and rotary tool. These systems
obtain representative sediments with precise vertical control and
volume. Penetration is nearly limitless, but the systems are costly
and at-sea operations require multipoint mooring of the support
vessel.

In cooperation with the Mississippi/Alabama Sea Grant College
Program, the Mississippi Mineral Resources Institute has recently
developed a powered cuttings-lift drill that retains many of the
capabilities of the basic churn drill system along with the advantages
of the lower operating cost sediment-lift systems. The drill includes
a heavy pneumatic vibrator that can drive as much as 15 meters of
dual-wall casing into sediment with a moderate range of grain size and
some induration. Water and air are used as a means of transporting
sections of disintegrated core, cut by a cutting shoe, up an eductor
hose to a shipboard processing table. Water is directed in the casing
annulus to jet ports in the shoe, while air from the vibrator exhaust
is dumped into the eductor line via a manifold. The operation
involves driving the drill down at intervals of 1/2 to 1 meter and
pausing to flush representative cuttings from that interval. Its
advantage is that the system can be used in any vessel of opportunity,
because it requires no specialized handling equipment. Since all
connections between vessel and drill are flexible, standard two point
anchoring is sufficient for vessel stabilization (Woolsey, 1985,
personal communication).

DISCUSSION AND CONCLUSIONS

Active placer mines and known deposits provide the stratigraphic
clues and the paleoenvironmental indicators useful to offshore
exploration programs. It is well known that coarse-grained deposits
occur commonly in channels, (Markewicz, et al., 1958), terraces,
(Griggs, 1945), beaches (Pirkle, et al., 1977), and downdrift
headlands (Komar, 1985, personal communication). Fine-grained
deposits (such as rock flour) of native metals gold and platinum, are
likely to be much dispersed in low concentrations over large areas

(Moore and Welkie, 1976). Available tools, described in this paper, seem to be capable of providing most of the kinds of information necessary to make informed interpretations and to find prospective targets for detailed exploration near the coast. However, once beyond the nearshore area and the relatively facile extension of terrestrial knowledge of this region is passed, exploration problems loom large. It becomes not so much a question of the tools or how to apply them, but rather where to apply them. The shelf area is vast and largely hidden from view. Where to begin will require using all skills and tools to shorten the scope of explorations to areas that have a high probability of placer occurrence. Farther offshore on shelves of sedimentary material, prospective sites also are farther from primary source rocks. Hence, the goal should be to search initially for those environments where processes are likely to have concentrated heavy minerals: overridden and buried barrier islands and beaches, headlands, and drowned delta fronts.

The obvious first phase of exploration must involve swath bathymetric mapping with interpretation (environmental classification) of the bedforms. Subsequent stages should employ those methods which would lead to a smaller search field, where one could then employ the various sampling technologies mentioned above for evaluating a prospective ore body and mine site. Analysis of large numbers of samples must follow in order to help narrow the prospective search field. For that purpose, computers and statistical techniques are basic tools. Depositional and process models are also significant and keys to narrowing exploration regions. The empirically derived model of an "exploration window" described by Moore and Welkie (1976), where certain key parameters (such as element to element ratios, element to grain size associations) exhibit positive correlation coefficients, offered promise. It is especially significant now that the parameters have been modified and conform to a tested geochemical basis (Owen, 1980).

What appears to be lacking presently is a tool or technique which would offer an efficient, rapid, and inexpensive means of surveying that gap which exists presently between the large scale rapid reconnaissance methods and the detail small scale precision survey and analytical methods.

REFERENCES

Archer, A. A., 1973, Economics of offshore exploration and production
    of solid minerals on the continental shelf; Ocean Management,
    v. 1, no. 1, pp. 5-40.

Crane, K., Hammond, S., Embley, R., and Malahoff, A., 1985,
    Distribution of geothermal fields in the Juan de Fuca Ridge; Jour.
    Geophys. Res., v. 90, pp. 727-744.

Dabrowski, A. H., 1982, Solid-state room-temperature energy-dispersive
    X-ray, In: Russ, J. C., Barrett, C. S., Predecki, P. K., and
    Leyden, D. E. (ed.); Advances in X-ray analyses; Plenum Press,
    v. 25, pp. 1-21.

Glenn, M. F., 1970, Introducing an operational multibeam array sonar;
    Int. Hydrographic Rev., v. 47, pp. 35-39.

Griggs, A. B., 1945, Chromite bearing sands of the southern part of the
    coast of Oregon; U.S. Geol. Surv. Bull. 945-E, pp. 113-150.

Kalafus, R. M., 1985, Differential GPS standards; Sea Technology,
    pp. 52-54.

Komar, P. D., and Wang, C., 1984, Processes of selective grain
    transport on the formation of placers on beaches; Jour. Geol.,
    v. 92, pp. 637-655.

Laughton, A. S., 1983, The first decade of GLORIA; Inter. Hydrographic
    Rev., v. 60, no. 1, pp. 13-45.

Ling, S. C., 1972, State of the art of marine soil mechanics and
    foundation engineering; U.S. Army Engineers, Waterways Exper.
    Station, Vicksburg, MI, Tech. Rept. S-72-11, 153 p.

Markewicz, F. J., Parrillo, D. G., and Johnson, M. E., 1958, Titanium
    sands of Southern New Jersey; New Jersey Bureau of Geol. and
    Topog., Dept. of Conserv. and Econ. Devel., 15 p.

Miller, J. M., Roberts, P. D., Seymour, G. D., Miller, N. H., and
    Wormald, M. R., 1977, A towed seabed gamma ray spectometer for
    continental shelf surveys, In: Nuclear Techniques and Mineral
    Resources; Int. Atomic Energy Agency, Vienna, Austria, pp. 465-
    496.

Mitchell, A. H. G., and Garson, M. S., 1981, Mineral deposits and
    global tectonic settings; Academic Press, London, 405 p.

Moore, D. G., and Palmer, H. P., 1968, Offshore seismic reflection surveys; Proc. Conf. on Civil Engineering in the Oceans, Am. Soc. Civil Engineers, pp. 780–806.

Moore, R. J., and Welkie, C. J., 1976, Metal-bearing sediments of economic interest, coastal Bering Sea; Proc. Symposium on Sedimentation, Alaska Geol. Soc., Anchorage, pp. K1–K17.

National Academy of Sciences, 1975a, Mineral resources and the environment; Committee on Min. Resources and the Environment, Commission on Natural Resources, National Research Council, Washington, D.C., 348 p.

National Academy of Sciences, 1975b, Mining in the outer Continental Shelf and in the deep ocean; The Marine Board, Assembly of Engineering, National Research Council, Washington, D.C., 119 p.

National Academy of Sciences, 1977, Priorities for Research in Marine Mining Technology; The Marine Board, Assembly of Engineering, National Research Council, Washington, D.C., 72 p.

National Advisory Committee on Oceans and Atmosphere, 1983, Marine minerals: an alternative mineral supply; National oceans goals and objectives for the 1980´s; U.S. Govt. Printing Office, Washington, D.C., 33 p.

Nelson, C. H., and Hopkins, D. M., 1972, Sedimentary processes and distribution of particulate gold in the northern Bering Sea; U.S. Geol. Surv., Prof. Paper 689, 17 p.

Noakes, J. E., Culp, R. A., and Spaulding, J. D., 1985, Continuous seafloor sediment sampler system for trace metal surficial sediment studies, Ch. 6: Mapping strategies in chemical oceanography; Adv. in Chem., Ser. 209, Amer. Chem. Soc., pp. 99–116.

Noakes, J. E., Harding, J. L., and Spaulding, J. D., 1974, Locating offshore mineral deposits by natural radioactive measurements, Marine Technol. Soc. Jour., v. 8, pp. 36–39.

Owen, R. M., 1980, Quantitative geochemical models of sediment dispersal patterns in mineralized nearshore environments; Marine Mining, v. 2, pp. 231–249.

Perry, R. B., 1985, Multibeam surveys of the U.S. Exclusive Economic Zone; Proc. Canadian Hydrographic Conf.–85, Halifax, Nova Scotia, (in press).

Pirkle, E.C., Pirkle, W.A. and Yoho, W.H., 1977, The highland
    heavy-mineral sand deposit in Trail Ridge in northern
    peninsular Florida; Florida Geol. Surv., Rept. of Invest.,
    no. 84, pp. 1-50.

Siapno, W.D., this volume, Nodule exploration: Accomplishments,
    needs and problems; pp. 244-254.

Sikorski, R.S., 1985, Positioning accuracy: An economic approach;
    Sea Technology, v. 24, pp. 28-32.

U.S. Geological Survey, 1984, A national program for the assess-
    ment and development of the mineral resources of the United
    States Exclusive Economic Zone; Proc. of a Symposium, Nov.
    15-17, 1983, Reston, Virginia, U.S. Geol. Surv. Circ. 929,
    308 p.

Wynn, J.C. and Grosz, A.E., 1983, Geophysical response of titanium-
    bearing deposits of placer heavy minerals in northeastern
    Florida; Proc. Expl. Geophys., 53rd Ann. Int. Meeting, Las
    Vegas, pp. 192-193.

THE DEVELOPMENT OF TECHNIQUES FOR MARINE GEOLOGICAL SURVEYS

D. A. Ardus
Marine Earth Sciences Research Program
British Geological Survey
Murchison House
West Mains Road
Edinburgh, EH9 3LA, U.K.

ABSTRACT. In the course of systematic mapping of the continental shelf of the United Kingdom new survey equipment has been developed and an integrated approach has been adopted for geophysical and geological operations.

A control system that provides independent programming of seismic sources, obtains optimum performance from high-resolution and deep-penetration reflection seismic profilers operated simultaneously with a side-scan sonar. Improved high-resolution profile data have been obtained by extending the tow depth of a boomer system to 1000 m.

A remotely operated vibracorer/rock-drill system retrieves cores as long as 6 m in unconsolidated sediments and as long as 5.5 m in hard rock at either outcrop or subcrop. Operations can be performed in more than 1500 m water depth. Drilling equipment, using wireline techniques, that can drill deeper, has improved the extent and quality of cores retrieved.

The importance of planning to achieve the right balance of information and to minimize the cost of data acquisition, together with the development and maintenance of a geological data bank are necessary aspects of developing a comprehensive regional interpretation of continental margin geology.

INTRODUCTION

The systematic geological mapping of the United Kingdom's Continental Shelf by the Continental Shelf Division of the Institute of Geological Sciences, now renamed the Marine Surveys Directorate of the British Geological Survey (BGS), commenced in 1966. Initially this work was funded by Department of Education and Science through the Natural Environment Research Council (NERC). Since the introduction of commissioned research in NERC, a large portion of the program has been supported by the Department of Energy.

Two objectives are pursued in this program. The first, undertaken in confidence for the Department of Energy by the Hydrocarbon Assessment Program, provides an overall synthesis of the offshore geology based on

81

P. G. Teleki et al. (eds.), Marine Minerals, 81–98.

confidential, commercial seismic and borehole data. The second, carried
out by the Marine Earth Sciences Research Program, is a reconnaissance
study of the Continental Shelf and upper slope. The primary product of
this activity is a series of maps at a scale of 1:250,000, encompassing
gravity and magnetics (conventionally based on aeromagnetic data but supplemented by marine surveys where the former are not available) as dye-
line copies, and maps of surficial sediments and bathymetry, Quaternary
geology and solid (hard rock) geology in the form of full color sheets.

Those maps that display solid geology and the geophysical parameters
provide full coverage of the land and shelf areas. In a few instances,
the Quaternary-age geological data are combined with data on surficial
sediments. This initial map production program is scheduled to be
completed by 1990.

The data derived by the BGS from this reconnaissance mapping
program are not confidential and provide a basis for research and
commercial activity in offshore areas.

CONVENTIONAL DATA ACQUISITION

Data are compiled at a scale of 1:100,000 and charts prepared
include bathymetry, geophysical tracks and sample- and drill-site
locations. Initial interpretations of data on seabed deposits,
Quaternary geology and solid geology are made at this scale. A
1:1,000,000-scale gravity map of the North Sea has also been prepared.

A 1:500,000-scale geophysical data atlas summarizes the geophysical
traverses run. Gravity, magnetic, echo sounder and navigation data are
routinely edited and one-minute sample intervals are digitized and
recorded on magnetic tape from which track and data-point-value maps
are generated.

A data bank of bottom sample, shallow core and borehole data has
been created, where every sample is uniquely identified and accurately
located in an unambiguous geographic coordinate system. This system is
based on sequential data acquisition in areas one degree square. Information concerning samples and core are logged and entered into a
computer onboard ship, and subsequently transferred to the onshore data
bank.

Grab samples are examined, classified, sub-sampled for geochemical
and micropalaeontological analyses and the material is stored for
subsequent onshore laboratory treatment, consisting primarily of particle size analysis and determination of carbonate content.

Gravity and vibracore sediment samples are recovered in plastic
liner tubes fitted in the core barrel and these are extracted, cut into
segments one meter long, capped and split lengthwise. The split cores
are photographed and logged with respect to observable sedimentary
features, visual classification of particle size, carbonate content
and color, and subsampled for detailed sediment size analysis, petrographic and micro-palaeontological studies, palaeomagnetic measurements
and occasionally for moisture content and liquid- and plastic-limit
determinations.

Geotechnical tests are also performed on clay cores immediately

after retrieval using a hand-held shear vane to determine undrained
shear strength, and with a pocket-penetrometer to measure unconfined
compressive strength. Following these tests, the half-section cores
are sealed in plastic tubing to preserve moisture content, and stored.

Cores derived from borehole drilling usually contain more consolidated and lithified material. These, therefore, are not split. Tests
performed on them are similar to those described above with the exception that selected sections of these cores are packed in waxed containers for subsequent laboratory geotechnical tests onshore.

Since 1967, 220,000 km of multiple seismic reflection traverses
have been gathered, samples and cores as long as 6 m have been obtained
from more than 25,000 locations. More than 500 boreholes, with as much
as 300 m penetration, have been drilled.

In addition to its own operations, BGS has commissioned multi-
channel seismic reflection surveys and participates in the British
Institutions Reflection Profiling Syndicate (BIRPS) for which it acts
as archivist and distributor for the 15- and 30-second two-way-time
reflection seismic data gathered.

NEW DEVELOPMENTS IN OFFSHORE OPERATIONS

The planning, execution and interpretation of the surveys are
undertaken on a collaborative basis by geophysicists and geologists,
whose aim is to achieve the most cost-effective pattern of operations.
Careful planning of the least number of sampling and drilling stations
necessary to calibrate the geophysical traverses is required to accomplish this aim.

The offshore operations follow appraisal of available and pertinent
information from hydrographic surveys and from academic and commercial
sources of bathymetric data. Operations commence with geophysical
traverses, commonly with a line spacing of 5 km. Following initial
shipboard interpretation of these data and their more comprehensive
onshore analysis, stations are occupied to obtain samples of seabed
material and shallow cores. Boreholes are also drilled to provide the
necessary subsurface control.

Navigation and position control of surveys are provided by satellite navigation, integrated Doppler sonar and a radio navigation system
using a land-based transponder network. Sampling usually uses only
the latter system. Drilling sites are located by radio navigation and
confirmed by satellite navigation. For underway geophysical traversing
this gives a navigation accuracy of the order of 200 m in shelf depths
and 400 m in deeper water. Drill sites are confirmed to 10 m position
accuracy.

Geophysics

The standard suite of equipment used in the geophysical investigations is comprised of a gravimeter, magnetometer, precision depth
recorder, side-scan sonar, high-resolution profiling system (pinger,

standard boomer or deep-tow boomer), sparker and airgun. These are
considered conventionally necessary in studies that ascertain the
geology to approximately one km below the seabed with the best possible
data resolution.

The deep-tow boomer system sound source produces acoustic energy
in the frequency range of 1-12 KHz with a peak around 5 KHz. At the
maximum power of 540 joules this gives about 70 m penetration in soft
sediments with a vertical resolution of approximately 0.25 m using a
single-element hydrophone. Most frequently, the boomer is used with a
short trailing hydrophone array which provides improved penetration;
its vertical resolution is slightly degraded but better than 1 m. The
pinger used by BGS operates at 3.5 KHz and does not penetrate as deeply
as the boomer; its vertical resolution is of the order of 1 m. The
multi-electrode sparker source produces most of its energy in the 200
Hz-1 KHz range. Operating at energies between 0.5 and 3 KJ it can
penetrate a subsurface depth of 500 m with a resolution approaching 5
m. The small airgun used most often, 40 cubic inch capacity, operates
in the 50-300 Hz range with a resolution of 20 m.

It is important to correlate high-resolution and deep-penetration
seismic data and side-scan sonar data especially when lateral variation
in geology is present. Therefore, these systems are, by necessity,
operated concurrently. However, many of these instruments emit signals
with overlapping frequency spectra. The side-scan sonar system, operating at frequencies between 50-200 KHz, does not produce interference
on the seismic profiles but suffers interference from the sparker and
airgun sources. Consequently, signal interference has to be taken into
account to achieve optimal performance in simultaneous operation of the
systems. BGS developed a control system that allows independent programming of each seismic source and graphic recorder. This programming
is used to achieve minimal acoustic interference and a maximum firing
rate for each source, compatible with varying depths of penetration, in
pursuit of the highest lateral resolution relative to achievable vertical resolution (Dobinson et al, 1982).

The mutual acoustic interference between the seismic systems is
minimized by synchronizing the firing cycles of the sources while permitting the graphic recorders to operate in a controlled asynchronous
mode. The control system provides flexibility to optimize performance
and penetration.

BGS has extended the depth of operation for the deep-tow boomer to
1000 m and successful records have been obtained throughout the full
tow-depth range. Penetration to 150 m has been achieved with resolution
better than 1 m (Fig. 1). The fish is equipped with an alternate
sparker source, which is limited to a maximum operating depth of 300 m,
that is used as a back-up system. The 800-kg tow-fish has a short
trailing hydrophone array and is deployed from a remotely controlled
winch on a 2.5 km long tow cable.

BGS, in conjunction with the U.K. Atomic Energy Research
Establishment's Harwell Laboratory, has developed a number of gamma-ray
spectrometers for seafloor surveys (Thomas et al., 1984). The instruments are housed, encased in plastic hose for protection from contact
with rocks or wreckage, in a towed 'eel'. The spectrometers measure

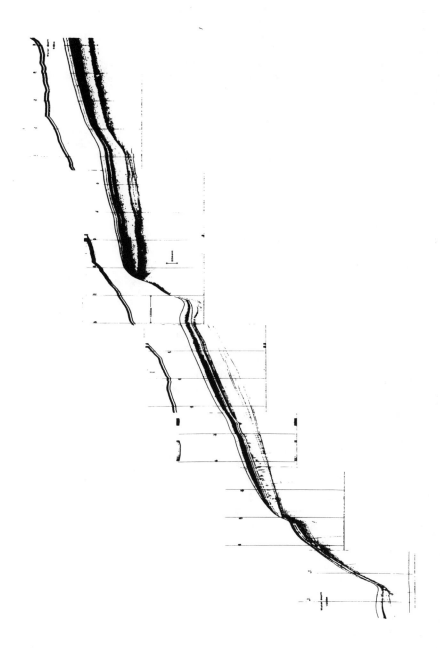

Figure 1. Deep-tow boomer record from the slope on the east of the Rockall Trough west of Lewis in water depths between 500 and 1500 m, illustrating the surface expression of faults in Quaternary sediments.

both the natural radioactive emission in constituents of sediments and
those of radionuclides introduced by man. Their use contributes to the
mapping of surficial sediments and solid geology and the measurements
are directly applicable to exploration for placer deposits, uranium-
bearing veins and phosphorites. Natural gamma-ray profiles have been
recorded along more than 15,000 km traversed in the past decade in
water as deep as 600 m. Additionally, the ability of the probe to
measure artificial radioactivity enables the system to map the distribution of effluents from nuclear installations.

A neutron source can be incorporated in the seabed probe to
determine transition elements from their instantaneous gamma-ray
emissions. Neutron interaction analysis has successfully measured the
grade and extent of manganese nodules off the west coast of Scotland
and the technique is also relevant to exploration for placers and cobalt-
rich crusts.

Remotely operated coring

For taking cores of fine-grained, unconsolidated sediments or soft
rock, such as marl, at an outcrop, gravity (Fig. 2) or piston coring is
the most cost-effective technique available. However, in the presence
of coarser sediments, pebbles, boulders or consolidated clays, subsurface penetration becomes increasingly difficult to ensure when using
these coring tools.

BGS has developed a composite tool for use in areas of the
Continental shelf and slope that are not amenable to gravity coring.
This tool is rated now for operation in 1500 m water depth. This unit
cores both sediment and rock, and its design has eliminated the need to
duplicate heavy lift handling systems, to move alternate kinds of
equipment to a single deployment point, or to restrict an operation to
a single retrieval procedure when the geology of the area being
investigated requires both sediment and hardrock coring facilities.
Further, it is capable of coring rock overlain by several meters of
overburden (Pheasant, 1984).

The equipment is based on BGS' electric vibracorer (Fig. 3), which
in standard form of usage takes a 6-m core, 85 mm in diameter, using a
barrel with a 100-mm outer diameter. The barrel is fitted with a
plastic (cellulose acetate buterate) liner. This tool generates a
vibration force of 6 tons at the frequency of 50 Hz, and is equipped
with a retraction winch capable of a 12-ton pull at the rate of 1.5 m per
minute. The rate of penetration and retraction is monitored acoustically
(Fig. 4) and the record of penetration rate usefully complements the
core recovered. A piston, designed to remain at the level of the
sediment surface during coring but fixed relative to the barrel during
retraction and recovery, minimizes chances of sediment compaction.

An electrohydraulic rotary drill, 5.5 m long, has been integrated
into this system (Fig. 5) to provide a capability to core rock. This
drill can retrieve a 45-mm diameter core in a double-walled barrel. It
is driven by a removable, split, hexagonal kelly bush rotating at speeds

Figure 2. The BGS gravity core system. The system uses a 1/2-ton corer chassis deployed from a flared trough, that incorporates a bowsing winch. A choice of a 5-ft barrel with 65-mm outer diameter or 10- and 20-ft barrels having outer diameters of 100 mm, and fitted with plastic liner, are available for use according to sediment characteristics. Alternatively, thick-walled barrels are available to core soft rocks.

Figure 3. The BGS electric vibracorer being deployed from the stern of
a support vessel by pivoting on the tension chains from a horizontal
position on deck to a vertical attitude for lowering and operation.

Figure 4. Detail of the base of the BGS vibracorer/rotary rock drill shows the 700 KHz acoustic transducer in the foreground. This is focussed on the vibrator housing and the reflected signal indicates the barrel extension with a resolution of 30 mm.

Figure 5. The BGS vibracorer/rock drill. This view of the base assembly from the head of the unit illustrates the hexagonal outer core barrel, the pressure–compensated power swivel and kelly–bush, the electro–hydraulic system, computer housing, acoustic penetrometer and the retraction winch.

controllable between 0 and 600 rpm and the bit is loaded by the passive
vibrator weight of 0.75 tons. Two-stage flushing of the tungsten
carbide or diamond bit allows a degree of jetting through the overburden
to counteract any tendency for the bit to block in pebbly sedimentary
layers. The 12-ton retraction force retrieves the barrel following
completion of coring.

The experience gained from use of the older, 1-m remotely operated
rock drill of BGS and the drill developed for CONSUB I (Fig. 6), the
remotely operated submersible built by British Aerospace and jointly
funded by BGS and Department of Industry (Eden et al., 1977), have both
contributed to the concept and the creation of the combined unit.
Unlike the previously used instruments, it does not require concurrent
television monitoring of the coring site to discriminate in-situ outcrop
from erratics, gravel and cobble lag deposits. This is primarily due
to its ability to penetrate deeper.

The additional reach and the competence to core through overburden
to subcrop have made significant areas of the shelf available for
investigation by this technique. These include Mesozoic basins off
Northern Scotland, where the cover of surficial sediments is relatively
thin.

The vibracore system and the combined vibracorer/rock drill are
both operated through a 2.5 km long torque-balanced, contra-helical
steel cable that contains internal power, instrumentation and command
conductors. The cable provides lift, power and control, the latter
through a microprocessor. Information presented, by both video display
unit and chart recorder, includes attitude, speed, flushing flow rate,
penetration depth and monitoring of retraction.

The drill is deployed on this single umbilical from a dynamically
positioned vessel which references its position using a transponder
fitted to the head of the equipment. This procedure has regularly
allowed the successful occupation of fifteen vibracore stations, 10
miles apart during 24-hour operational days. Such activity took place
in 300 m water depth west of Shetland and was sustained in, and
occasionally above, Beaufort-scale Force 6 conditions. Further
development of this system to operate in waters as deep as 2.5 km is
planned.

Drilling

The penetration depth for offshore boreholes of BGS are constrained
by regulations to a maximum of 300 m below seabed and are designed to
provide lithological, stratigraphical and geotechnical control for the
seismic interpretations used in the regional reconnaissance mapping
program. Although the dynamically positioned drilling vessels are now
competent to operate in water as deep as 1000 m, sites considered so far
are in water depths of only 500 m.

In 1977, BGS initiated the development of a drill-barrel system
capable of pursuing continuous coring from the seabed into unconsolidated
sediments and solid rock (Ardus et al., 1982). This is a wireline
system operated through a seabed template that permits re-entry, and it

Figure 6. CONSUB I, BGS' remotely operated vehicle, showing the rock drill and its advance mechanism. These are mounted, together with a television camera and a stereo camera system, on the pan and tilt unit.

is designed for use with a heave-compensated, 5 1/2-inch outer diameter
drillstring, bored out to have a 4-inch internal diameter. Continuous
refinement has improved the system, as well as the range of bits
available to use with it. The latter are designed to achieve stability
in the borehole and to cut core cleanly, and to this end, a series of
stepped wing bits, with pilot inserts, has been extended. A total-loss
mud system continues to be used for hole stability and bit flushing.

Natural gamma logging is used routinely to compensate for any
incomplete recovery of core by providing a continuous record from the
borehole, indicating the occurrence of clay minerals in the stratigraphic sequence.

USE OF THE DATA IN GEOLOGICAL INTERPRETATIONS

The benefits of an integrated approach to data acquisition and
interpretation are shown by the results of studies of Quaternary
sediments in the North Sea (Fig. 7) because they provide data fundamental to offshore site investigations for pipeline and structure design
and installation (Fannin, 1980; Stoker et al., 1985). The different
ages, facies and stress histories of Quaternary sediments are major
factors in the determination of regional geotechnical variations. The
following examples illustrate aspects of studies from two different
geological settings facilitated by the coherent survey coverage and
good quality data.

Studies of seismic stratigraphy calibrated by drilling and coring
have shown a more complete Pleistocene sequence than was previously
known in the North Sea, reflecting a continuation of the Tertiary basinal subsidence with the succession attaining its greatest thickness,
of the order of 600 m in the Central Graben area (Fig. 8). Paleomag-
netic studies have revealed, by the presence of a reversely magnetized
zone thought to represent the Matuyama epoch, the presence of early
Pleistocene strata (Stoker et al., 1983).

The Lower Pleistocene section is characterized by sediments of
temperate marine facies. The earliest recognizable glaciogenic sediments are Middle Pleistocene age. The remainder of the Middle and
Upper Pleistocene succession is represented by a complex sequence of
marine and glaciomarine facies, overlain in turn by a veneer of Holocene
sediments (Stoker et al., in press).

Stoker and Fannin (1984) have shown that, on the slope into the
Faroe-Shetland Trough, in an area with slope angles generally less than
1.5°, considerable evidence for sediment movement exists as deep as
1200 m water depth (Fig. 9). Coarse sands and gravels are commonly
found to a depth of 500 m with glacial dropstones occurring throughout,
and clasts of stiff clay from near the shelf break occur in a flow
matrix far down slope.

On the upper slope in this area linear patches of recent largely
biogenic sand move over winnowed coarse gravelly sand, and iceberg
grounding marks are discernible in present day water depths to 350 m
and possibly as deep as 450 m. Below this zone, sediment slumping,
possibly by small rotational shears, occurs above a broad mound of

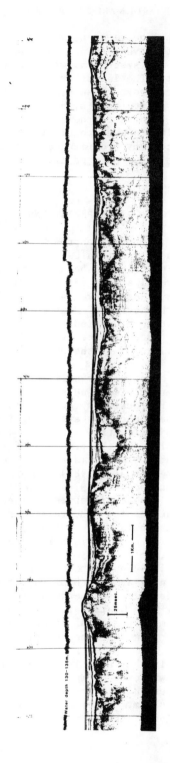

Figure 7. A complex Quaternary sequence in the North Sea east of Orkney illustrated by a deep-tow boomer record.

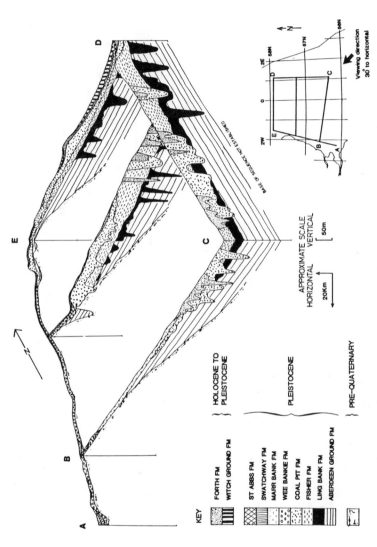

Figure 8. Stratigraphic relationships of the Quaternary formations in the central North Sea (after Stoker et al., in press).

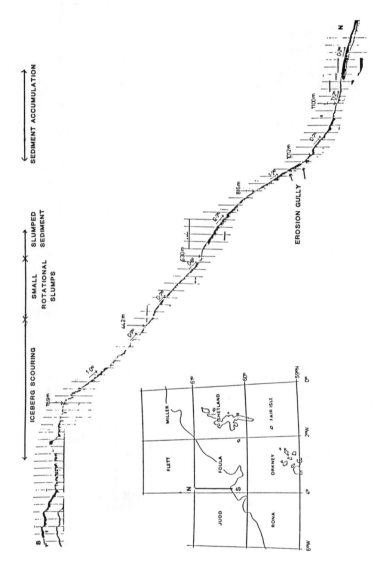

Figure 9. A standard deep-tow boomer record from the south-eastern slope of the Faroe-Shetland Trough, west of Shetland, illustrates iceberg scouring and sediment slumping. The location diagram shows the 200 m isobath. Water depths (m) and slope angles are shown on the traverse profile (after Stoker and Fannin, 1984).

structureless sediment. Between 700 and 1050 m depth the upper 10 m of surficial sediments exhibit parallel reflectors. These are cut by erosional gullies perpendicular to the slope. The gullies are straight, about 20 km long, 20 m deep and several hundred meters wide, and supply sediment to an area beyond the 1050-m isobath where structureless sediment buries the acoustically layered material to subsurface depth of 20 m.

This erosion and post-depositional sediment movement may arise from gravitational loading, storm loading or seismicity, separately or in combination.

CONCLUSION

Regional offshore mapping provides a basis for detailed geological studies. These include applied objectives such as the identification of potential resources, or as a fundamental prerequisite to site investigation for offshore structure design, emplacement and long-term monitoring.

Satisfactory interpretation will only be achieved if the various aspects of the survey are matched to the geological setting and the appropriate complement of instruments are deployed.

The BGS seismic control unit, the deeper-towed boomer system, the composite vibracorer/rock drill and the wireline drilling barrel enhance the utility of the conventional equipment used. They, together with the towed seabed spectrometer, are directly relevant to the study of the mineral concentration and resource assessment not only in shelf areas but also in deep water locations.

ACKNOWLEDGEMENTS

Published with the permission of the Director, British Geological Survey, (Natural Environment Research Council).

REFERENCES

Ardus, D.A., Skinner, A., Owens, R. and Pheasant, J., 1982, Improved coring techniques and offshore laboratory procedures in sampling and shallow drilling; Proc. Oceanology International, 1982, Brighton, U.K, 18 p.

Dobinson, A., Roberts, P.R. and Williamson, I.R., 1982, The development of a control system for the simultaneous operation of seismic profiling equipment; Proc. Oceanology International, 1982, Brighton, U.K., 16 p.

Eden, R.A., McQuillan, R. and Ardus, D.A., 1977, UK experience of the use of submersibles in the geological survey of the continental shelves; In: Geyer, R.A. (ed.) Submersibles and their uses in oceanography and ocean engineering; Elsevier, Amsterdam, pp. 235–278.

Fannin, N.G.T., 1980, The use of regional geological surveys in the
   North Sea and adjacent areas in the recognition of offshore hazards;
   In: Ardus, D.A. (ed.) Offshore Site Investigation; Graham and
   Trotman, London, pp. 5–22.

Pheasant, J., 1984, A microprocessor controlled seabed rockdrill/
   vibracorer; Quart. Jour. Underwater Technology, v. 10, no. 1, pp.
   10–44.

Stoker, M.S. and Fannin, N.G.T., 1984, A geological framework for the
   northwest UK Continental Shelf and Slope; Brit. Geol. Surv. Marine
   Geology Research Programme, Internal Rept. 84/21, 15 p.

Stoker, M.S., Long, D. and Fyfe, J.A., in press, A revised Quaternary
   stratigraphy for the Central North Sea; Brit. Geol. Surv. Rept.

Stoker, M.S., Long, D., Skinner, A.C. and Evans, D., 1985, The
   Quaternary succession of the Northern UK Continental Shelf and
   Slope: implications for regional geotechnical investigations; In:
   Advances in Underwater Technology and Offshore Engineering, v. 3,
   Offshore Site Investigation: Graham and Trotman, London, pp. 45–61.

Stoker, M.S., Skinner, A.C., Fyfe, J.A. and Long, D., 1983, Palaeomagnetic evidence for early Pleistocene in the central and northern
   North Sea; Nature, v. 304, no. 5924, pp. 332–334.

Thomas, B.W., Miller, J.M. and Malcolm, A., 1984, Radiometric surveys
   of the seabed; Proc. Oceanology International, 1984, Brighton,
   U.K., 11 p.

MODEL OF TERTIARY PHOSPHORITES ON THE WORLD'S CONTINENTAL MARGINS

Stanley R. Riggs
Department of Geology
East Carolina University
Greenville, North Carolina 27834
U.S.A.

ABSTRACT. Shallow subsurface Tertiary sediments of mid-latitude,
modern continental margins of the world contain a tremendously
extensive and poorly known Neogene sediment sequence that has a high
theoretical potential for containing major deposits of unconsolidated
phosphate sediments. Reworked phosphates occurring in thin, Holocene
surface sediments are important tracers for these shelf deposits which
may crop out or more commonly occur in the shallow subsurface. This
hypothesis should be tested by combining techniques of high-resolution
seismics with detailed networks of vibracores on continental shelf
areas with anomalous phosphate concentrations in the surface sediments.
Ultimately, deeper core drilling is required to adequately understand
the third dimension. Future mining and production will undoubtedly
combine existing technology from the offshore petroleum industry with
new technology being developed within the mineral industry such as
hydraulic slurry mining.

INTRODUCTION

Most marine sediments and rocks contain less than 0.3% P_2O_5.
However, periodically through geological time phosphorites,
(containing 5% P_2O_5 or greater) formed on the seafloor in response to
specialized oceanic conditions and accumulated in sufficient
concentrations to produce major deposits of regional extent. Marine
phosphate formation and deposition represents periods of low rates of
sedimentation in combination with large supplies of nutrient
phosphorus derived through upwelling on shallow continental slope and
shelf environments in low- to mid-latitudes. Cold, nutrient-enriched
upwelling currents are considered to be prerequisite mechanisms
supplying essential elements for production of large volumes of
organic matter to the sediments. Phosphorus is then concentrated by
various mechanisms, possibly bacterial, at either the sediment-water
interface or within interstitial pore waters. This process leads to
the primary formation and growth of phosphate grains which may remain
where they formed, or be transported as clastic particles within the

P. G. Teleki et al. (eds.), Marine Minerals, 99–118.
© 1987 by D. Reidel Publishing Company.

environment of formation.   During subsequent periods of time, some
primary phosphate grains may be physically reworked into another
sediment unit in response to either changing or different
environmental processes.

Some phosphate formed during all major sea-level transgressions
during the 67 million year Cenozoic history, however, some periods
were more important than others with respect to producing large
volumes of phosphorites and preserving them in the geologic column.
During the Paleocene and Eocene several major episodes of
phosphogenesis occurred within the major east-west ocean which
included Tethys, producing extensive phosphorites throughout the
Middle East, Mediterranean, and northern South American regions.   By
the Neogene this circum-global ocean had been destroyed by plate
tectonic processes, and the north-south Pacific and Atlantic Oceans
dominated global circulation patterns.   Upper Cenozoic phosphogenesis
(Table I) occurred along the north-south ocean-ways which now contain
the modern continental margins.   On the basis of the extent of known
phosphate deposits, the Miocene was by far the most important episode
of phosphate formation during the Upper Cenozoic (Table I).

TABLE I. Episodes of Upper Cenozoic Phosphogenesis

GEOLOGIC PERIOD	AGE *	DURATION *
Quaternary	<3 ma	10 ta
Pliocene	5- 4 ma	1 ma
Miocene	19-13 ma	6 ma
Oligocene	29-25 ma	4 ma

* ma = million years ago; ta = thousand years ago

In the southeastern United States, phosphate formation began in
the Carolina phosphogenic province during the late Oligocene producing
the Cooper Formation.   Miocene phosphate formation began by at least
late early Miocene and continued cyclically into late middle Miocene
throughout the southeast producing the Hawthorn and Pungo River
Formations (Riggs, 1984; Riggs et al., 1985).   Local concentrations of
primary phosphorite in the lower Yorktown and Bone Valley Formations
formed during the very short Pliocene second-order sea-level cycle.

Some Holocene and late Pleistocene phosphate on the shelves off
Peru/Chile (Burnett, 1977), Namibia (Baturin, 1982), and east
Australia (O'Brien et al., 1981) is primary and has formed during one
or more brief Quaternary phosphogenic episodes which coincided with
the glaciation/deglaciation cycles.   In other shelf areas such as
offshore of North Carolina (Luternauer et al., 1967; Pilkey et at.,
1967; Riggs et al., 1985) local concentrations of phosphate in the
Pleistocene and Holocene sediments are largely reworked from
underlying units of Tertiary age.

The early to middle Miocene was a major period when anomalous
concentrations of phosphate formed contemporaneously throughout a
major portion of the world's continental margins with latitudes lower

than 45° (Sheldon, 1964; Cook et al., 1979; Baturin, 1982; Riggs, 1984). Table II outlines the extensive distribution of known Miocene sediments containing major phosphorites along the modern continental margins. If these Miocene phosphate deposits are integral responses to common processes associated with an oceanographic episode of global extent, there should be large-scale temporal and geological similarities among the deposits. However, details of the sediment patterns within each deposit should be dependent upon the local geologic setting interacting with regional oceanographic and climatic conditions. Each of these known Miocene deposits contain similar patterns of phosphorite sedimentation which are products of global events and changing paleoceanographic conditions through time.

TABLE II. Distribution of known Miocene sediments containing phosphorites along the modern continental margins.

CONTINENTAL MARGIN	REGION
East Atlantic:	Portugal, Northwest Africa through South Africa, and Agulhas Bank
West Atlantic:	North Carolina through Florida, Cuba, Venezuela, and Argentina
East Pacific:	California through Baja California, Mexico, and Peru through Chile
West Pacific:	Sakalin Island, Sea of Japan, Indonesia, Chatham Rise east of New Zealand, and East Australian shelf.

Most known Miocene phosphate deposits such as in North Carolina, Florida, Venezuela, California, Baja California, and Peru occur on the emerged coastal plain and are geologically well known. In several cases, these emerged deposits are only the updip limit of a larger Miocene section that extends seaward beyond the coastal plain and constitute large portions of the upper sediment regime which have built the modern continental shelves. Snyder (1982), Riggs (1984), and Riggs et al. (1985) demonstrated that the Miocene was characterized by fourth-order sea-level cyclicity (100,000 to 1 million years duration) which repeatedly moved the depositional and erosional processes of shelf sedimentation across the shelf system to produce abnormally thick Miocene sections. Locally, these Miocene sediments are exposed on the seafloor, however, generally they are buried below thin covers of Plio-Pleistocene and Holocene surface sediments. Thus, the potential for discovering new phosphate deposits within the Miocene sediments on the world's continental shelves is very high for the following reasons: 1) phosphate genesis is known to occur throughout the shelf and upper slope environments, 2) thicker and more extensive sequences of Miocene sediments occur on the shelves

than on adjacent coastal plains, and 3) the shallow subsurface Neogene
geology of most of the world's shelves is poorly known.

Two important sediment relationships have been developed
concerning reworked phosphate in surficial sediments on the North
Carolina continental shelf: 1) the distribution of phosphate in
surface sediments closely reflects the distribution within underlying
Miocene sediment units and 2) the process of reworking significantly
dilutes the concentration of phosphate in the surficial sediments
(Riggs et al., 1985). These relationships have also been recognized
on the shelves of northwest and southwest Africa and could represent
important exploration tools for richer Tertiary phosphorites occurring
within the shallow subsurface on many continental shelves of the world.
Thus, the presence and distribution of anomalous phosphate
concentrations in surficial shelf sediments not characterized by
upwelling and modern phosphate formation, may represent an important
sedimentologic tracer for older subsurface phosphate sequences.

Paleogene phosphate deposits may be locally important on the
continental margins such as on the northwestern African shelf;
however, the potential is considerably lower than that of the Neogene
for several reasons. First, Paleogene phosphogenesis appears to have
been largely related to equatorial upwelling associated with the
east-west oceanic system dominant at that time (Sheldon, 1980); large
portions of this Tethyan Province now occurs on land masses with minor
areas on modern continental margins. Second, on most modern margins,
the Paleogene sediments are deeply buried beneath an extensive
sequence of Neogene sediments; only regionally have local tectonic
processes preserved the Paleogene section at a depth shallow enough to
represent a potential phosphate target on continental shelves, even
then these sediments are often too indurated to be recovered easily.
Thus, Paleogene phosphorites on continental shelves generally have a
low probability, and only locally can they have a high potential as a
source of future marine resources.

PHOSPHOGENESIS

The Mineral Components

Phosphate forms and occurs within specific lithofacies with very
distinctive mineral associations. Riggs (1979a, 1980) defined the
"phosphate sediment package" as a unique assemblage of authigenic
sediments which formed in response to specialized environmental
conditions in specific geologic settings during abnormal periods of
sedimentation. This aberrant assemblage of authigenic sediments
(Table III) occurs in many contemporaneous marine sediments deposited
on the world's continental margins during major phosphogenic episodes,
such as occurred during the Miocene.

Structural Control

Riggs (1979b, 1984) defined first-order structural highs as
large-scale morphological or structural features which determine the

location of major phosphogenic provinces. Within each province, second-order structural or topographic highs control specific sites of phosphate deposition. These second-order highs diverted upwelled waters into adjacent embayments where phosphate formed along mid-slope environments. During transgression, phosphate depocenters migrated farther onto the inner shelf of second-order embayments as oceanic currents increasingly impinged on the continental margin.

TABLE III. The aberrant authigenic sediment suite associated with episodes of phosphogenesis during the Miocene on the Atlantic continental margin of southeastern United States (modified from Riggs, 1980, 1984).

SEDIMENT FACIES	FACIES RELATIONSHIPS
1. PHOSPHATE	Inner to outer shelf; Associated with first-order structural highs.
2. DOLOMITE	Seaward of phosphate facies and off shelf edge; Increases southward.
3. DIATOMACEOUS MUDS (Chert and porcellanite)	Seaward of phosphate facies and off shelf edge; Increases northward.
4. ORGANIC MATTER (Black shales)	Occurs with phosphate, diatomaceous, and dolomite facies.
5. MAGNESIUM-RICH CLAY MINERALS (Attapulgite, sepiolite, etc.)	Inner to outer shelf; Increases southward and into first-order embayments.
6. MANGANESE and FERROMANGANESE	Seaward and downslope of plateau phosphate facies.
7. IRON SEDIMENTS (Iron silicate: glauconite) (Iron sulfide: minor pyrite)	North of phosphate facies; Within phosphate facies.

Stratigraphic Control

Riggs (in press) found that major episodes of phosphogenesis and the associated authigenic suite of sediments (Table III) do not form within extensive sequences of uniform sediments. Rather, most phosphogenic sequences occur in the transition zone between two distinctly different lithologic groups of sediments; that is in transition zones between thicker and more uniform sections of normal carbonates, siliciclastics, or volcanoclastic sequences or vice versa. Throughout the geologic column, this transition zone within the

sediments occurs in consort with changes in sea level, and it is
apparent in both the large-scale framework of first- and second-order
sea-level fluctuations and in the smaller-scale framework associated
with local facies and unconformities resulting from third- and
fourth-order fluctuations. Riggs (in press) proposes that major
episodes of aberrant authigenic sedimentation reflect global changes
in tectonism and/or climates which affected and modifed
paleoceanographic conditions.

Along the continental margin of the southeastern United States, a
major phosphogenic episode with its associated aberrant authigenic
sediment sequence occurred in the transition zone between
predominantly limestone sediments of the Eocene to late Oligocene and
predominantly siliciclastic sediments of the Pliocene and Quaternary.
The phosphogenic transition period began during the late Oligocene
second-order sea-level transgression (about 29 my ago) and continued
into the middle Miocene (about 13 my ago). In this period, an
abnormal concentration and volume of phosphorus was deposited with
contemporaneous facies of glauconite, diatomaceous muds, organic
matter, dolomite, and Mg-rich clays (Table III).

Cyclic Deposition

Snyder (1982), Snyder et al. (1982), Riggs (1984), and Riggs et
al. (1985) established that multiple depositional sequences formed
within the Miocene and Pliocene section of North Carolina in response
to established third-order cycles of eustatic sea-level change (Vail
et al., 1979). At a still smaller scale, each third-order
depositional sequence is composed of multiple sediment units, each
deposited in response to transgression and regression associated with
fourth-order sea-level cycles. Complex and changing patterns of
deposition and erosion were produced by variations in the Gulf Stream
processes through each transgression. Regression and low stand
portions of sea-level cycles were characterized by nondeposition,
scouring and channeling, and diagenetic alteration. These
destructional processes severely complicated the depositional patterns
of the Neogene sediments.

In North Carolina each fourth-order sea-level event is
represented by either the following idealized lithogic cycle or
regional and temporal variations upon this theme (Riggs et al., 1982;
Riggs, 1984). Siliciclastic sedimentation was dominant during the
early-stage of each transgression and within shallower, landward
environments; carbonate sedimentation was dominant during the
late-stage of each transgression and within deeper, seaward
environments. If an aberrant sequence of authigenic sediments formed,
including phosphorites, its formation began during the early-stage and
continued through the mid-stage of transgression; it occurred within
the transition zone between dominantly siliciclastic and dominantly
carbonate sedimentation, both laterally and vertically.

SOUTHEASTERN UNITED STATES MODEL

Carolina Phosphogenic Province

Miocene sediments were deposited in a large Neogene basin off the east flank of the Mid-Carolina Platform High (Figure 1), both major first-order structures (Riggs, 1979b, 1984). Cape Lookout High, a second-order east-west oriented paleotopographic high (Snyder, 1982; Snyder et al., 1982; Riggs, 1984; Riggs et al., 1985), subdivided the Neogene basin into two second-order regional embayments. North of Cape Lookout High is the Aurora Embayment, which has producing phosphate mines in the Aurora phosphate district. South of Cape Lookout High is Onslow Embayment, which has the recently discovered Northeast Onslow Bay and Frying Pan phosphate districts.

Miocene sediments of the Pungo River Formation occur as highly truncated sequences of apparent offlapping depositional units which strike southwest and dip east and southeast off the flank of the Mid-Carolina Platform High (Figure 1). The sediments thicken from the western updip limit to over 500 meters at the shelf edge (Riggs et al., 1985). Miocene sediments on the coastal plain occur totally in the subsurface, extend onto the continental shelf where they occur in shallow subcrop, and occur in the subsurface below a thin Plio-Pleistocene carbonate sequence at the shelf edge (Blackwelder et al., 1982; Snyder, 1982; Popenoe, 1985).

Minor amounts of phosphate ($<1\%$ P_2O_5) occur everywhere in the surface sediments within Onslow Bay (Figure 1) where the Pungo River Formation crops out below a very thin and variable Holocene surficial sand blanket. However, areas of higher concentration of phosphate grains were first described by Pilkey et al. (1967) and Luternauer et al. (1967), who mapped two areas of high concentration in the surficial sediments which contained more than 14% and 7% phosphate grains. Riggs et al. (1985) have demonstrated that these two "hot spots" occur over major beds of Miocene phosphate within the Frying Pan and Northeast Onslow Bay phosphate districts, respectively.

Within the Frying Pan phosphate district, Holocene phosphorite gravelly sands contain 5% to 75% phosphate grains (3.2% to 21.7% P_2O_5) which are continously being diluted by the addition of new biogenic debris and biocorroded rock fragments. Sand thickness ranges up to 3 m thick and occurs largely between the carbonate rock-capped mesas. The 3% P_2O_5 contour corresponds to the limit of the Holocene sediment which contains significant phosphate and almost exactly mimics the outcrop pattern of a very rich (4.8% to 22.9% P_2O_5) phosphate bed of the Miocene Pungo River Formation (Riggs et al., 1985). Outside the 3% contour, phosphate concentrations decrease rapidly to less than 1% P_2O_5.

On the southern flank of the Mid-Carolina Platform High (Figure 1), the distribution and character of associated phosphate sediments is controlled by second-order topographic features. The Beaufort, South Carolina deposits, with generally low phosphate concentrations, occur in the Ridgeland Trough west of the Beaufort High. These sediments increase in phosphate concentration southward into the

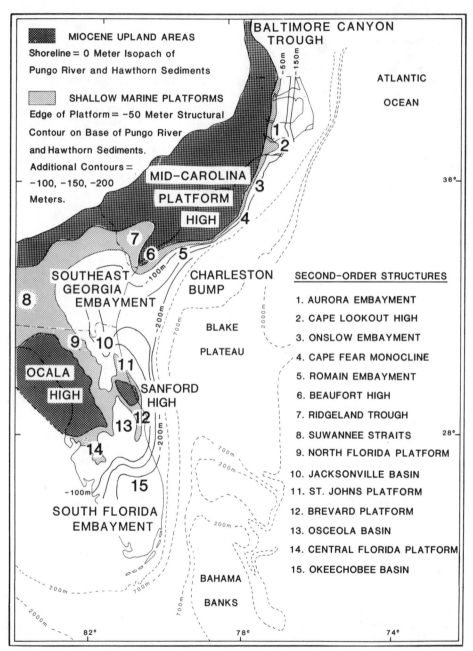

FIGURE 1. Map of southeastern United States showing the general
Miocene shoreline and major first- and second-order structural and
topographic features which controlled Neogene phosphate sedimentation
(from Riggs, 1984).

Chatham, Georgia deposits, which occur on and around the southern flank of the Beaufort High, a paleotopographic feature which controlled phosphate formation and deposition during the Miocene (Riggs, 1984). Phosphorites extend east and north across the inner continental shelf as the Savannah deposits. Little is known about other second-order features and phosphorite sequences across the middle and outer shelf on the southern flank of the Mid-Carolina Platform High for lack of detailed studies. However, surface sediment mapping by Pevear et al. (1966) and Gorsline (1963) demonstrate significantly increased P_2O_5 contents within the Holocene sediments on the inner continental shelf, where Woolsey (1976) demonstrated that phosphate-rich Miocene units occur in subcrop.

Florida Phosphogenic Province

The Ocala High (Figure 1) is a first-order feature with a core of Eocene Ocala Limestone which controlled deposition of all Neogene sediments (Riggs, 1979b). Two first-order sediment basins occur in association with the Ocala High. The southern portion of the Florida Peninsula is occupied by the large Okeechobee Basin which accumulated over 225 meters of phosphatic carbonate sediments during the Miocene. To the north, the Ocala High is terminated in Georgia by the Southeast Georgia Embayment and Suwanee Straits.

Miocene phosphorites were deposited in association with second-order features around the perimeter of the Ocala High (Figure 1). The southern end of the Ocala High forms the broad and extensive Central Florida Platform which plunges gently into the subsurface; the crest and adjacent flanks of this structure contain the world's largest producing phosphate districts. Along the eastern side of the Ocala High is a broad and irregular North Florida Platform. The southern half of this Platform contains only a thin sequence of Miocene sediments with minor amounts of phosphate, while the northern half contains an irregular section of Miocene sediments which are as much as 25 meters thick, with high concentrations of phosphate within the second-order embayments. Downdip, the Miocene phosphatic sands and clays increase to over 150 meters in thickness in the Jacksonville Basin.

Another positive element of extreme importance to Neogene sedimentation was the Sanford High (Figure 1). This smaller second-order feature has long and narrow plunging platforms off the north and south ends, which contain thick and extensive sequences of phosphate-rich sediments. These phosphorites extend down the flanks of the platforms into the adjacent Okeechobee, Osceola, and Jacksonville Basins, as well as extend eastward onto the present Florida continental shelf. Samples from a core hole 45 km offshore of Jacksonville, Florida, contain thick and very phosphate-rich downdip facies of this Miocene stratigraphic sequence (Charm et al., 1969; Hathaway et al., 1976). Holocene surface sediments from the southern Georgia and northern Florida continental shelf contain anomalously high amounts of phosphate (Gorsline, 1963; Pevear et al., 1966). This surface enrichment probably represents reworked material from the underlying Miocene phosphate-rich section.

The western margin of the Ocala High occurs on the western
Florida continental shelf in the Gulf of Mexico (Figure 1), where the
subsurface Tertiary geology is poorly known. However, Neogene
phosphorites should exist on the shelf and should have a similar
distribution and relationship to first- and second-order structures as
elsewhere within the Florida and Carolina phosphogenic provinces
(Riggs, 1984). Birdsall (1978) found minor concentrations of phophate
in the surface sediments throughout much of the western Florida
continental margin suggesting that major Neogene phosphate deposits
might exist in the shallow subsurface.

OTHER EXAMPLES OF KNOWN CONTINENTAL MARGIN DEPOSITS

Caribbean Basin

In addition to the well-known Miocene phosphate deposits that
exist from North Carolina through Florida, similar deposits occur on
the Atlantic coastal plains of Cuba (Ilyin et al., in press) and
Venezuela (Bellizzia, in press). Analysis and comparison of these,
apparently geologically related, phosphate deposits suggest that the
Miocene phosphate province could extend throughout the entire
Caribbean region. However, much of the Miocene section occurs in deep
and insufficiently drilled basins on the islands and on adjacent
continental shelves below thin and variable Plio-Pleistocene surface
sediment cover. The latter is presently either poorly known or not
known at all, because of lack of shallow subsurface marine research on
these continental margins.

Northwest Africa

An extensive belt of phosphatic surface sediments occurs on the
continental shelf of northwest Africa extending from Casablanca to
Agadir (Nutter, 1969; Summerhayes, 1970; Summerhayes et al., 1972;
Bee, 1973; and McArthur, 1974). Phosphate in the unconsolidated
surface sediments ($<8\%$ P_2O_5) was derived from erosion of underlying
Upper Cretaceous, Paleocene, Eocene and Miocene phosphorites with no
evidence of modern phosphate formation. More specifically, phosphate
in the surface sediment occurs as sub-angular clastic fragments that
are lithologically similar to phosphate in the adjacent outcrops, and
that concentrations in surface sediments are richest near exposed and
eroding, older phosphatic sedimentary units. Some sediment dispersal
has taken place by longshore drift during previous times of lowered
sea-level.

Bee (1973) summarized the specific distribution of phosphorites
in the surface sediments on the shelf as follows. 1) The inner shelf
phosphate is being derived from outcropping Cenomanian strata by
coastal erosion. 2) The mid-shelf phosphate is concentrated near
outcrops of Upper Cretaceous and Eocene phosphatic units. Phosphate
in the two previous categories consists of highly abraded grains with
margins cutting across the original textural fabric. 3) On the outer

shelf and upper slope, the area of anomalous phosphate correlates directly to the distribution of exposed phosphatic Miocene strata. Reworked phosphate and glauconite grains, occurring almost entirely as whole, unabraded foraminiferal tests, have been derived from Miocene strata well below the outer limits of sea-level regressions by submarine erosion. 4) Low, but still anomalous phosphate values also occur within modern beach and dune sands, old beach terraces, and in river sands and clays which are deposited in the river-mouth sediments on the inner shelf.

Reconnaissance surveys in the shelf area of southern Morocco (between Cap Juby and Agadir) indicated that, the phosphatic facies in surface sediments were definitely related to exposed Upper Cretaceous and Tertiary phosphorites (Bee, 1973). However, on account of low concentrations of phosphate in the surface sediments, Bee concluded that either the phosphate outcrops are not as widespread or the rocks are not as rich in phosphate as on the Moroccan shelf to the north.

Cretaceous and Paleogene deposits that occur in the shallow subsurface on the Moroccan shelf, are the same age as the units that contain the world's richest and most extensive phosphate deposits on the adjacent coastal plain. These phosphorites were generated as a consequence of major episodes of phosphogenesis associated with the equatorial Tethys Ocean, and are similar to deposits that occur throughout the Mediterranean region and the Middle East. Considering that in areas where modern phosphorites are not forming today, processes of modern shelf erosion and sedimentation are generally processes of phosphate dilution by terrigenous sand, rock fragments, and biogenic debris, one can only imagine what the real potential might be for these older beds in the subsurface.

Central-west Africa

The coastal plain of Gabon, Congo, and Cabinda (Angola) contain phosphates of Maastrichtian, late Paleocene, early Eocene, and Miocene age (Giresse, 1980; Barusseau et al., this volume). In Cabinda, the Paleocene to Eocene sediments consist of interbedded phosphorite sands (12% to 25% P_2O_5), white marly limestones, and silicified units. Miocene sediments are exposed along the coast of Cabinda; they are locally phosphatic, contain bone beds, and extend north onto the continental shelf of Congo and south Gabon. These Miocene phosphorites occur as flat-lying, shallow, unconsolidated sediments, which have been interpreted to be reworked from Eocene and Paleocene phosphates, which they overlie. According to Barusseau et al. (this volume) post-Paleogene sea-level fluctuations severely dissected the Paleocene-Eocene surface producing submarine platforms or ridges with 7 to 8 m of relief and reworked the Paleogene sediments, concentrating the coarser phosphate grains into Miocene to Quaternary gravel accumulations, 15 to 50 cm thick, around the platforms. Gamma radiometric surveys of the surface sediments, have extended the surficial phosphates southwest onto the Cabinda shelf.

The importance of this example is not necessarily in the volume of phosphate or in the economic potential of these surficial sediments.

Rather, it is what these sediments are indicating about the resource
potential of the Paleocene-Eocene strata directly beneath the
surficial sands, or even of the Miocene section further offshore where
it forms a very thick sediment wedge (Giresse et al., 1976).

Southwest Africa

Extensive areas of phosphatic rocks and phosphate-rich surface
sediments occur on the continental margin off the south coast (Agulhas
Bank area) and west coast of South Africa (Birch, 1979). Middle
Eocene to upper Miocene phosphatic rocks, containing greater than 5%
P_2O_5, generally form a rocky relief on the outer shelf.
Unconsolidated surface sediments, containing greater than 1% P_2O_5, are
coincident with the distribution of phosphatic rocks on the outer
shelf; the richest phosphatic sediment occurs in areas where the
richest phosphate rock (averaging 17.8% P_2O_5) has been recovered,
southwest of Saldahna Bay, South Africa (Birch, 1979). The adjacent
coastal plain contains isolated remnants of Miocene and Pliocene
phosphate units of the Elandsfontyn and Varswater Formations,
respectively (Hendy, 1981), which are presently being mined for
phosphate.

To date, no evidence has been found to suggest any formation of
modern phosphate on the shelf of South Africa (Birch, 1979). However,
north off the Namibian coast, is one of the areas where there has been
extensive documentation of modern phosphorite formation in the surface
sediments (Baturin et al., 1972; Veeh et al., 1974; Baturin, 1982).

Baja California, Mexico

An extensive sequence of phosphorites occur in the surface
sediments on the Pacific shelf west of Baja California, Mexico
(D'Anglejan, 1967). These deposits extend from Cedros Island and
Viscaino Bay, southward to Cabo San Lazaro and Bahia Magdelena.
Phosphate occurs in fine sands, in waters less than 100 m deep, where
locally, areas contain phosphate concentrations between 15% and 40% of
the total sediment. Surface concentrations persist down cores to a
depth of slightly over 1 m. Similar phosphate concentrations occur in
the adjacent modern coastal system in the Santo Domingo area,
including the beaches, dunes, and associated estuaries, and within
Pleistocene deposits on the outer coastal plain where preparations are
presently underway to mine low-grade phosphate sand deposits.

Salas (1978) and Quintus-Bosz (1980) described two Neogene units
on the Baja Peninsula which contain major phosphorites. The Miocene
Monterrey Formation, a sequence of black shales, diatomites,
phosphorites, and carbonates is being mined for phosphate at San Juan
de la Costa. The second unit is the extensive Pliocene Salada
Formation.

D'Anglejan (1967) described the shelf as an erosion platform cut
into more than 4,000 m of Tertiary sediments; he dredged samples of
the Monterrey Formation from rock scarps, where the Miocene was
exposed on the shelf. His lack of success in demonstrating the

presence of a modern component in the phosphate within surficial
sediments, its occurrence as a continuous unit with the coastal plain
deposits, close parallelism in textural characteristics between
phosphate and non-phosphate components, and presence of fossil
phosphatic brachiopod fragments lead him to the conclusion that at
least a major portion of the phosphate in the surface sediments, if
not all of it, has been reworked from older sedimentary units.

Peru/Chile

    The Peru/Chile continental margin has significant potential for
containing extensive Neogene phosphate deposits in the subsurface. It
has been demonstrated that this modern shelf area, which is a region
of coastal upwelling, is an area of modern phosphogenesis (Baturin et
al., 1972; Veeh et al., 1973; Burnett et al., 1977; Baturin, 1982;
Burnett et al., 1982). The modern nodules and pellets form
phosphorites in association with organic-rich diatomaceous oozes at
the sediment-water interface, and apparently form, preferentially,
near the upper and lower boundaries of the interception of the
oxygen-minimum zone with the continental margin (Burnett, 1977;
Burnett et al., 1982).
    Processes of phosphogenesis have existed on the Peru/Chile
continental margin in the past, and probably have persisted through
much of the Neogene. Miocene deposits occur in the Sechura Desert on
the northern coast of Peru. The pelletal phosphate deposits occur as
1 to 1.5 meter thick units that contain 20% P_2O_5, interbedded with 3
to 20 meter thick beds of diatomite which contain 1 to 7% P_2O_5 (Cheney
et al., 1979), and are highly uniform over large areas of the Sechura
Desert. The entire phosphatic sequence of middle Miocene sediments
ranges from 135 to 215 meters thick and consists of at least 14
interbedded sequences of diatomite and phosphorite units. These
cyclic couplets begin with diatomite and grade irregularly with
abundant interlaminated stringers into the overlying phosphorites,
which have sharp and planar upper boundaries with the overlying
diatomite. Recent work has demonstrated that similar Neogene deposits
occur in numerous localities further south such as the Pisco Basin in
Peru (Baker et al., in press) and Mejillones Peninsula in Chile
(Valdebenito, in press).
    These examples represent small areas of the continental margin
where Miocene sediments are located on the outer coastal plain and
were influenced by local tectonic events. There is no reason to
believe that similar phosphogenic processes have not persisted across
larger portions of the Peru/Chile shelf cyclically throughout the
Neogene. If this has been the case, then the submarine shelf should
contain extensive stratigraphic sequences of phosphorite sediments,
and their specific locations should be controlled by the regional
structural framework and former variable paleoceanographic conditions.
Consequently, this region would represent an excellent area to test
the general hypothesis of this paper; several core holes drilled
through the Neogene section, and the results tied into a network of
high-resolution seismic profiles, could define the relationship of

phosphogenesis to global paleoceanographic episodicity throughout the upper Cenozoic.

SUMMARY

The future potential for development of marine phosphate resources from the continental margins of the world is very large. This potential is greatest for the Neogene deposits, particularly the Miocene, and decreases with sediments of Paleogene age. The present continental margins of the world were dominated by Cenozoic sedimentation. The last major period of growth was during the Miocene; only locally do Pliocene and Pleistocene sediments constitute a major accumulation. In stable shelf environments the Miocene is dominantly at or near the surface except in areas of very rapid and active sedimentation (i.e. large deltas, active tectonic areas, etc.). In areas of active tectonics, Paleogene sediments may occur at or near the surface of the continental shelves (i.e. Northwest African shelf).

Phosphate forms below 45° to 50° latitude. Paleogeography of the Neogene was similar enough to todays geography that these latitudes represent realistic exploration limits for Neogene deposits on the world's present continental margins. This is not true for the Paleogene, particularly in areas such as Australia and India.

Phosphate forms during major episodes when critical oceanographic conditions change in response to changes in global climates. During these episodes phosphogenesis is far more extensive over broader regions than during other periods. The Miocene was the last major phosphogenic episode, since the Pliocene and Pleistocene episodes were too brief in duration (1 million and 10,000's of years, respectively) to provide sufficient time for the formation of major deposits.

These Miocene sediments only occasionally occur on coastal plains; however, when they do, they often contain major phosphate deposits (i.e. southeastern U.S., Cuba, Venezuela, California, Baja California, Peru, Chile, Argentina, South Africa, etc.). Such coastal plain deposits are often the landward pinchout of an extensive submarine Miocene wedge of sediments, most of which underlies the adjacent continental shelves and slopes. Sometimes the Miocene is exposed at the surface, but more often it is buried below a thin and varible skin of Plio-Pleistocene sediments.

Modern phosphates are known to be forming today in only a few locations on continental margins (i.e. Peru/Chile and Namibia). If the modern era does not really represent a major phosphogenic episode, imagine how much phosphate could have formed in these regions during a major phosphogenic episode such as the Miocene. The few scattered Miocene occurrences that exist on the adjacent coastal plains of each of these countries contain major phosphate deposits.

CONCLUSIONS

What is the realistic potential of phosphate resource development from the world's seafloors in the near future? Based upon major advances in our scientific understanding of the genesis of marine phosphorites during the past two decades, it can be demonstrated that there is a very large potential which could be economically viable in the near future. However, before this happens, a sophisticated three-dimensional evaluation of the world's continental margins, with the aid of high-resolution seismic profiling and shallow drilling, will have to be undertaken. This type of research is only beginning; until recently, marine scientific efforts associated with phosphate research have dealt primarily with surficial and shallow subsurface sediments and their geologic framework. But there is an important third-dimensional component; geologic time has allowed for a great volume of continental-margin sediments to build wide shelf areas during the Neogene. Glimpses into these extensive sediment packages are possible wherever updip portions of the Neogene sediments occur on adjacent coastal plains. The surficial sediments of the world's margins are little more than the "tip of the iceberg" and, in most cases, these sediments are not what will contain the economic phosphate resources of the oceans in the near future.

However, in most places Miocene sediments will be buried below thin but variable amounts of Plio-Pleistocene sediments. These overlying Plio-Pleistocene sediments will contain reworked phosphate, derived during former sea-level fluctuations associated with glaciation and deglaciation. It is likely that much of the phosphate, that has been sampled and described in the surface sediments from continental margins around the world, reflects extensive Tertiary phosphate beds in the shallow subsurface, as previously demonstrated for the southeastern U.S. and west African margins. Thus, phosphate in the surface sediments, which often occurs in very low concentrations, can be used as a tracer indicating considerably richer deposits in the shallow subsurface.

The Miocene represents a major global phosphogenic episode and a period of major growth of the continental margins. This period produced a thick pile of sediments that are commonly unconsolidated to poorly indurated. These factors make the Miocene sediments an important target for phosphates which would be potentially available to recovery by slurry mining techniques from the shallow subsurface in the marine environment.

ACKNOWLEDGEMENTS

This article is based upon work supported by 1) NSF grants OCE-7908949, OCE-8110907, and OCE-8400383; 2) University of North Carolina Sea Grant College Program, 3) the phosphate research program at East Carolina University; and 4) International Geologic Correlation Program (IGCP) 156 on Phosphorites co-sponsored by UNESCO and IUGS. Specific acknowledgements go to W.C. Burnett, A.C. Hine, R.P. Sheldon, S.W. Snyder, S.W.P. Snyder, and my colleagues and students at East Carolina University.

REFERENCES

Barusseau, J.P., and Giresse, P., this volume, Some mineral resources
   of the west African continental shelves related to Holocene
   shorelines: phosphorite (Gabon, Congo), Glauconite (Congo), and
   ilmenite (Senegal, Mauritania); pp. 133-153.

Baker, P.A., and Allen, M.R., in press, Origin of dolomite in
   continental margin organic-rich sediments, In: Burnett, W.C. and
   Riggs, S.R. (eds.), Genesis of Neogene and Recent phosphorites;
   Cambridge Univ. Press, England.

Baturin, G.N., 1982, Phosphorites on the sea floor; Elsevier,
   Amsterdam, 343 p.

Baturin, G.N., Merkulova, I.I., and Chalov, P.I., 1972, Radiometric
   evidence for recent formation of phosphatic nodules in marine
   shelf sediments; Marine Geology, v. 13, pp. 37-41.

Bee, A.G., 1973, The marine geochemistry and geology of the Atlantic
   continental shelf of central Morocco; unpub. Ph.D. dissert.,
   Imperial College, Univ. London, 263 p.

Bellizzia, A., in press, The phosphorites of the Falcon and Tachira
   areas, Venezuela, In: Notholt, A.J., Sheldon, R.P., and Davidson,
   D. (eds.), World Phosphate Resources; Cambridge Univ. Press,
   England.

Birch, G.F., 1979, Phosphatic rocks on the western margin of South
   Africa; Jour. Sed. Petrology, v. 49, pp. 93-110.

Birdsall, B.C., 1978, Eastern Gulf of Mexico, Continental Shelf
   phosphorite deposits; M.S. Thesis, Univ. South Florida, St.
   Petersburg, 87 p.

Blackwelder, B.W., I.G. MacIntyre, and O.H. Pilkey, 1982, Geology of
   the continental shelf, Onslow Bay, North Carolina, as revealed by
   submarine outcrops; Amer. Assoc. Petroleum Geol. Bull., v. 66,
   pp. 44-56.

Burnett, W.C., 1977, Geochemistry and origin of phosphorite deposits
   from off Peru and Chile; Geol. Soc. America Bull., v. 88,
   pp. 813-823.

Burnett, W.C., Beers, M.J., and Roe, K.K., 1982, Growth rates of
   phosphate nodules from the continental margin off Peru; Science,
   v. 215, pp. 1616-1618.

Burnett, W.C., and Veeh, H.H., 1977, Uranium-series disequilibrium
   studies in phosphorite nodules from the west coast of South
   America; Geochimica et Cosmochimica Acta, v. 41, pp. 755-764.

Charm, W.B., Nesteroff, W.D., and Valdes, S., 1969, Detailed
    stratigraphic description of the JOIDES cores on the continental
    margin off Florida; U.S. Geol. Surv. Prof. Paper 581-D, 13 p.

Cheney, T.M., McClellan, G.H., and Montgomery, E.S., 1979, Sechura
    phosphate deposits, their stratigraphy, origin, and composition;
    Econ. Geology, v. 74, pp. 232-259.

Cook, P.J., and McElhinny, M.W., 1979, A reevaluation of the spatial
    and temporal distribution of sedimentary phosphate deposits in the
    light of plate tectonics; Econ. Geology, v. 74, pp. 315-330.

D'Anglejan, B.F., 1967, Origin of marine phophorites off Baja
    California, Mexico; Marine Geol., v. 5, pp. 15-44.

Giresse, P., 1980, The Maastrichtian phosphate sequence of the Congo,
    In: Bentor, Y. (ed.), Marine phosphorites; Soc. Econ. Paleontol.
    and Miner. Spec. Publ. 29, pp. 193-207.

Giresse, P., and Cornen, G., 1976, Distribution, nature et origine des
    phosphates miocenes et eocenes sous-marins des plates-formes du
    Congo et du Gabon; Bull. du Bureau de Recherches Geol. et Min.,
    sect. IV, no. 1-1976, pp. 5-15.

Gorsline, D.S., 1963, Bottom sediments of the Atlantic shelf and slope
    of the southern United States; Jour. Geology, v. 71, pp. 422-440.

Hathaway, J.C., J.S., Schlee, C.W., Poag, 1976, Preliminary summary of
    the 1976 Atlantic margin coring project of the U.S.G.S.; U.S.
    Geol. Surv. Open File Rept. no. 76-844, 218 p.

Hendy, Q.B., 1981, Geological succession at Langebaanweg, Cape
    Province, and global events of the late Tertiary; S. African Jour.
    Sci., v. 77, pp. 33-38.

Ilyin, A.V., and Ratnikova, G.T., in press, Miocene phosphorites of
    Cuba; In: Burnett, W.C. and Riggs, S.R. (eds.), Genesis of Neogene
    and Recent Phosphorites; Cambridge Univ. Press, England.

Luternauer, J.L., and O.H. Pilkey, 1967, Phophorite grains; their
    application the interpretation of North Carolina shelf
    sedimentation; Marine Geol., v. 5, pp. 315-320.

McArthur, J.M., 1974, The geochemistry of phosphorite from the
    continental margin off Morocco; unpub. Ph.D. dissert., Imperial
    College, Univ. London, 200 p.

Nutter, A.H., 1969, The origin and distribution of phosphate in marine
    sediments from the Moroccan and Portugese continental margins;
    unpub. D.I.C. dissert., Imperial College, Univ. London, 158 p.

O'Brien, G.W., Harris, J.R., Milnes, A.R., and Veeh, H.H., 1981, Bacterial origin of east Australian continental margin phosphorites; Nature, v. 294, pp. 442-444.

Pevear, D.R., and Pilkey, O.H., 1966, Phosphorite in Georgia shelf sediments; Geol. Soc. America Bull., v. 77, pp. 849-858.

Pilkey, O.H. and J.L. Luternauer, 1967, A North Carolina shelf phosphate deposit of possible commercial interest; Southeastern Geol., v. 8, pp. 33-51.

Popenoe, P., 1985, Seismic stratigraphy and Tertiary development of the North Carolina continental margin, In: Poag, C.W. (ed.), Geological evolution of the U.S. Atlantic Continental Margin; Hutchinson and Ross, Stroudsburg, Pennsylvania, pp. 125-188.

Quintos-Bosz, R., 1980, Petrology and distribution of phosphate in the lower Salada Formation, Santa Rita, Baja California Sur, Mexico; unpub. M.S. thesis, Colorado School of Mines, Golden, 126 p.

Riggs, S.R., 1979a, Petrology of the Tertiary phosphate system of Florida; Econ. Geol., v. 74, pp. 195-220.

Riggs, S.R., 1979b, Phosphorite sedimentation in Florida--a model phosphogenic system; Econ. Geol., v. 74, pp. 285-314.

Riggs, S.R., 1980, Tectonic model of phosphate genesis; In: Sheldon, R.P., and Burnett, W.C. (eds.), Fertilizer Mineral Potential in Asia and the Pacific; East-West Resource Systems Inst., East-West Center, Honolulu, Hawaii, pp. 159-190.

Riggs, S.R., 1984, Paleoceanographic model of Neogene phosphorite deposition, U.S. Atlantic continental margin; Science, v. 223, no. 4632, pp. 123-131.

Riggs, S.R., in press, Phosphogenesis and its relationship to exploration for Proterozoic and Cambrian phosphorites; In: Cook, P.J., and Shergold, J.H. (eds.), Proterozoic and Cambrian phosphorites; Cambridge Univ. Press, England, v. I, chap. 7.

Riggs, S.R., Lewis, D.W., Scarborough, A.K., and Snyder, S.W., 1982, Cyclic deposition of Neogene phosphorites in the Aurora area, North Carolina, and their possible relationship to global sea-level fluctuations; Southeastern Geol., v. 23, no. 4, pp. 189-204.

Riggs, S.R., Snyder, S.W.P., Hine, A.C., Snyder, S.W., Ellington, M.D., and Mallette, P.M., 1985, Geologic framework of phosphate resources in Onslow Bay, North Carolina continental shelf; Economic Geology, v. 80, pp. 716-738.

Salas, G.P., 1978, Sedimentary phophate deposits in Baja California, Mexico; Amer. Institute Mining Eng., preprint 78-H-75, 21 p.

Sheldon, R.P., 1964, Palaeolatitudinal and palaeogeographic distribution of phosphate; U.S. Geol. Surv. Prof. Paper 501-C, pp. C106-C113.

Sheldon, R.P., 1980, Episodicity of phosphate deposition and deep ocean circulation--a hypothesis, In: Bentor, Y. (ed.) Marine phosphorites; Soc. Econ. Paleontol. and Miner., Spec. Publ. 29, pp. 239-248.

Snyder, S.W.P., 1982, Seismic stratigraphy within the Miocene Carolina Phosphogenic Province: chronostratigraphy, paleotopographic controls, sea-level cyclicity, Gulf Stream dynamics, and the resulting depositional framework; M.S. Thesis, Univ. of North Carolina, Chapel Hill, N.C., 183 p.

Snyder, S.W.P., A.C. Hine, & S.R. Riggs, 1982, Miocene seismic stratigraphy, structural framework and sea-level cyclicity: North Carolina Continental Shelf; Southeastern Geol., v. 23, no. 4, pp. 247-266.

Summerhayes, C.P., 1970, Phosphate deposits on the northwest African continental shelf and slope; unpub. Ph.D. dissert., Imperial College, Univ. London, 282 p.

Summerhayes, C.P., Nutter, A.H., and Tooms, J.S., 1972, The distribution and origin of phosphate in sediments off northwest Africa; Sedimentary Geol., v. 8, pp. 3-28.

Vail, P.R., and Mitchum, R.M., Jr., 1979, Global cycles of relative changes of sea level from seismic stratigraphy, In: Watkins, J.S., Montadert, L., and Dickerson, P.W. (eds.), Geological and geophysical investigations of continental margins; Am. Assoc. Pet. Geol. Mem. 29, pp.469-472.

Valdebenito, E.M., in press, The Tertiary phophorites of Mejillones Peninsula, Chile, In: Notholt, A.J., Sheldon, R.P., and Davidson, D. (eds.), World phosphate resources; Cambridge Univ. Press, England.

Veeh, H.H., Burnett, W.C., and Soutar, A., 1973, Contemporary phosphorite on the continental margin of Peru; Science, v. 181, pp. 844-845.

Veeh, H.H., Calvert, S.E., and Price, N.B., 1974, Accumulation of uranium in sediments and phosphorites on the south west African shelf; Marine Chemistry, v. 2, pp. 189-202.

Woolsey, J.R., 1976, Neogene stratigraphy of the Georgia coast and
    inner continental shelf; unpub. Ph.D. dissert., Univ. of Georgia,
    Athens, 222 p.

OPEN—OCEAN PHOSPHORITES —— IN A CLASS BY THEMSELVES?

W.C. Burnett
Department of Oceanography
Florida State University
Tallahassee, Florida 32306
USA

D.J. Cullen
New Zealand Oceanographic Inst.
P.O. Box 12346
Wellington, North
NEW ZEALAND

and

G.M. McMurtry
Hawaii Inst. Geophysics
University of Hawaii
Honolulu, Hawaii 96822
USA

ABSTRACT.   Most   phosphorites   occurring   on   the   crests of seamounts,
ridges, and other   elevated portions of   the sea floor   appear to be   of
marine origin, although at least some appear to have originated from the
submergence   of   an   insular—type   phosphate   deposit.   The   close
relationship between iron—manganese crusts and "open—ocean" phosphorites
may provide important insights into the mode of deposition of both Fe/Mn
crusts and the phosphatic deposits.

INTRODUCTION

In   1950,   when   the   Mid—Pacific   Mountains   were first sampled by
dredging, it was discovered that the summit rocks contained phosphatized
limestone   (Hamilton,   1956).   Subsequently,   further   expeditions have
established that   phosphorites are   not uncommon   features of   seamounts
throughout   the   world's   oceans.   Besides   the   Mid—Pacific Mountains,
seamount phosphorites have also been found in the north Pacific Ocean on
Sylvania Guyot near   Bikini Atoll (Hamilton and Rex,   1959), on the Wake
and Geisha   Seamounts (Heezen et   al., 1973), on   numerous guyots in the
Marcus—Necker   Ridge system between Marcus   Island and the Japan Trench,
on the   Milwaukee Bank   in the   southern part   of the   Emperor Seamounts
(Baturin, 1982), on   the Necker Ridge and nearby seamounts (Hein et al.,
1985a) and   the Musicians   Seamounts (Moberly   and Sinton,   pers. comm.,
1982).   Reported occurrences   in the   south Pacific   Ocean include those
from seamounts in the Southern Line   Islands (Haggerty et al., 1982), on
seamounts   in the Tasman Sea (Slater   and Goodwin, 1973), occurrences on
the slope of the Manihiki Plateau (Baturin, 1982), on isolated seamounts

119

P. G. Teleki et al. (eds.), Marine Minerals, 119–134.

near the Tokelau Islands and the Northern Cook Islands and on guyots on the North Fiji Plateau (Cullen and Burnett, 1986).

The phosphorite found on seamounts is typically phosphatized limestone, including rudist, bioclastic—calcarenite, and nannoplanktonic foraminiferal phosphatized limestones, both of which are often coated with iron—manganese crusts (Bezrukov et al., 1969; Hamilton, 1956; Heezen et al., 1973). The mineralogy and major element chemistry is similar to continental margin deposits. Amounts of P_2O_5, CaO, CO_2, and F are found to coincide with the degree of phosphatization and the composition of the original material. In contrast to shelf deposits, seamount phosphorites are geochemically characterized by very low concentrations of organic carbon and uranium. There is also an absence of pyrite sulfur and a relative enrichment of the rare earth elements (REE) in the heavy lanthanoids (Baturin, 1982). The seamount phosphorite REE pattern closely resembles that of seawater (Kolodny, 1981; Burnett et al., 1983), while continental margin deposits tend to display "shale—like" REE patterns (McArthur and Walsh, 1985).

Recent awareness of the seamount type of phosphorite has been stimulated by the keen interest in the Co—rich iron—manganese crusts which are also found in this environment. Several recent reports have pointed out that phosphorite is often present as a substrate and/or admixture together with the iron—manganese crusts (Halbach et al., 1982; Halbach and Manheim, 1984; Hein et al., 1985a; Hein et al., 1985b).

What is the significance of these open—ocean phosphorites? Do they, like continental margin deposits, imply a depositional setting involving upwelling and high biological productivity? If not, how did they form? Is there any significance to their common association with iron—manganese crusts? What is their areal extent and what possible economic value do these deposits have? The purpose of this paper is to briefly address these questions and recommend further areas of research which should improve the understanding of open—ocean phosphorites.

For the purposes of this paper, open—ocean phosphorites are defined as those that occur on relatively isolated seamounts, plateaus, ridges, and other elevated portions of the sea floor. The Chatham Rise and the Blake Plateau deposits, which have geological and geochemical affinities quite unlike the deposits to be considered here, are specifically excluded. The Blake Plateau iron—manganese and phosphate deposits have been covered in considerable detail by Manheim and his colleagues (Manheim et al., 1980; Manheim et al., 1982). The Chatham Rise deposits have also been studied in quite some detail because of possible mining interests (Cullen, 1980; Kudrass and Cullen, 1982; and von Rad, 1984).

DISTRIBUTION AND COMPOSITION

Because the sampling coverage has been limited and non—recovery of phosphatic material does not necessarily eliminate the possibility of its occurrence, it is impossible to make many generalizations about the areal extent of seamount phosphorites (Table I). Besides the occurrences in the Pacific Ocean as described herein, phosphorite from seamounts in the Atlantic have been reported from Annan Seamount in the eastern equatorial Atlantic (Jones and Goddard, 1979), the New England Seamounts (Manheim, 1972), near the northern boundaries of the Romanche Fault Zone (Bonatti et al., 1970), a guyot near the Aves Swell in the Caribbean (Marlowe, 1971), and on the Jan Mayen Ridge in the north Atlantic (Kharin, 1974, cited in Baturin, 1982). Known occurrences in the Indian Ocean are limited to seamounts in the West Australian

Table I   Documented examples of seamount phosphorite deposits from the Atlantic and Pacific Oceans.

Location	Reported Age	Reference
-- ATLANTIC OCEAN --		
Aves Swell	Miocene-Holocene	Marlowe (1971)
Annan Seamount	Eocene	Jones and Goddard (1979)
New England Seamounts		Manheim (1972)
Jan Mayen Ridge		Baturin (1982)
Romanche Fracture Zone		Bonatti et al. (1970)
-- PACIFIC OCEAN --		
Albert Henry Seamount	Late Pliocene	Cullen and Burnett (1986)
Colahan Seamount	--	Hein et al. (1985a)
Line Islands	Late Cretaceous	Haggerty et al. (1982)
		Halbach and Manheim (1984)
Kalolo Seamount	--	Cullen and Burnett (1986)
Marcus-Necker Ridge	Cretaceous	Baturin (1982)
Mid-Pacific Mountains	Late Cretaceous/Eocene	Hamilton (1956)
	Pre-middle Miocene	Halbach et al. (1982)
		Halbach and Manheim (1984)
		Hein et al. (1985a)
Milwaukee Bank	Neogene	Baturin (1982)
Musicians Seamounts	Miocene and older (?)	Burnett et al. (1983)
		DeCarlo et al. (1986)
Necker Ridge	--	Hein et al. (1985b)
North Fiji Plateau	--	Cullen and Burnett (1986)
Sylvania Guyot	Cretaceous	Hamilton and Rex (1959)
S.P. Lee Guyot	--	Hein et al. (1985b)
Tasman Sea Guyots	Late Pliocene/ Pleistocene	Slater and Goodwin (1973)
Wake and Geisha Seamounts	Late Cretaceous/Eocene	Heezen et al. (1973)

Basin and the Cocos Ridge, investigated by Russian scientists (Baturin, 1982).

Although details and coverage are sketchy, it is clear that deposits occur in all three major oceans in a wide latitudinal range. It is possible that if reconstructed to their positions during phosphatization there could be a low-latitude bias. Sheldon (1980) has already pointed out that the north Pacific phosphorites on seamounts described by Heezen et al. (1973) were all in equatorial positions when they formed during the Late Cretaceous and Eocene. Hein et al. (1985b) also suggest a possible relationship between the presence of carbonate fluorapatite, the main mineral phase of sedimentary phosphorite, and proximity to the equator. These authors have implied that a genetic relationship may well be expected to exist because of the enhanced productivity and related phenomena associated with equatorial upwelling.

Because of the increased sampling of seamount summits and flanks during the last few years as a result of the Co-crust exploration programs, it has become increasingly clear that seamount phosphorites are rather typical rock types in this environment. Based on our sampling of seamounts in the Pacific, we feel that phosphorites are quite common on or near the summits of seamounts. Because the phosphatic rock is often indurated, while the associated iron-manganese crusts, hyaloclastites, altered volcanics, limestones, etc. may not be, dredge sampling may bias the rock-type distribution substantially. In many situations, we have observed that continued sampling on any one seamount will eventually result in recovery of phosphorite. Clearly, improved sampling techniques, including the *in situ* recovery of outcrop samples by means of manned submersibles, will improve this situation.

The major element chemistry of seamount phosphorites is very similar to that of other sedimentary phosphorites (Table II). Averages of microprobe results (omitting the volcaniclastic samples KK-80: 20-5 and 22-12) are shown together with averages for Pacific island phosphate deposits and a modern shelf deposit from the continental margin of Peru-Chile (Table III). Unfortunately, the data on fluorine obtained during the microprobe analyses were too erratic to be useful. Chemical analyses of samples Q551a and Q556, both from the southwest Pacific, display normal amounts of F relative to phosphorus with F/P_2O_5 mass ratios of 0.124 and 0.127 respectively. Both samples also contained about 5.5 - 6.0% structural CO_2 as determined by X-ray diffraction analysis (Gulbrandsen, 1970).

Although the Pacific seamount phosphorites vary little from other sedimentary phosphorites in terms of their major element geochemistry, they are quite distinct geochemically from Pacific island phosphorites. The average reported in Table III represents 24 analyses of 5 insular samples, 2 from Nauru Island, 2 from Fanning Island, and 1 from Washington Island. The samples were similar chemically, with the exception that the 2 samples from Fanning Island were not completely phosphatic; that is, small amounts of unphosphatized calcium carbonate were still present. The island samples were shown to have F/P_2O_5 ratios close to that of pure fluorapatite (0.089) and low structural CO_2 concentrations of approximately 1%. The unsubstituted nature of the insular apatite relative to the seamount samples is also reflected in

Table II   Microprobe   analyses of polished sections of seamount
             phosphorites from the north and south Pacific Ocean.

Location	------- Musicians Seamounts -------					-- Tokelaus --		North Cook Is.
Seamount	Wagner	Wagner	Schubert	Brahams	Chopin	Kalolo	Kalolo	A.Henry
Sample #	8-1	9-13	20-5	22-12	35-13	Q551	Q551A	Q556
	(4)*	(3)	(2)	(3)	(3)	(4)	(3)	(3)
SiO_2	1.39	1.20	39.28	27.63	3.42	0.07	0.97	4.88
Al_2O_3	0.60	0.41	11.59	8.11	1.04	0.02	0.10	1.19
FeO	0.25	0.25	8.46	14.67	0.62	0.07	1.44	1.01
MgO	0.27	0.25	1.75	1.67	0.45	0.67	0.58	0.46
CaO	52.10	52.48	13.01	9.49	49.29	50.97	47.22	33.36
SrO	0.12	0.10	0.11	0.03	0.08	0.26	0.31	0.10
Na_2O	0.88	0.72	3.01	1.21	0.89	0.87	1.26	0.90
K_2O	-**	-	2.31	2.20	-	-	-	-
P_2O_5	29.86	31.26	7.59	5.75	29.47	26.65	25.93	21.23
SO_3	1.49	1.47	0.40	0.38	1.37	1.90	2.00	1.20
CaO/P_2O_5	1.74	1.68	1.71	1.58	1.67	1.92	1.82	1.57
MgO/P_2O_5	0.009	0.008	0.040	0.570	0.015	0.025	0.022	0.021
Na_2O/P_2O_5	0.029	0.023	0.400	0.930	0.030	0.033	0.049	0.042
SO_3/P_2O_5	0.050	0.047	0.053	0.066	0.046	0.071	0.077	0.057

Note. Analyses performed at the U.S.G.S. microprobe facility in
Denver, Colorado.
* Number of spots analyzed per sample.
** Below detection limit.

the elemental ratios of MgO, Na_2O, and SO_3 to P_2O_5 (Table III). The
significance of these chemical differences concerning the origin of
seamount phosphorite will be discussed later in this paper.
     One of the most striking differences between the seamount and shelf
deposits concerns the very low concentration of uranium (<9 ppm) in
isolated seamount-phosphorites compared to the values (50-100 ppm)
reported in shelf deposits. Arrhenius (1963) attributed this difference
to a reflection of the depositional environment of the respective
deposits; oxidizing in the case of seamounts, reducing in the case of
the shelf deposits.   A recent analysis of the oxidation state of
uranium, however, in seamount phosphorites has shown it to be entirely
in the reduced, tetravalent form (Halbach, pers. comm., 1982). Besides,
the summits of seamounts containing phosphate deposits are often bathed
in mid-depth, low-oxygen waters rather than in highly oxidizing
conditions. The difference in concentration, therefore, may be caused
by the mode of formation or other factors, as well as the oxidation
state of the depositional environment.

Table III   Averages of microprobe   analyses of phosphorite   sample sets
             from Pacific seamounts,   Pacific islands, and the Peru/Chile
             margin.

	Pacific Seamounts* (20)***	Pacific Ocean Islands** (24)	Peru/Chile Margin (66)
SiO_2	1.99	0.17	4.46
Al_2O_3	0.56	0.19	1.37
FeO	0.61	0.06	1.04
MgO	0.45	0.30	1.28
CaO	47.57	49.35	44.28
SrO	0.16	0.27	0.25
Na_2O	0.92	0.06	1.15
K_2O	< 0.10	< 0.10	0.32
P_2O_5	27.40	34.19	28.69
SO_3	1.70	0.17	2.10
CaO/P_2O_5	1.74	1.44	1.54
MgO/P_2O_5	0.016	0.008	0.045
Na_2O/P_2O_5	0.034	0.002	0.040
SO_3/P_2O_5	0.058	0.005	0.073

Note. Island and Peru/Chile shelf data previously unpublished. All
analyses performed at the U.S.G.S. microprobe facility in Denver,
Colorado.
*   Average of data from   Table II excluding samples   20-5 and 22-12.
**   Nauru, Fanning, and Washington Islands, Central Pacific Ocean.
*** Number of spots analyzed.

     REE also display some interesting differences   between shelf
deposits and those from seamounts. Burnett et al. (1983) pointed out
that north Pacific seamount phosphorites have high concentrations and a
distinct seawater-like pattern, unlike phosphorites from the Peru-Chile
shelf which have very low concentrations and a shale-like REE pattern.
McArthur and Walsh (1985) have   recently confirmed that modern shelf
deposits have low abundances of REE and flat patterns when normalized to
shale. The distinct cerium depletion and enrichment of the heavier REE
in seamount samples suggests that these elements (and perhaps phosphorus
as well) were derived from seawater which has essentially the same
pattern. Many ancient phosphorites have REE patterns and concentrations
similar to the seamount samples (Altschuler, 1980) rather than to the
phosphorites from modern upwelling zones. The conclusion, therefore, is
that REE must be taken up secondarily from the environment during

exposure, or that the modern phosphorites analyzed for REE are not typical of those preserved in the geologic record. Recent analysis of modern phosphate pellets separated from mud collected on the Peru shelf indicates that they have concentrations and patterns which more closely resemble the geologic deposits than do the indurated nodules which were analyzed earlier (Piper et al., 1986).

PHOSPHORITE/IRON-MANGANESE CRUST ASSOCIATIONS

Fe/Mn crusts and phosphorites occur on seamounts both separately and in close association. Seamount crusts often occur on phosphorite or phosphatic limestone substrates but more commonly they have been found on hyaloclastite, volcanic breccia, or basalt. The volcanic materials, however, are often cemented, impregnated, or replaced by phosphorite. Additionally, the inner, older parts of the crusts themselves are commonly impregnated by carbonate fluorapatite (Hein et al., 1985b).

As part of the survey of the Co-enriched iron-manganese deposits in the Hawaiian Islands, De Carlo et al. (1986) have pointed out that phosphorite is a common substrate and/or admixture together with the Fe/Mn crusts on seamounts in that region. Phosphatic material within the metallic crusts was observed to be higher on the off-axis seamounts. The crusts are also thicker on the off-axis areas; most likely because these seamounts are significantly older, probably Late Cretaceous.

Phosphorite associations with iron-manganese crusts were observed during a study of the crusts which occur on Necker Ridge, Horizon Guyot, and the S.P. Lee Guyot (Hein et al., 1985a; Hein et al., 1985b). An interesting aspect of these observations was that the relative amount of carbonate fluorapatite contained within the iron-manganese crusts themselves increased with decreasing latitude. It is tempting to relate this observation to present-day oceanographic parameters in these areas, such as the depth and extent of the oxygen-minimum zone in the respective areas, but until these phosphorite occurrences are dated and the seamount positions backtracked to their sites of phosphatization, it is difficult to evaluate what oceanographic parameters were important.

Phosphorite as a substrate was also found more commonly in low-latitude stations in a recent survey of seamounts in the southern hemisphere (Fig. 1). Cullen and Burnett (1986) report that iron-manganese encrustations were discovered over phosphorite on two isolated seamounts north of $10°$ S latitude -- Kalolo Seamount (northwest of the Tokelau Islands) and Albert Henry Seamount (north of the Northern Cook Islands). Only mildly phosphatic material was found in the more southerly stations. The sample from Albert Henry Seamount is quite young, at about 2.8 - 3.0 m.y. (late Pliocene), while the phosphorite from Kalolo Seamount (Fig. 2) must be older because of its 40 mm thickness of iron-manganese material. At an assumed manganese accumulation rate of 2 mm/10^6, the substrate phosphorite would have to be at least 20 m.y. old (early Miocene).

The Hawaiian Archipelago studies have shown that phosphorite substrates underlying Fe/Mn crusts are quite common on at least two of the southern Musicians Seamounts, Mendelssohn and Schumann. Some

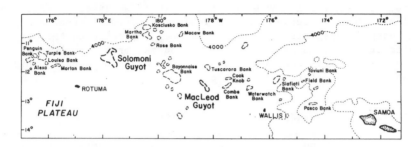

Fig. 1. Index map of shallow banks and deeper guyots along the North Fiji Plateau. Phosphorite in association with dolomite was sampled from both Solomoni and MacLeod Guyots. (From Cullen and Burnett, 1986.)

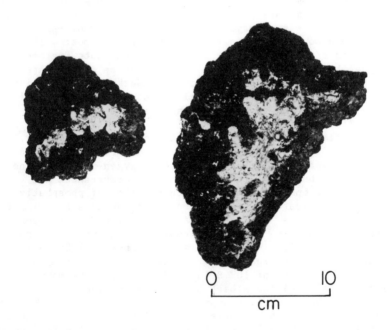

Fig. 2. Iron—manganese crust on a phosphatized limestone substrate from Kalolo Seamount, South Pacific Ocean. The crust thickness approaches 4 cm.

samples show distinct evidence of multiple growth or depositional episodes. According to De Carlo et al. (1986) these same samples generally contain significant quantities of carbonate fluorapatite, both in a dispersed form throughout the crust, and as a discrete layer separating metal oxide zones. A crust from Schumann Seamount, for example, shows two distinct oxide-growth layers of crustal development separated by a very thin layer of phosphorite (Fig. 3). Judging by the thickness of the crust overlying the phosphorite zone, the age of the phosphatic material is probably Miocene. This probably is not an isolated occurrence, as apparent hiatuses in iron-manganese crust development, marked by paper-thin layers of phosphorite separating different generations of crust material, have been reported from several areas in the Pacific (Halbach et al., 1982; Halbach and Manheim, 1984; Hein et al., 1985b).

Hein and his colleagues noted that all the thicker crusts on Horizon Guyot displayed distinct inner and outer layers, the apparent result of two periods of crust formation. The two layers are separated by a very thin phosphorite layer and the inner crust is impregnated with as much as 12% carbonate fluorapatite. Crusts from S.P. Lee Guyot were also reported to display two distinct generations in the thicker crusts with as much as 11% apatite in the inner zones only.

The significance of the two-layer Fe/Mn – phosphorite "sandwich" is usually taken to represent two periods of hydrogenous metal oxide deposition, separated by a hiatus during which time the phosphorite formed. In at least some cases, the phosphorite appears to be the result of phosphatizing coccolith ooze, so the actual sequence of events may be periods of non-deposition of carbonate sediment (during which time the hydrogenous metal oxides are accumulating) at the crust site separated by a short interval of ooze accumulation. Perhaps the accumulation "event" could occur during a relaxation of local currents. For whatever reason, some time after the accumulation of carbonate sediment, a period of phosphatization occurs. It would be extremely useful if these events could be dated. Halbach et al. (1982) estimate that inner crusts from the mid-Pacific (which average about 2.0 cm in thickness) are about 10 m.y. old. The phosphatization period, therefore, may have occurred sometime in the Middle Miocene –– a time of worldwide deposition of sedimentary phosphorites. In one sample from the Central Pacific Basin reported by Halbach and Manheim (1984), preserved coccoliths from the phosphatic nuclei of a seamount nodule were separated and tentatively identified as belonging to a group "...of Oligocene to Miocene age...", thus setting an upper limit for the period of phosphatization somewhere in the middle Tertiary.

The suggestions of Halbach et al. (1982), and more recently of Hein et al. (1985b), are correct in that the origin of the phosphorite, and perhaps of the Fe/Mn crusts themselves, is somehow linked to a mid-water phenomenon in the ocean involving the oxygen-minimum zone. Expanded, more intensified oxygen-minimum zones probably occurred several times during the geologic past when productivities, caused by high sea level (Fischer and Arthur, 1977), were enhanced on the enlarged continental shelves. The "oceanic anoxic events" of the Upper Cretaceous do not seem to correspond exactly to periods of widespread phosphorite deposition on

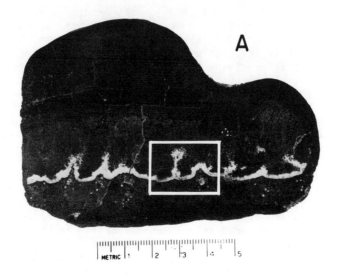

Fig. 3A.   Iron-manganese crust phosphorite "sandwich" from Schumann
Seamount in the southern Musicians Seamounts, north of Hawaii.   The
vein material is almost pure carbonate fluorapatite.

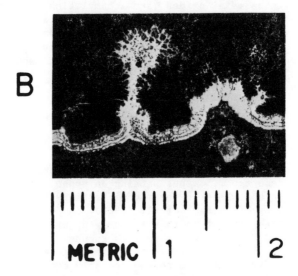

Fig. 3B. Detail of boxed area of sample in which alternating layers
of Fe/Mn oxides and phosphorite may be seen within the vein.

continental margins around that time, but they may correspond with the occurrences of Upper Cretaceous/Eocene phosphorites that occur on seamounts in the north Pacific Ocean (Arthur and Jenkyns, 1981). During these periods, phosphate concentrations in the mid-water column may have been high enough to convert fine-grained calcium carbonate to carbonate fluorapatite.

Seamount crusts and phosphorites may well be linked genetically, insofar that the origin of each, at least in part, is related to the oxygen-minimum zone in the ocean (Frakes and Bolton, 1984). Studies on the Peru shelf have shown that modern phosphorite tends to occur near the boundaries of the intersection of the oxygen-minimum zone with the continental margin (Burnett, 1977; Burnett et al., 1983). Oxygen-deficient waters are not only rich in phosphate and other nutrients, but contain elevated levels of manganese and cobalt as well (Klinkhammer and Bender, 1980; Knauer et al., 1982). Inasmuch as phosphorite is probably not forming continuously in the open-ocean environment, it is suspected that the relative amounts of apatite contained in Fe/Mn crusts are mainly controlled by a "phosphate-switching" mechanism related to regional or worldwide variations in such factors as equatorial upwelling and associated biological productivity which affects the oxygen-minimum zone. The iron-manganese crust, on the other hand, may have a relatively constant rate of hydrogenous precipitation, punctuated by these infrequent phosphatizing hiatuses that are phosphatizing episodes.

DROWNED ISLANDS, PHOSPHATE, AND BIRDS

There is little, if any, doubt after the pioneering work of Hamilton and his colleagues in the early 1950's that many guyots and seamounts of the Central Pacific Basin are submerged islands. Did some of these islands contain guano-derived phosphate deposits before they submerged? Is this the mechanism by which seamount phosphorites originate? This suggestion was made some time ago by Bezrukov (1973) as grounds for exploring for phosphorite on seamounts south of Christmas Island in the Indian Ocean. The fact that phosphorites were found there as well as in the vicinity of other well-known phosphate islands supports Bezrukov's hypothesis.

It may seem likely that a few seamount phosphate deposits did form by submergence of phosphate islands, but it is unlikely that the majority of them originated in this fashion. Although birds originated in the mid-Mesozoic, their major evolutionary advances and global dispersion took place in the early Cenozoic, probably accomplishing the major steps by the end of the Eocene (Beerbower, 1968). Thus, at least a few seamount deposits appear to be too old to have originated from bird droppings. The former islands along the Marcus-Necker Ridge, for example, were probably emergent until some time in the Late Cretaceous. In addition, the concept of drowned guano deposits is inconsistent with what is known about the history of some of the deposits. The phosphate on several northwestern Pacific seamounts, for example, is thought to have had two stages of phosphatization, the latter of which must have occurred in at least a relatively deep pelagic environment because of the texture of the phosphatized sediment (Heezen et al., 1973).

Cullen and Burnett (1986) recognized two distinct types of phosphorite during their work in the south Pacific Ocean. Based on petrographic observations, sedimentary associations, and geochemical criteria, they suggested that the phosphorite found on some isolated seamounts was marine in origin while phosphorite found on guyots associated with a series of shallow banks on the North Fiji Plateau was probably of insular origin. The suspected submerged phosphate was associated with dolomite, another common diagenetic mineral of phosphate islands (Fig. 4). Another example of a submerged "insular-type" phosphate deposit was discovered in 1976 by a group led by Andre Rossfelder in the lagoon of Mataiva Atoll near Tahiti (Rossfelder, 1984). This deposit was apparently formed during one or more former glacial periods when sea level was lower and the deposit was subsequently drowned and covered by a veneer of carbonate sediments during high stands of the sea (Roe and Burnett, 1985).

In dealing with these relationships, another side of the issue is whether emergent seamounts containing marine phosphate deposits may result in insular phosphate deposits. Whether this has ever happened has not yet been established but it certainly must be considered a possibility.

Fig. 4. Scanning electron microscope photograph of small botryoidal aggregates of carbonate-fluorapatite (light material) coating subhedral dolomite crystals in a sample from MacLeod Guyot, North Fiji Plateau (Cullen and Burnett, 1986). Dolomite crystals approximately 20-30 μm across.

CONCLUSIONS

Based on our present knowledge of the distribution and geologic setting of open—ocean phosphorites, the following conclusions are reached:

(1) Phosphorites and phosphatic rocks are commonly occurring lith- ologies on summits of many seamounts in the world's oceans.

(2) Most of these deposits are marine in origin, although some of them are clearly derived by submergence of "phosphate islands."

(3) The present distribution of these deposits suggests that they may form preferentially in low latitudes, perhaps in response to equatorial upwelling and associated phenomena (high biological productivity and consequent development and maintenance of an oxygen minimum zone, etc.).

(4) The association of phosphorite with iron—manganese crusts on seamounts in the Central Pacific Ocean is probably related to the enhanced concentrations of phosphate and metals within the zone of minimum dissolved oxygen which bathes the summits of many seamounts.

ACKNOWLEDGEMENTS

The authors wish to express their gratitude to the officers and crew of the R/V TANGAROA (New Zealand Oceanographic Institute) and the R/V KANA KEOKI (University of Hawaii) for their efforts during the collection of many of the samples discussed in this paper. We also thank E. Maughan of the U.S. Geological Survey for arranging for the use of the electron microprobe. Ms. Sheila Heseltine (Florida State Univer- sity) provided extremely useful editorial assistance on this manuscript, as did Barbara Whitehouse—Jones (University of Hawaii). Financial support to the senior author was provided by grants from the National Science Foundation (OCE8317181 and INT8111783). This paper is a contribution to IGCP Project 156 (Phosphorites). Hawaii Institute of Geophysics contribution 1701.

REFERENCES

Altschuler, Z.S., 1980, The geochemistry of trace elements in marine
    phosphorites - Part I. Characteristic abundances and enrichment.
    In: Bentor, Y.K. (ed.), Marine Phosphorites; Soc. Econ. Min.
    Paleont. Spec. Publ. 29, pp. 19-30.

Arrhenius, G., 1963, Pelagic sediments, In: Hill, M.N. (ed.), The Sea;
    John Wiley & Sons, New York, v. 3, pp. 655-727.

Arthur, M.A. and Jenkyns, H.C., 1981, Phosphorites and paleoceanography;
    Proc. 26th Inht. Geol. Cong., Geology of Oceans Symposium, Paris,
    July 7-17, 1980, In: Oceanologica Acta, pp. 83-96.

Baturin, G.N., 1982, Phosphorites on the Sea Floor; Elsevier Publ. Co.,
    New York, 343 p.

Beerbower, J.R., 1968, Search for the Past, 2nd Ed.; Prentice-Hall,
    Inc., Englewood Cliffs, 512 p.

Bezrukov, P.L., 1973, Main scientific results of the 54th voyage of the
    RV VITYAZ in the Indian and Pacific Oceans (February-May 1973);
    Okeano- logiya, no. 5.

Bezrukov, P.L., Andrushenko, P.F., Murdmaa, I.O., and Skornyakova, N.S.,
    1969, Phosphorite on the floor of the central part of the Pacific
    Ocean; Dokl. AN SSSR, v. 187, no. 4.

Bonatti, E., Honnorez, J., and Ferrara, R., 1970, Equatorial
    Mid-Atlantic Ridge: petrologic and Sr-isotopic evidence for an al-
    pine type rock assemblage; Earth and Planet. Sci. Lett., v. 9, no.3.

Burnett, W.C., 1977, Geochemistry and origin of phosphorite deposits
    from off Peru and Chile; Geol. Soc. Am. Bull., v. 88, pp. 813-823.

Burnett, W.C., Roe, K.K. and Piper, D.Z., 1983, Upwelling and
    phosphorite formation in the ocean, In: Suess, E. and Thiede, J.
    (eds.), Coastal Upwelling: Its Sediment Record, Proc. NATO Adv. Res.
    Inst.; Plenum Press, New York and London, v. 10, pp. 377-397.

Cullen, D.J., 1980, Distribution, composition and age of submarine
    phosphorites on Chatham Rise, east of New Zealand, In: Bentor, Y.K.
    (ed.), Marine Phosphorites; Soc. Econ. Min. Paleont. Spec. Publ. 29,
    pp. 139-148.

Cullen, D.J. and Burnett, W.C., 1986, Phosphorite associations on
    seamounts in the tropical southwest Pacific Ocean; Marine Geology,
    in press.

DeCarlo, E.H., McMurtry, G.M., and Kim, K.H., 1986, Geochemistry of
    ferromanganese crusts from the Hawaiian Archipelago Exclusive

Economic Zone, I: Northern Survey Area; Submitted to Deep—Sea Research.

Fischer, A.G. and Arthur, M.A., 1977, Secular variations in the pelagic realm; Soc. Econ. Min. Paleont. Spec. Publ. 25, pp. 19—50.

Frakes, L.A. and Bolton, B.R., 1984, Origin of manganese giants: sea-level change and anoxic—oxic history, Geology, v. 12, pp. 83—86.

Gulbrandsen, R.A., 1970, Relation of carbon dioxide content of apatite of the phosphoria formation to regional facies; U.S. Geol. Surv. Prof. Paper 700—B, pp. B9—B13.

Haggerty, J.A., Schlanger, S.O., and Silva, I.P., 1982, Late Cretaceous and Eocene volcanism in the Southern Line Islands and implications for hotspot theory; Geology, v. 10, pp. 433—437.

Halbach, P., Manheim, F.T., and Otten, P., 1982, Co—rich ferromanganese deposits in the marginal seamount regions of the Central Pacific Basin — results of the MidPac '81; Erzmetall, v. 35, no. 9, pp. 447—453.

Halbach, P. and Manheim, F.T., 1984, Potential of cobalt and other metals in ferromanganese crusts on seamounts of the Central Pacific Basin; Marine Mining, v. 4, pp. 319—336.

Hamilton, E.L., 1956, Sunken Islands of the Mid—Pacific Mountains; Geol. Soc. America, Memoir 64, Boulder, 97 p.

Hamilton, E.L. and Rex. R.W., 1959, Lower Eocene phosphatized ooze from Sylvania Guyot; U.S. Geol. Surv. Prof. Paper 260—W.

Heezen, B.C., Mathews, J.L., Catalano, R., Natland, J., Coogan, A., Tharp, M., and Rawson, M., 1973, Western Pacific Guyots; Init. Rept., DSDP, Leg 20, U.S. Govt. Printing Office, Washington, D.C., pp. 653—702.

Hein, J.R., Manheim, F.T., Schwab, A.S., Daniel, C.L., Bouse, R.M., Morgenson, L.A., Sliney, R.E., Clague, D., Tate, G.B., and Cacchione, D.A., 1985a, Geological and geochemical data for seamounts and associated ferromanganese crusts in and near the Hawaiian, Johnston Island, and Palmyra Island Exclusive Economic Zones; U.S. Geol. Surv. File Rpt. 85—292, Menlo Park, U.S.A., 129 p.

Hein, J.R., Manheim, F.T., Schwab, W.C., and Davis, A.S., 1985b, Ferro-manganese crusts from Necker Ridge, Horizon Guyot, and S.P. Lee Guyot: geological considerations; Marine Geology, in press.

Jones, E.J.W., and Goddard, D.A., 1979, Deep—sea phosphorite of Tertiary age from Annan Seamount, eastern equatorial Atlantic; Deep—Sea Research, v. 26A, pp. 1363—1379.

Klinkhammer, G.P. and Bender, M.L., 1980, The distribution of manganese in the Pacific Ocean; Earth Planet. Sci. Lett., v. 46, pp. 361-384.

Knauer, G.A., Martin, J.H., and Gordon, R.M., 1982, Cobalt in northeast Pacific waters; Nature, v. 297, pp. 49-51.

Kolodny, Y., 1981, Phosphorites, In: Emiliani, C. (ed.), The Sea; John Wiley and Sons, New York, pp. 981-1023.

Kudrass, H.R. and Cullen, D.J., 1982, Submarine phosphorite nodules from the central Chatham Rise off New Zealand — composition, distribution, and reserves (VALDIVIA cruise 1978); Geol. Jahrb., v. D-51, pp. 3-41.

McArthur, J.M. and Walsh, J.N., 1985, Rare-earth geochemistry of phosphorites; Chemical Geology, v. 47, pp. 191-220.

Manheim, F.T., 1972, Mineral resources off the northeastern coast of the United States; U.S. Geol. Surv. Circular 669, 28 p.

Manheim, F.T., Pratt, R.M., and McFarlin, P.F., 1980, Composition and origin of phosphorite deposits of the Blake Plateau, In: Bentor, Y.K. (ed.), Marine Phosphorites, Soc. Econ. Min. Paleontl. Spec. Publ. 29, pp. 117-138.

Manheim, F.T., Popenoe, P., Siapno, W., and Lane, C., 1982, Manganese-phosphorite deposits of the Blake Plateau, In: Marine Mineral Deposits — New Research Results and Economic Prospects, Proc. of the Clausthaler Workshop, Repub. of Germany; Verlag Gluckauf, pp. 9-44.

Marlowe, J.I., 1971, Dolomite, phosphorite, and carbonate diagenesis on a Caribbean seamount; Jour. Sed. Petrol., v. 41, pp. 809-827.

Piper, D.Z., Baedecker, P.A., Crock, J.G., Loebner, B.J., and Burnett, W.C., 1986, Rare earth elements in apatite concretions from the Peru shelf; Submitted to Marine Geology.

Roe, K.K. and Burnett, W.C., 1985, Uranium geochemistry and dating of Pacific Island apatite; Geochim. Cosmochim. Acta, v. 49, pp.1581-1592.

Rossfelder, A., 1984, Phosphate and Facts;Pacific Islands Monthly, p.11.

Sheldon, R.P., 1980, Episodicity of phosphate deposition and deep ocean circulation — an hypothesis, In: Bentor, Y.K. (ed.), Marine Phosphorites, Soc. Econ. Min. Paleont. Spec. Publ. 29, pp. 239-248.

Slater, R.A. and Goodwin, R.H., 1973, Tasman Sea guyots; Marine Geology, v. 14, pp. 81-99.

von Rad, U., 1984, Outline of SONNE cruise SO-17 on the Chatham Rise phosphorite deposits east of New Zealand; Geol. Jahrb., v. D-65, pp. 5-23.

SOME MINERAL RESOURCES OF THE WEST AFRICAN CONTINENTAL SHELVES RELATED
TO HOLOCENE SHORELINES: PHOSPHORITE (GABON, CONGO), GLAUCONITE (CONGO)
AND ILMENITE (SENEGAL, MAURITANIA)

J.P. Barusseau and P. Giresse
Laboratoire de Recherches de Sedimentologie Marine
Université de Perpignan
66026 Perpignan, France

ABSTRACT.   In marine environments, Recent and fossil shorelines are
special targets of interest for hard mineral resources exploration on
account of their close relationship with both their mineral constituents
and the physical forces responsible for the accumulation of these
minerals.   Such occurrences of heavy minerals, associated with littoral
features (shoreface, shore and dune), are displayed on the West African
coast of Gabon–Congo and Senegal–Mauritania.
    Holocene phosphatic coprolites of the Congolese and Gabonese shelves
are reworked from Miocene and Eocene submarine outcrops.   The water
depth in the area ranges uniformly from ~ 30 to 50 m.   The phosphatic
deposits are flat-lying, shallow and largely unconsolidated.   Their
high grade is related to the winnowing and sorting of the Holocene
shoreline deposits for the last 8,000 years.   In the Republic of Congo,
there are large deposits of muddy glauconite and glauconitic carbonate
mud on the continental shelf adjacent to a 18,000 years old shoreline.
On the adjacent land, vast stretches of very poor, sandy soils could
potentially benefit from having the glauconite applied as an unprocessed
fertilizer.   Trial applications have shown that glauconite produces
good yields and preserves an adequate level of soil fertility despite
considerable leaching.   Pleistocene and Holocene beaches of West African
coasts contain numerous layers rich in Ti- and Zr- minerals.   In the
shallow waters of the Ivory Coast (Vridi–Fresco), Senegal (Mbour–Joal;
Cayar–St. Louis) and Mauritania (Aftout es Saheli; South Cap Timiris)
concentrations of ilmenite and zircon occur in sedimentary beds.   These
heavy minerals are reworked from old littoral dunes or dune massifs by
erosion and deposited along the present shore by waves.

PHOSPHATIC DEPOSITS OF THE CONTINENTAL SHELVES OFF CONGO AND GABON

    Much as the offshore phosphatic deposits on the western margin of
Africa (Morocco: Summerhayes et al., 1973; Tooms and Summerhayes, 1968;
South–Africa: Parker and Siesser, 1972; Birch, 1979), the submarine
phosphates of Congo and Gabon belong to sedimentary basins, where
several phosphatic sequences are known to occur on land (Fig. 1).

135

P. G. Teleki et al. (eds.), Marine Minerals, 135–155.
© 1987 by D. Reidel Publishing Company.

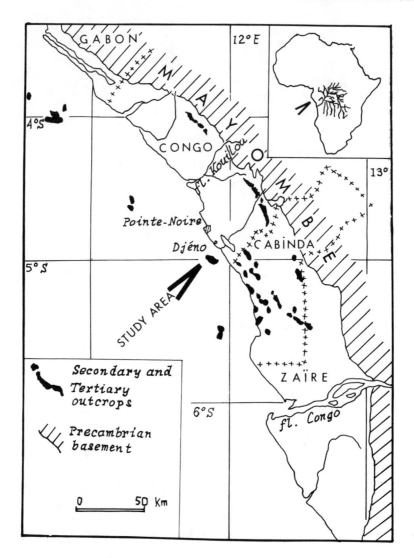

Figure 1.   Location map of the phosphate study area.

The oldest phosphatic units exposed in the basin are of
Maestrichtian age and occur in a narrow belt trending NW-SE parallel to
the Precambrian Massif of Mayombe.   Down-faulting prevented their erosion (Giresse, 1980).   The age of the highly altered upper beds is
Paleocene.   Other phosphatic horizons, usually decalcified and silicified,
are reported to occur in the Upper Cretaceous beds along the southern
coast of Gabon (Furon, 1932).   On land, along the coast of Cabinda

(Angola), phosphatic deposits are exposed parallel to a N 110°–120° structural trend, which is controlled by Eocene moderate compression. These significant, but irregularly distributed phosphates have been discovered along the present coast in the Ypresian and Montian–Thanetian beds and in Maestrichtian beds farther east (Cunha Gouveia, 1960). Miocene sequences also merit attention because of the presence of bone beds, but their content of P_2O_5 is modest. In addition, in the course of petroleum exploration drilling, phosphate layers of Late Cretaceous and Paleogene age had been encountered in the subsurface.

In 1973 (Cornen et al.) and in 1976 (Giresse and Cornen), Eocene phosphates redeposited in the Miocene and Quaternary were discovered on the submarine shelf of Southern Congo and Southern Gabon. The deposits are flat-lying, shallow and essentially unconsolidated, and contain successively reworked material from Ypresian age phosphatic outcrops (that are an extension of those on land in Cabinda) or, more frequently, from Miocene beds which overlie the Ypresian reefs.

The first geophysical and vibracore reconnaissance survey (Phoscap mission) was conducted in an area offshore Southern Gabon (Horn, 1978). The preliminary evaluation of phosphate resources on the inner and outer shelves proved to be disappointing (Malounguila-NGanga, 1983). By contrast, offshore of Djeno, Southern Congo, phosphorite occurrences appeared to be more promising, particularly those located between the 35 m and 50 m bathymetric contours (Fig. 2). This area was studied during two successive geophysical and drilling/sampling surveys, sponsored by the United Nations Revolving Fund for Natural Resources Exploration and the French Bureau de Recherches Geologiques et Minieres and performed by Zellars–Williams Inc., Lakeland, Florida (Woolsey et al., 1984). The resulting observations were partially presented by Bargeron (1984), including those of earlier studies, and together give a reasonably complete description of the deposits. An evaluation of the resource potential was not made.

On land in Cabinda, Ypresian age beds are composed of white marls and marly limestones, frequently silicified and relatively rich in P_2O_5 (12% to 25%) and phosphatic sands with more than 26% P_2O_5 (Cunha Gouveia, 1960). The northern extension of the hardest beds (phosphatic and silicified) outcrop on the present continental shelf. Differential erosion during Oligocene had dissected these beds and narrow terraces remained, which have a hummocky surface and a relief of 7–8 m. The outcropping beds were subsequently strongly reworked by transgressions of the sea during the Miocene. The Miocene shoreline can be traced from 80 m water depth offshore of Southern Gabon to –50 to –60 m offshore of northern Congo, to –35 to –40 m offshore of Djeno and, finally, on land in Cabinda area (Landana). This progressive uplift of the Miocene shoreline seems have been caused by upward epeirogenic movements controlled by tensional tectonics increasing in scale from north to south during the Neogene and the Quaternary.

The coastal facies of the Miocene includes a conglomerate with brown and glazed coprolites of Selacians containing a calcitic and apatitic cement. This conglomerate resulted from in-situ reworking of Ypresian reefs in a Miocene age breaker zone. The facies is present within the entire study area, although it is particularly well preserved

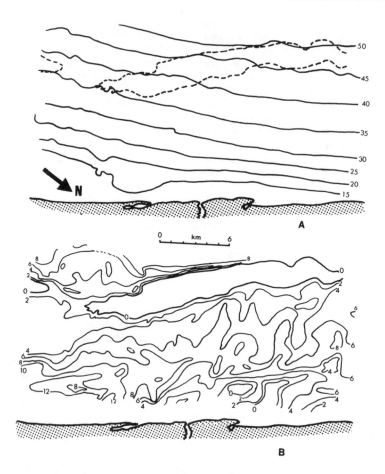

Figure 2.   Bathymetric (A) and isopach (B) maps of the Djeno
phosphatic bank region offshore of Congo.   Contours are in meters.
The dashed line in the upper sketch is the outline of the bank.
Isopachs in the lower sketch represent the distribution of
Holocene deposits (phosphatic unit and fine-grained sediments).

offshore.   The Miocene-age conglomerate and some of the Ypresian beds
were exposed several times during Quaternary regressions of the sea,
and during the intervening transgressive periods became truncated and
reworked.   In the process, Miocene phosphatic coprolites were liberated,
which then accumulated and became concentrated in the areas neighboring
the outcrops.   The Holocene transgression controlled the last redeposi-
tion.   Offshore of Southern Gabon, occurrences of Miocene conglomerate
are rare and located in deep water (−50 to −90 m), where the sea transgressed rapidly.   Offshore of Djéno, by contrast, the transgression

reached the Tertiary outcrops between −35 to −50 m approximately 9,000 years ago and the barrier was crossed or bypassed quite quickly. About 7,000 years ago, when the rate of transgression had slowed down (Giresse et al., 1984) and reached a static stage of sea-level, the wave action caused intensive reworking and winnowing of the deposits to a depth of 15 to 30 m (Fig. 3).

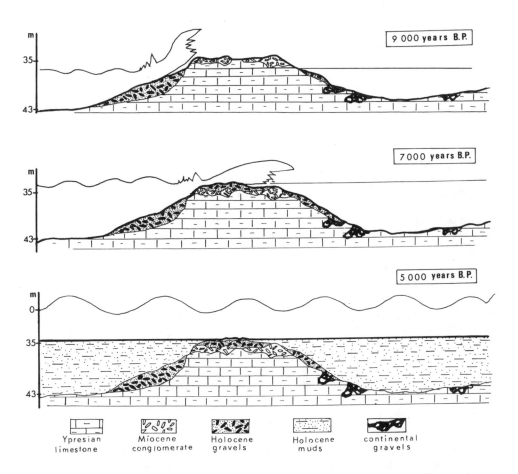

Figure 3. Steps of mechanical concentration of phosphate on the bank. Top: 9,000 years B.P., the winnowing action of the waves is located seaward of the bank. Middle: 7,000 years B.P., intensive reworking and winnowing action during the time when transgression was slowing down. Bottom: 5,000 years B.P., phosphorite gravels become thinly covered by barren muds.

About 5,000 years B.P., the transgression reached the present elevation and the winnowing action of the waves reached areas where water depths presently range from 35 to 50 meters. Bedrock and the

and the phosphorite gravels became thinly covered by silty muds eman-
ating from the Congo River mouth. Here, the only phosphorite layer
that has any areal continuity is the basal one. Its thickness averages
about 15 cm with a few pockets 50 cm deep.

The phosphorite-rich gravels that outcrop just offshore Djéno
are estimated to contain 4,000,000 tons of phosphatic sands. However,
this estimate is preliminary, and awaits further investigations in an
area that is covered by as much as 3 meters of mud. Bedrock, or its
alteration products, were reached in about 45% of the cores taken.
Gravel accumulation off the bank on the upslope of the "fossil" wave
breaker zone could also be of economic interest. Another target is the
northern extension of the outcrop zone where longshore currents might
have accumulated and concentrated coarse phosphatic particles 7,000
years ago. A gamma radiometric survey of the surface sediments (Fig.
4) showed that the phosphatic deposit on the Cabinda shelf extends
southwest. This extension must be taken into account in the regional
evaluation of the deposits, and in subsequent economic considerations.

It is apparent that even with the most favorable tonnage, grade,
geometry and relationship to overburden and bedrock, these deposits
would not be competitive on the world phosphate market. However, with
the present growth of agricultural efforts in the Republic of Congo,
the deposit could prove valuable as a domestic phosphate resource.
Attention has focussed upon its potential as a directly applicable
fertilizer, which would not require the costly treatment associated
with the production of super-phosphates. In its favor is that the
deposit is not cemented, lies in modest water depths and is very near
the Pointe-Noire harbor. Moreover, being less soluble in the very
acidic podzolic or ferralitic soils that characterize this tropical
area, the phosphorite is able to remain active in the soil much longer
than super-phosphates can. Because the phosphatic gravels contain
carbonate fragments of tests from the bedrock epiphyte fauna, these
further enhance their value as an efficient fertilizer for use on
low-pH acidic soils. The residual muddy component of these gravels
could also be made useful if organic matter would be added to it,
because this would create bonding in the highly siliceous soils.

GLAUCONITE DEPOSITS OF THE CONTINENTAL SHELF OF CONGO

The accumulation of glauconitic grains in significant amounts
reflects mainly two genetic processes. The first is a geochemical/
mineralogical evolution, the second is mechanical concentration.
Description of both processes is useful to the understanding of the
genesis of glauconites.

Authigenesis

Generally speaking, most sediments of the world's continental
shelves and slopes are authigenic. One of the necessary conditions
for authigenesis is the availability of a substrate (fecal pellets,

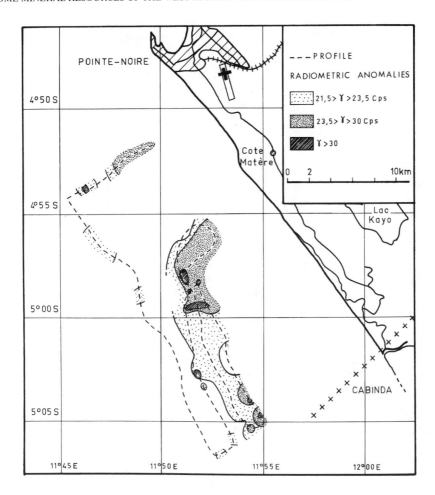

Figure 4. Distribution of phosphatic sediments as measured by gamma-ray. Values are given in counts per second (Cps).

foraminiferal tests, porous surfaces or plates, micaceous layers), which also play a prominent role in the geometric definition of a geochemical microenvironment different from that of the surrounding sediment. The substrates and the matrix are separated by an oxic-anoxic boundary. During the first stages of the authigenesis, the role of iron is most important. In mobile form (ferrous iron), it can leave the ambient medium and become incorporated and concentrated in the microenvironment.

The mineralogical composition of the microenvironment of this substrate appears to be significant. The nearer to glauconite the composition is, the more receptive the substrate will be. For example, this is the case for clay minerals such as illite, smectite (particularly

nontronite) or biotite particles. In the case of biotite, glauconite
genesis can presumably appear by a simple transformation process. Ion
exchange across the oxic-anoxic boundary is all the more enhanced when
the ambient environment is organic-rich and reducing and able to mobi-
lize the ferrous iron. This is today the case along much of the seafloor in tropical zones offshore of river mouths.

Whenever iron is being depleted, conditions will be unfavorable
for glauconite formation. Various data from modern continental shelves
lead to conclude that three controlling factors are important to the
primary accumulation of glauconite and in evaluating future prospects of
economical interest:

a) In a vertical depositional sequence the distribution of organic
matter is irregular and controlled by the rate of sedimentation.
Higher concentrations are present in layers of fast sedimentation, from
which iron can be extracted and concentrated in the layers of slower
rate sedimentation. In sediment layers of centimetric or decimetric
dimensions, iron would be displaced towards the oxygenated water-
sediment interface. This process is similar to that observed in the
genesis of manganese-oxide deposits. These observations support the
model of glauconitization proposed by Ireland et al. (1983). The
greater the interaction time at the interface, the higher the glauconite
concentration.

b) Iron can be delivered bound to its original mineralogical
substrate (biotites, pyroxenes, hornblends); thus, it can reach the
deepsea bottom directly. This was observed on the active margin of
western Canada (Bornhold and Giresse, 1985). This process implies a
rapid delivery and/or the presence of unaltered particles of "first
cycle" sediments such as those derived from periglacial environments.

c) Iron can be related to intense hydrolysis present in tropical
continental environments, and it is accumulated offshore of several
river mouths such as those of the Congo, the Niger and the Orinoco, to
mention only the most important ones, where iron is associated with a
relatively high content of organic matter. The influx of iron is
controlled by several factors that depend on the lithology of the river
catchment area. Ferralitic sequences of soils take up iron, and this
is stocked until a rhexistasic event takes place; this is occurring
today in the Amazon Basin, whose continental shelf is poor in glauconite.
By contrast, in a predominantly sandy basin, podzolization processes
are related to the amount of iron carried to the ocean. The basin
drained by the Congo River is a good example, and the glauconitic
accumulations offshore are significant (Giresse, 1983). However, even
in this very favorable case, the relatively high content of glauconite
observed in Recent surficial muds is still less than 20% of the total
volume of sediment, and commonly it is less than 10%.

Thus, post-sedimentary processes are required to form the "green
sands" with a concentration that would elicit any economical interest
in their exploitation.

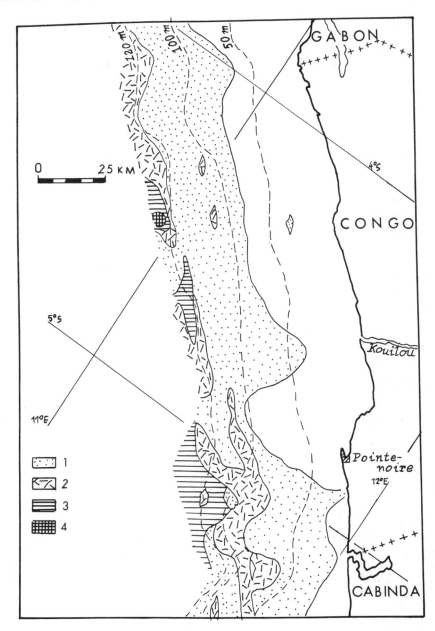

Figure 5. Distribution of the magnetic fraction in the surficial sediments of the shelf offshore of Congo. 1: 5-25%, 2: 25-75%, 3: 50-75%, 4: more than 75%.

The concentration process

A very significant, controlling aspect of glauconitization is the
availability and continuing production of proper substrates. Whether
such substrates can be produced in a given area depends on certain
paleoenvironmental factors that governed:
　　1)　organic primary production of calcareous tests, porous substrates favorable to various epigenesis (as on the outer Continental
Shelf of Spain, off Asturia, during the Miocene; Lamboy, 1976);
　　2)　mechanical reworking or bioturbation of semi-consolidated muds
(outer shelf of the Western Canadian Continental margin; Bornhold and
Giresse, 1985);
　　3)　high production of fecal pellets by deposit-feeder organisms
such as Polychaetes. This process, present on most of the present
continental shelf of West Africa, is controlled by the quantity of
nutritive matter in the sediment, which is at peak production during
high rates of upwelling. Offshore of Congo, the regressive period
approximately 18,000 years ago was particularly favorable to such
production (Giresse et al., 1981).
　　Mechanical sorting and concentration, by successive reworking, is
significant in the genesis of green-sand deposits. This takes place
even on those margins where the rate of glauconite genesis is particularly low today, such as off Morocco (Tomms and Summerhayes, 1968) or
Western Spain (Lamboy, 1976), where accumulations are largely related
to Pliocene events.
　　Some conclusions can be drawn from the study of more recent glauconite accumulations on the western shelves of Africa, especially
offshore of Congo where relatively high contents have been observed
(Fig. 5 and 6). On these shelves, a semi-consolidated marine mud
outcrops in many places. This deposit, whose age is mid-Wisconsin, has
a very low glauconite content, thus, it is indicative of the starting
point of the subsequent glauconitization. During the last period of
lowered sealevel (18,000 years B.P.) the shoreline retreated to approximately the present 110-m bathymetric contour. The shelf was a narrow
strip where upwelling was preponderant. The coastline and the morphology of the narrow platforms were both controlled by monoclinal "cuestas"
of Miocene outcrops, which controlled the configuration of the shoreline
and whether areas were sheltered or exposed to waves. Fecal pellets
provided the microsurfaces for mineralization. In a 6-m-thick vertical
sequence, alternating, rather pure green sands and glauconitic muds
occur. On rocky substrates, the cover is nearly pure glauconite, but
it is only about ten centimeters in thickness. At most locations, fine
particles had been winnowed and removed from the area of 10 to 20 m
water depth and transported directly to the upper slope. This mechanical process separated the authigenic grains from their muddy matrix.
Without this process, glauconite could not have accumulated on the
West-African margin in recent times.
　　During the post-Wisconsin transgression, the shoreline rapidly
moved landward from the green-sand zone. As a result, the deposits,
being far from any terrigenous sediment supply, remained uncovered
during this period. The warm transgressive waters induced a rapid

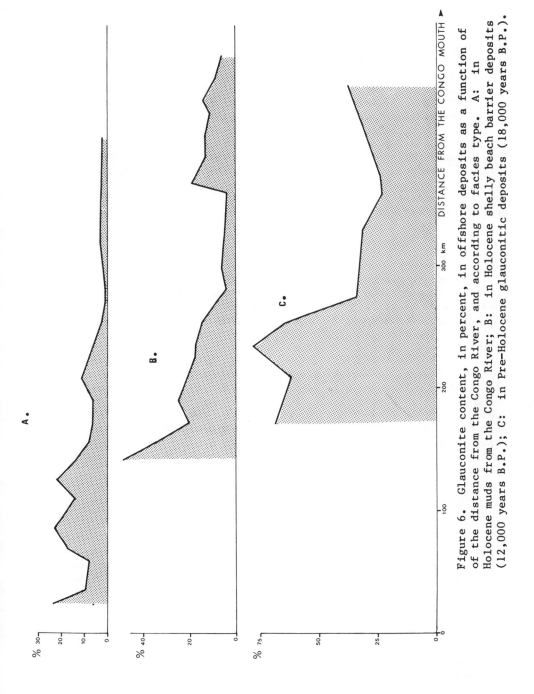

Figure 6. Glauconite content, in percent, in offshore deposits as a function of of the distance from the Congo River, and according to facies type. A: in Holocene muds from the Congo River; B: in Holocene shelly beach barrier deposits (12,000 years B.P.); C: in Pre-Holocene glauconitic deposits (18,000 years B.P.).

production of invertebrate calcareous tests which accumulated in water
depths of 95 to 105 m during a stable phase of the shoreline position.
The resulting calcareous substrate, unlike the fecal pellets, was not
as favorable a medium for glauconite genesis.

After 12,000 years B.P., the transgression proceeded rapidly
shoreward and eventually reached the present coast. The middle and the
inner shelves became areas of deposition for finer clastics, such as
mud delivered largely by the Congo River.

Although authigenesis of glauconite occurred at a higher rate in
the organic and ferrugineous muds near the mouth of the rivers, where
Polychaete fecal pellets were particularly abundant, concentrations of
glauconite by mechanical means also took place on the sea bottom near
the high-energy coast and near the subaqueous Cretaceous outcrops
located in waters 30 m deep. However, in none of these Holocene muds
does glauconite reach concentrations greater than 20%.

Resources

The entire glauconitic mass, accumulated during the last 18,000
years on the Congolese margin, represents about 59×10^9 tons. If the
evaluation is restricted to the green sands of the outer shelf, (about
57×10^9 tons), the average K_2O content of this accumulation is estimated
to be 6%. The green sands have been tested in agricultural field trials
to determine whether they could increase soil fertility (Giresse and
Jamet, 1982). The results, following a full growth cycle of the Cassava
plant (18 months), show that glauconitic sediments produce good agricultural yields and preserve an adequate soil fertility despite considerable
leaching. This is significant, because in the Republic of Congo, vast
areas of very poor sandy soils are present on the entire coastal plain
and in "Serie des Cirques" continental sandy deposits. These marine
deposits, therefore, could be used on fields locally and more economically than commercial fertilizers that need prolonged and expensive
transportation.

ILMENITE DEPOSITS OF PRESENT AND ANCIENT SHORELINES OF THE SENEGAL-
MAURITANIA COAST

Heavy mineral placers, with high ilmenite content, have been known
for a long time on the coastline of Senegal and Mauritania (Schwarz,
1931). These deposits exist within the component parts of the littoral
prism, including the dunes.

Hebrard (1981) reported that such accumulations also occur along
Ivory Coast. Littoral placer deposits of upper Holocene age along the
shore overlie the so-called "Arca senilis beach," whose location is
described by Elouard (1967). This deposit is dated to be 5,500 years
old ("Nouakchottian"). Recent offshore surveys have revealed a third
occurrence at 25-30 m depth.

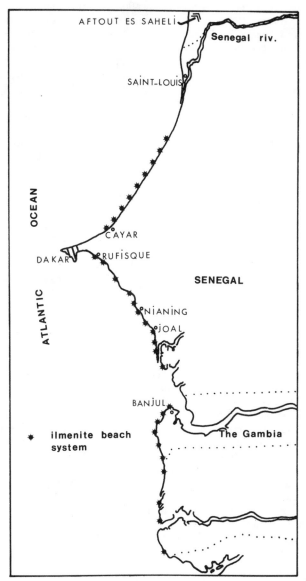

Figure 7. Littoral accumulation of ilmenite within the beach and dune system of the coast of Senegal.

Modern Deposits

From north to south, Holocene placers occur in the vicinity of the Baie d'Arguin islands (Blanchot, 1946; Schwarz, 1931), along the Mauritania coast between Nouakchott and Cape Timiris (Vogt, 1956), and

north of mouth of the Senegal River ("Aftout es Saheli"; Hebrard, 1981).
Along the Senegalese coast (Fig. 7), numerous placer occurrences are
known to occur between Lompoul (South St. Louis) to Cayar and, South of
Cap Vert, from Rufisque to Joal ("Petite Côte"). Minor occurrences,
rather poorly surveyed, have also been reported for the Gambia and the
Casamance coasts (Bureau de Recherches Géologiques et Minières, 1973).

      Ilmenite is the main component of a heavy mineral assemblage,
with lesser amounts of zircon and rutile. The latter enhance the
economic value of these deposits, that are composed of layers of
millimetric to centimetric thickness (a few of them decimeters thick),
often as lenticular beds within the beach sand prism. Littoral drift
is the primary agent in offshore and beach-face concentrations, and
aeolian reworking is responsible for concentrating ilmenite in the
dunes. In the first mode, deposits result from a two-fold mechanism of
littoral transport alongshore and in the offshore. Most likely, the
sediment originates from erosion of the large Akchar, Agneitir and
Azefal dune fields in Mauritania ("Ogolian" arid phase, coeval with the
Wisconsin glacial period), although it is possible that aeolian deposits
that covered the northern and central parts of Senegal, could also be
contributing to the supply. This process, along with the reworking of
aeolian sands, ensures the availability of a rather steady supply of
sand material. The sediments are composed principally of fine sand
with a major size distribution mode at 0.08 mm and, at times, show a
second modal value at 0.2 mm (Dropsy, 1943). Because the littoral
transport direction is southward, the sedimentary drift crosses the
Cayar sub-marine valley, whose head is in the active surf zone. The
canyon deflects part of the sediment load seaward, and this modifies
the sedimentary budget of the southern coast resulting in coastal
erosion in this part of Africa (i.e., one meter per year on the south
Mauritania coast). Furthermore, the process of onshore-offshore transport is also active, whose offshore component is prominent, whereby the
removal of winnowed light quartz particles from the nearshore area is
constant. A recent veneer of fine sand on the inner shelf is a consequence of this winnowing action.

      ·As a result of these processes, heavy minerals and especially
ilmenite grains become concentrated as lag deposits (Barusseau, 1985).

      Various appraisals (Blanchot, 1946; Vogt, 1956; and Hebrard, 1981)
indicate that the deposits along the southern coast of Mauritania contain between 0.15×10^6 and 10×10^6 m^3 of ilmenite. This range is
so broad, that these estimates could only be considered as preliminary.

Fossil deposits

      Known ilmenite deposits older than those forming today are those
of late and early Holocene age. Late Holocene placers (5,500 years
old) were deposited during the transgressive "Nouakchottian" phase,
during which the post-glacial sea reached its uppermost level (2 to 3 m
above the present M.S.L.). As a result of transgression, numerous
embayments and rias were formed which make for an irregular coastline.
Subsequent climatic decay ("Tafolian stage"; 4,000-2,000 years B.P.)

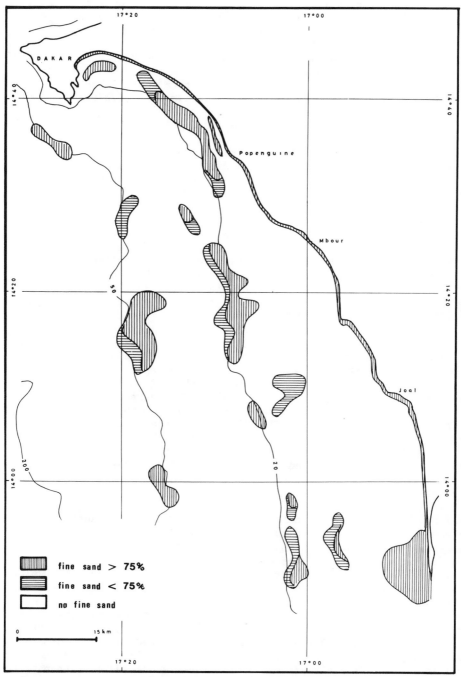

Figure 8. Fossil and present coastal sedimentary systems south of Cap Vert (Dakar).

imparted greater aridity (Hebrard, 1972) and simultaneously a slight
lowering of sea level took place (Eisele et al., 1974; 1977; Pinson-
Mouillot, 1980). During this phase, long beach barriers closed the
indentations of the coastline.

The Holocene accumulations are located at the northernmost part of
the region (El Memghar-Zreif near Cape Timiris) and at the southernmost
part (Nianing, Mbodiene, Joal). High concentrations of heavy minerals
have been noted at Nianing, of which 8% is ilmenite, 1.5% is zircon and
0.1% is rutile (Hebrard, 1981).

Early Holocene shorelines can be distinguished easily on the shelf
by modal grainsize analysis of the sediment. South of Cap-Vert, three
shorelines have been mapped at 80 to 90 m; 45 to 55 m and 25 to 30 m
water depths (Masse, 1968; Barusseau 1984; Barusseau et al., 1984) that
correspond to a slowing down or pause in the transgressive motion of
the post-glacial sea level (Fig. 8). During these periods, extensive
erosion of dune field segments emplaced on the emerged shelf during the
prior regression should have taken place. Consequently, large dune
barriers were built southward on account of the littoral drift and a
subsequent formation of beach-rocks, presently submerged, occurred.

North of Cap Vert, the early Holocene paleo-shorelines are also
present (Pinson-Mouillot, 1980). During an exploration for ilmenite
(Bureau de Recherches Géologiques et Minières, 1973), seismic surveys
and gamma-ray spectrometry showed that a deposit existed between the 10
and 45 m isobaths, whose highest concentrations were between 25 and 30 m
(Fig. 9). The linear extent of this deposit clearly shows the close
tie between this pattern and the 8,500 to 7,000 years B.P. shoreline.

In these ilmenite-rich sediments, the median grainsize of the light
fraction ranges between 150 and 250 μm (north Cap Vert), values which
do not compare well with the main mode of the original aeolian material.
Heavy minerals are known to have a smaller median grainsize on account
of differences in hydraulic equivalences of light and heavy particles.
Computations of the Shields criterion for the quartz and ilmenite
fractions give a value of 2.25 for the ratio of the diameters which is
in agreement with the median grainsize of ilmenite north of Cap Vert
(100-125 μm). The equivalence in this case is not questionable.

Discussion

Finally, modern and ancient ilmenite-rich sands of Mauritania and
Senegal are clearly related to present or fossil shorelines, each
having a recognizable pattern. However, these patterns are not in
good agreement with the shape of the indented transgressive coast.
Furthermore, in spite of numerous evidences of erosion of large Ogolian
dune fields, the material of present and ancient shorelines does not
evince an obvious relation with this source.

Both discrepancies may be solved if two additional hypothesis are
raised about the nature of mechanical processes acting on this part of
the African coasts: 1) longshore drift contributed to establish a
regular lay-out of the shoreline whenever the sea level was stable, by
building long barriers beaches; 2) onshore-offshore sediment transport
from the littoral zone to both the shoreface and the inner shelf was

responsible for the observed sorting of light minerals whose fine
components were winnowed in an offshore direction while heavy particles
were concentrated. This process was largely governed by the slope of
the shelf (0.2 to 0.3%) between Cape Timiris and Casamance with steeper
slopes (2.2%)·off Cap Vert. The same set of factors have resulted in
the same mechanical concentrations within the submerged (ancient)
shoreface, beach and dune of each paleoshoreline and within the present
coastal sedimentary prism.

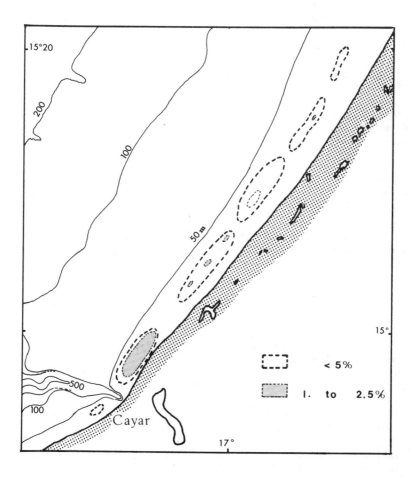

Figure 9. Fossil ilmenite deposits along a submarine Holocene
shoreline north of Cap Vert. Concentrations are shown in percent.

SUMMARY AND IMPLICATIONS OF FINDINGS

Phosphorite, glauconite and ilmenite deposits off the West African coast (Congo, Gabon, Senegal and Mauritania) have been identified by using sedimentological techniques. Exploration, however, has taken place only on the Djeno phosphatic bank, using gamma-ray spectrography, vibracore sampling and seismic reflection profiling.

The comparison of depositional conditions emphasizes the common characteristics of the three types of sedimentary deposits:

1) mechanical sorting and concentration of continental shelf mineral resources must be considered wherever ancient shoreline positions can be identified;

2) there are few paleo-shorelines on the continental shelf because littoral features were not formed between standstills of the sea-level due to the fast rate of transgression;

3) the position of ancient shorelines may have fluctuated because of tectonic movements and, therefore, it is impossible to relate a specific depth to a given ancient shoreline;

4) it is not realistic to believe that beach barriers, with their mechanically sorted sands, have moved shoreward during a transgression. More likely longshore processes played a major role in the forming of pre-Holocene morphology;

From an economical point of view, development of small mineral deposits in developing countries may be valuable. Phosphorites of Congo shelf seem especially attractive in this respect.

ACKNOWLEDGEMENTS

The authors gratefully acknowledge Dr. Paul Teleki for his contributions to improving the author's original manuscript and for his thorough and constructive review of the paper. Dr. William Burnett also made many valuable comments and we would particularly like to thank him.

REFERENCES

Bargeron, D.L., 1984, An economic, mineralogic and chemical investigation of the offshore phosphate deposit, People's Republic of the Congo, West Africa; unpubl. M.S. thesis, University Mississippi, 82 p.

Barusseau, J.P., 1984, Analyse sédimentologique des fonds marins de la "Petite Côte" (Sénégal); Doc. Sc. C.R.O.D.T., no. 94, 22 p.

Barusseau, J.P., 1985, Evolution de la ligne de rivage en République Islamique de Mauritanie; unpubl. rept., UNESCO, 104 p.

Barusseau, J.P., Diop, S., Faure, H., Giresse, P., Lezine, A.M., Masse, J.P. and Saos, J.L., 1984, Environnements sédimentaires marins au

cours du Quaternaire récent sur la marge atlantique de l'Afrique;
Proc. 5th Eur. Congr. on Sedimentology, Marseille, pp. 43–44.

Birch, G.F., 1979, Phosphatic rocks on the western margin of South
Africa; Jour. Sed Pet., v. 49, pp. 93–110.

Blanchot, A., 1946, Étude des sables noirs littoraux de la Mauritanie
(du Cap Timiris à Saint-Louis du Sénégal); Unpubl. rept., Arch,
Dir. Mines Afr. Occid., 14 p.

Bornhold, B. and Giresse, P., 1985, Conditions de glauconitisations
profondes et abondantes sur une marge active de l'Océan Pacifique
Nord (Vancouver, Canada); Compte rendu Acad. Sci. Paris, v. 300,
Ser. II, no. 11, pp. 517–522.

Bureau de Recherches Géologiques et Minières, 1973, Recherche d'ilmenite
au large des côtes du Sénégal (Operation Rosilda); B.R.G.M. Dept.
Géol. Marine, Orléans, Rept. No. 73 SGN 228 MAR, 122 p.

Cornen, G., Giresse, P. and Odin, G.S., 1973, Découverte de dépôts
phosphates néogènes sous-marins sur les plateaux continentaux du
Sud du Gabon et du Nord du Congo; Bull. Soc. Géol. France, v. 15,
pp. 9–11.

Cunha Gouveia, J.A., 1960, Notas sobre os fosfatas sedimentares de
Cabinda; Serv. Geol. y Minas Angola, Rept./ 1, pp. 49–65.

Dropsy, M.U., 1943, Etude granulométrique sur quelques sables de
Mauritanie; Bull. Soc. Gr. Mineral, v. 66, pp. 251–263.

Einsele, G., Herm, D. and Schwarz, H.U., 1974, Sea level fluctuation
during the past 6000 years at the coast of Mauritania; Quaternary
Research, v. 4, pp. 282–289.

Einsele, G., Elouard, P., Herm, D., Kogler, F.C. and Schwarz, H.U.,
1977, Source and biofacies of late Quaternary sediments in relation
to sea level on the shelf off Mauritania, West Africa; Meteor
Forsch. Ergebnisse, v. C-26, pp. 1–43.

Elouard, P., 1967, Eléments pour une définition des principaux niveaux
du Quaternaire sénégalo-mauritanien, 1: Plage à Arca senilis;
Bull. de l'IFAN, v. 29, sér. A., no. 2, pp. 822–836.

Furon, R., 1932, Les roches phosphatees de la cote du Gabon; Compte
rendu Acad. Sci. Paris, pp. 1959–1960.

Giresse, P., 1980, The Maastrichtian phosphate sequence of the Congo;
Soc. Econ. Paleont. Min., Spec. Publ. no. 29, pp. 193–207.

Giresse, P., 1983, La succession des sédimentations dans les bassins
marins et continentaux du Congo depuis le début du Mésozoique;

Sci. Géol. Bull. Strasbourg., v. 35, no. 4, pp. 183-206.

Giresse, P. and Cornen, G., 1976, Distribution, nature et origine des
     phosphates miocènes et éocènes sous-marins des plates-formes du
     Congo et du Gabon; Bull. Bur. Recherches Géol. et Min., v. 4, no.
     1, pp. 5-15.

Giresse, P., Jansen, F., Kouyoumontzakis, G. and Moguedet, G., 1981,
     Les fonds du plateau continental congolais et le delta sous-marin
     du fleuve Congo-Bilan de huit années de recherches sédimentologiques,
     paléontologiques, géochimiques et géophysiques; Tr. Doc. Orstom,
     no. 38, pp. 13-45.

Giresse, P. and Jamet, R., 1982, Essais de fertilisation de la culture
     du manioc par les sédiments marins glauconieux du Congo; Cahiers
     ORSTOM, s. Pedol., v. 19, no. 3, pp. 238-292.

Giresse, P., Barusseau, J.P., Malounguila-Nganga, D. and Wiber, M.,
     1984, Les phosphates au large du Congo et du Gabon. Nature
     géochimique et conditions mécaniques d'accumulation; Proc., 2nd
     Int. Sem. Offshore Mineral Resources, Brest, Germinal, pp. 315-326.

Hebrard, L., 11972, Un épisode quaternaire en Mauritanie (Afrique
     occidentale) a la fin du Nouakchottien: le Tafolien, 4000-2000
     ans B.P.; Bull. ASEQUA, no. 33/34, pp. 5-15.

Hebrard, L., 1981, Gisements de minerais de titane et de zirconium des
     plages fossiles du littoral ouest-africain; Unpubl. rept., 2 p.

Horn, R., 1978, La géophysique marine. Application à la reconnaissance
     du plateau continental africain; Compte rendu, Sem. Ressources
     minérales sous-marines, Bur. Rech. Geol. Min., Orléans, pp. 185-
     316.

Ireland, B.J., Curtis, C.D. and Whiteman, J.A., 1983, Compositional
     variation within some glauconites and illites and implications
     for their stability and origin; Sedimentology, v. 30, pp. 769-786.

Lamboy, M., 1976, Géologie marine et sous-marine du plateau
     continental au Nord-Ouest de l'Espagne. Genèse des glauconies
     et des phosphorites; These Doct. es Sci. Nat., Rouen, 285 p.

Malounguila-Nganga, D., 1983, Les environments sédimentaires des
     plate-formes du Nord Congo et du Sud Gabon au Quaternaire
     superieur d'après les données de vibrocarottages; Unpubl.
     thesis, University of Toulouse, 169 p.

Masse, J.P., 1968, Contribution à l'étude des sédiments actuels du
     plateau continental de la région de Dakar; Trav. Lab. Géol.
     Dakar, no. 23, 81 p.

Parker, R.V. and Siesser, W.G., 1972, Petrology and origin of some phosphorites from the South African continental margin; Jour. Sed. Pet., v. 42, pp. 434-440.

Pinson-Mouillot, J., 1980, Les environnements sédimentaires actuels et quaternaires du plateau continental sénégalais; Unpubl.

thesis, Univ. Bordeaux, 106 p.

Schwarz, L., 1931, Contribution to the geology of Mauritania; The Geogr. Jour., pp. 61-73.

Summerhayes, C.P., Birch, G.F., Rogers, J. and Dingle, R.V., 1973, Phosphate in the sediments of South Western Africa; Nature, v. 243, pp. 509-511.

Tooms, J.S. and Summerhayes, C.P., 1968, Phosphatic rocks from the northwest African continental shelf; Nature, v. 218, pp. 1241-1242.

Vogt, J., 1956, Reconnaissance préliminaire d'indices d'ilménite de la côte de Mauritanie; unpub. rept., Gouv. Gen. d'AOF, Dir. Gen. Mines et Geologie, 23 p.

Woolsey, R., Ferrante, M. and Le Lann, F., 1984, Exploration for phosphate in the offshore territories of the Congo; Proc., 2nd Int. Sem. Offshore Mineral Resources, Brest, Germinal, pp. 32-338.

EXPLORATION AND GENESIS OF SUBMARINE PHOSPHORITE DEPOSITS FROM
THE CHATHAM RISE, NEW ZEALAND - A REVIEW

U. von Rad and H.-R. Kudrass
Bundesanstalt für Geowissenschaften
und Rohstoffe (BGR),
Postfach 510153,
D-3000 Hannover 51, FRG

ABSTRACT. The Chatham Rise phosphorites are one of the best surveyed
deposits of the condensed replacement type.

The exploration methods used during the 1981 ship-borne survey with
R/V SONNE included acoustic transponder positioning, underwater television and photography, deep-towed side-scan sonar and high-resolution
seismic profilers, as well as a new, large, power-driven grab sampler,
a large hopper, and a vibrating sieving device. Statistically, the
phosphorite resources were difficult to assess, because of a high
regional variability in the distribution and thickness of the phosphorite deposits. Both the surveying and the resource assessment should be
improved in future years and could be applied for the prospecting of
similar phosphorite deposits.

The phosphorite nodules are partially phosphatized chalk pebbles
overlying a Miocene hardground. The time of phosphatization (8–11 Ma)
was ascertained by the age of the youngest phosphatized foraminiferal
assemblage (lower Late Miocene), the age of phosphatized whale bones
(Late Miocene or younger), and the radiometric age of the glauconitic
rim of phosphorite nodules. The highest amount of phosphorite formed on
top of elevated areas by morphologically controlled, multistage phosphatization, especially in a broad saddle of approximately 400 m water
depth on the central Chatham Rise. We present two models for the mode
of phosphatization: (1) $CaCO_3$ replacement in the pore water of organic-
rich anoxic sediments, where phosphate is released through bacterial
activity, and (2) replacement of $CaCO_3$ by direct uptake of phosphate
dissolved in sea water, followed by physico-chemical or microbial
mineralization. The latter model is favored.

INTRODUCTION

The phosphorite gravel on the central Chatham Rise east of New
Zealand (Fig.1) was first described by Reed and Hornibrook (1952) and
investigated in more detail by Norris (1964). After a commercial
reconnaissance survey (Pasho, 1976), the most phosphorite-rich area of

157

P. G. Teleki et al. (eds.), Marine Minerals, 157–175.
© 1987 by D. Reidel Publishing Company.

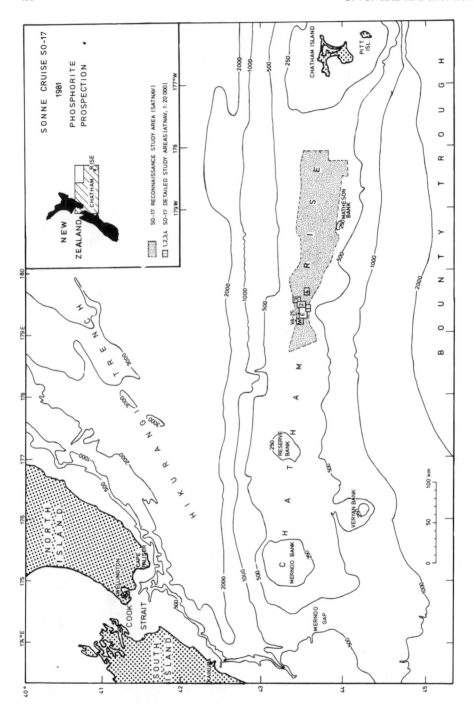

Fig. 1. Area of investigation for phosphorite deposits on the Chatham Rise (VA 25/1978 and SO-17/1981 cruises)

the Rise between 179°E and 180° was visited by four New Zealand Oceanographic Institute (NZOI) cruises to establish the age, distribution,
petrology and geochemistry of the phosphorites (Cullen and Singleton,
1977; Cullen,1978; Cullen,1980). The Chatham Rise is a 150 km wide and
1000 km long, submerged microcontinent with a minimum water depth of
about 400 m, extending from the South Island of New Zealand to the
Chatham Islands (Fig. 1). At the crest of the central portion of the
Rise there is loose gravel, consisting of partly phosphatized chalk
pebbles, intermixed with glauconitic, foraminiferal sandy muds, and
resting on Oligocene chalk. The P_2O_5 content of the phosphorite is a
function of nodule size and ranges from about 19% (large nodules) to 23%
(small nodules).

Inasmuch as New Zealand, the largest per-capita consumer of
phosphate in the world, has no phosphorite deposits on land, it depends
fully on the import of superphosphate fertilizer for its very important
agricultural and grassland farming economy. The Chatham Rise phosphorite
could replace, in part, the guano phosphate, imported from Christmas,
Nauru and Ocean Islands, where the reserves are rapidly dwindling
(Burnett and Lee, 1980). Therefore attempts were made already in the
1970's to assess the extension of the Chatham Rise phosphorite deposits
and to estimate its reserves (Karns, 1976; Cullen, 1979). In 1978, these
investigations were taken a significant step forward by a joint NZOI-BGR
project, employing sophisticated navigational techniques and a major
sampling program with the R/V VALDIVIA (Kudrass and Cullen, 1982). The
encouraging results of this cruise were considered sufficiently
interesting for a major New Zealand company to apply for a prospecting
license covering much of the Chatham Rise. In 1981, BGR and the New
Zealand Department of Scientific and Industrial Research (DSIR)
conducted another 2-month-long cruise with the R/V SONNE to the Chatham
Rise (von Rad, 1984). As a consequence, the Chatham Rise accumulations
are one of the best surveyed submarine phosphorite deposits of the
world.

In the following sections of this paper, we describe the exploration methods used during the SONNE survey, in view that these methods
might also be applicable to exploring for similar phosphorite deposits
elsewhere. Following this, we discuss briefly the genesis, as well as
the resource assessment and development of the Chatham Rise phosphorites.

EXPLORATION METHODS

Positioning

A navigational positioning accuracy of 10-50 m is necessary for the
sampling locations on the seafloor relative to the geophysical (seismic,
side-scan) survey lines, and subsequently for the statistical treatment
of the data derived from the samples. However, many submarine phosphorite deposits occur hundreds of kilometers from shore, and the distances
(as in the case of the Chatham Rise: about 400 km) seriously hinder the
use of land-based navigation systems. Instead, an underwater acoustic

navigation was deployed consisting of an array of as many as 8 transponders moored to the seafloor. This time-consuming deployment procedure which resulted in a positioning and location accuracy of 30-100 m in a relatively small area of about 10 x 10 km will in the future be replaced by the new global satellite positioning system (GPS).

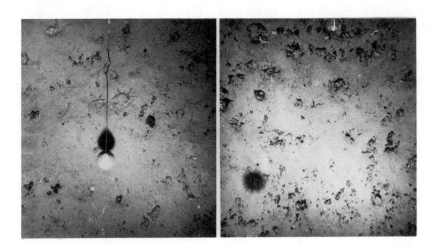

Figure 2.   Sea floor photographs from the central part of the Chatham Rise (in about 390 m water depth) showing abundant large and small phosphorite nodules, sea urchins, and burrowing small crustaceans (Galathea-type crabs). Note outcrop of white Oligocene chalk near large sea urchin. Weight diameter = 5 cm.

Underwater television and photography

     Although underwater TV and photography can be used to obtain information directly on the small-scale variability of the spatial distribution of phosphorite nodules on the sea floor, it is of limited value for the assessment of these resources, because, without coring, the thickness of the deposit or the abundance of phosphorite nodules hidden under a thin sedimentary cover remain unknown. For recording and measuring environmental parameters, however, such as the distribution of epibenthic communities, bioturbation features indicative of endobenthos, and the presence of fishes and spawning grounds, still and video photography continues to be of paramount importance (Dawson, 1984). For biologists, studying the activity of near-bottom communities, high-quality photographs are necessary. For the design of a mining device, an underwater TV/photo-system can also provide first information about the speed and direction of bottom currents (e.g. by means of scour marks), and about the micromorphology and possible submarine outcrops.

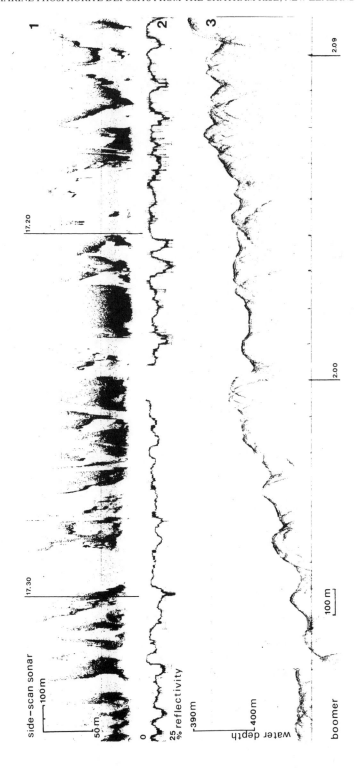

Fig. 3. Sidescan sonar (starboard side = 1), Huntec boomer record (=3) with acoustic reflectivity data (=2). Profiles (1 and 2/3) are 100-150 m apart and adjusted to the same scale along the ship's track. The frequency of of high- and low reflectivity patches on the boomer record is comparable to the frequency of dark and light patches, and high-reflectivity areas are presumably caused by phosphorite nodules. White-coloured, straight, 5-50 m wide stripes are thought to be iceberg scours (after Kudrass and von Rad, 1984a).

Photographs taken on SONNE-cruise traverses, over a total distance
of 20 km, revealed a highly bioturbated sea floor with a patchy distri-
bution·of near-surface phosphorite nodules (Fig. 2). Patchy phosphorite-rich areas, 360 m in diameter, were found by the aid of photography; these patches could not be correlated, however, with the micro-
relief of the seafloor (Kudrass and von Rad, 1984a).

Side-scan sonar surveys

Much as underwater television and photography, side-scan sonar
surveys are also not suitable for the estimation of three-dimensional
phosphorite resources. The method, however, is useful to show the extent
of high-concentration surficial phosphorite deposits on the seafloor
and to image and record the location of possible hard-rock outcrops. The
recognition of prominent sediment transport directions is also possible
by interpreting sonar imagery.
During the SONNE cruise, a side-scan sonar was towed at an
elevation of about 50 m above the sea floor, in order to record the
seafloor features encountered. On the Chatham Rise, side-scan sonar
records showed the existence of high-reflectivity patches, which we
interpreted to be areas of phosphorite gravel covering the seafloor
(Kudrass and von Rad, 1984a). These patches were found to be dissected
by numerous, 10-20 m wide furrows. The furrows are thought to be iceberg
scour marks impressed during a glacial low sea-level stand (Fig. 3). By
redistributing the phosphorite gravel, these scour marks are also
responsible for a high variability of the phosphorite coverage (Kudrass,
1984).

High-resolution reflection-seismic profiling

High-resolution seismic profiling equipment towed at a water depth
of about 100 m provided excellent detail of the sedimentary sequences,
because of its high energy (560 joules), the high frequency of the
seismic signal (1.4 kHz), short pressure pulse (0.2 ms), and high firing
rate (120 shots/min.). During the survey of the Chatham Rise, we
achieved a penetration of several decameters with a vertical resolution
of 0.3 m at a maximum water depth of 550 m. The "boomer" records also
provide information indirectly about the phosphorite distribution, as
several "seismic facies" types, characterized by diffraction hyperbolae
and varying seafloor relief, could be distinguished (Hansen, 1984).
Because these "seismic facies" types could be correlated with the
phosphorite coverage, as known from detailed sampling (Fig. 4A, B), we
used this method to "map" phosphorite-rich areas regionally and to
define targets for subsequent, more detailed seismic and sampling
surveys (Falconer et al., 1984; Kudrass, 1984). The "boomer" also records
show also the thickness of overlying sediments and the relation of the
deposit to the underlying strata.

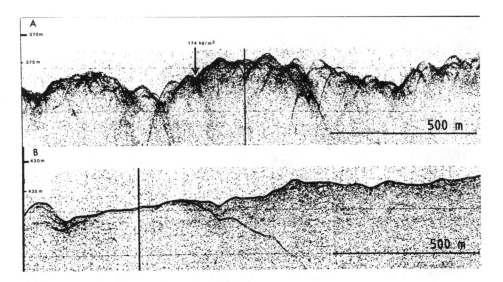

Figure 4. Correlation of high-resolution seismic (boomer) record with phosphorite yield (kg/m^2).
A. Continuous overlapping diffraction hyperbolae with hummocky surface (high to intermediate relief), without bottom reflector. Samples from this "seismic facies type" have usually a high phosphorite yield.
B. Smooth to irregular, rolling bottom topography without diffraction hyperbolae, but with distinct subbottom reflectors. Usually, no phosphorite was found in areas with this "seismic facies".

In contrast to side-scan sonar and high-resolution seismic surveys, the measurement of acoustic reflectivity (Parrot et al., 1980) cannot be used to map the distribution of phosphorite nodules, because reflectivity is mainly influenced by the small-scale topography of the sea floor and by the varying depth and morphology of the Tertiary chalk underlying the phosphorite-bearing sand (Fig. 3; Kudrass and von Rad, 1984a).

Sampling

Because phosphorite usually occurs as a deposit of gravel and coarse sand mixed with finer-grained, non-phosphoritic sediments, the quantity of a representative sample must be fairly large. On the VALDIVIA cruise (1978), we used a heavy grab sampler with a maximum sample weight capacity of 150 kg, whereas during the SONNE cruise (1981) we obtained more reliable results with a very heavy, large grab sampler, which had pneumatically operated jaws driven by compressed air (von Rad, 1984). The grab recovered as much as 1.8 tons (0.8 m^3) of sediment (including

a   b

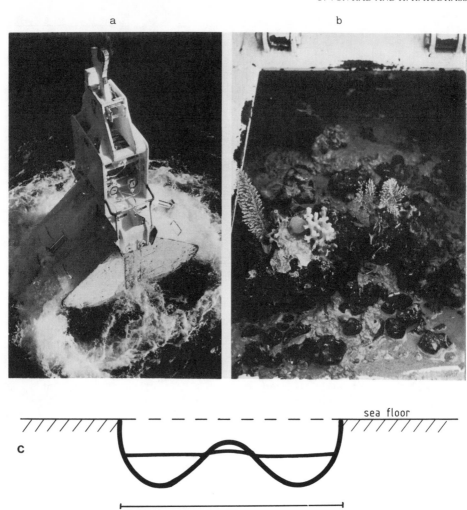

Fig. 5. Large grab sampler with pneumatic drive (PREUSSAG-PEINER A.G.)
a: Filled grab sampler being raised onboard
b: Undisturbed sediment surface in grab sampler: abundant phosphorite
   nodules, over-grown by corals and sponges
c: Estimated profile of seafloor, sampled by large grab sampler. Thick
   line shows th recovered cross-section with a deep initial penetration; thin line shows the section, if only a reduced penetration
   was reached due to higher phosphorite concentration. The area
   covered by the opened grab sampler is 190 cm x 100 cm.

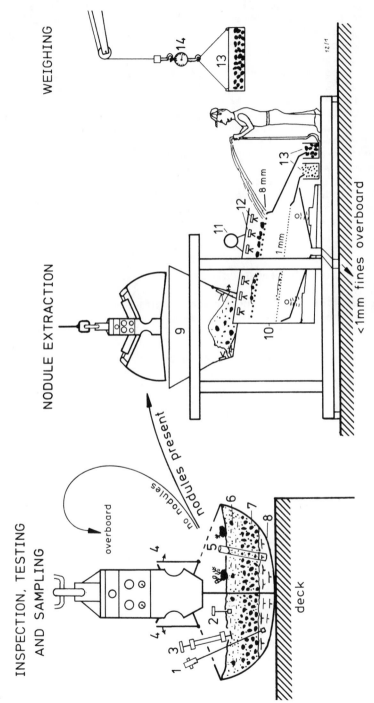

Fig. 6. Sketch of shipboard processing of large grab sampler. 1,2 = hand shear vanes, 3 = plate-bearing test, 4 = grab doors, 5 = sampling tube, 6 = soft, muddy glauconite-foraminiferal sand, 7 = same, with phosphorite nodules, 8 = nannoplankton ooze/chalk, 9 = hopper, 10 = nannoplankton ooze/chalk, 11 = vibrating motor, 12 = water jets to wash nodules free from mud and ooze, 13 = rectangular containers for nodule fractions >8 mm and 1-8 mm, 14 = spring scale.

as much as 0.5 tons of phosphorite), and the sampled area ranged from 2 to 1.6 m². The high closing power (1.5 t) prevented an incomplete closure of the jaws by breaking enclosed nodules or other rocks. Even when phosphorite concentrations reached 30-40% by weight, the grab could penetrate to a subsurface depth of 50 cm and reach, in most cases, the Oligocene chalk below the phosphorite deposit. When the Oligocene chalk was not reached, the quantitative evaluation of the results proved difficult, as the penetration of the grab depends on the phosphorite or gravel content of the sediment (Fig. 5). This is a general problem with all grab samplers, and it has to be taken into consideration in all quantitative assessments.

Another common problem is the handling and processing of huge amounts of sediment, after the grab sampler is retrieved. All samples containing phosphorite were washed onboard using a specially developed large hopper and vibrating sieving device (Fig. 6). Altogether we recovered about 40 tons of phosphorite during the SONNE cruise which could be used for shore-based bulk analysis, e.g. for fertilizer field trials.

GENESIS OF THE CHATHAM RISE PHOSPHORITE

Fig. 7 shows that the Neogene evolution, genesis, diagenesis and reworking of the Chatham Rise phosphorites comprise a complicated multi-stage process (Pasho, 1976; Kudrass and Cullen, 1982; Kudrass and von Rad, 1984b). This process started with the formation of a chalk hardground surface during mid-to late Miocene times (Fig. 7A). This hardground was bored and fragmented during a long non-depositional or erosional hiatus, forming chalk pebbles which became marginally ferruginized (Fig. 7A,H). The determination of the age of the partial phosphatization of the foraminiferal chalk pebbles by slow diagenetic replacement of CO_3 by PO_4 (von Rad and Rösch, 1984) is difficult, because radiometric dating by the U-series is only possible for phosphorites younger than 700,000 years. However, by studying (1) the foraminiferal age of the phosphatized chalk (Zobel, 1984), (2) determining the age of phosphatized whale bones associated with the nodules (Fordyce, 1984) and (3) the K-Ar age of the glauconitic rim of phosphorite nodules (Kreuzer, 1984), it was possible to date the time of phosphatization indirectly: the date is lower late Miocene (about 7-10 Ma ago) because (1) predates, and (2) and (3) postdate that "event" (Fig. 8; Kudrass and von Rad, 1984b).

A puzzling feature of the Chatham rise phosphorite is the juxtaposition of phosphorite-rich areas with an average coverage of 66 kg/m² and phosphorite-poor areas with an average of only 11 kg/m², although the nodules of both areas have the same chemical composition and grain-size distribution. Kudrass and von Rad (1984b) reasoned that this was the result of a morphologically controlled, multistage phosphatization of the surficial chalk pebbles during several late Miocene phases of submarine erosion and hardground formation (Fig. 9). Phosphatization was

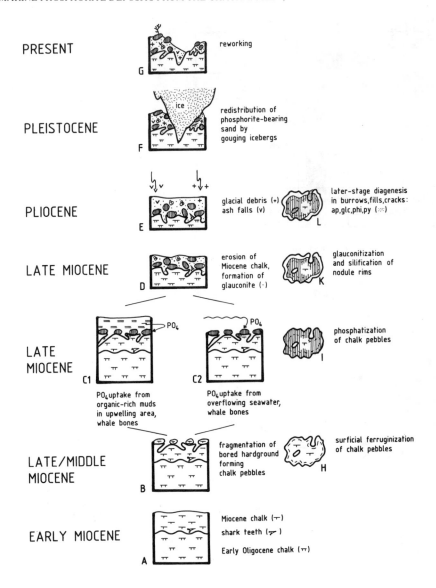

Fig. 7. Schematic sketch showing Neogene evolution, genesis and diagenesis of phosphorite and associated sediment on the central Chatham
Rise (C_1 and C_2 are alternative models for phosphatization process, see
text). Evolution of chalk pebbles to phosphorite nodules is shown in
four steps. From Kudrass and von Rad (1984b).

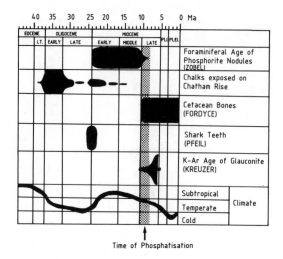

Time of Phosphatisation

Fig. 8. Timing of significant events on the crest of the Chatham Rise. Age of phosphatization of Chatham Rise phosphorite indirectly inferred from foraminiferal age of nodules, whale bones, and K-Ar age of glauconite rims. Climatic curve is constructed after data from Shackleton and Kennett (1975), and Knox (1980). From Kudrass and von Rad (1984b).

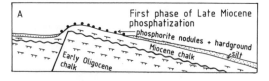

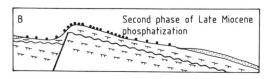

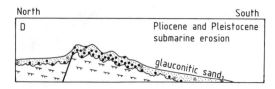

Fig. 9. Model of a morphological control of phosphatization with repeated cycles of partial burial, phosphatization, erosion, and hardground formation. Note the preferred formation of phosphorite nodules on uplifted parts of fault block. From Kudrass and von Rad (1984b).

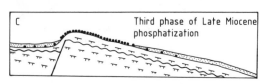

favored on uplifted parts of fault blocks, because they were better exposed to overflowing water masses and erosional processes than deeper-lying areas, where the rate of pelagic accumulation was slightly higher. Each erosion-phosphatization cycle added new phosphorite nodules to the elevated areas. Chalk pebbles were phosphatized to a size-dependent saturation level ranging from 17 to 21% P_2O_5. Most of the lesser phosphatized chalk pebbles were probably removed during the erosional phases. These processes resulted in the fairly uniform P_2O_5 concentration of the pebbles, irrespective of their provenance from phosphorite-rich or phosphorite-poor areas.

For the mode of phosphatization we discuss two models:
(1) CaCO$_3$ replacement in the pore water of organic-rich anoxic sediments, where P is released through bacterial activity (Fig. 7-C1). Today, phosphatization occurs principally on the upper continental slope and outer shelf off the western coasts of South America (Peru-Chile) and Africa (Namibia) by apatite precipitation/replacement from anoxic interstitial solutions, influenced by the high productivity in these upwelling areas (Bremner, 1980; Burnett and Roe, 1983). The fact that no phosphatized epibenthos or burrowing bivalve shells have been observed on the Chatham Rise nodules, might indicate that they were buried during phosphatiza-tion. Also the content of the Chatham Rise phosphorite (ca. 200 ppm) is much higher than that from seamount phosphorite (ca. 5 ppm), formed from "overflowing normal sea water" (W. Burnett, pers. commun., 1985). The abundance of whale bones may indicate nutrient-rich surface waters, possibly induced by upwelling effects with organic-rich muds being accumulated below a high-productivity zone. On the other hand, hardly any organic matter and very little biogenic silica (opal), both being important indicators of phospho-rites from high-productivity upwelling environments, is associated with the Chatham Rise phosphorites.

(2) Replacement of CaCO$_3$ by direct uptake of phosphate dissolved in seawater, followed by physico-chemical or microbial mineralization ("condensed sequences-plateau/rise model": Fig. 7-C2). Absorption of phosphate onto organic coating of the chalk may enhance this replacement. (Suess, 1973). According to O'Brien and Veeh (1983), apatite in contemporary phosphorites at the "non-upwelling" eastern Australian margin formed during post-mortem alteration of organic P originally present in bacterial cells; this happened at very low sedimentation rates, when the upper slope environment became quite restrictive and nutrient supply was drastically limited. According to Riggs (pers. commun., 1985), these deposits might have been geochemically recycled from older phosphorites and therefore have Holocene radiometric ages. Bacterially precipitated microcrystalline phosphorite mud was also described from Miocene shallow-water occurrences in Florida (Riggs, 1979). Direct uptake of phosphate dissolved in seawater seems also to be responsible for the phosphatization of the tops of seamounts by overflow processes (Burnett and Roe, 1983). We favor the "non-upwelling western

boundary current model" for the Chatham Rise, because of four lines
of evidence: (1) SEM photos of the collophane in phosphatized chalks
from the Chatham Rise show a microglobular apatite matrix with
spherules (about 1-3µ in diameter) suggestive of phosphatized
bacterial cells. (2) The δ^{34}S concentration, measured in Chatham
Rise phosphorites, is similar to that of sulfate in evaporites of
the same age, which indicates that the phosphatization medium was
open-sea water (Benmore et al., 1983). (3) The pronounced Ce defi-
ciency with respect to shale might indicate direct derivation of the
phosphate from seawater which has a similar REE (rare earth element)
pattern (Burnett and Roe, 1983; O'Brien and Veeh, 1983. (4) During the
"Terminal Miocene (Cooling) Event" (Kennett, 1982, a significant glacial expansion of the Antarctic ice sheet caused higher latitudinal
temperature gradients between polar and equatorial regions,
increased wind velocities and surface circulation, and enhanced
dynamic upwelling at the newly formed Subtropical Convergence which
formed over the Chatham Rise and separates cool subantarctic from
warmer subtropical water masses (Heath, 1981; Kennett, 1982, p.
739). Turbulent mixing at the boundary of these two distinct water
masses might have been the cause of increased silicoplankton productivity and the phosphatization of chalk, especially in areas where
the cool, nutrient- and phosphate-rich water masses flowed over
saddles in the rise and came in contact with exposed chalks during
extended times of nondeposition and erosion. Since the richest
phosphorite occurrences lie in a saddle position on the Chatham Rise
(Falconer et al., 1984), a somewhat channelized overflow of
phosphorus-rich water masses over extended time intervals might
indeed have been the cause of the phosphatization of the chalk
pebbles.

During later diagenetic stages the rims of the phosphorite nodules
were glauconitized and partly silicified by opaline silica (Fig. 7K).
Glauconite pellets formed about 7 million years ago in the associated
pelagic foraminiferal sands (Kreuzer, 1984). Later, the fillings of
burrows and cracks in the nodules were cemented by apatite, phillipsite,
or calcite (Fig. 7L). During the Pliocene and Pleistocene, submarine
erosion removed most of the non-phosphoritic parts of the Miocene chalks
and created a karst-like topography. During this period, ice-rafted
terrigenous debris and air-borne volcanic ashes were added to the
surficial lag deposit (Fig. 7E). During the Pleistocene, gouging
icebergs, indicated by straight, several-m deep furrows, considerably
disturbed the presumably uniform original distribution of phosphorite
nodules (Fig. 7 F; Kudrass, 1984).

RESOURCE ASSESSMENT AND DEVELOPMENT

Estimation of phosphorite resources

Phosphorite-rich portions of the SONNE-survey area were defined by the results of boomer and side-scan sonar interpretation. Mean phosphorite coverage (kg/m^2) of all samples within these areas were calculated after Krige (1962). The phosphorite-rich areas covering an area of 140 km^2 contain, on the average, 54 kg/m^2 with a concentration of 12.6 weight % phosphorite in the glauconitic sand. Adding to these resources the resources of the previously surveyed neighboring areas (Kudrass and Cullen, 1982), the total resources are 25 million tons of phosphorite, spread over an area of 378 km^2, with an average coverage of 66 kg/m^2.

In addition, an assessment of phosphorite resources was attempted using a statistical technique, known as kriging, by which a model of the spatial distribution of the ore grade can be generated (David, 1977). However, the regional variability of the phosphorite coverage was too high to generate a reliable model.

Mining aspects

The gouging by Pleistocene icebergs considerably complicates the mining of the Chatham Rise phosphorites, as thickness, overburden, concentration, and consolidation of the phosphorite-bearing sediments were highly disturbed. Large ice-rafted blocks may also be severe mining obstacles, although they seem to be fairly rare.

The design of an efficient mining device (collector with cutter or scraper, sieving system, hydraulic lifting system) must take into consideration the following steps: uptake of phosphorite-bearing sediments to a maximum depth of about 50 cm, separation of nodules from glauconitic foraminiferal sand and underlying sticky ooze or chalk, separation of oversized nodules, flint, and erratic pebbles, and, finally, lifting of the phosphorite nodules onboard. The mining companies which were involved in this SONNE survey have designed a prototype for such a mining system (Meyer et al., 1985) and plan to continue the present prospecting phase toward a mining test, which might lead to a joint mining venture.

Environmental biological studies

The importance of the gravelly phosphorite occurrences for the epibenthic fauna and for mobile scavengers (e.g. fish) is obvious, but its relation to the whole living community is not established. In order to find out, whether use of phosphorite-rich areas for both submarine mining and fishing is possible, a preliminary environmental impact study was made following the SONNE cruise (Dawson, 1984). According to the results of this study, it is necessary to put the concepts of resilience (the ability of an ecosystem to persist in the presence of perturbations) and the concept of stability (the ability of a system to return to an

equilibrium state after temporary disturbance) to a practical use in
monitoring operations.

Marine phosphorite as direct-application fertilizer

Some of the special geochemical and mineralogical aspects of marine
phosphorite for fertilizer use, e.g. P_2O_5, F, U contents; CO_3: PO_4 ratio
in apatite; maturity, reactivity and agronomic effectiveness for soil,
were also studied (von Rad and Rösch, 1984). Long-term field trials of
crushed, unprocessed phosphorite fertilizer in comparison with conventional superphosphate have been performed for a variety of soil types
(Mackay et al., 1980).

ACKNOWLEDGEMENTS

We are very grateful for the excellent cooperation with the technicians and scientists from BGR, the New Zealand Oceanographic Institute,
the New Zealand Geological Survey, Preussag A.G., Salzgitter A.G., and
Fletcher Challenge Ltd. during and after the SONNE cruise. The SONNE-17
cruise was financed by the Federal Ministry of Research and Technology
(BMFT, Bonn), with contributions by BGR, DSIR and Fletcher Challenge
Ltd.. W. Burnett (Tallahassee, Florida) and S. Riggs (Greenville, North
Carolina) kindly reviewed the manuscript. We appreciate the critical
comments by P. Teleki and David Davidson who considerably helped to
improve the style and contents of our paper.

REFERENCES

Benmore, R.A., Coleman, M.L. and McArthur, J.M., 1983, Origin of
    sedimentary francolite from its sulphur and carbon isotope compo-
    sition; Nature, v. 302, pp. 516-518.

Birch, G.F., 1979, Phosphorite pellets and rock from the western conti-
    nental margin and adjacent coastal terrace of South Africa; Mar.
    Geol., v. 33, pp. 91-116.

Bremner, J.M., 1980, Concretionary phosphorite from SW Africa; Jour.
    Geol. Soc. London, v. 137, pp. 773-786.

Burnett, W.C. and Lee, A.I.N. 1980, The phosphate supply system in the
    Pacific; GeoJournal, pp. 423-435.

Burnett, W.C. and Roe, K.K., 1983, Upwelling and phosphorite formation
    in the ocean; In: Suess, E. and Thiede, J. (eds.), Coastal upwel-
    ling, its sediment record; NATO Conf. Series, IV, v. 10A, Plenum
    Press, New York, London, pp. 377-398.

Cullen, D.J., 1978, The distribution of submarine phosphorite deposits on central Chatham Rise, east of New Zealand; 2. Sub-surface distribution from cores; New Zealand Ocean. Inst., Ocean. Field Rep., no. 12, pp. 1-29.

Cullen, D.J., 1979, Mining minerals from the sea floor: Chatham Rise phosphorite. - New Zealand Agricult. Sci. v. 13, pp. 85-91.

Cullen, D.J. 1980, Distribution, composition and age of submarine phosphorites on Chatham Rise, east of New Zealand; Spec. Publ., Soc. Econ. Paleont. Minereral., no. 29, pp. 139-148.

Cullen, D.J. and Singleton, R.J., 1977, The distribution of submarine phosphorite deposits on central Chatham Rise, east of New Zealand; New Zealand Ocean.Inst. Field Rep. no. 10, pp. 1-24.

D'Anglejan, B.E., 1967, Origin of marine phosphorites off Baja California, Mexico; Mar. Geol., v. 5, pp. 15-44.

David, M., 1977, Geostatistical ore reserve estimation; Developments in Geomathematics, v. 2, Elsevier, Amsterdam, 364 p.

Dawson, E.W., 1984, The benthic fauna of the Chatham Rise: an assessment relative to possible effects of phosphorite mining; Geol. Jahrb., v. D-65, pp. 209-231.

Falconer, R.K.H., von Rad, U. and Wood, R., 1984, Regional structure and high-resolution seismic stratigraphy of the central Chatham Rise (New Zealand); Geol. Jahrb., v. D-65, pp. 29-56.

Fordyce, R.E. 1984, Preliminary report on Cetacean bones from Chatham Rise (New Zealand); Geol. Jahrb., v. D.-65, pp. 117-120.

Hansen, D., 1984, High-resolution seismic results of the detailed SONNE-17 survey areas (Chatham Rise, New Zealand); Geol. Jahrb., v. D-65, pp. 57-67.

Heath, R.A., 1981, Physical oceanography of the waters over the Chatham Rise; New Zealand Ocean. Inst., Ocean. Summary, no. 18, pp. 1-15.

Karns, A.W., 1976, Submarine phosphorite deposits of the Chatham Rise near New Zealand; Am. Assoc. Petrol. Geol., Mem., no. 25, pp. 395-398.

Kennett, J.P., 1982, Marine Geology; Prentice-Hall, Englewood Cliffs, New Jersey, 813 p.

Knox, G.A., 1980, Plate tectonics and the evolution of intertidal and shallow-water benthic biotic distribution patterns of the southwest Pacific; Paleogeogr. Palaeoclimat. Palaeoecol., v. 31, pp. 269-297.

Kreuzer, H., 1984, K-Ar dating of glauconite rims of phosphorite nodules (Chatham Rise, New Zealand); Geol. Jahrb., v. D-65, pp. 121-127.

Krige, D.G., 1962, Statistical applications in mine valuation. Part I; Jour. Inst. Min. Surv. South Africa, v. 12, pp. 45-84.

Kudrass, H.R., 1984, The distribution and reserves of phosphorite on the central Chatham Rise (SONNE-17 cruise 1981); Geol. Jahrb., v. D-65, pp., 179-194.

Kudrass, H.R. and Cullen, D.J., 1982, Submarine phosphorite nodules from the central Chatham Rise off New Zealand - composition, distribution, and reserves (Valdivia-cruise 1978); Geol. Jahrb., v. D-51, pp. 3-41.

Kudrass, H.R. and von Rad, U., 1984a, Underwater television and photography observations, side-scan sonar and acoustic reflectivity measurements of phosphorite-rich areas on the Chatham Rise (New Zealand); Geol. Jahrb., v. D-65, pp. 69-89.

Kudrass, H.R. and von Rad, U., 1984b, Geology and some mining aspects of the Chatham Rise phosphorite deposits - a synthesis of SONNE-17 results; Geol. Jahrb., v. D-65, pp. 233-252.

Mackay, A.D., Gregg, P.E.H. and Syers, J.K., 1980, A preliminary evaluation of Chatham Rise phosphorite as a direct-use phosphatic fertilizer; New Zealand Jour Agricult. Res., v. 23, pp. 441-449.

Meyer, K.W., Kudrass, H.R. and von Rad, U., 1985, The Chatham Rise, New Zealand, marine phosphorite deposit - geology, reserves and development aspects; Proc. Int. Fert. Conf., London British Sulfur Corp., pp. 165-173.

Norris, R.M., 1964, Sediments of Chatham Rise; Bull. New Zealand Dept. Sci. Indust. Res., v. 159, pp. 1-39.

O'Brien, G.W. and Veeh, H.H., 1983, Are phosphorites reliable indicators of upwelling? In: Suess, E. and Thiede, J. (eds.), Coastal upwelling, its sediment record, NATO Conf. Series IV, v. 10A, Plenum Press., New York, London, pp. 399-418.

Parrott, D.R., Dodds, D.J., King, L.H. and Simpkin, P.G., 1980, Measurement and evaluation of the acoustic reflectivity of the sea-floor; Can. Jour. Earth Sci., v. 17, no. 6, pp. 722-737.

Pasho, D.W., 1976, Distribution and morphology of Chatham Rise phosphorites; Mem. New Zealand Ocean. Inst., no. 77, pp. 1-27.

Reed, J.J. and Hornibrook, N. de B., 1952, Sediments from the Chatham Rise; New Zealand Jour. Sci. Technol., v. 34B, pp. 173-188.

Riggs, S.R., 1979, Phosphorite sedimentation in Florida – a model phosphogenic system; Econ. Geol., v. 74, pp. 285–314.

Riggs, S.R., 1984, Paleoceanographic model of Neogene phosphorite deposition, U.S. Atlantic continental margin; Science, v. 223, pp. 123–131.

Shackleton, N.J. and Kennett, J.P., 1975, Paleotemperature history of the Cenozoic and the initiation of Antarctic glaciation: oxygen and carbon isotope analyses in DSDP Sites 277, 279 and 281; In: Kennett, J.P., Houtz, R.E. et al., Init. Repts Deep Sea Drilling Project, v. 29, U.S. Gov. Print. Off., Washington, D.C., pp. 743–755.

Suess, E., 1973, Interaction of organic compounds with calcium carbonate; Geochim. Cosmochim. Acta, v. 37, no. 11, pp. 2435–2447.

von Rad, U., 1984, Outline of SONNE cruise SO–17 on the Chatham Rise phosphorite deposits east of New Zealand; Geol. Jahrb., v. D–65, pp. 5–23.

von Rad, U. and Rösch, H., 1984, Geochemistry, texture, and petrography of phosphorite nodules and associated foraminiferal glauconite sands (Chatham Rise, New Zealand); Geol. Jahrb., v. D–65, pp. 129–178.

Zobel, B., 1984, Foraminiferal age of phosphorite nodules from the Chatham Rise (SO–17 cruise); Geol. Jahrb. v. D–65, pp. 99–105.

CONTROLS ON THE NATURE AND DISTRIBUTION OF MANGANESE NODULES IN THE WESTERN EQUATORIAL PACIFIC OCEAN

D.S. Cronan,
Marine Mineral Resources Programme,
Applied Geochemistry Research Group,
Department of Geology,
Imperial College,
London, SW7 2BP, U.K.

ABSTRACT. Three main factors appear to govern the location of potentially economic manganese nodules in the western equatorial Pacific: elevated biological productivity, mainly over $50 gCm^{-2}y^{-1}$, depth near or below the CCD, and an absence of turbidite sedimentation. These factors indicate the possibility that potentially economic nodules may be present in parts of the East Central Pacific Basin, West Central Pacific Basin and the North Penrhyn Basin.

INTRODUCTION

In exploration for potentially economic manganese nodules, knowledge of the conditions under which the deposits form will aid in the search for them, and exploration programs can be directed to those environments in which conditions for formation of economic varieties of the deposits are optimal. Thus a deductive approach to manganese nodule exploration is essential.

Potentially economic manganese nodules can be defined as those which satisfy first generation mine site requirements. It is generally agreed that the average abundance of nodules required for a first generation mine site is $10 kg/m^2$ or more, with a minimum of $5 kg/m^2$ (Bastien-Thiery et al., 1977; Archer, 1979). Compositionally, average and cut-off contents of 2.27% and 1.18% combined Ni and Cu, the two elements of greatest economic importance in manganese nodules, are deemed to be necessary to support a first-generation mining operation (Kildow et al., 1976).

Potentially economic manganese nodules are known to be best developed in the northeastern equatorial Pacific Ocean, between the Clipperton and Clarion fracture zones, where nodules contain as much as 3% Ni+Cu+Co, are rich in Mn, have a high Mn/Fe ratio, and the principal mineral phase is represented by todorokite. The nodules are located along the northern margin of the equatorial zone of high biological productivity, largely in the region bounded by the 50 and 100 grams-of-carbon-per-square-meter-per-year ($gCm^{-2}y^{-1}$) productivity isolines (Fig. 1), below the depth at which calcium carbonate dissolves (CCD), on sediments composed largely

P. G. Teleki et al. (eds.), Marine Minerals, 177–188.
© *1987 by D. Reidel Publishing Company.*

of siliceous ooze.

These high-grade nodule deposits are thought to form as a result of the extraction of manganese and other metals from seawater by organisms and organic material and their transport to the seafloor in organic remains. On the ocean bottom, organic decay liberates the metals into the interstitial waters of the sediments, and the metal concentrations gradually become higher. By complex chemical reactions, the manganese and other metals migrate to centers of nucleation and precipitate to form manganese nodules. The quantity of sinking organic remains is sufficient to supply the valuable metals in the nodules, and the organic decay, which takes place on the seafloor, is able to drive the diagenetic reactions that concentrate the metals in the nodules.

It is evident from the environmental conditions under which high-grade nodules form in the Clarion-Clipperton Zone that, in exploration for potentially economic nodules in the western equatorial Pacific, two principal environmental parameters have to be taken into account. First, deposits should be sought beneath areas of high biological productivity, ideally more than $50\,gCm^{-2}y^{-1}$, which ensures that enough organic material is transported to the seafloor to drive the diagenetic reactions leading to the formation of todorokite and an accompanying enrichment of metals. Secondly, deposits should be sought below the lysocline, and most probably in the vicinity of, or below, the CCD, because it is only here that sedimentation rates are sufficiently low to enable the metal-rich organic phases to become sufficiently concentrated at the sediment/water interface; conditions that promote the formation of high-grade nodules. These reduced sedimentation rates at and below the CCD are also conducive to promoting high concentrations of nodules on the sea-floor. Thus, under the most favourable conditions, a strong inter-relation between high nodule grades and high abundances should exist, in contrast to the occasionally reported negative correlation between these attributes. However, even where grades are high and abundances are seemingly low, because grade tends to vary proportionately much less than abundance in a given area, detailed search in such an area could still locate deposits with greater than average abundance. Grade would thus appear to be more important than abundance in initial explor-atory investigations for potentially economic manganese nodules.

AREAS OF THE WESTERN EQUATORIAL PACIFIC WHERE CONDITIONS FOR THE FORMATION OF POTENTIALLY ECONOMIC MANGANESE NODULES ARE MET

Based on the considerations discussed above, it is possible to rank the various regions of the western equatorial Pacific Ocean in terms of their likelihood of containing potentially economic manganese nodules.

The necessity of having elevated biological productivity in surface waters that creates a sufficient flux of organic matter to the seafloor to induce the formation of potentially economic nodules, also restricts the likely occurrence of these nodules to equatorial and sub-equatorial areas. Examination of the isolines of primary productivity in the Pacific (Fig. 1) show an expansion of the zone of high productivity in

an easterly direction, and thus, the range of latitudes in which ore
grade nodules are likely to occur should also be expected to expand
eastwards. This is supported by the data of Aplin and Cronan (1985),
which show that the area containing both high Mn/Fe ratios and an
abundance of todorokite, characteristics of ore grade nodules, widens
eastwards. However, the axial region of the equatorial zone of high
productivity where it lies above the CCD, and where carbonate may
inhibit nodule formation by acting as a diluent of the labile metal-rich
phases, contains few nodules. Thus nodules are best sought in those
areas immediately to the north and south of the most highly productive
waters, between the 50-100 g$Cm^{-2}y^{-1}$ productivity isolines, although they
are not necessarily absent under areas of higher productivity.

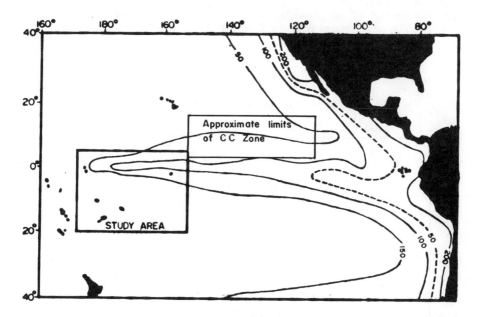

Fig. 1   Study area and approximate limits of the Clarion-Clipperton
Zone, together with isolines of primary productivity from Schmithusen
(1976).

Beneath the area of high biological productivity, the most favourable environment for ore-grade nodule development will be below the
lysocline and near or below the CCD. A plot of $CaCO_3$ against depth in
surface sediments (Fig. 2) from the S.W. equatorial Pacific (Aplin 1983)
shows the CCD to be below about 5100 m in the western Pacific between
the equator and 20°S, which is in good agreement with other estimates.
Within this area, the CCD is likely to be at greater depths in the
north than in the south. Bathymetric charts of this region show large
expanses of the seafloor deep enough to be near or below the CCD. Hence,
the 5500 m isobath can probably be taken as the maximum depth for the

common occurrence of high-grade nodules in the region, although Exon
(1983) has reported deeper occurrences, and the 4500 m isobath as the
minimum depth. Highest grades appear to be centered approximately
between 4900 and 5500 m. This maximum depth is probably determined by
the fact that, with increasing depth below the CCD, decay of organic
material in the water column will proportionately increase relative to
that on the seafloor, and, thus, will reduce the effectiveness of diagenetic nodule-forming processes in the sediments. However, it is
important to remember that the CCD can vary somewhat throughout the
region, as noted by Pautot and Melguen (1979) in the area immediately
to the east of that considered here; hence, the critical depth for
highest nodule-grades is likely to vary slightly from basin to basin
throughout the study area.

Another factor of importance in determining the occurrence of ore
grade nodules in the study area is the influence of terrigenous and
bioclastic sedimentation. So far in the present work, discussion of
sedimentation has been restricted to particle by particle sedimentation,

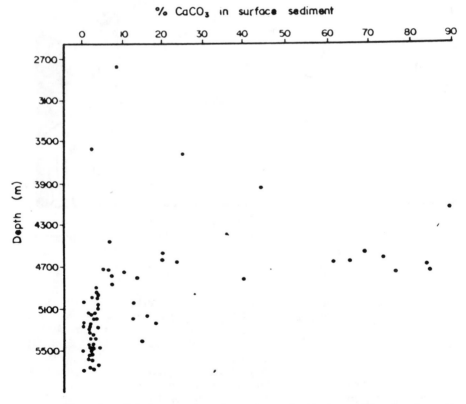

Fig. 2. Calcium carbonate versus depth in sediments from the
S.W. Equatorial Pacific (from Aplin, 1983)

principally of biogenic origin. However, in the vicinity of the islands, sedimentation rates may be locally enhanced due to slumping of either biogenic (calcareous turbidites) or volcaniclastic detritus (Orwig, 1981). This reduces the abundance of nodules on the seafloor, and inhibits the formation of high-grade varieties, because the organic material required for high-grade nodule forming reactions becomes diluted. These processes may be responsible for the low grades reported by Exon (1983) in the region at depths and latitudes that might, otherwise, be expected to contain high-grade nodules.

ENVIRONMENTS OF POTENTIALLY ECONOMIC DEPOSITS OF NODULES WITHIN THE STUDY REGION

On the basis of the above, it is possible to define certain parts of the study area to be promising for the occurrence of potentially economic nodules.

East Central Pacific Basin

To the west, the East Central Pacific Basin is bounded by the Howland-Tokelau island chain, to the east by the Line Islands, and to the south by the Manihiki Plateau and North Penrhyn Basin (Fig. 3). Water depths in much of this area are below 5000 m, and in its northern part below 5500 m. According to Exon (1982), the southern part of the basin is floored by calcareous clay, which in view of the water depths present, suggests that it must lie just above the CCD. In contrast, the northern part of the basin is floored with siliceous ooze, indicating that it is below the CCD (Aplin, 1983). Among a total of approximately 140 survey stations in the East Central Pacific Basin, Exon (1982) recorded that sixty-eight percent contained nodules, which is a very high percentage for a western equatorial basin. Nodule abundances varied from near zero to more than 30 kg/m^2 and the highest concentrations were found to occur between 5250 and 5750 m, namely at and below the CCD. Nodule grades, based on the composite of Ni+Cu+Co, also varied considerably; the highest grades of more than 2% combined Ni+Cu+Co occurred between 5100 and 5600 m. Thus, to a considerable extent, the zones of greatest abundance and grade coincide, and are near and below the CCD. It would appear, therefore, that on the basis of existing data, the CCD should be taken as the upper limit of exploration for potentially economic nodules in the East Central Pacific Basin . These observations are in agreement with the factors determining nodule grade and abundance, discussed earlier in this paper.

The geographic distribution of nodules with abundances of more than 10 kg/m^2 and grades of more than 2% combined Ni+Cu+Co in the East Central Pacific Basin are shown in Fig. 3. Biological productivity isolines from Schmithusen (1976) are also shown. It is evident from the examination of this distribution that such grades and abundances are rather widespread throughout the East Central Pacific Basin, notwithstanding that much of the basin has not been sampled, and that a higher proportion of them fall between the 50 and 100 $gCm^{-2}y^{-1}$ productivity isolines than either above or below these limits. A larger

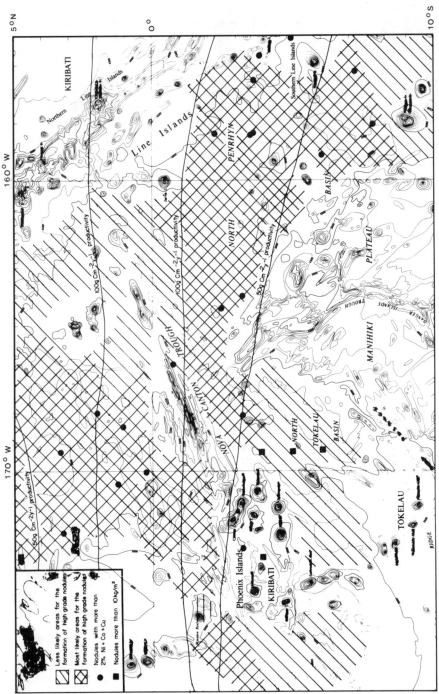

Figure 3.   Areas of known or potential nodule occurrence in the East Central
Pacific Basin and adjacent areas.

number of stations exhibit acceptable grades than acceptable abundances, but this may have resulted from having recorded more data on grades than on abundances. Abundance is nonetheless more variable locally than grade; therefore, high abundances may also occur at or in the vicinity of areas with acceptable grades.

It is interesting to note that some high grade nodules occur either within or adjacent to the equatorial high productivity maximum of over $100 \, g\,Cm^{-2} y^{-1}$ (Fig. 3). This maximum is often regarded as unfavourable for the occurrence of potentially economic nodules because it is accompanied by significantly high calcareous sedimentation rates. However, the higher flux of calcareous organisms to the seafloor is balanced by the water depth in this part of the basin that is greater than elsewhere (more than 5500 m) and which is well below the CCD. Consequently even under the zone of maximum biological productivity, potentially economic nodules can form as long as the seafloor is deeper than the CCD.

Samples from a group of stations near the southern Line Islands, centered on about $3^{\circ}S$, $157^{\circ}W$, exhibit high nodule grades (Fig. 3). Abundance data for these stations are not available. Lying on the southern flank of the equatorial zone of high biological productivity, where the seafloor is very close to the CCD, conditions in this area appear to be ideal for the formation of potentially economic nodules, barring the influence of the turbidite sedimentation.

Within and in the vicinity of the Phoenix Islands, high nodule abundances have been noted at several stations, two of which exhibited high grades. In this area, sediment slumping from the islands is likely, and could cause sedimentation rates to be quite variable. However, where basins are sheltered from mass movement in the vicinity of the 5000 m isobath, potentially economic nodules may be present.

Most remaining areas of the East Central Pacific Basin are either very poorly sampled, or not sampled at all. However, by using productivity and bathymetric data, their likelihood for containing deposits of potentially economic nodules can still be assessed.

The area between 2°-$5^{\circ}N$, west of the northern Line Islands (Fig. 3) could possibly be prospective below the 5000 m isobath. However, Orwig (1981) has shown that this area is subject to the influence of chanelled turbidite sedimentation, resulting in locally enhanced sediment accumulations which would not be conducive to forming potentially economic nodules. Any nodule deposits in this area are likely to have a limited areal extent, being confined by turbidite flows; thus this area would not be listed among high priority exploration areas.

The area northeast of Howland and Baker Islands (Fig. 3) and north of the Phoenix Group may also be subject to turbidite deposition. Moreover water depths in most of this area are less than 5000 m which, considering the position of the equatorial high productivity maximum, should further reduce any likelihood that this area contains potentially economic nodules.

Immediately north of the Phoenix Islands is a largely unsampled area located between the 50 and $100 \, g\,Cm^{-2} y^{-1}$ productivity isolines. This area extends as far east as the Nova Canton Trough, and water depths are mostly at or below 5000 m. Based on this setting, the area

is likely to contain potentially economic nodules. Rosendahl et al.
(1975) have reported that an acoustically transparent layer composed
largely of siliceous ooze is widespread in the region below 5000 m. By
analogy with other areas, this should be a likely setting for the occurrence of potentially economic nodules. However, according to Rosendahl
et al. (1975), pelagic turbidites are also present in the area, which
would limit the size of nodule fields.

The area extending from the longitude of the Phoenix Islands past
the Nova Canton Trough and the northern margin of the Manihiki Plateau
and into the Line Islands and North Penrhyn Basin areas to date has not
been sampled adequately. However, during the Wake-Tahiti Transect (Usui,
1984) a few stations were occupied, some of which yielded nodules with
more than 2% combined Ni+Cu+Co. Because it lies beneath the equatorial
zone of high biological productivity, with depths that are largely below
the CCD, this area's potential for containing economic nodules should be
high. However, the proximity of the Manihiki Plateau on its southern
margin, which could supply turbidites, could have a locally deleterious
effect on nodule genesis, and seamounts near the equator could exert a
similar influence.

South and southeast of the Phoenix Islands is an area largely
deeper than 5000 m. It is south of the $50 \, g \, Cm^{-2} y^{-1}$ isoline, and its
prospectivity is likely to be lower than that of the more biologically
productive areas north of it.

West Central Pacific Basin

The West Central Pacific Basin is bounded on the east by the
Howland, Phoenix and Tokelau Islands, on the west by the Gilbert and
Ellice Islands, and on the south by the North Melanesian Borderland
(Fig. 4). Water depths in much of this region are below 5000 m,
and reach 6000 m in the central and southern sectors (Fig. 4).
Information on sediments for this region is limited. McDougall and
Eade (1981) recorded "siliceous brown clay" at several stations near
the equator east of the Gilbert Islands, which indicates that the CCD
is above the seafloor. To the south, as the seafloor deepens, the CCD
is further above the bottom. Aplin (1983) showed that in the east of
the area, near the Phoenix Islands, siliceous remains comprise 10-25%
of the sediments to latitudes of 5° S and decrease to 2-10% as far as
8° S. Considering these data and the position of productivity isolines,
it would appear that the equatorial sector of the area falls under the
zone of high productivity. As would be expected, the productivity is
lower, however, in the southern sector of this area.

Sample data from about 50 stations in the West Central Pacific
Basin compiled by Exon (1982), showed that 45% contain nodules. However,
data on abundances are available for only two stations and on grade for
only 14. Grades vary from 0.28% combined Ni+Cu+Co to a maximum of
1.58%.

In spite of the limited nodule data available for the West Central
Pacific Basin, it is possible on the basis of productivity and bathymetric considerations to state generally where potentially economic
nodules might occur. The area on the southern boundary of the high
productivity zone immediately to the west of the Phoenix Islands is

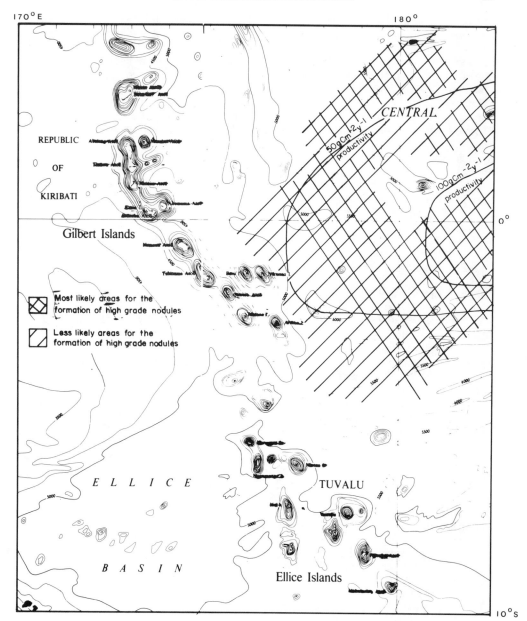

Figure 4.   Areas of known or potential nodule occurrence in the
West Central Pacific Basin.

below 5000 m water depth;therefore, its prospectivity for potentially economic nodules on grounds of both productivity and bathymetry is high. However, turbidite sedimentation from the islands or from seamounts may reduce the possibility of potentially economic nodules being present. The area southwest of the Phoenix Islands must be regarded as having lower prospectivity for potentially economic nodules considering available biological productivity data, although bathymetrically suitable for their occurrence. The area due east of the Gilbert Islands is floored by siliceous ooze/brown clay sediments, falls largely between the 50-100 g$Cm^{-2}y^{-1}$ productivity isolines, and is below the CCD. Hence by analogy with similar conditions elsewhere, this area is expected to contain potentially economic nodules.

The area to the west of the Gilbert Islands (Fig. 4) although barely investigated, would seem to have a low likelihood for the occurrence of potentially economic nodules, because it is west of the high productivity area and much of it is shallower than 4500 m. Additionally, it is likely to have been sedimented by turbidite material slumped from the islands. None of the four stations in the area considered by Exon (1982) showed any presence of nodules.

North Penrhyn Basin

The North Penrhyn Basin is bounded on the west by the Manihiki Plateau, on the east by the Southern Line Islands, on the south by the South Penrhyn Basin, and on the north by the East Central Pacific Basin (Fig. 3). Water depths generally range from 5000 to 5500 m. According to Aplin (1983) sediments in the North Penrhyn Basin are mostly $CaCO_3$-poor clays. This suggests that most of the basin floor is below the CCD, in accord with its latitudinal position and average depth greater than 5000 m. Small amounts of siliceous debris are common north of 8°S, reflecting elevated productivity in surface waters north of this latitude.

In data from about 35 stations in the North Penrhyn Basin, compiled by Exon (1982), 70% recorded the presence of nodules, the highest percentage in any of the basins studied. Nodule abundance was found to vary from near zero to just over 10 kg/m^2, with nodules being most abundant in the depth interval of 5200 to 5600 m. However, this observation was based on a very small sample population, only 11 samples, because abundance data were not available for many of the stations in the North Penrhyn Basin that were considered by Exon (1982). Nodule grades in the North Penrhyn Basin were found to vary from under 1% combined Ni+Cu+Co to just over 2%. The grades show little relationship to depth, probably because of the small sample population analysed, although the highest grades were found to occur below 5000 m.

The geographic distribution of nodules with abundances of more than 10 kg/m^2 or grades of more than 2% Ni+Cu+Co (Exon, 1982; Aplin, 1983) are shown in Fig. 3. North of 10°S in the North Penrhyn Basin there are no stations where more than 10 kg/m^2 of nodules have been recorded, and only three where grades exceed 2% combined Ni+Cu+Co. These three stations are near the Southern Line Islands, adjacent to the prospective area in the south-eastern East Central Pacific Basin mentioned previously.

In spite of the limited data available on nodule grade and abundance in the North Penrhyn Basin, it is possible to predict likely occurrences of potentially economic nodules on the basis of productivity and bathymetric data. The area to the north, east and immediately south and southeast of Starbuck Island impinges on the prospective area of the East Central Pacific Basin, falls under the zone of elevated productivity and is near or just below the CCD. Thus, its prospectivity should be high. The area between Starbuck and Penrhyn Islands falls south of the main region of high productivity but the presence of siliceous organisms in its sediments that extends to about $8^{o}S$ indicates a moderate prospectivity at least this far to the south. Further southward, the likelihood for containing potentially economic nodules decreases.

CONCLUSIONS

Based on analogy with the Clarion-Clipperton zone, three main factors appear to influence the occurrence of potentially economic nodules in the study area, (i) elevated biological productivity mainly above $50\,g\,Cm^{-2}y^{-1}$, (ii) seafloor depth near or exceeding that of the CCD, (iii) an absence of turbidite sedimentation. These conditions are fulfilled in parts of the following basins, where available data indicate that high grade and sometimes abundant nodules do occur:

i) the East Central Pacific Basin west of the Line Islands extending to the East Central Pacific Basin north and east of the Phoenix Islands

ii) the West Central Pacific Basin west and northwest of the Phoenix Islands

iii) the North Penrhyn Basin north of Penrhyn Island

Similarities between the environmental conditions under which potentially economic nodules occur in the Clarion-Clipperton zone and in the areas listed above are noteable. Nonetheless, such similarities may relate only to present day conditions. During the relatively long time needed for nodule formation, environmental conditions may have varied, and could certainly have been different from those prevailing today. However, the deposits in the two areas show similarities which suggests that the same genetic model can be applied to both. Possibly, any changes in environmental conditions during the period of formation of the deposits has also similarly affected both areas. Such postulates can only be tested through a much more thorough evaluation of the evolution of nodule deposits in both areas.

ACKNOWLEDGEMENTS

This work developed from a project initiated by CCOP/SOPAC and has been supported by the SERC. The article is published with the permission of CCOP/SOPAC and the United Nations. The opinions expressed are not necessarily endorsed by either organisation.

REFERENCES

Aplin, A., 1983, The Geochemistry and Environment of Deposition of some
    Ferromanganese Oxide Deposits from the South Equatorial Pacific,
    unpubl. Ph.D. thesis, University of London, 348p.

Aplin, A. and Cronan, D.S., 1985, Ferromanganese oxide deposits from
    the Central Pacific Ocean, II, Geochim. Cosmochim. Acta., v.49,
    pp.437-451.

Archer, A., 1979, Resources and potential reserves of nickel and copper
    in manganese nodules, In: Manganese Nodules: Dimensions and
    Perspectives, D. Reidel Co., Dordrecht, Netherlands, pp.71-87.

Bastien-Thiery, H., Lenoble, J.P. and Rogel, P., 1977, French exploration
    seeks to define mineable nodule tonnages on Pacific floor, Eng. Min.
    Jour., July 1977, v.171, pp.86-87.

Exon, N.F., 1982, Manganese nodules in the Kiribati region, equatorial
    western Pacific, South Pacific, Mar. Geol. Notes, v.2(6), CCOP/SOPAC,
    Suva, Fiji, pp.77-102.

Exon, N.F., 1983, Manganese nodule deposits in the Central Pacific Ocean
    and their variation with latitude. Marine Mining, v.4, pp.79-107.

Kildow, J.T., Bever, M.B., Dar, V.K. and Capstaff, A.E., 1976, Assess-
    ment of economic and regulatory conditions affecting ocean minerals
    resource development. Mass. Inst. Technol. Rept. (Unpubl.) 1146p.

McDougall, J.C. and Eade, J.V., 1981, Manganese nodules in western
    Kiribati (Gilbert Islands). South Pacific. Mar. Geol. Notes, v.2(5)
    CCOP/SOPAC, Suva, Fiji, pp.67-75.

Orwig, T.L., 1981, Channelled turbidites in the eastern Central Pacific
    Basin. Marine Geol. v.39, pp.33-58.

Pautot, G. and Melguen, M., 1979, Influence of deep water circulation
    and seafloor morphology on the abundance and grade of Central South
    Pacific manganese nodules, In: Bischoff, J.L. and Piper, D.Z. (eds.)
    Marine geology and oceanography of the Pacific Manganese nodule
    province, Plenum Press, New York, pp.621-649.

Rosendahl, B.R., Moberly, R., Halunen, J., Rose, J. and Kroenke, L.,
    1975, Geological and geophysical studies of the Canton Trough region,
    Jour. Geophys. Res. v.80, pp.2565-2574.

Schmithusen, J., 1976, Atlas zur Biogeographie, Bibliographisches Inst.,
    Mannheim/Wien/Zurich; 74p.

Usui, A., 1984, Regional variation of manganese nodule facies on the
    Wake-Tahiti Transect: morphological, chemical and mineralogical
    study, Marine Geology, v.54, pp.27-51.

GROWTH HISTORY AND VARIABILITY OF MANGANESE NODULES OF THE EQUATORIAL NORTH PACIFIC

U.von Stackelberg
Bundesanstalt für Geowissenschaften und Rohstoffe
Postfach 510153
3 Hannover 51, Federal Republic of Germany

ABSTRACT. "Seeding" of coarse particles at marked horizons is responsible for the origin of manganese nodule fields. Such "seeding" occurred at hiatuses which were formed by the erosional capacity of the currents of the Antarctic Bottom Water during the Neogene. Two kinds of nodules are prevalent, one of which owes its existence to these hiatuses. This is the main type of the equatorial North Pacific. The other nodule type originated at the surface of consolidated and bioturbated ash layers.

A steady process of biogenic lifting has prevented most of the nodules from burial. Occasionally encountered buried nodules prove, however, that nodules form at a distinct horizon, as well as reveal their progressive growth stages.

Pelagic accumulation rates are directly related to the rate of decay of organic matter within near-surface sediments. This decay correlates positively with the Mn-flux within the interstitial water in seafloor sediments. The Mn-flux, in turn, is responsible for the diagenetic growth of nodules. The amount of organic matter available on the seafloor positively influences how intense the activity of burrowing organisms is, and it is this activity that keeps nodules from being buried below the sedimentary surface. Sediment accumulation rates are mainly determined by the velocity of bottom currents, which, in turn, are strongly influenced by bottom morphology.

Mn-nodule deposits in the equatorial North Pacific Ocean represent accumulations at least during the last 15 million years. From site to site, however, sedimentation rates, geochemical and oceanographic conditions differ, and this may explain the great variability of manganese nodule facies.

INTRODUCTION

The occurrence of manganese nodules on the seafloor is worldwide. Deposits of high abundance, that is, containing more than 10 kg nodules per m, occur in the Pacific and Indian Oceans and around Antarctica (Cronan, 1980). The metal content of most of these

P. G. Teleki et al. (eds.), Marine Minerals, 189–204.
© *1987 by D. Reidel Publishing Company.*

deposits, however, is low and not of economic interest. Such deposits, with a Cu + Ni + Co content of more than 2.5% exist in the central Indian Ocean and in the South Pacific, and in the equatorial North Pacific between 5° and 10°N. The area represented by this so-called nodule belt approximately equals half the area of Europe, on account of which a very large potential of metal resources is assumed to exist within this region.

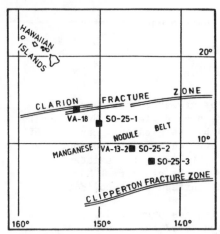

SURVEY AREAS

Figure 1. Survey areas of VALDIVIA and SONNE cruises in the Pacific manganese nodule belt.

Because of this potential economic interest, several research expeditions in the past surveyed this belt, situated between the Clarion- and Clipperton Fracture Zones. The survey areas of several West German cruises, which are discussed in this paper, also lie within this zone (Fig. 1). Area VA-18 is located at the northern edge within the Clarion Fracture Zone. The three areas of SONNE cruise SO-25 lie in the northern, central and southern part of the nodule belt. Area VA-13-2 generally coincides with that of area SO-25-2. From a total of 5 surveys, various kinds of samples were recovered at 845 stations and more than 900 m of sediment cores and manganese nodules from 560 stations were collected. All nodules were classified and their dimensions recorded. Of the 9,000 nodules, so obtained, 1,400 were cut, polished and studied in detail.

Based on published data and the examination of this large quantity of material, general statements can be derived: 1) manganese nodules occur nearly exclusively on the surface of the seafloor, as only a few nodules are found buried in the sediment; 2) the nodule facies is uniform with respect to geochemistry, mineralogy and nodule shape within the limited perspective provided by any one individual bottom photo-graph or sample, both of which usually show only one type of nodule. Significantly, not only the shape, but the size of nodules is also consistently uniform, that is, their size distribution is unimodal. Sample-to-sample or photograph-to-photograph, however, there is a

large variability in nodule types, which differ in shape, surface
structure, and composition.

The three main types are: spherical rough nodules, mostly small
and ellipsoidal smooth nodules, which frequently grow together forming
polynodules, and large ellipsoidal nodules with asymmetric growth,
with smooth tops and rough bottoms. A change from one nodule type
to the other may be found within a distance of a few 100 meters.
(von Stackelberg, 1982).

In the following sections the occurrence and variety of nodules is
described in more detail.

1 cm

Figure 2. Polished radial section of a large ellipsoidal nodule with
asymmetric growth from area VA-13/2. The nucleus (dashed line) is
composed of consolidated sediment. White arrows indicate overlapping
and wedging of dark-coloured layers at the equatorial zone which proves
that nodules are turned over from time to time.

GROWTH STRUCTURES

The section of a nodule normally shows a concentrically layered
structure (Fig. 2). In the center we always find a core of foreign
material, such as a fragment of consolidated sediment or frequently
fragments of volcanic rocks or segments of fragmented older nodules,
as well as organic remains such as sharks' teeth. Nodule growth

cannot start without having a nucleus. Within 1177 nodule sections
studied, 2068 nuclei were identified, which showed the following
distribution of types: 54% sediments, 26% nodule fragments, 8% basalt,
5% organic remains, 4% pumice.

The asymmetric growth pattern of nodules is also illustrated in
Fig. 2. At the top it is layered and dense (appears dark in reflected
light); at the bottom it is irregular and porous (appears light grey).
The boundary between the two types of growth, called the equatorial
zone, coincides with the sediment-water boundary, whereby the upper
part of the nodule is exposed above the sedimentary surface. The
nodule in Fig. 2 was accreting material apparently on both sides,
although in a different manner below and above this zone. Mineral-
ogically there is also a significant difference in the kind of Mn-oxide
minerals found on the upper and lower surfaces. At the top it is
composed mostly of δ-MnO$_2$, at the bottom it is todorokite. The
growth below the sediment-water interface is diagenetic; above it is
hydrogenetic (Calvert and Price, 1977). The change of dark and
light layers indicates that growth conditions change frequently.

Isotopic dating has shown that the speed at which manganese
nodules grow is astonishingly low (Heye, 1978), generally not more
than a few mm per million years. Accordingly, it must be assumed that
deep-sea manganese nodules are several million years old.

LIFTING PROCESS

If it is true that nodules are millions of years old, how can it
be explained that they mainly occur at the surface of Recent sediments,
when the accumulation rate of these deep-sea sediments is three orders
of magnitude higher than the growth rate of nodules? Some force must
be assumed to exist which repeatedly raises the nodules and prevents
them from being buried. This mystery has concerned many researchers
and has led to a number of hypotheses, among which are microearthquakes
(Glasby, 1977) and gas ascending from the ocean crust (Malmquist and
Kristiansson, 1981). These assumptions have not been proved by
observations. It is much more reasonable to refer to erosion by
bottom currents as a possible cause. However, all typical indirect
indicators of strong currents on the seafloor, such as sediment
ripples and current marks, are missing. Traces of organisms are,
nonetheless, found everywhere in and on top of the sediments (Fig. 3).
Manganese nodules surrounded by soft deep-sea sediment, are a
favored substrate of sessile benthic organisms. Grazing organisms
feed on these sessile types and may move the nodules. Holothurians
or fish, searching the seafloor for food, are capable to move and
even turn nodules over.

Polished sections of large basin-nodules show distinct wedging
in the dark-colored layers at the equatorial zone (Sorem et al.,
1979). The upper half of the hydrogenetically grown layer overlaps
the lower one (Fig. 2). Apparently, there are two hemispheres of
different age. The lower one formed when the nodule was turned upside
down. It is assumed that even polynodules, that show very finely

laminated growth structures, are constructed from quasi-hemispheres
and not from spheres, because a part of the nodule must have rested
in the surface sediment. Frequent turnover, however, produces a quasi-
concentric growth.

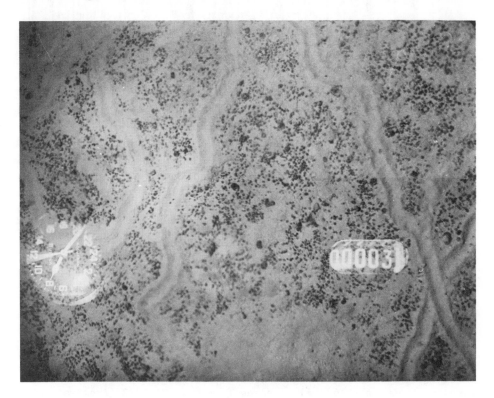

Figure 3. Bottom photograph of manganese nodules and traces of
benthic organisms from area SO-25-3.

    Grazing organisms and sediment feeders, therefore, may contribute
to the movement and upward lifting of the nodules (Menard, 1964). With
a sedimentary accumulation rate of a few mm per thousand years, dislodgement every 100 years is sufficient to keep the nodules at the surface.

FORMATION OF "HIATUS NODULES"

    If manganese nodules are old and biogenic lifting has kept them at
the sediment surface, the question arises immediately: when, where and
how did nodules form? To answer this question, several aspects of nodule formation must be examined. In most of the sediment cores, as illustrated in Fig. 4, hiatuses were found (von Stackelberg, 1979).

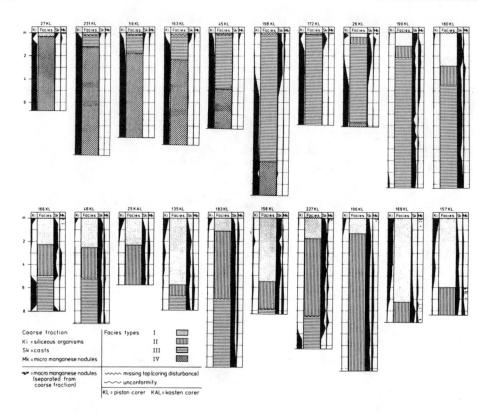

Figure 4 . Lithologic facies and concentration of selected components
such as casts of siliceous organisms and micromanganese nodules within
the coarse fraction (> 63 m) of cores from area VA-13/2 (from von
Stackelberg, 1979). Facies I: Light-brown siliceous ooze to siliceous
mud. II. Dark-brown phillipsitic clay. III: Reddish-brown phillipsitic clay; mainly light-brown radiolarian-bearing nannofossil ooze.

    The schematic diagram of sediment cores (Fig. 5) shows that sedi-
ments of the time interval between Early Miocene and Late Pliocene are
absent, which means that there is a hiatus of about 10 million years
between 15 and 4 million years before present. Near such a hiatus, the
tests of siliceous organisms disappear, because they become dissolved
through time. Calcareous tests do not exist at all below the Carbonate
Compensation Depth (CCD). . Chemically resistant particles, such as fish
teeth do, however, concentrate. The abundance of siliceous organisms
correlates negatively with authigenic minerals, and the latter form
partly at the expense of these planktonic tests. Smectites, zeolite
minerals and micro-manganese nodules are found within radiolarian casts.

Adjacent to the hiatus, a concentration of volcanogenic mineral grains, such as feldspars and magnetites, can commonly be observed. These are chemically resistant relicts of basaltic rocks or pumice. This concentration of volcanogenic minerals, however, is not a result of enhanced volcanism, but of greater submarine weathering and condensation of residual components.

Processes of dissolution and formation of authigenic minerals lead to accumulations of coarser sediments on the seafloor. Diagenetic processes may also result in the cementation of the sedimentary surface. The consolidated sediment is subsequently burrowed by benthic organisms, and coarse fragments of the matrix become detached.

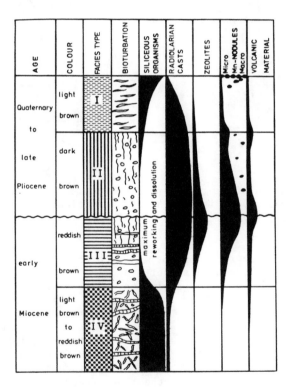

Figure 5. Schematic core diagram from area VA-13/2. Wavy line indicates a hiatus (from von Stackelberg, 1982).

Figure 6. Distribution of manganese nodule deposits (1), and pathway of
Antarctic Bottom Water (2), in the Pacific and Atlantic Ocean (after
Dangeard and Vanney, 1978).

     All observations at the seafloor indicate that bottom currents have
a strong impact on the textural properties of bottom sediments, through ·
the processes of reworking and dissolution. Fig. 6 shows the distribution
of the Antarctic Bottom Water Current at present. Within the nodule belt
its influence is generally weak. Current velocities increase from South
to North. About 13 million years ago, however, there was a stronger flow
than today of Antarctic Bottom Water toward the North, triggered by a
wide-ranging expansion of the Antarctic ice cover (van Andel et al.,
1975). Furthermore, uplift of Central America disrupted the circum-
equatorial circulation and enhanced the flow of South-to-North bottom
water currents at that time (Keller and Barron, 1983). These Tertiary
bottom currents may have followed the same pathways as today, but with
much higher intensity. The result was the Tertiary hiatus described above.
     This completes the discussion of the characteristics of a hiatus and
how it was formed. But what significance does this hold for the origin of
nodules? During the hiatus the supply of dissolved metals increased.
Primarily, however, the increased supply of coarse relict components,
authigenic minerals and consolidated sediment particles was the crucial

impulse for the origin of manganese nodules. The hiatal horizon was the
starting line for nodule growth. The high percentage (54%) of nuclei,
that are composed of indurated sediment, are strong indicators of the
importance of hiatal horizons supplying such fragments.

     The nodule cross-section of Fig. 2 clearly shows the principles of
growth history. The nucleus is surrounded by hydrogenetically grown Mn-
oxide (dark), which indicates that the initial growth is not yet influenced
by sedimentation as bottom currents prevented the deposition of sediments.
The nodule remained uncovered, deriving metal ions directly from seawater.
Only later, after the start of post-hiatal sedimentation, did diagenetic
growth appear and that has prevailed since.

     A great number of manganese nodules from the equatorial North Pacific
Ocean show a similar growth sequence, therefore, have probably originated
at former hiatuses too. The course of the Tertiary Antarctic Bottom
Water might be responsible for the E-W configuration of this nodule belt.

FORMATION OF "ASH NODULES"

     Manganese nodules form not only at hiatuses but also at any distinct
horizon produced by other geological processes. Site VA-18 lies at the
northern margin of the nodule belt on a horst between two grabens of the
Clarion Fracture Zone (Fig. 1).

     Cores from that area characteristically show a sequence of three
sedimentary units (Fig. 7). The upper unit includes three layers of
volcanic ash with a distinct, thin lamination which is disturbed by
bioturbation only in the upper parts (von Stackelberg, 1982). Paleo-
magnetic measurements and study of radiolaria give this ash sequence a
probable age of $1-2 \times 10^6$ years (Rehm et al., 1980). Chemical investigations support the assumption that the ashes were supplied from Maui,
Hawaii. Buried nodules were found only in unit II and the lower part of

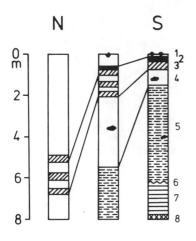

Figure 7. Schematic diagram of
cores from area VA-18 (after von
Stackelberg, 1982).
1) Manganese nodules;
2) manganese crust   } unit I
3) volcanic ash layers;
4) dark brown clay;
5) dark reddish-brown clay; = unit II
6) hiatus;
7) laminated clay;   } unit III
8) laminated chert.

unit I. These nodules, with diameters as wide as 5 cm, reveal two
growth layers similar to those in nodules from area VA-13/2, and probably originated during a hiatal period that occurred between the deposition of unit II and III. The nearly undisturbed lamination of the ash
layers indicates that nodules, after the first "rain" of ash, were not
dislodged further by benthic organisms.

Nodules on the surface are generally spheroidal and remarkably
small (max. diameter = 1.2 cm.). Their nuclei consist mostly of tiny
fragments of consolidated ash or manganese crust. Most of the ash
layers must have been indurated very early. Furthermore, it was the
uppermost part of the youngest ash layer that served as a favorable
habitat for benthic organisms. The burrows excavated by these organisms
became subsequently encrusted by manganese oxide (Fig. 8). Burrowing
was destructive so that fragments of consolidated ash and manganese
crust, which served as seeds for a new generation of nodules, were
detached. Growth started immediately after seeding during a period of
normal-rate sedimentation, unlike that which occurred during the hiatal
period observed in area VA-13/2. Therefore, only diagenetic Mn-oxide
growth is found. The occurrence of the small nodules is restricted to
areas where sedimetation rate is low and the uppermost ash layer is
encrusted with manganese. The relatively short growth period of 1-2 x
10^6 years available is the reason for the small size of the nodules.

5 cm

Figure 8. Surface of extensively burrowed volcanic ash layer from area
VA-18 by encrusted manganese.

Lastly in area VA-18 and SO-25-1, layers of coarse foreign material were found, supplied from seamounts or ridges. These layers have a similar bearing on nodules as ash layers (von Stackelberg et al., 1984). Here, old nodules also became buried and a new generation of nodules has started to form at the consolidated surface.

Thus, the hiatus model is not the only one that explains the origin of nodule fields. For the nodule belt, however, it probably is the most important model.

GROWTH HISTORY OF BURIED NODULES

The concepts put forth above, on the origin and lifting of nodules, must be treated as mere assumptions as long as information on intermediate growth stages is lacking. This information actually can be found in buried nodules. However, only a few buried nodules had been studied.

If we take the rules of statistics into account, biogenic lifting should not be the fate of every nodule. A few individual nodules should actually become buried. On the SONNE cruise in 1982, approximately 200 buried nodules were collected; all were investigated in detail.

Fig. 9 shows a sediment core, in which 39 buried nodules were found, and in which two hiatuses can be seen. Beneath the main hiatus, which is Early Miocene to Late Pliocene age, there is a second hiatus of Early Miocene to Late Eocene age. Nodules originated at the older hiatal level (A). Burrowing organisms separated fragments of a zeolitic indurated horizon; furthermore, coarse residual components were concentrated at this level. These particles served as nuclei for nodule growth. The thicknesses of outer manganese-oxide layers (layer A) growing around the nucleus increase from one buried nodule to the next, upward throughout unit 3. Most nodules became fragmented while exposed later at the second, i.e. the principal hiatus, and were covered, in turn, by a younger layer (B). The thickness of this younger layer also increases upwards throughout units 1 and 2.

It is recognized that nodule genesis actually starts at a distinct horizon and that the products are subject to a long and variable growth history until they reach the stage observed in nodules found on the surface. Buried nodules reveal various intermediate steps of growth. They are, in some way, the earlier stage "deceased" versions of surface nodules. Consequently there is a strong support for the assumption that manganese nodules are old, indeed. Furthermore, analyzing similar core sections containing buried nodules from marginal basins has proved that polynodules in such settings did not slump from hills, but originated and grew in place.

Recent investigations using Scanning Electron Microscope (SEM) photography revealed that the characteristic brittleness of the surface of buried nodules is not caused by dissolution but by a final, abnormal diagenetic growth.

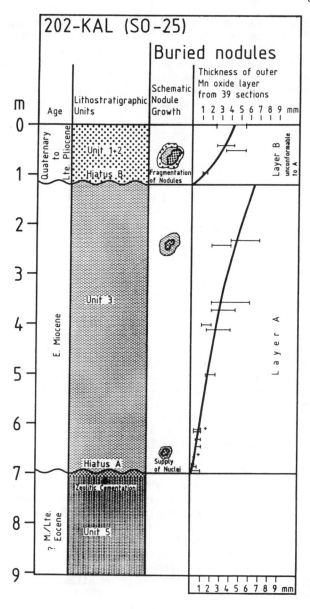

Figure 9. Growth history of buried nodules from sediment core 202-KAL
from area SO-25-2 (from von Stackelberg, 1984).

VARIABILITY OF SURFACE NODULES

The previous discussion describes how manganese nodules originated, what process held them at the sedimentary surface and how the character of their growth changes as soon as the sedimentary environment changes. This was essentially a description of the growth history along the vertical dimension. Obviously there is a need to review briefly the growth conditions along the horizontal dimension too, and to discuss the areal and geomorphologic variability of nodule types.

Sedimentation within the nodule belt is characterized by rapidly changing accumulation rates. Fig. 10 shows areas of accumulation, non-deposition and erosion. Erosion occurs on top of hills and within basins. Maximum accumulation rates are found in marginal basins adjacent to hills. Based on isotopic dating, basin accumulation rates were determined to be as high as 20 mm/thousand years (Mangini and von Stackelberg, in press). However, accumulation rates discussed in this paper are only valid for post-hiatal sedimentation (Plio-Pleistocene).

As it regards their distribution, nodules are absent in basins where erosion prevails. They occur, however, and are highly abundant on tops of hills, where erosional processes are active and sediments are abraded, in the form of smooth polynodules. Quite similarly shaped smooth polynodules also occur in marginal basins, where sedimentation rates, as well as nodule abundance are high, the latter as much as 15 kg/m^2.

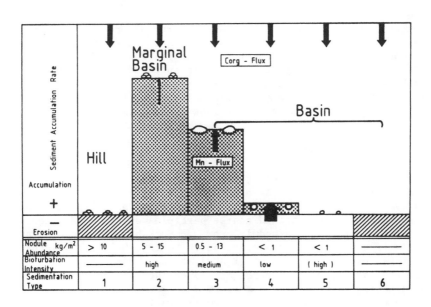

Figure 10. Correlationships among manganese nodule type and abundance, sediment accumulation rate, flux of organic carbon from the sea water and manganese-flux from interstitial water.

Asymmetrically growing nodules are associated with medium-high sedimentary accumulation rates. At very low rates, rough spherical nodules form (von Stackelberg, 1979, 1982).

These observations contradict the usual concept that, at high sedimentary accumulation rates, a nodule should be covered and mainly grow diagenetically, and conversely at low accumulation rates the growth is hydrogenetic (Calvert and Price, 1977). How can this contradiction be resolved?

The precipitation rate of organic matter from surface plankton, within any relatively small area is more or less uniform (Fig. 10), and its distribution in surface sediments is also relatively uniform. The sediment cores, however, reveal distinct variations with depth. In areas of high sedimentary accumulation rates, the content of organic matter in sediments slowly decreases with core depth (Müller and Mangini, 1980). Consequently, relatively high content is still found in deeper parts of the sedimentary column. Within marginal basins characterized by high influx of reworked old material, the young organic matter is rapidly covered and buried, and that prevents its oxidation. In areas of low accumulation rates the reverse is true. Here, the oxidizing influence of the bottom water is stronger, and most of the organic matter decays. This decay changes the micro-environment. Manganese is gradually dissolved and transported upwards by the interstitial water, which may explain the occurrence of rough, diagenetically grown nodules. In marginal basins, however, the manganese flux should be less, hence, the nodules grow hydrogenetically and have smooth surfaces.

Organic matter, preserved in deeper parts of the sedimentary column in marginal basins, is sought by mud-feeders who burrow into the sediment. Consequently, the intensity of bioturbation is high, biogenic lifting becomes especially strong, and in spite of relatively high sedimentary accumulation rates, nodules are kept at the surface. Where little food is available in the sedimentary mass, bioturbation is weak, and nodules are covered by sediment most of the time.

SUMMARY AND CONCLUSIONS

During the Neogene, the flow of the Antarctic Bottom Water was much more intensive than today. As a result, areally extensive hiatuses were formed by erosion. This erosion, together with diagenesis and bioturbation produced and concentrated coarse particles on the ocean floor, which served as nuclei for nodule growth. Such hiatal nodules are the main type present in the equatorial North Pacific Ocean. Two other nodule types exist, one that originated at the surface of indurated and bioturbated ash layers, the other at the surface of sedimentary layers composed of reworked foreign material.

Foremost, it is the seeding of coarse particles at a distinct horizon that is responsible for the generation of new nodules. Most nodules, having originated at deeper horizons, are found now on or near the sedimentary surface. A steady process of biogenic lifting has prevented the majority of them from being buried, except for the few that had escaped dislodgement and lifting. Sediment cores containing buried nodules prove that this assumption about the formation of nodules at distinct horizons is correct and demonstrate well their progressive growth stages from the Miocene to the present day.

Sediment accumulation rates and content of organic matter are two important factors in nodule growth. Where sedimentary accumulation rates are low, most of the organic matter decays within, at, or just below the surface sediments. This leads to the generation of a strong manganese flux. The resulting growth of the nodules is mainly diagenetic. Bioturbation is weak as only a limited amount of food is available to the benthic organisms in deeper parts of the sedimentary column. Therefore, biogenic lifting is diminished and nodules remain covered by sediment most of the time. In areas of higher accumulation rates, especially in marginal basins adjacent to hills, conditions are reversed. There is a strong correlation between nodule type and sediment type, and both may vary within short distances. Within any area of limited size, accumulation rates are mainly determined by velocities of bottom currents which, in turn, are strongly influenced by bottom morphology.

The genesis of nodule facies is a function of several interacting geological processes, the combined effect of which differ from one location to another. As some 15 million years of geological history is involved, each nodule has its own unique past. This may be one explanation for the great variability of manganese nodule types in the world's oceans.

REFERENCES

Calvert, S.E. and Price, N.B., 1977, Geochemical variation in ferromanganese nodules and associated sediments from the Pacific Ocean; Mar. Chem., v. 5, pp. 43-74.

Cronan, D.S., 1980, Underwater Minerals; Academic Press, London, 362 p.

Dangeard, L. and Vanney, J.R., 1978, Environment physiographique géologique et biologique des pavements profonds de nodules polymétalliques; Revue de géogr. phys. géol. dynam., v. 20, no. 5, pp. 371-387.

Glasby, G.P., 1977, Why manganese nodules remain at the sediment-water interface; New Zealand Jour. Sci., v. 20, pp. 187-190.

Heye, D., 1978, Growth conditions of manganese nodules. Comparative studies of growth rate, magnetisation, chemical composition and internal structure; Prog. Oceanography, v. 7, pp. 163-239.

Keller, G. and Barron, J.A., 1983, Paleoceanographic implications of Miocene deep-sea hiatuses; Bull. Geol. Soc. Am., v. 94, pp. 590-613.

Malmquist, L. and Kristiansson, K., 1981, Microflow of geogas -

a possible formation mechanism for deep-sea nodules; Mar. Geol.,
v. 40, pp. M1- M8.

Mangini, A. and von Stackelberg, U., in press, Depositional history
in the Central Equatorial Pacific during the past 250,000 years;
Geol. Rundschau.

Menard, H.W., 1964, Marine Geology of the Pacific; McGraw-Hill,
New York, 271 p.

Müller, P.J. and Mangini, A., 1980, Organic carbon decomposition
rates in sediments of the Pacific manganese nodule belt dated by
^{230}Th and ^{231}Pa; Earth Planet. Sci. Lett., v. 51, pp. 94-114.

Rehm, E., Halbach, P. and Raschka, H., 1980, Geochemische und mineralo-
gische Untersuchungen von Basalten und vulkanischen Aschenlagen aus
der Clarion-Bruchzone (Zentralpazifik); Berliner Geowissen-
schaftliche Abhandlungen, v. A-19, pp. 184-186.

Sorem, R.K., Fewkes, R.H., McFarland, W.D. and Reinhart, W.R., 1979,
Physical aspects of the growth environment of manganese nodules
in the "Horn region," East Equatorial Pacific Ocean; Coll. Int.
du C.N.R.S., No. 289, La genese des nodules de manganese,
pp. 61-76.

van Andel, Tj.H., Heath, G.R., and Moore, T.C., Jr., 1975, Cenozoic
history and paleoceanography of the Central Equatorial Pacific
Ocean; Geol. Soc. Am. Mem., v. 143, pp. 1-134.

von Stackelberg, U., 1979, Sedimentation, hiatuses and development of
manganese nodules: VALDIVIA Site VA-13/2, Northern Central
Pacific; In: Bischoff, J.L. and Piper, D.Z. (eds.) Marine geology
and oceanography of the Pacific Manganese Nodule Province, Plenum
Press, New York, pp. 559-586.

von Stackelberg, U., 1982, Influence of hiatuses and volcanic ash rains
on the origin of manganese nodules of the equatorial North Pacific
(Valdivia cruises VA-13/2 and VA-18); Marine Mining, v. 3, no. 3-4,
pp. 297-314.

von Stackelberg, U. Beiersdorf, H. und Riech, V., 1984, Zur Mangan-
knollengenese im äquatorialen Nordpazifik, SONNE-Fahrt SO-25,
In: Herminghaus, C. and Kessler, C. (eds.) Projektleitung
Rohstofforschung der KFA, Statusbericht 1984; Jülich, pp. 444-458.

CHEMISTRY AND GROWTH HISTORY OF CENTRAL PACIFIC MN-CRUSTS AND THEIR ECONOMIC IMPORTANCE

A. Mangini[1], P. Halbach[2], D. Puteanus[2], M. Segl[3]

[1]Heidelberg Academy of Sciences Dating and Paleoclimate Project
c/o Institut fur Umweltphysik der Universität Heidelberg,
Im Neuenheimer Feld 366, D-6900 Heidelberg, F.R.G.

[2]Institut fur Mineralogie und Mineralische Rohstoffe,
Technische Universität Clausthal,
D-3392 Clausthal-Zellerfeld, F.R.G.

[3]Institut fur Umweltphysik der Universität Heidelberg
Im Neuenheimer Feld 366, D-6900 Heidelberg, F.R.G.

ABSTRACT. Hydrogenetic ferromanganese deposits from seamount areas of the Central Pacific Basin should be seriously evaluated as sites for economic development. MIDPAC exploration cruises in 1981 and 1984 have shown that crust thickness varies between 0.5 and 10 cm and that best qualities of metals of primary interest such as Mn (average 26 %), Co (average 0.9%) and Ni (average 0.55 %) and trace concentrations in Pt (average 0.5 ppm) are found between depths of 1,000 and 2,000 meters. Two generations of crust growth occur: an apatite phase is present either in an impregnated form in the older crust generation, or as discrete apatite layer in the older crust generation, and in between the older and younger generations. ^{10}Be dating helps better understanding of the origin and growth history of the Mn-crusts.

INTRODUCTION

The prime mineral resources of potential importance in the deep sea are abyssal ferromanganese nodules and crusts. These deposits contain, besides Mn, as much as 2 % combined Ni, Co and Cu, and on account of this could become a source of raw materials in the future. It is assumed that they form by precipitation of dissolved metal ions and hydrated metal-oxide compounds around nuclei of volcanic, biogenic or nodule fragments, or on exposed and altered rock surfaces. Ferromanganese deposits accumulate basically by two different accretion processes, by diagenetic growth and by hydrogenetic growth. The accumulation mechanisms lead to genetically different types of deposits.

P. G. Teleki et al. (eds.), Marine Minerals, 205–220.

DIAGENETIC PROCESSES

The diagenetic nodules are supplied by pore-water transport of hydrous divalent cations, such as Mn, Ni, Cu, Zn, and Mg, within the peneliquid surface layers of pelagic sediments. The main authigenic phase in diagenetic nodules, the 10-Å manganate (identical to todoro-kite; Burns and Burns, 1979), can be formed thermodynamically under oxidizing conditions in the deep-sea. Halbach et al. (1982) calculated the corresponding Gibbs Free Energy Δ Gr to be ≤ -6.6 kcal/Mol. Diage-netically formed nodules display primary growth features, such as den-dritic microtextures that consist of alternating laminae of crystalline 10-Å manganate and amorphous material. The latter is composed of an intimate mixture of ferric hydroxide, silicate and MnO_2 (Burns and Brown, 1972, Halbach et al., 1982). Burns and Brown (1979) suggested that autocatalytic oxidation of the 10-Å manganate may proceed in con-tact with seawater to yield a particular horizon of higher oxidation grade. This horizon is characterized by a partial or complete loss of the 10-Å basal structure.

Halbach et al. (1982) postulated that the rhythmic sequences of 10-Å manganate and amorphous material are created by physico-chemical changes in the microenvironment of accreting surfaces. The formation of the 10-Å manganate starts with release of H^+ ions, thus, the pH level becomes considerably depressed near the accreting deposit. This may accelerate precipitation of Fe-bearing silicate gel and, as a con-sequence, inhibit further formation of 10-Å manganate. Since the oxidation of Mn^{2+} continues, δMnO_2 particles are precipitated as well, producing the material appearing as amorphous under X-rays.

Hydrogenetic nodules and crusts accumulate by direct supply of colloidal, hydrous, metal-oxide particles from near-bottom seawater,i.e., the process of precipitation is controlled by the transport of these particles to the surface and by surface chemical reactions. The prin-cipal Mn-phase in the hydrogenetic substance is δMnO_2, which is a poor-ly crystalline, two-line manganate in X-ray diffraction, that is iden-tical to the mineral phase "vernadite" (Burns and Burns, 1979). This phase is intimately intergrown with FeOOH x nH_2O, amorphous under X-rays, and Fe-bearing aluminosilicates. The hydrogenetic material contains several microlayers that appear grey. The fine lamination suggests that, the precipitates have grown in sequential accumulation, in response to chemical/thermal conditions of the surrounding ocean waters. Boudreau and Scott (1978) have shown that the rate of flux of Mn from seawater to the surface of a nodule matches the radiometric rates surprisingly well, assuming that, diffusion had controlled transfer of dissolved Mn from seawater across the benthic boundary layer to the nodule. Their studies also proved that the very low (0.02 ppb), ambient Mn-concentration of seawater, is sufficient to account for all manganese in hydrogenetic nodules. These findings lead to the hypothesis that fast bottom currents encourage hydrogenetic growth, not only by lowering sedimentation rates through erosion, but also by in-creasing the flux of manganese to nuclei and substrate rocks. Obvious-ly the pH of the microenvironment and the oxidation grade should be very sensitive to changes in current velocities at the site of the deposition as well. A higher current velocity will also enhance diffusion of H^+ and O_2 through the benthic boundary layer at the water-nodule interface.

The trace-element content of seawater influences directly the chemical composition of the Mn-deposits. As suggested earlier by Cronan (1972) and Bonatti et al. (1972), one of the reasons for the presence of different trace-element composition in ferromanganese deposits at different localities is the vertical and lateral inhomogeneity of the ocean in regard to its content of trace-elements, which have short residence times. Trace-elements with residence times smaller than 10^2 years (such as Fe, Mn, Co, Th), that is, much shorter than the mixing time-scale of the ocean (about 10^3 years), will be concentrated close to their sources (Mangini et al., 1986). Dissolved manganese is, for example, highly enriched in the water column close to hydrothermally active areas, as well as in the oxygen-minimum zone of a reducing environment (Klinkhammer and Bender, 1980). Hydrogenetic crusts grown in relatively shallow water depths beneath the oxygen-minimum zone are generally richer in Mn than hydrogenetic deposits in abyssal water (Halbach and Puteanus, 1984a). Bonatti et al. (1972) attempted to classify genetically Mn-deposits on the ocean-floor, and suggested that four types of deposits exist:

1) hydrogenous (or hydrogenetic) deposits:slowly precipitating from seawater in oxidized environments;

2) hydrothermal deposits: precipitation from hydrothermal solutions in areas of volcanism and/or high heat flow. Higher contents of Mn and Fe than in abyssal deposits. High to extremely high growth rates, low to very low trace-element contents;

3) halmyrolytic deposits: Mn is supplied by submarine weathering of basaltic debris;

4) diagenetic deposits:precipitation results from diagenetic remobilization of Mn in the sedimentary column;very high Mn/Fe ratios and growth rates; trace-element contents are low. A good example is a Mn-nodule from the Peru Basin that grows at a rate of 168 mm/Ma (Reyss et al., 1982).

A widely distributed abyssal nodule type, known as the mixed-type, forms by accretion. The process in a mix of hydrogenesis and diagenesis, whereby the upper part of the nodule is under the influence of near-bottom seawater, whereas the lower part is supplied from pore water.

Based on new data from the MIDPAC expeditions (Halbach and Puteanus, 1984b) and new results from [10]Be dating (Segl et al., 1984a) the hydrogenetic class could be subdivided further into:

a) shallow pelagic types:precipitation of Mn beneath the oxygen-minimum zone;lowest growth rates, highest trace-element content;end-member for hydrogenetic growth. An example of this type is Mn-crusts from the Central Pacific (MIDPAC cruises 81 and 84).

b) abyssal pelagic types:lower Mn/Fe ratios than in shallow pelagic types because of the dissolution of carbonates (which deliver significant amount of Fe and cause low Mn/Fe ratios);

growth rates are generally slightly higher than that of shallow
deposits.

GROWTH RATES

At abyssal water depths sedimentation rates are low. But nodules
grow at rates that are even lower than those of the surrounding sediment, by several orders of magnitude. This implies that the processes
of Mn-nodule genesis and accumulation have been active for millions of
years. Mechanisms suggested for keeping Mn-nodules at the sediment
surface are a) biological uplift (von Stackelberg, 1979 and this
volume) and b) periodical events such as faster deep-water circulation
which may erode and redistribute sediments (Moore et al., 1981; Mangini
and Kuehnel, 1986).

The growth rates of open-ocean deposits range from 1 to about 5
mm/Ma. These growth rates were determined making use of radioactive
decay of natural isotopes that are incorporated into nodules at the
time of their formation. Most commonly applied isotopes are ^{10}Be
(half-life:1.5 Ma) and ^{230}Th (7.5×10^4a). A compilation of results is
given in Segl et al. (1984a). The detection of ^{10}Be by the recently
developped accelerator mass spectrometry has enabled absolute dating
of a large number of samples. The detection limit of this technique is
14-15 Ma. One of the outstanding results was finding a number of event
markers in ferromanganese deposits, that represent either changes in
growth rate or in mineralogical structure. These markers coincide with
well-established paleoceanographic events. The most obvious ones in
nodules and crusts date 1.2 Ma, 3.2 Ma, 6.2 Ma and 14.0 Ma (Segl et al.,
1984a and Segl et al., 1984b). This suggests that ferromanganese deposits reflect past changes in paleoceanographic conditions, as such
they might be valuable to reconstructing global paleoceanographic
events (Table I).

TABLE I : Ages of important time markers recorded in pelagic ferro-
          manganese deposits (from Segl et al., 1984b)

Age	Event
1.2 Ma	Single event unclear - Beginning of Mid-Pleistocene oscillations of the global ice volume.
3.3 Ma	Initiation of the glaciation in the Northern Hemisphere
6.2 Ma	Shift in δ^{13}C in benthic forams. Beginning of modern bottom water circulation. Drop of the CCD (Calcium Carbonate Compensation Depth) in Central Pacific.
14.5 Ma	Increase of the δ^{18}O in benthic forams attributed to enhanced bottom water flow caused by the onset of major Antarctic glaciation. Rise of the CCD.

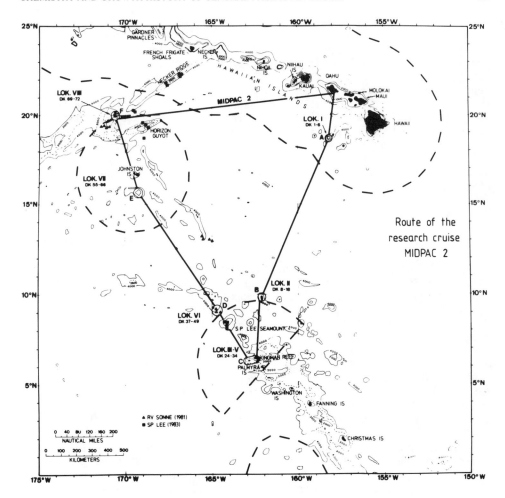

Figure 1 : Location map of the study sites and samples obtained
during the research cruises: MIDPAC 81 (RV SONNE 1981)
and MIDPAC 2 (1984). The sample sites denoted A, B, C,
D, E, and F are also listed in Table III.

DEPOSITS OF THE CENTRAL PACIFIC

Intensive research during the past 20 years has also focussed on the economic potential of ferromanganese deposits. From such view-point, the metal content of Mn-crusts, recovered during the MIDPAC cruises to the Central Pacific 1981 and 1984 seem very promising for future extraction, especially if Co associated with Mn, could be bene-ficiated. Thus a description of the area of the deposits and their composition and economic potential is warranted.

The objective of the MIDPAC 81 (1981) and MIDPAC 2 (1984) expe-ditions to the Mid-Pacific Mountains and Line Islands was to map the distribution of cobalt in these provinces and to improve the under-standing of how they formed (Fig. 1).

Distribution of deposits

The origin of the Line-Islands Ridge and the Mid-Pacific Moun-tains is principally related to Late Cretaceous basaltic volcanism. The first geologic event of a basaltic eruption in this area could be as old as 100 to 106 Ma (Clague, 1981). Jackson and Schlanger (1976) proposed that the entire region underwent synchronous volcanism and uplifted the preexisting shields into shallow water prior to 80 to 85 Ma. Subsequently, 50 - 60 Ma, the islands began to submerge and the epoch of intense volcanism ended.

The topography of the area is characterized by high-relief sea-mounts and seamount chains. The depth of the water varies between 4,900 and 1,100 m. A few of the seamount slope investigated extend from a depth of 2,000 m to 3,000 m, with intervening small terraces, others range only between 1,000 and 1,500 m. The tops of the sea-mounts are often flat (plateaus), several nautical miles in diameter, which had formed during subsiding post-volcanic marine or subaerial erosional conditions. Typical cliff-and-terrace structures exist on several seamount slopes. In the intervening valleys, ferromanganese deposits are scarcer. In a few valleys, whose surface is covered by calcareous sediments at depths of 2,600 to 2,700 m, asymmetric ripples were observed, which indicate that strong bottom currents are present. Ferromanganese encrustations are common on seamount slopes and summit plateaus between 3,000 and 1,100 m water depth, covering exposed rocks (Halbach et al., 1982). On plateaus and flat terraces, nodules lie on top of partly consolidated calcareous sediments.

The principal types of substrates for the crusts are: (1) alkali basalt and its alteration products, (2) hyaloclastites with fragments of highly vesicular basaltic rock, that suggest a relatively shallow water or subaerial origin, (3) yellowish-green smectite rock, (4) in-durated phosphorite, partly formed by replacement of calcareous ooze, sometimes enclosing fragments of the older generation of ferromanga-nese crust, and (5) occasionally claystone.

The availability of ferromanganese crust or rock fragments in plateau regions promotes the formation of nodules. In seamount regions, the bare basement rocks serve as nucleating surface for Mn-crusts. One important prerequisite to promote nucleation by hydro-

TABLE II : Chemical composition (values in wt.% of dried substance) of seamount encrustations from the MIDPAC 81 survey area (see Fig. 1).

Stat.	Long. W	Lat. N	Water depth [m]	Mn	Fe	Co	Ni	Cu	SiO_2	Al_2O_3	Ca	P_2O_5	TiO_2	Mn/Fe
31/1	164 48'	09 09'	2100	28.1	13.2	1.13	0.56	0.06	3.37	0.47	2.6	0.33	1.07	2.13
43/4	165 02'	13 08'	3350	24.1	14.8	0.63	0.48	0.15	7.75	1.42	3.6	0.54	1.62	1.63
57	165 29'	13 05'	4380	18.9	14.3	0.62	0.27	0.10	5.92	0.71	2.8	0.50	1.18	1.32
75/2	171 04'	19 22'	1880	23.4	14.6	0.83	0.38	0.09	7.84	1.23	3.6	0.60	1.73	1.60
76/3	170 59'	19 22'	1190	29.0	15.6	1.38	0.51	0.03	3.44	0.47	4.3	0.81	1.17	1.86
31/4	164 48'	09 09'	2100	20.1	9.6	0.53	0.36	0.07	1.92	0.19	8.4	4.70	0.82	2.09
58/3	165 28'	13 12'	1510	22.1	8.0	0.71	0.56	0.05	3.82	0.71	8.5	6.05	1.05	2.76
76/2	170 59'	19 21'	1190	21.6	7.9	0.50	0.62	0.11	2.92	0.52	8.4	4.7	1.50	2.73
111/4	170 38'	20 05'	1240	24.6	7.3	0.75	0.82	0.18	1.42	0.42	8.7	4.55	1.18	3.37

younger crust (stations 31/1, 43/4, 57, 75/2, 76/3)

older crust (stations 31/4, 58/3, 76/2, 111/4)

TABLE III : Minima, maxima, and average metal values (in wt.% or ppm of dried substance) of the areas investigated on the research cruise MIDPAC 2 (1984; see Fig. 1).

Area		Mn %	Fe %	Ni %	Co %	Cu %	Zn ppm	Ti %
A	Min.	8.81	9.6	0.16	0.23	0.048	609	0.55
	Max.	23.50	21.6	0.54	0.79	0.252	923	1.82
	Aver.	17.7±3.7	10.7±3.5	0.33±0.10	0.51±0.13	0.120±0.04	772±77	1.1±0.34
B	Min.	10.5	6.2	0.29	0.33	0.029	646	0.55
	Max.	26.7	15.8	1.35	1.52	0.116	1568	1.26
	Aver.	22.8±5.5	11.6±2.7	0.62±0.23	0.98±0.39	0.057±0.02	942±197	0.88±0.23
C	Min.	16.6	8.6	0.37	0.54	0.019	710	0.83
	Max.	25.8	19.5	0.62	1.97	0.106	1319	1.32
	Aver.	21.8±2.9	15.0±4.0	0.48±0.07	0.89±0.46	0.056±0.03	905±191	1.04±0.20
D	Min.	10.2	4.8	0.42	0.46	0.019	701	0.41
	Max.	29.8	16.3	0.84	1.81	0.097	1134	1.36
	Aver.	25.6±4.5	11.1±2.4	0.67±0.09	1.23±0.38	0.047±0.01	934±88	0.94±0.22
E	Min.	17.7	6.0	0.35	0.30	0.019	572	0.58
	Max.	30.0	18.4	0.94	1.75	0.145	1208	2.99
	Aver.	25.2±2.7	12.9±1.9	0.64±0.10	1.02±0.24	0.073±0.02	864±82	1.13±0.18
F	Min.	19.9	7.2	0.36	0.41	0.029	618	0.62
	Max.	25.2	18.6	0.79	1.07	0.194	1245	1.49
	Aver.	22.5±1.9	14.6±4.1	0.53±0.16	0.83±0.20	0.082±0.05	853±207	1.1±0.34

genesis is to prevent sediment deposition on substrate rocks. In the
areas investigated the AABW (Antarctic Abyssal Bottom Water) or other
deep-water currents are sufficiently erosive to prevent the accumulation of pelagic material.

Chemical composition

The elemental composition of crust samples obtained during the
MIDPAC expeditions presents a wide range of variation in concentrations
(Table II and III), and for Mn and Fe these are noticeable. The Mn-
content varies between 15 and 30%, the Fe content between 7 and 21%.
The Fe concentrations lie in the same range as that reported for other
oceanic seamount deposits, whereas the Mn-values are higher than those
reported by Cronan (1977). The Mn/Fe ratio varies between 1.0 and 3.4
and increases with decreasing water depth. This indicates that hydrogenetic precipitates from shallower water depth have Mn/Fe quotients
higher than 2.5. Overall, however, no significant compositional
differences could be distinguished between crusts and hydrogenetic
nodules from adjacent seamount regions.

In samples of crust from summits and top parts of slopes (water
depth less than 1,500 m) Co-content reaches 2%, Ni values lie between
0.7 and 0.8% , and Cu averages 0.05%. The Mn/Fe ratio ranges from 2.0
to 3.4. The average concentrations of all crust samples investigated
are 0.9% Co, 0.5% Ni, and 0.1% Cu. Co, Ni, and Mn/Fe ratio show inverse relationship to water depth (Table IV), whereas Cu increases
with water depth. Higher Cu concentration was observed in hydrogenetic
polynodules from deep-water stations (4,800 - 4,900 m), where the Cu-
values increase to 0.4 %. Here Ni is 0.5% and Co 0.3%. The significant positive correlation of Mn with Co and Ni (Table V) proves that
the increasing abundance of δMnO_2 in the ferromaganese material controls the enrichment of Co and Ni. The positive correlation of Co
and Ni with Mn can be explained by a specific adsorption of hydrous
CO^{2+} and Ni^{2+} ions by MnO_2.

TABLE IV : Average metal concentrations of ferromanganese
           seamount deposits (MIDPAC 81) from different
           ranges of water depth (wt.% of dried substance)
           (Halbach et al., 1982)

water depth m	Mn	Fe	Co	Ni	Cu	Mn/Fe
4400 - 4000	19.7	16.7	0.67	0.24	0.10	1.17
4000 - 3000	20.5	18.0	0.63	0.35	0.13	1.41
3000 - 2400	20.5	19.5	0.69	0.18	0.09	1.05
2400 - 1900	25.5	16.1	0.88	0.41	0.07	1.58
1900 - 1500	24.7	15.3	0.90	0.42	0.06	1.61
1500 - 1100	28.4	14.3	1.18	0.50	0.03	1.99

The concentration of Pt are surprisingly high and vary between 0.2 and 1.2 ppm (average 0.5 ppm). Older crusts are richer in Pt (average 0.6 ppm) than younger ones. Highest Pt values, between 0.8 and 1.2 ppm, were obtained for older crusts located in water depths of 1,100 to 1,300 m (Halbach et al., 1984a). Compared to the concentration of Pt in seawater of 0,0002 ppb (Hodge et al., 1985), the crust samples show an enrichment factor of about 2×10^6.

Fe shows no significant relationship to Mn, Cu and Co. Ni, however, related inversely to Fe at the 99% level of confidence (Table V). The indifferent behavior of Fe versus Mn suggests that, Fe may have a dual bond, one part of the Fe being bound to the hydrated ferromanganese oxide phase, the other to a silicate phase. Ti-content varies between 0.7 and 2.3% TiO_2 in younger crusts, and between 0.2 and 1.5% TiO_2 in older crusts. SiO_2 and Al_2O_3 show a positive correlation (r = 0.95, n = 17), which indicates that both components are bound to an Al-silicate phase. Ca values above 4% and P_2O_5 above 2% indicate that a discrete apatite phase is present, in an impregnated form in the older ferromanganese crusts (Table II). The apatite content varies between 8 and 15% in the phosphatized samples investigated. Calculated on an apatite-free basis, it should be noted that, the ferromanganese material of older crusts contains less Co, higher Ni and Pt, and somewhat higher Mn/Fe ratios than those of younger encrustations. In general, the older crust material, deposited before phosphorites formed, is poorer in Fe than the younger encrustation layers (Table II).

TABLE V : Interelement relationships in seamount crusts and plateau nodules from the Central Pacific MIDPAC 81 survey area; 46 samples (after Halbach et al., 1982).

	Fe	Ni	Cu	Co
Mn	0.08	0.52[*]	−0.51[*]	0.75[*]
Fe		−0.60[*]	−0.13	−0.03
Ni			−0.07	0.54[*]
Cu				−0.60[*]
Mn/Fe		0.83[*]	−0.26	0.49[*]

[*] Level of confidence 99%

Elemental fluxes and growth rates

The most important factor governing the Co-enrichment is the growth rate; no local source is needed to explain the observed Co-concentrations that reach 2% in samples from summit areas (Halbach et al., 1983). The Co-flux is near constant (2.95 ± 0.6 $\mu g/cm^2$ 1,000 a) over the observed range of growth rates of 0.8 to 2.7 mm/Ma. This directly implies that the Co-concentration in hydrogenetic ferro-manganese deposits might be inversely related to the growth rate. In contrast the proportionality between Mn and the growth rate is trivial, as Mn is the main component of the precipitate. Fluxes of Co, Mn, Ni, Fe and ^{230}Th calculated to reach Mn deposits in the MIDPAC 81 area are summarized in Table VI. For comparison, fluxes into pelagic clays are shown in the second line of this table. The latter were derived assuming average values of 55 ppm (Co), 6,000 ppm (Mn), 200 ppm (Ni) and 6% (Fe) from data of Bischoff et al. (1979), an average sediment accumulation rate of 2 mm/1,000a, and an average in-situ sediment density of 0.5 g/cm^3. This comparison shows that the MIDPAC 81 depo-sits have incorporated only a minor fraction of these particle-reactive trace elements available in the water column.

TABLE VI : Fluxes of Co, Mn, Ni, Fe and ^{230}Th into Mn-deposits at MIDPAC 81 survey area (see Fig. 1).

Deposit types	Element fluxes [$\mu g/cm^2 ky$]				
	Co	Mn	Ni	Fe	^{230}Th[*]
Crusts	2.95 ± 0.6	92 ± 15	1.83 ± 0.7	54 ± 9	10
Pelagic clays	5.5	600	20	6000	99
Flux ratio clays / crusts	1.9	6.5	11	111	10

[*] % of production

Paleoceanographic control of ferromanganese accumulation

When considering the paleoceanographic evolution of the Pacific Ocean it should be pointed out that only specific conditions of the water column promote the formation of ferromanganese seamount crusts. Halbach and Puteanus (1984a) have proposed that the formation of hydrogenetic ferromanganese deposits is initiated by marine carbonates and their dissolution. An increasing rate of dissolution is linked to an enhanced supply of colloidal $Fe(OH)_3$ x nH_2O in the oceanic water column.

According to results of age determinations on the basis of ^{230}Th and ^{10}Be (Halbach et al., 1983; Segl et al., 1984a; and Segl et al., 1984b), younger crusts, whose average thickness is 2.4 to 2.8 cm, are not older than 11 Ma. The period during which younger crusts formed probably corresponds to the phase of decreasing current activity of the AABW and to a decreasing calcite dissolution rate which started around 8 Ma. The oldest layers of the younger crusts are therefore very rich in Fe, indicating a maximum in Fe-flux. The upper layers, i.e., the more recent ones, are poorer in Fe and show increasing concentrations of Mn, Co and Ni. The period of genesis for older crusts corresponds to the phase of increasing current activity of the AABW initiated by a major global cooling that started in the early Middle Miocene; their growth period may have lasted from about 18 to 12 Ma. Therefore, the intercalated apatite formation and discrete layers of the older crust growth period took place during the Mid-Miocene phosphogenic event, which is dated to 13 - 19 Ma (Riggs, this volume).

Resource potential of Mid-Pacific Mountain ferromanganese deposits

Estimates of the resource potential of Mn-crusts from the Mid-Pacific Mountain area is shown on Table VII. It is based on recent estimates by Halbach and Manheim (1984). The estimate is based on the following assumptions: 1) the average thickness of the crusts is taken to be 2.5 cm. This is a conservative value, ignoring local increments caused by curvature, top-and-bottom encrustations and discrete nodules, 2) nearly the entire area of the upper sections of seamount slopes and the tops are encrusted with ferromanganese material.

The numbers in Table VII suggest that the estimated values for the investigated seamount areas equal the values for prime deep-water nodule areas. Using the assumption that 40% of the exploitable slopes and tops are in water less than 2,600 m deep, and are covered by oxide crust as described in Table IV, approximately 5 million tons is calculated to cover each seamount. Many seamounts have surfaces larger than 300 km^2 in this area (Chase et al., 1971). Thus, many seamounts may meet the tonnage requirements indicated to serve as a minimum for commercial sites, namely 500 km^2 of prime nodule tract, corresponding to about 4 million tons of nodule recoverable during one year of effort (Halbach and Fellerer, 1980). Co, Ni and Mn are viewed as recoverable metals. Whether Pt might be of additional commercial interest is unknown, insofar that no pyrometallurgical or hydrometallurgical method has been developed so far to recover Pt from ferromanganese ore.

CONCLUSIONS

Marine ferromanganese deposits form basically by two different accretion processes, by diagenetic growth supplied by pore-water transport of divalent cations and by hydrogenetic growth. The accumulation mechanisms lead to genetically different types of deposits.

TABLE VII : Metal grades and values for seamount crusts and abyssal nodules *

Component	Price ($/kg)	MPM crusts				Line Is. crusts				MIDPAC				Nodule (NB)			
		% dry	% wet	$/t wet	$/m²	% dry	% wet	$/t wet	$/m²	% dry	% wet	$/t wet	$/m²	% dry	% wet	$/t wet	$/m²
Co	27.56	0.83	0.54	148.82	2.98	1.00	0.65	179.14	3.58	0.79	0.51	140.56	2.81	0.24	0.16	44.10	0.44
Ni	4.98	0.49	0.32	15.94	0.32	0.55	0.36	17.93	0.36	0.49	0.32	15.94	0.32	1.21	0.79	39.34	0.39
Cu	1.77	0.07	0.05	0.89	0.02	0.07	0.05	0.89	0.02	0.07	0.04	0.71	0.01	1.00	0.65	11.51	0.12
Mo	10.58	0.06	0.04	4.23	0.08	0.06	0.04	4.23	0.08	0.06	0.04	4.23	0.08	0.04	0.03	3.17	0.03
**	1.52	25.0	16.25	247.00	4.94	25.0	16.25	247.00	4.94	24.6	15.99	243.05	4.86	25.2	16.38	248.98	2.49
***	0.58	25.0	16.25	94.25	1.89	25.0	16.25	94.25	1.89	24.6	15.99	92.74	1.85	25.2	16.38	95.00	0.95
Total																	
**				416.88	8.34			449.19	8.98			404.49	8.08			347.10	3.47
***				264.13	5.29			296.44	5.93			254.18	5.07			193.12	1.93

* MPM refers to crusts from the Mid-Pacific Mountains and Line Islands Ridge, at water depths of 2,600 m or less. MIDPAC refers to samples from all depths encompassed in the MIDPAC 81 cruise (n = 61;Halbach et al., 1982). NB refers to abyssal manganese nodules from the nodule belt area in the NE Pacific. Approximate ore abundance is 20 km m⁻² for the crust deposits and 10 kg/m² for the abyssal nodules. Molybdenum metal values for the concentration are taken from previous studies (Craig et al., 1982). Metal prices are for May 83 (Source: Erzmetall 36 (1983) p. 397). Decimals are for computation only. Recent trace element determinations of ferromanganese crust samples from MIDPAC 81 have shown that platinum may reach concentrations of up to 1.0 g/t (average content 0.5 g/t; range 0.3-1.0 g/t). This might be of additional commercial interest since presently mined platinum-bearing ores have contents of 1-2 g/t of platinum.

** Manganese (99.95%).

*** Ferromanganese (78% Mn) recalculated to 100 % Mn basis.

Hydrogenetic nodules and crusts accumulate by direct supply of colloidal, hydrous, metal-oxide particles from near-bottom seawater.

The trace-element concentration of seawater and its vertical and lateral variation influence directly the chemical composition of these deposits.

Based on new data and results the hydrogenetic class could be subdivided further into: (a) shallow pelagic type and (b) abyssal pelagic type. Growth rates of open-ocean ferromanganese deposits range from 1 to about 5 mm/Ma. Using [10]Be-dating method a number of event markers in ferromanganese crust layers could be identified; these markers coincide with well-established paleoceanographic events.

MIDPAC expeditions to the Central Pacific have shown that Co-rich and Pt-bearing ferromanganese crusts are common on seamount slopes and summit plateaus between 3000 and 1100 m water depth, covering exposed bare substrate rocks. One prerequisite to promote nucleation by hydrogenesis is to prevent sedimentation. The AABW or other deep-water currents are sufficiently erosive to cause non-sedimentation.

Hydrogenetic crusts from shallower water depth have higher Mn/Fe quotients and higher Co, Ni, and Pt contents, in contrast Cu increases with water depth.

Two growth generations of ferromanganese crusts have been identified. The younger crust is not older than 11 Ma, the growth period of the older one may have lasted from 18 to 12 Ma. Both crust generations can be related to specific conditions of the water column promoting the hydrogenetic precipitations.

Considerations of the resource potential of crust show that the metal composition and the quantity might be of commercial interest.

REFERENCES

Bischoff, J.L., Heath, G.R. and Leinen, M., 1979, Geochemistry of deep-sea sediments from the Pacific, Manganese Nodule Province: DOMES sites A, B and C; In:Bischoff, J.L. and Piper D.Z. (eds.), Marine geology and oceanography of the Pacific Manganese Nodule Province, Plenum Press, N.Y., pp. 397-436.

Bonatti, E., Kraemer, T. and Rydell, H., 1972, Classification and genesis of submarine iron-manganese deposits; In: Horn, D.R. (ed.), Ferromanganese deposits on the ocean floor, Nat.Sci.Found., Washington, D.C., pp. 149-165.

Boudreau, B.P. and Scott, M.R., 1978, A model for the diffusion controlled growth of deep-sea manganese nodules; Am.Jour.Sci., V. 278, pp. 903-929.

Burns, R.G. and Brown, B.A., 1972, Nucleation and mineralogical controls on the composition of manganese nodules; In:Horn, D.R. (ed.), Ferromanganese deposits on the ocean floor, Nat.Sci.Found., Washington, D.C., pp. 59-61.

Burns, R.G. and Burns, V.M., 1979, Manganese oxides; Burns, R.G. (ed.),
    In: Marine minerals, Mineral.Soc.Am., Short Course Notes 6,1.

Chase, T.E., Menard, H.W. and Mammerickx, J., 1971, Topographic map of
    the North Pacific; Scripps Inst. Ocean., La Jolla, California,
    Tech. Rept. TR-13.

Clague, D.A., 1981, Linear island and seamount chain, aseismic ridges
    and intraplate volcanism: results from the Deep-Sea-Drilling-Pro-
    ject; Soc.Econ.Paleont. and Min., Spec. Publ., 45, 127 p.

Cronan, D.S., 1972, Regional geochemistry of ferromanganese nodules in
    the world Ocean; In: Horn, D.R. (ed.), Ferromanganese deposits on
    the ocean floor, Nat.Sci.Found., Washington, D.C., pp. 19-29.

Cronan, D.S., 1977, Deep Sea Nodules, distribution and geochemistry;
    In:Glasby, G.P. (ed.), Marine manganese deposits, Elsevier,
    Amsterdam, pp. 11-44.

Halbach, P. and Fellerer, R., 1980, The metallic minerals of the
    Pacific seafloor; Geol.Jour., v. 4, pp. 407-420.

Halbach, P., Giovanoli, R. and von Borstel, D., 1982, Geochemical pro-
    cesses controlling the relationship between Co, Mn, and Fe in
    early diagenetic deep-sea nodules; Earth.Planet.Sci.Lett., v. 60,
    pp. 226-236.

Halbach, P. and Manheim, F.T., 1984, Potential cobalt and other metals
    in ferromanganese crusts on seamounts of the Central Pacific Basin;
    Marine Mining, v. 4, no. 4, pp. 319-336.

Halbach, P. and Puteanus, D., 1984a, Platinum concentrations in ferro-
    manganese seamount crusts from the Central Pacific; Naturwissen-
    schaften, v. 71, pp- 577-579.

Halbach, P. and Puteanus, D., 1984b, The influence of the carbonate
    dissolution rate on the growth and composition of Co-rich ferro-
    manganese crusts from Central Pacific seamount areas; Earth.Planet.
    Sci.Lett., v. 68, pp. 73-87.

Halbach, P., Segl, M., Puteanus, D. and Mangini, A., 1983, Co-fluxes
    and growth rates in ferromanganese deposits from Central Pacific
    seamount areas; Nature, v. 304, pp. 716-719.

Heath, G.R., 1979, Burial rates, growth rates and size distribution
    of deep sea manganese nodules; Science, V. 205, pp. 903-904.

Hodge, V.P., Stallard, M., Koide, M. and Goldberg, E.D., 1985, Platinum
    and the platinum anomaly in the marine environment; Earth.Planet.
    Sci.Lett., v. 72, pp. 158-162.

Jackson, E.D. and Schlanger, S.O., 1973, Regional synthesis, Line Island Chain, Tuamotu Island Chain and Manihiki Plateau, Central Pacific Ocean; Initial Reports of the Deep Sea Drilling Project, Nat.Sci. Found., Washington, D.C., v. 33, pp. 915-925.

Klinkhammer, G.P. and Bender, M.L., 1980, The distribution of manganese in the Pacific Ocean; Earth.Planet.Sci.Lett., v. 46, pp. 361-384.

Mangini, A. and Kuehnel, U., in press, Depositional history in the Clarion-Clipperton Zone during the last 250,000 years; In: v. Stackelberg, U. and Beiersdorf, H.H. (eds.), Relationship of Mn nodules and sediments in the Equatorial North Pacific.

Mangini, A., Segl, M., Kudrass, H., Wiedecke, M., Bonani, G., Hofmann, H.J., Morenzoni, E., Nessi, M., Suter, M. and Wölfli, W., in press, Diffusion and supply rates of [10]Be and [230]Th radioisotopes in two manganese encrustations from the South China Sea, Geochim. Cosmochim.Acta.

Moore, T.C., Pisias, N.G. and Keigwin Jr., L.D., 1981, Ocean basin and depth variability of oxygen isotopes in Cenozoic benthic foraminifera, Mar. Micropaleont., v. 6 (5/6), pp. 465-481.

Piper, D.Z., Leong, K. and Cannon, W.F., 1979, Manganese nodule and surface sediment compositions: DOMES sites A, B and C; In: Bischoff, J.L. and Piper, D.Z. (eds.), Marine Geology and Oceanography of the Pacific Manganese Nodule Province, Plenum Press, New York, pp. 437-474.

Reyss, J.L., Marchig, V. and Ku, T.L., 1982, Rapid growth of a deep-sea manganese nodule, Nature, v. 295, pp. 401-403.

Riggs, S.R., this volume, Model of tertiary phosphorites on the world's continental margins, pp. 96-116.

Segl, M., Mangini, A., Bonani, G., Hofmann, H.J., Nessi, M., Suter, M., Wölfli, W., Friedrich, G., Plüger, W.L. Wiechowski, A., and Beer, J., 1984a, [10]Be dating of a Mn-crust from central North Pacific and implications for ocean paleocirculation, Nature, v. 309, pp. 540-543.

Segl, M., Mangini, A., Bonani, G., Hofmann, H.J., Morenzoni, E., Nessi, M., Suter, M. and Wölfli, W., 1984b, [10]Be dating of the inner structure of Mn encrustations applying the Zürich Tandem Accelerator; Nucl.Instr.Meth., v. B5, pp. 359-364.

v. Stackelberg, U., 1979, Sedimentation, hiatuses and development of manganese nodules: VALDIVIA site VA-13/2, Northern Central Pacific, In: Bischoff, J.L. and Piper, D.Z., (eds.), Marine geology and oceanography of the Pacific Manganese Nodule Province, Plenum Press, New York, pp. 559-586.

GEOCHEMICAL METHODS IN MANGANESE NODULE EXPLORATION

H. Kunzendorf
Risø National Laboratory
P.O. Box 49
DK-4000 Roskilde
Denmark

ABSTRACT. The general strategy for a geochemical survey of manganese nodule fields includes reconnaissance free-fall grab sampling from a relatively large research vessel and analysis of nodule samples by shipboard X-ray fluorescence methods. Nodules are classified geochemically, based on a diagram of Mn/Fe versus Ni+Cu, as samples with good economic potential (Mn/Fe > 1.5) or low potential (Mn/Fe < 1.5). Detailed sampling has to be continued in areas where manganese nodules are considered to be economic to exploit. A geochemical exploration campaign can terminate at either the reconnaissance or the detailed scale.

INTRODUCTION

Since their discovery by the "Challenger" expedition (1873-1876), ferromanganese nodule occurrences have been frequently reported. Prominent data on their characteristics can be found in the comprehensive work of Mero (1965), Horn et al. (1973), Glasby (1977), Cronan (1980) and McKelvey et al. (1983). The recent review given by McKelvey et al. (1983) is based on about 2400 chemical analyses of manganese nodules gathered in the Scripps Institute of Oceanography's Sediment Data Bank (Frazer and Fisk, 1980). Chemical composition of nodules varies considerably, and data on average values are difficult to interpret, mainly because of irregular and widely-spaced sampling stations. Even for the Pacific Ocean, sampling density is not better than 1 sample per 1000 km^2.

Although Mn-nodule occurrences are well-known by now, any systematic exploration of these occurrences is, at best, still in a progressive state, because only a few private firms have sufficiently detailed information available from exploration cruises carried out in the 1970's. However, as the surficial extent of the mineralized areas is enormous, nodule-mining consortia at this time probably have difficulties to arrive at sound economic evaluations of necessary large mining sites. Also, nodule variability prevents choosing "the" profitable area especially if metal markets are taken into

221

P. G. Teleki et al. (eds.), Marine Minerals, 221–234.

consideration. Thus, while the location of mineral deposits are known, their metal potential and areal extensions are only known adequately in a few cases.

One of the important methods used to explore manganese nodule sites involves exploration geochemistry, which can lead to estimating the grade of nodules. In this way, areas that have, at present, low mining potential on account of low metal content, can be excluded. In the following, geochemical techniques for manganese nodule characterization are described, including sampling large quantities of nodules and the necessary analytical steps. Emphases are given to determining the metals Ni, Cu, Co, Mn and Fe at their present value, and to other chemical elements of importance in these evaluations, and to genetic considerations of mineralization. Other chemical elements in nodules are, however, discussed only briefly. Furthermore, a possible shipboard geochemical classification scheme for the nodules is presented.

SAMPLING METHODS

A proposed exploration strategy for manganese nodules has recently been described (United Nations, 1984). The strategy consists of a sequence of activities, including a regional resource appraisal, target area refinement, identification of one or more economic deposits and delineation of these deposits. However, in practice exploration of manganese nodule fields in deep water is reduced to only two steps, the regional appraisal and the refinement of targets. This, in part, is the result of the high cost of operating large exploration vessels, which require that exploration strategies be economized. In short, instead of a long period of exploration, during which the four proposed steps are all carried out, greater emphasis is given to evaluation and analytical work directly onboard the ship. A move in this direction is now possible with advances made in computer technology.

Manganese nodule exploration in the deep ocean requires a relatively large research vessel. The vessel is usually equipped with instrumentation for mapping the ocean floor (to a depth of 10 km). A possible, and perhaps practical, exploration strategy for manganese nodules involving bathymetric mapping, sampling, sample analysis and evaluation is presented in Table I.

Presently, exploration is focussed on known fields whose nodules contain more than 1.5 to 2 weight percent (dry weight) combined Ni and Cu. In the most prospective area of the Pacific Ocean, between the Clarion and Clipperton Fracture Zones (CCFZ), the Ni+Cu content is often 2.5%, with nodule abundance (areal coverage) of approximately 5 to 10 kg/m^2.

Exploration of the preselected area is usually initiated by a multibeam echosounder (see Spiess, this volume). At present, few multibeam-based bathymetric maps of nodule areas are available, mainly because the multibeam echosounding technique for use in civilian oceanographic research became available only in the mid-1970's, and also because installation of the equipment is still considerably expensive. Nonetheless, the detailed bathymetric mapping of every exploration area is regarded vital.

Table I. Proposed exploration strategy for the evaluation of manganese nodule fields.

Step	Work	Remarks
0	Basic research	Select known information on manganese nodule project area; budget for exploration survey(s)
1	Selection of ship and logistics	Compare available research vessels with basic equipment for mapping and navigation; laboratory facilities; multibeam echosounder system; satellite navigation; sampling devices
2	Bathymetric survey	Multibeam echosounding survey of preselected project area; some reconnaissance sampling
3	Sampling survey	Grid sampling with mainly free-fall grabs with mounted still camera; use of dredges and box corers; future sampling by ROV
4	Classification and analysis	Evaluation of still-camera photos; nodule classification according to mainly size and morphology; shipboard analytical program by XRF; metals determined: Mn, Fe, Co, Ni, Cu and Zn; TV and/or photographic inspection while analyzing
5	Geochemical characterisation	Evaluation of analytical survey; nodule chemistry in the Mn/Fe vs. Ni+Cu diagram; if exploration time is left, start further sampling
6	Detailed analysis and data evaluation	In the home laboratory; by a number of analytical and computer techniques; if necessary select new target area(s) and start at step (0); if detailed survey is needed start at (3) and continue
7	Project evaluation	Determine grade and amount of nodules in selected project area; probable economic evaluations

After the bathymetry has been determined, preliminary seafloor maps can be constructed onboard ship. Some reconnaissance sampling is often initiated, as is the execution of certain special surveys (sidescan sonar, visual observation techniques). The seafloor large-scale maps, with contour intervals of 20 m, are important to prepare, as they provide a basis for detailed sampling and geochemical surveys of smaller target areas.

At the reconnaissance stage of exploration, sampling is commonly carried out on a preselected grid, whereby distances of several tens to hundreds of kilometers between sampling stations have to be tolerated. In the follow-up stage, less than 1 km between stations can be achieved. Exploration surveys carried out in the 1970's indicated the simple free-fall grab to be the most reliable and effective sampling device for manganese nodules. The grabs cover only about 0.1 m^2 of the seafloor but by having several closely spaced grab stations (about 200 m between grabs) along pre-defined lines or grids, a reasonable first approximation of nodule occurrences of a nodule field can be obtained. With a grab's drifting speed of about 13 minutes per kilometer, retrieving a sample in water depths of about 5 km can be accomplished within 2 to 3 hours. This means, that a free-fall grab survey involving launching and recovering the grabs at 5 to 7 grab stations, can be carried out within one working shift (6 hours). Grabs are usually equipped with a still camera, radio beacons for day-time use and flash beacon for night operation. The free-fall grab is particularly important to local small-grid sampling. In the future, it would be desirable to extend the photographed area of sampling by other visual inspection methods.

Other sampling devices for manganese nodules include chain-bag and box dredges, and box corers. As is the case with all cable-mounted sampling devices, only one sample can be recovered within a 2 to 3 hour period, clearly a shortcoming. During the operation of such devices, the ship is generally not available to carry out other activities, such as surveys. Remotely operated vehicles (ROV), with sampling capabilities, will probably replace some of the shipboard-operated grab samplers in the near future.

Conventionally, recovery with a grab sampler is limited to 1 to 10 kg (assumption: 50% coverage with nodule diameters as much as 4 cm) but relatively large pieces of crust have also been recovered when grabs did not close fully. Nodule abundances are often between 5 and 10 kg/m^2.

Manganese nodules retrieved are classified mainly according to size and morphology. Each grab load is photographed onboard ship for visual documentation purposes. A practical size classification is 20 mm, 20–40 mm, 40–60 mm, 60–80 mm, and 80 mm. Aside from these, there is no generally accepted classification scheme in use.

ANALYTICAL TECHNIQUES

The metals usually contained in ferromanganese nodules are manganese, iron, cobalt, nickel, copper and zinc. For their rapid and precise determination a reliable and relatively simple analytical method is required, which should also be employable onboard ship during an exploration survey.

The most widely used analytical technique in terrestrial exploration is atomic absorption spectrophotometry (AAS), with emission spectroscopy (ES) and colorimetry as secondary analytical techniques (Fletcher, 1982). However, measurements by these techniques are difficult to carry out onboard an exploration vessel, partly because samples dissolved are difficult to weigh accurately in a ship's laboratory. For the analysis of manganese nodules, a number of recommended analytical methods are compiled and shown in Table II.

It is not within the scope of this paper to describe these techniques in detail; instead, the reader is referred to the available literature on these methods.

Table II. Analytical techniques used preferentially in manganese nodule exploration. Elements in parenthesis are not determined routinely by the method indicated.

Method	Elements determined	Remarks
X-ray fluorescence (XRF)	Mn, Fe, Co, Ni, Cu, (Zn) major and other elements	Sophisticated instruments installed onboard ship
Energy-dispersive XRF (EDXRF)	Mn, Fe, Co, Ni, Cu other elements	Similar to XRF; portable units exist
Atomic absorption spectrophotometry (AAS)	Mn, Fe, Co, Ni, Cu, Zn many other elements	Difficult to use onboard ship
Instrumental neutron activation (INAA)	About 30 elements, but not Ni and Cu	Time consuming and expensive; requires nuclear reactor
Others: Emission spectroscopy (ES, ICP-ES)	Many elements	Complicated instrumentation
Delayed-neutron counting (DNC)	U, Th	Nuclear research reactor required

Manganese nodule samples are usually dried before analysis, so that elemental contents can be reported on a dry-weight basis. Drying at 105° C for at least 5 hours is the standard procedure, followed by grinding to at least -100 mesh size. Standardized sample preparation techniques are necessary on account of the hydroscopic behavior of the powder of manganese nodules.

By far the most important analytical technique in manganese nodule analysis is X-ray fluorescence (XRF), a purely instrumental technique that does not require sample dissolution and weighing operations. Only sample preparation in the form of grinding and/or pelletizing of samples is needed. XRF instrumentation, coupled to a microcomputer is capable of accomodating a large number of samples with usually good analytical precision. Detection limits for important elements in manganese nodules are at or lower than the 10-ppm level with precision of measurements generally less than $\pm$ 3%.

XRF is based on wave-length dispersion of X-rays. A modification of XRF is the energy-dispersive type (EDXRF) which works on the same physical principles but employs a nuclear X-ray detector system instead of the single-crystal system (Kunzendorf, 1973; 1979). By replacing the excitation system (X-ray tube) with a small radioisotope, portable EDXRF units can be constructed. These have been used onboard ship in manganese nodule exploration (e.g. Lüschow, 1973).

In the home laboratory, a large number of analytical techniques may be employed. XRF is often replaced by atomic absorption spectrophotometry (AAS, a more inexpensive method) and by several other instrumental techniques. For base metals AAS and XRF are by far the most frequently used techniques. A multi-element technique often applied in geochemical studies is instrumental neutron-activation analysis (INAA). This method, however, is time consuming and expensive compared to the other techniques. The elements that can be differentiated by INAA include rare and precious elements, Fe, Co, Zn, As, Sb, Ba, etc. Because INAA requires access to a nuclear research reactor, multi-element analysis is increasingly performed by emission spectroscopy using an inductively coupled plasma source (ICP-ES). Routine analysis for uranium is carried out by delayed-neutron counting (Kunzendorf et al., 1980).

For standardization purposes, the U.S. Geological Survey's manganese-nodule standards, Nod-A-1 and Nod-P-1, should be used (Flanagan and Gottfried, 1980; Kunzendorf and Gwozdz, 1984).

There have been several attempts in the past to carry out analyses of nodules using towed geochemical sensors that measure properties in-situ, thus, require no retrieval of samples. These techniques have recently been reviewed by Noakes and Harding (1982), but none have been used routinely in deep oceanic waters yet. As is the case with all towed heavy instrumentation, there is always the risk of losing instrumentation during towing operations. A modern system prototype employing INAA techniques was lost during test operations (Borcherding et al., 1977). At present, there is little need to use such techniques in deep waters. Future improvements in ROV techniques, however, may lead to the development of more successful in-situ equipment for the analyses of manganese nodules.

GEOCHEMICAL CHARACTERIZATION

As mentioned earlier, the most important metals in manganese nodules are Mn, Fe, Co, Ni, Cu, and Zn. The presence and quantity of metals can be detected directly onboard ship; therefore, a rough geochemical characterization of nodules and nodule fields can already be established during the exploration phase of a program.

As regards the origin of nodules, Bonatti et al. (1972) have proposed a characterization plot of Mn, Fe, and Ni+Cu+Co, leading to well-defined nodule fields of hydrothermal, hydrogeneous or diagenetic origin. Most of the known deepsea manganese nodule occurrences fall within the hydrogeneous field.

From an exploration point of view, it is sufficient to determine Mn, Fe, Ni and Cu and classify the nodules in a Mn/Fe vs. Ni+Cu diagram (Fig. 1). The Ni+Cu sum could also be extended to include, e.g., Co and/or Zn. Plotting the ratio of Mn and Fe usually reduces systematic errors and enables intercomparing the performance of different kinds of analytical equipment.

Globally, most nodule data fit into the field outlined in Fig. 1, which represents analytical results of more than 2000 samples obtained during several surveys in the Pacific (Friedrich et al., 1974; Kunzendorf et al., 1983). Fig. 1 also includes the data reported by Dymond et al. (1984) from the MANOP expedition, the averaged data reported by McKelvey et al. (1983), and the averaged data of Haynes et al. (1983) for the most promising area in the Clarion-Clipperton

Fracture Zone (CCFZ) as well as other areas.

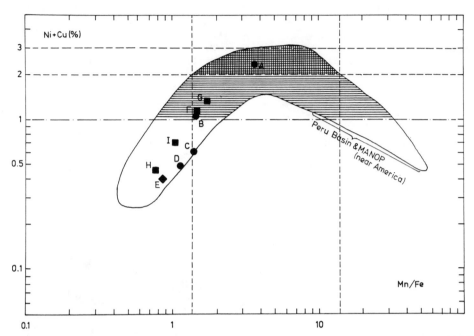

Figure 1. Geochemical classification plot based on more than 2000 manganese nodule analyses from mainly samples of the Clarion-Clipperton Fracture Zone area. Averaged data from other areas are also plotted. A: Clarion-Clipperton Fracture Zone (CCFZ), B: other abyssal plains, C: Mid-Pacific seamounts, D: other seamounts, E: fossil nodules from Timor. A to D from Haynes et al. (1983), E from Margolis et al. (1978). Indicated are also averaged data from: F: World average, G: Pacific stations, H: Atlantic stations, I: Indian stations, all from McKelvey et al. (1983).

     The scatter of data points is caused by sampling errors, and/or differing analytical methods and nodule standards are reflected. In the field plotted in Fig. 1, economic-grade manganese nodules are considered, at present, to be above the 2% Ni+Cu line. Such nodules also have relatively high Mn/Fe ratios (> 1.5). Although this plot is purely empirical and will likely be modified in the future, it displays the mineralogical variations in nodules well, among which are the common Fe-rich and Mn-rich oxyhydroxides, vernadite (δ-MnO$_2$), todorokite (10Å manganite), birnessite (7Å manganite), and amorphous Fe-oxides. The lower left part of the distribution shown in Fig.1 is related to nodule growth from mainly seawater terminating in Fe-rich manganese mineral phases (vernadite). Nodules high in Ni+Cu are usually found in areas beneath surface waters high in biogenic productivity, where Ni and Cu (and Mn) are supplied additionally by diagenetic

processes (todorokite, middle part of distribution in Fig. 1). Nodules
with high Mn and faster growth rates near continents (e.g. in the Peru
Basin) show higher occurrences of birnessite with generally lower Ni+Cu
(right-hand of the distribution in Fig. 1).

Most importantly, Fig. 1 shows that very few manganese nodules
exist that have Ni+Cu in excess of 3%, which is important to consider
in their economical evaluations as a resource. Although Ni and Cu is in
plentiful supply in ionic phases in, e.g., the CCFZ area, it appears
that todorokite cannot accomodate more than about 4% combined Ni+Cu.
This is the level of Ni+Cu observed in micro-manganese nodules too, and
in loose material around nodules resting in the sediment (Friedrich et
al., in press).

Economic grade nodules have high Mn/Fe ratios. However, at Mn/Fe
ratios of greater than about 15, manganese nodules very seldom achieve
Ni+Cu grades greater than 2%.

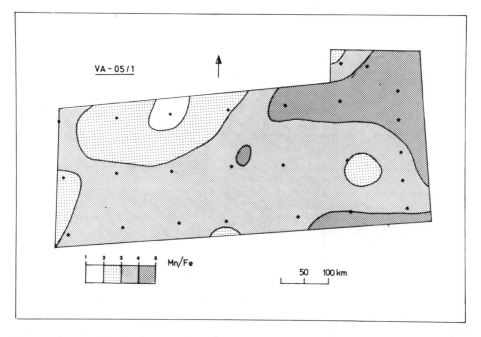

Figure 2. Geochemical map of Mn/Fe ratios in nodules from the CCFZ area
in the Pacific. Sampling density is approximately 1 sample per 5000
km^2. Note that there are no location coordinates shown because the data
are kept confidential.

As an example, chemical results of a regional reconnaissance
survey in the CCFZ area involving approximately one sampling location
per 5000 km^2 (with several sampling stations per location) are
presented in Fig. 2. The map is based on data from about 30 sampling
locations and represents about 200 nodules that have been obtained and
analyzed onboard ship. A double-linear, quadratic and distance-

weighting   computer-based procedure was applied to generate   the
isolines. The resulting illustration shows that the northeastern area
has Mn/Fe ratios that are relatively high, and that is indicative also
of high Ni+Cu content. However, because the areal density of sampling
was very low, omitting in the process many nodule fields between
stations, this large area could not be evaluated properly.

Another manganese nodule sampling survey, with approximately one
sampling station per 2 km^2 (Fig. 3), shows that Mn/Fe ratios (and high
Ni+Cu) clearly are connected with featureless seafloor areas, inbetween
abyssal hills. Sediment thicknesses are much greater in these areas
(von Stackelberg, 1979) than on abyssal hill slopes; therefore, nodule
chemistry is thought to be influenced by seafloor topography.   The
results of this survey also suggest that there might be some local
effect, in that the Mn/Fe ratio in nodules, southeast of the two
abyssal hills, is higher than in the northwestern area. Oxygenated
bottom water flow may be responsible for this,   although more
investigations are needed to prove this hypothesis.

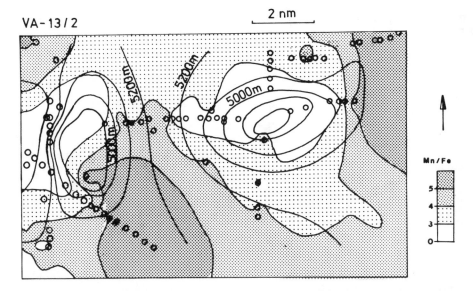

Figure 3. Geochemical map of Mn/Fe ratio in nodules from a subarea in
the CCFZ area (1 sample per 2 km^2). Concentric solid lines are
isobaths, circles represent sampling stations. Note that the
coordinates are left out for reasons that data are still confidential.

The possible geochemical classification of nodules is mainly based
on their Mn and Fe contents and, as mentioned earlier, this represents
the two principal ferromanganese mineral phases. A number of other
elements are also connected with these phases, most of which have been
discussed in detail by Haynes et al. (1983). Selected elemental groups
occurring in manganese nodules are shown in Table III. The amount of
minor and trace elements vary widely in the manganese nodules, although

it should be noted that these have not been studied adequately and the available data base may be yet strongly influenced by samples from the CCFZ area. More detailed information on these elements in nodules can be found in, e.g., Calvert and Price, (1977) and Dymond et al. (1984).

TABLE III. Selected elements occurring in manganese nodules ascribed to the major ferromanganese mineral phases. Those in parenthesis cannot be associated with certainty with the phases.

Element type	Mn-rich ferromang. mineral phase	Fe-rich ferromang. mineral phase	Other mineral phases
Metals of immidiate econ. interest	Ni, Cu, Zn, (Co) (V, Mo)	Co, (V)	
Major elements	(Mg, Ca, Na)	Ti, Ca	Al, Ca, Mg, K, Si, Na, (Ti)
Minor elements		Zr, Sr	Sr, Ba
Environment. related elements	Cd, (Ba)	As, Pb, (Hg)	(Se, Cr)
Rare and precious metals		rare earths	(rare earth, Au, Pt-group elements)
Radioactive elements		U, Th	(Th)

The grouping in the Table III is somewhat empirical and many of the precious, rare and other trace elements have not been sufficiently investigated to be relatable to the major oxyhydroxide phase. Although recent technological trends point toward the extractability of Co, Mo, V and possibly precious metals too from the nodules, it will be prevailing economics at the time of future exploitation that will determine whether these metals will be viewed as valuable or as a hindrance to the refining processes. In view of the requirements for mining of large tonnages of manganese nodules, future geochemical studies should also take into account those elements that are detrimental to the environment in concentrated form (e.g. Ba, As). In some areas environmental studies have been made already such as for the DOMES project (Bischoff and Piper, 1979) and MANOP.

CONCLUSIONS

Geochemical methods, applied to terrestrial mineral exploration, can be successfully applied to marine manganese-nodule exploration. Exploration geochemistry is one of the methods that is well-suited to shipboard adaption.

A manganese nodule survey should be composed of:

1) Sampling by mainly free-fall grab samplers; in the future possibly by remotely operated vehicles (ROV).
2) Analysis of large numbers of samples by X-ray fluorescence onboard ship for Mn, Fe, Co, Ni, Cu and Zn.
3) Geochemical classification of nodules according to the Mn/Fe vs. Ni+Cu diagram, followed by detailed sampling in areas of favorable grade (> 1.5% Ni+Cu).
4) Final evaluation of grade and areal extent of nodule fields in the survey area.

At present, there is still only a limited amount of information on other metals of economic potential in nodules. Such metals as Mo, V, As (and others) should be investigated further to permit to evaluate whether these metals could be reasonably beneficiated from the nodules extracted.

ACKNOWLEDGEMENTS

Most of the samples considered in this text were collected during cruises with research vessels of the Federal Republic of Germany. The cooperation with colleagues from the Rheinisch-Westfälische Technische Hochschule Aachen, F.R. Germany, during the surveys and regarding some of the analytical work is greatly appreciated.

REFERENCES

Bischoff, J.L. and Piper, D.Z. (eds.), 1979, Marine geology and oceanography of the Pacific Manganese Nodule Province; Plenum Press, New York, 842 pp.

Bonatti, E., Kramer, T. and Rydell, H.S., 1972, Classification and genesis of submarine iron-manganese deposits, In: D.R. Horn (ed.) Ferromanganese deposits on the ocean floor; National Science Found.-IDOE Washington, D.C., pp. 149-166.

Borcherding, K., Döbele, R., Eberle, H., Erbacher, I., Hauschild, J., Hübener, J., Lange, J., Müller, G., Rapp, W., Rathjen, E., Rusch, K.D., Schäf, A. and Tamm, U., 1977, Manganknollen-Analysensystem MANKA; KFK 2537, 194 pp.

Calvert, S.E. and Price, N.B., 1977, Geochemical variation in ferromanganese nodules and associated sediments from the Pacific Ocean; Mar. Chem., v. 5, pp. 43-74.

Cronan, D.S., 1980, Underwater minerals; Academic Press, London, 362 p.

Dymond, J., Lyle, M., Finney, B., Piper, D.Z., Murphy, K., Connard, R. and Pisiasis, N., 1984, Ferromanganese nodules from MANOP sites H, S, and R: Control of mineralogical and chemical composition by multiple accretionary processes; Geochim. Cosmochim. Acta, v. 48, pp. 931-949.

Flanagan, F.J. and Gottfried, D., 1980, USGS rock standards, I - II: manganese nodule reference samples USGS-Nod-A-1 and USGS-Nod-P-1; U.S. Geol. Surv. Prof. Paper 1155, 39 p.

Fletcher, W.K., 1981, Analytical methods in geochemical prospecting; Elsevier, Amsterdam, 255 p.

Frazer, J.Z. and Fisk, M.B., 1980, Availability of copper, nickel, cobalt and manganese from ocean ferromanganese nodules (III); Bur. Mines OFR 140(2)-80, 112 pp.

Friedrich, G.H.W., Kunzendorf, H. and Plüger, W.L., 1974 Ship-borne geochemical investigations of deep-sea manganese nodule deposits in the Pacific using a radioisotope energy-dispersive X-ray system; Jour. Geochem. Explor., v. 3, pp. 303-317.

Friedrich, G., Plüger, W.L. and Kunzendorf, H., in press, Chemical composition of manganese nodules, In: Halbach, P., Friedrich, G. and von Stackelberg, U. (eds.) The manganese nodule belt of the Pacific Ocean: scientific, technical and economic aspects; Enke, Stuttgart.

Glasby, G.P. (ed.), 1977, Marine Manganese Deposits. Elsevier, Amsterdam, 312 p.

Haynes, B.W., Law, S.L. and Barron, D.C., 1983, Mineralogical and elemental description of Pacific manganese nodules; U.S. Dept. of the Int., Bur. Mines, Inform. Circ. 8906, 59 p.

Horn, D.R., Horn, B.M. and Delach, M.N., 1973, Factors which control the distribution of ferromanganese nodules and proposed research vessels track North Pacific, Phase II Ferromanganese Program; Tech. Rept. no. 8 Nat. Sci. Found.- IDOE, Washington, D.C., 40 p.

Kunzendorf, H., 1973, Non-destructive determination of metals in rocks by radioisotope X-ray fluorescence instrumentation, In: Jones, M.E. (ed.) Geochemical Exploration 1972; Inst. Mining and Metallurgy, London, pp. 401-414.

Kunzendorf, H., 1979, Practical experiences with automated radioisotope energy-dispersive X-ray fluorescence analysis of exploration geochemistry samples. Risø Nat. Lab., Roskilde, Denmark, Risø-Rept. no. 407, 24 p.

Kunzendorf, H., Løvborg, L. and Christiansen, E.M., 1980, Automated uranium analysis by delayed-neutron counting; Risø Nat. Lab., Roskilde, Denmark, Risø-Rept. no. 429, 38 p.

Kunzendorf H., Plüger, W.L. and Friedrich, G.H., 1983, Uranium in Pacific deep-sea sediments and manganese nodules; Jour. Geochem. Explor., v. 19, pp. 147-162.

Kunzendorf, H. and Gwozdz, R., 1984, U, As and W in USGS manganese nodule standards Nod-A-1 and Nod-P-1; Geostandards Newsletter, v. 8, no. 2, pp. 169-170.

Lüschow, H.-M., 1973, Non-dispersive X-ray spectrometric analysis of manganese nodules: procedures and results, In: Morgenstein, M. (ed.), The origin and distribution of manganese nodules in the Pacific and prospects for exploration; Proc. Symp., Hawaii Inst. Geophysics, Honolulu, pp. 103-108.

Margolis, S.V., Ku, T.L., Glasby, G.P., Fein, C.D. and Audley-Charles, M.G., 1978, Fossil manganese nodules from Timor: Geochemical and radiochemical evidence for deep-sea origin; Chem. Geol., v. 21, pp. 185-198.

McKelvey, V.E., Wright, N.A. and Bowen, R.W., 1983, Analysis of the world distribution of metal-rich subsea manganese nodules; U.S. Geol. Surv. Circ. 886, 55 p.

Mero, J.L., 1965, The mineral resources of the sea; Elsevier, Amsterdam, 312 p.

Noakes, J.E. and Harding, J.L., 1982, Nuclear techniques for seafloor mineral exploration; Proc. Oceanology Int., paper OI82 1.3, 16 p.

United Nations, 1984, Analysis of exploration and mining technology for manganese nodules; United Nations, Ocean Economics and Technology Branch, 140 p.

Spiess, F.N., 1986, Deep-ocean near-bottom surveying and sampling techniques; this volume, pp. 255-271.

Usui, A., 1979, Minerals, metal contents, and mechanism of formation of manganese nodules from the central Pacific Basin (GH76-1 and GH77-1 areas), In: Bischoff, J.L. and Piper, D.Z. (eds.) Marine geology and oceanography of the Pacific Manganese Nodule Province; Plenum Press, New York, pp. 651-679.

von Stackelberg, U., 1979, Sedimentation, hiatuses, and development of manganese nodules: Valdivia site VA-13/2, northern central Pacific,

In: Bischoff, J.L. and Piper, D.Z. (eds.) Marine geology and oceanography of the Pacific Manganese Nodule Province; Plenum Press, New York, pp. 559-586.

ANALYSIS AND METALLURGY OF MANGANESE NODULES AND CRUSTS

Benjamin W. Haynes and Michael J. Magyar
U.S. Department of the Interior
Bureau of Mines, Avondale Research Center
Avondale, Maryland U.S.A.

ABSTRACT. This paper presents an overview of recent research completed by the Bureau of Mines for manganese nodules and crusts collected from the Pacific Ocean. Included are elemental and mineralogical compositional analyses, methods of analysis for the minerals and wastes from processing, and process metallurgy. Data are presented for 16 major and minor elements and the three major mineral phases present in nodules and crusts. Methods of analysis for nodules and crusts and their tailings include physical, chemical, mineralogical, and leaching test procedures. Results of processing show varying recoveries of the metals cobalt, copper, manganese, and nickel.

INTRODUCTION

Since the discovery of marine ferromanganese oxides during the cruises of the H.M.S. Challenger (Murray and Renard, 1891), interest in these marine oxides has changed from a scientific curiosity to a potential resource of value metals. Interest initiated by Mero (1965) was partially responsible for the formation of several multi-national consortia to investigate and develop mining methods for manganese nodules and subsequent recovery of cobalt, copper, nickel, and, in some processes, manganese. Recent research by Hein et al. (1985) has shown that manganese crusts on seamounts contain cobalt several times the concentration of Pacific manganese nodules of the Clarion-Clipperton fracture-zone area. The interest in these cobalt-enriched ferromanganese crusts is further enhanced by their location within exclusive economic zones (EEZ's) of the United States and other countries.
    The Bureau of Mines has studied various aspects of manganese nodule and crust composition and processing. Several reports on nodules (Haynes et al., 1982, 1983, 1983a, 1985) were published describing the mineralogical and elemental content, primary processing flowsheets for recovery of value metals, methods for determining physical and chemical properties of nodules and their processing wastes, and results of laboratory processing and characterization to

P. G. Teleki et al. (eds.), Marine Minerals, 235–246.
© *1987 by D. Reidel Publishing Company.*

determine potential processing waste disposal problems.

This paper presents information on composition of nodules and crusts, chemical and physical methods of analysis, results of processing, and the impact of these parameters on resource assessment analyses.

MINERALOGY AND ELEMENTAL CONTENT

Mineralogy

Manganese nodules and crusts consist of a complex mixture of materials, including organic and colloidal matter, nucleus fragments, and crystallites of various minerals of hydrogenous, detrital, and authigenic origins. The phases in nodules and crusts are fine grained, commonly metastable, poorly crystalline, and intimately intergrown so that it is difficult to extract a homogeneous, single-phase sample for study. The minerals are characterized by structural defects, essential vacancies which may not be ordered, domain intergrowths, extensive solid solutions, and cation exchange properties that lead to nonstoichiometry and inhibit long-range ordering of the crystals.

The phases in manganese nodules and crusts are the oxide minerals of manganese and iron, and the accessory minerals. The isotropic oxides in manganese nodules and crusts are not easily classified by optical means. Standard X-ray diffraction techniques generally cannot determine more than one or two of the manganese minerals and several accessory minerals. Good results are obtained by scanning electron microscopy (SEM) and transmission electron microscopy (TEM) studies. Recent work by Turner and Buseck (1979, 1981), Turner et al. (1982), and Siegel (1981) determined the structure of todorokite in manganese nodules. Chukhrov et al. (1976, 1979, 1980) studied the iron oxides in nodules. The dehydration and oxidation of both the manganese and iron phase minerals occurs readily upon removal from seawater.

Manganese forms a large number of oxide minerals, but only the low-temperature oxides that can be formed in water are relevant to the mineralogy of nodules and crusts. Tetravalent manganese predominates in nodules and crusts, but the presence of Mn^{2+} and Mn^{3+} ions in some minerals has been inferred from crystallographic and thermodynamic data. Todorokite, birnessite, and vernadite (δ-MnO_2) are the predominant manganese-oxide minerals, occurring as cryptocrystalline phases in manganese nodules. The todorokite and birnessite appear to contain variable linkages of edge-shared (MnO_6) octahedra and are characterized by numerous structural defects, cation vacancies in the chains or layers of linked octahedra, domain intergrowths of mixed periodicities, cation exchange properties, and extensive substitution of Ni^{2+}, Cu^{2+}, Zn^{2+}, Mg^{2+}, or other divalent ions for Mn^{2+} (Burns, 1970). The formation of vernadite, sometimes catalyzed by microorganisms, results in a poorly crystalline to amorphous phase with a high surface area and cation adsorption properties, concentrating cobalt by substitution of Co^{3+} for Mn^{4+} (Burns, 1976; Murray and Dillard, 1976).

Studies have shown that vernadite (δ-MnO$_2$) is the major manganese mineral in crusts (Hein et al., 1985). The highest cobalt contents occur in crusts at water depths of 800 to 2,500 meters. The oxide, oxyhydroxide, and hydrated oxide phases of iron relevant to manganese nodules and crusts are structures of close-packed oxygens containing Fe^{3+} and/or Fe^{2+} ions in various octahedral interstices forming different assemblages of edge-shared (FeO$_6$) octahedra. Certain iron (III) oxyhydroxides are isostructural with manganese (IV) oxides, with (FeO$_6$) and (MnO$_6$) octahedra edge-shared in various arrangements. The larger ionic radius of Fe^{3+} compared with that of Mn^{4+} results in larger spacings for the ($10\bar{1}0$) and ($11\bar{2}0$) planes of the hexagonal close-packed systems (approximately 2.50 to 2.56 Å and 1.48 to 1.54 Å, respectively), and Fe-Fe interatomic distances across edge-shared (Fe(O,OH)$_6$) octahedra ranging from 2.95 to 3.05 Å. The most commonly observed iron-bearing phases in manganese nodules and in crusts are goethite, lepidocrocite, and feroxyhyte (Haynes et al., 1982).

In crust samples from the Necker Ridge and other areas of the Pacific, no iron-phase minerals were identified (Hein et al., 1985). This is as expected, because of the extremely fine-grained nature of the marine iron oxides. Accessory minerals observed include plagioclase, quartz, apatite, and calcite for crusts and apatite, plagioclase, phillipsite, quartz, and smectite for the substrates.

Elemental content

Many reports have been published on the elemental content of manganese nodules with Haynes et al. (1982) and McKelvey et al. (1983) being the most comprehensive. Analyses of the elemental composition of manganese crusts are limited compared to manganese nodules. The most recent information published by Hein et al. (1985) detailed the 1983 U.S. Geological Survey cruises to the central Pacific. Ongoing studies by the Hawaii Institute of Geophysics supported by the Minerals Management Service have analyzed crust samples from the Hawaiian Archipelago. Table I gives concentraton values for 16 elements in nodules and crusts from selected areas of the Pacific Ocean. A comparison of metal concentrations of nodules from the mid-Pacific seamounts region (Haynes et al., 1982) to crusts from the Necker Ridge and northern Line Islands (Table I) give similar concentrations for metals such as manganese, copper, cobalt, iron, and nickel.

METHODS OF ANALYSIS

The ability to analyze marine minerals, substrates, and processing wastes by rapid, precise, and accurate methods is essential in obtaining a reliable resource assessment. Attention must be given to shipboard chemical analysis and laboratory physical, chemical, and mineralogical analysis of marine minerals; to physical and chemical properties of the substrate materials of crusts; and to physical and chemical properties of the process tailings.

Table I.   Comparison of elemental composition of manganese nodules and manganese crusts from various areas of the Pacific Ocean. Values given in weight percent.

| Elements | Manganese nodules[1] | | | | Necker Ridge (36) | | Manganese crusts[2] | | | |
| | Pacific Ocean | | CC-Zone[3] | | | | Hawaiian Archipelago (124) | | Kingman Reef (14) Palmyra Atoll | |
	Mean	Range	Mean	Range	Mean	Range	Mean	Range	Mean	Range
Al	2.8	<0.25 -10.0	2.9	0.5 - 8.0	1.48	0.40- 3.79	1.58	0.26- 6.55	0.57	0.28- 1.00
Ba	0.24	<0.005- 0.80	0.28	<0.01 - 0.76	0.23	0.16- 0.38	0.28	0.03- 1.50	0.26	0.15- 0.43
Ca	2.12	<0.05 -25.0	1.7	<0.5 -18.0	2.99	1.70- 6.22	2.26	1.06 -15.9	3.08	2.50- 5.68
Co	0.44	<0.05 - 2.5	0.24	<0.1 - 0.9	0.67	0.25- 1.26	0.66	0.07- 1.75	1.27	0.75- 3.02
Cu	0.66	<0.01 - 2.00	1.02	0.1 - 2.0	0.12	0.03- 0.29	0.07	0.03- 0.32	0.06	0.02- 0.10
Fe	10.4	1 -25	6.9	1 -25	18.5	11.2 -22.7	16.8	1.82 -24.0	15.5	8.85 -19.8
Mg	1.53	<0.25 - 5.0	1.65	<0.25 - 3.0	1.14	1.00- 1.55	1.37	0.84- 4.01	1.29	1.01- 2.29
Mn	21.6	1 -40	25.4	1 -39	22.4	12.0 -32.6	21.3	1.20 -45.6	28.6	22.6 -38.8
Mo	0.041	<0.005- 0.150	0.052	<0.005- 0.12	0.05	0.02- 0.09	0.04	0.005- 0.09	0.08	0.05- 0.11
Ni	0.89	0.10 - 2.0	1.28	0.1 - 2.0	0.36	0.07- 0.64	0.36	0.06- 1.16	0.56	0.34- 1.00
P_2O_5	0.23	<0.01 - 2.2	0.23	<0.01 - 2.2	1.34	0.60- 3.58	1.07	0.16 -13.2	1.43	0.83- 3.88
Pb	0.072	0.005- 0.470	0.045	0.005- 0.18	0.20	0.11- 0.32	0.18	0.02- 0.37	0.18	0.11- 0.26
Sr	0.083	<0.005- 0.30	0.045	<0.005- 0.16	0.17	0.09- 0.24	0.15	0.03- 0.23	0.19	0.15- 0.22
Ti	0.73	<0.05 - 2.20	0.53	0.10 - 2.20	1.41	0.83- 2.63	1.17	0.28- 3.26	1.14	0.81- 1.77
V	0.51	<0.005- 0.30	0.047	<0.005- 0.08	0.07	0.04- 0.10	0.06	0.02- 0.16	0.08	0.06- 0.10
Zn	0.11	<0.05 - 1.00	0.14	<0.05 - 0.95	0.08	0.05- 0.16	0.05	0.01- 0.38	0.09	0.06- 0.15

[1]From Haynes et al. (1982).
[2]Values in parentheses for crusts indicate number of samples analyzed.
[3]CC-Zone: Clarion-Clipperton fracture-zone area.

The physical properties of the marine minerals and substrates are important in considering process type, design, and equipment. The physical properties of the tailings affect disposal options and subsequent cost. Lewis and Tandanand (1974) recommended procedures for rock testing that could be used on crust substrate as well as on manganese nodules and crusts. The American Society for Testing and Materials (ASTM) soil mechanics procedures have been successfully used to determine the physical characteristics of coal refuse (Busch et al., 1975) and manganese nodule tailings (Haynes et al., 1985). Table II lists the recommended ASTM procedures for testing wastes.

Table II. Suggested test procedures for determining physical properties of manganese nodule and crust tailings (ASTM, 1977)

Property	Procedure
Grain size distributions:	
Plus 200 mesh	ASTM D422-63
Minus 200 mesh	Allen (1975)
Specific gravity	ASTM D854-58
Triaxial shear	ASTM D2850-70
Permeability	ASTM D2434-68[1]
Maximum density	ASTM D698-78
Minimum density	ASTM D2049-69
Atterberg limits:	
Liquid	ASTM D423-66
Plastic	ASTM D424-59
Soil class	ASTM D2487-69
Slurry density	ASTM D2216-71

[1]Using Bureau of Reclamation Earth Manual Procedure E13.

The determination of mineralogical and chemical composition of nodules has been described by Haynes et al. (1983). The use of optical and electron microscopic techniques in conjunction with X-ray diffraction is required. Chemical compositions can be determined by many methods including atomic absorption spectroscopy (AAS), inductively coupled plasma atomic emission spectroscopy (ICP), neutron activation analysis (NAA), X-ray fluorescence spectrography (XRF), and ion chromatography (IC). Detailed discussions of these methods, including their uses and limitations can be found in Haynes et al. (1983).

Another important aspect of analysis is determining whether tailings generated from the processing of manganese nodules and crusts should be considered hazardous wastes. Depending on whether disposal is on land or at sea, three tests are currently available to determine the impact of disposal: the U.S. Environmental Protection Agency's

(EPA) extraction procedure (EP) toxicity test (Federal Register, 1980), the ASTM shake extraction test (ASTM, 1982), and the EPA-Corps of Engineers seawater elutriant test (U.S. EPA, 1977). Details of the procedures for these tests are described in Haynes et al. (1983).

METALLURGY

Processing routes to recover the potentially economic metals manganese, copper, cobalt, and nickel from manganese nodules were developed by several consortia in the late 1960's and in the 1970's. Two recent studies (Dames and Moore, 1977; Haynes et al., 1983a) detailed the conditions for processing and presented flowsheets for five technically feasible processes. Four of those five processes are based on adaptations of processes for land-based nickeliferous laterites, and the other process is unique to manganese nodules. The five processes are the gas reduction and ammoniacal leach, Cuprion ammoniacal leach, high-temperature and high-pressure sulfuric acid leach, reduction and hydrochloric acid leach, and smelting and sulfuric acid leach. All of these processes are amenable to processing manganese crusts as well as nodules because of the similarity of their manganese and iron mineral structures.

The first two processes listed are based on the Caron process as practiced at Nicaro, Cuba (Agarwal et al., 1979; Alonso and Daubenspek, 1960; Caron, 1924; Graaf, 1979, 1980; Szabo, 1976). The first of the two processes is almost a direct adaptation except for the metals separation and purification section. The Cuprion process, however, uses a near ambient reduction in aqueous ammonia-ammonium carbonate solution with cuprous copper and carbon monoxide (Agarwal et al., 1979; Szabo, 1976). The high-temperature and high-pressure H_2SO_4 leach process is based on the metallurgy as developed by the Freeport Nickel Co. and practiced at Moa Bay, Cuba (Carlson and Simons, 1961; Duyvesteyn, 1979; Neuschutz, 1977). The separation and purification steps for the metals are different for nodules than for nickeliferous laterites. The HCl process represents relatively newly developed technology. The smelting process has many similarities with several nonferrous smelting processes and has been tested for manganese nodules (Halbach et al., 1977; Haynes et al., 1985; Montanteme et al., 1978; Septier et al., 1979; Sridhar et al., 1977; Wilder et al., 1981). For nodules processed in the laboratory (Haynes et al., 1985), the ammoniacal leach processes recovered >90% of the copper and nickel, and 50% of the cobalt, but rejected most of the iron and manganese. The H_2SO_4 pressure leach process recovered >92% of the copper, nickel, and cobalt, and rejected most of the iron and manganese. The smelting process recovered >95% of the copper, cobalt, and nickel, and rejected >98% of the manganese.

Recent work on one manganese crust sample has been completed for three processes: Cuprion ammoniacal leach, high-temperature and high-pressure H_2SO_4 leach, and the smelting and H_2SO_4 leach. Results obtained from the processing of the manganese crust sample were similar to those obtained in processing nodules for the latter two processes.

The mineralogical, chemical, and physical properties of the crust
tailings were also similar to the nodule tailings. Whereas the Cuprion
process was good for nodules, it was somewhat less effective for the
cobalt crust sample used. The more promising processes for the crust
sample were the high-temperature and high-pressure H_2SO_4 process and
the smelting and sulfuric acid process. For the crust sample, the
Cuprion process extracted, in a single stage, 60% of the cobalt and 50%
of the nickel. The sulfuric acid pressure leach process extracted 90%
of the cobalt, copper, and nickel in a single stage. The smelting
process separated 98% of the cobalt and nickel, >94% of the copper, and
92% of the iron into a metal alloy while rejecting 98% of the manganese
into the slag phase as designed. An important result of the work
completed on manganese nodules and crusts is that the tailings and
slags generated in the laboratory were a nonhazardous waste based on
the EPA EP-toxicity test criterion. The other two test procedures
produced leachates with equal or lower metal concentrations than the EP
toxicity test.

RESOURCE ASSESSMENT CONSIDERATIONS

    Elemental composition and mineralogy, analysis methodology, and
metallurgy all play important roles in the resource assessment of
manganese nodules and crusts. The information for manganese nodules is
quite comprehensive and some of this information can be applied
directly to the assessment of manganese crusts. Methodology for
physical, chemical, and mineralogical determinations are available and
can be used on crusts with little variation.

    The lack of significant data on the grade and abundance of crusts
reflects the early stage of crust exploration. The major emphasis in
further studies needs to be on the systematic gathering of samples for
determining metal content, coverage, thickness, and substrate type. An
important difference of manganese crusts as compared to nodules is in
the accessory mineral composition. Whereas Pacific manganese nodules
contain moderate to low levels of phosphorus as phosphate (see Table
I), manganese crusts that have higher cobalt content also contain
significantly more phosphate. This can have a major impact on
processing if pyrometallurgical techniques are used. This gangue
material for crusts would also include any substrate recovered with the
crust. The substrate could be as high as 50% of the amount of material
recovered and must be removed prior to processing to reduce the
processing plant size. Knowledge of the physical properties and
composition of the various substrate types associated with crusts will
be necessary to determine optimum methods of separation. This in turn
will influence any detailed resource assessment analysis.

    In recovering samples of crusts and evaluating potential mine
sites, detailed bottom topography (seamount slopes and microtopography)
and bathymetry must be obtained. The terrains in which crusts are
found are vastly different from the abyssal plains where Pacific
nodules are found. This will radically change the present design of
the seabed mining machines in respect to locomotion, stability

compensation, and collector heads, thereby affecting the economics of mining.

From a metallurgical viewpoint, manganese nodules and manganese crusts will react similarly in hydrometallurgical and pyrometallurgical processes. However, the concentration differences of metals in nodules (high Cu and Ni and lower Co) as compared to those in crusts (high Co, moderate Ni, and low Cu) will mandate changes in the metallurgy developed for manganese nodules if it is to be applied to crusts.

Chloride levels of marine minerals are unique and are not generally encountered in land-based ore processing. Chloride is known to degrade gold blow-out protection disks in high-pressure processing of manganese nodules (Haynes et al., 1985), but the use of tantalum has been effective in preventing this rapid degradation. The greater corrosion caused by chloride could affect capital and maintenance costs of equipment. The higher phosphate concentrations in crusts will require greater control in pyrometallurgical processing which will be reflected as increased operating costs needed to meet the alloy specifications.

CONCLUSIONS

Although many parameters must be considered and be accounted for to obtain a reliable resource assessment of manganese nodules and crusts, only those related to composition, analysis, and metallurgy are addressed in this paper.

The amount of grade and abundance data for manganese nodules within certain areas of the Clarion-Clipperton zone is sufficient to provide reliable economic resource assessment appraisals.

Analysis methodology for manganese nodules can be applied directly to manganese crusts with little or no change.

Data needed to evaluate manganese crust resource potential include elemental composition, physical properties of crusts and substrates, topography, bathymetry, and composition of the gangue materials.

The quantity and quality of data required for crusts should approach the levels acquired for nodules to make resource assessment meaningful.

Mining machines designed for manganese nodules are not suitable for crust collection because of the adhering properties of crusts to substrate and the differences in seafloor topography.

The adaptations of land-based laterite processes used for manganese nodules can be applied to manganese crust processing.

Variations in Co, Ni, and Cu concentrations between manganese crusts and nodules are a major factor in considering metallurgical options.

Phosphate concentrations in crusts are higher than those

in nodules and will influence any pyrometallurgical approach.

ACKNOWLEDGMENTS

The authors acknowledge the support and cooperation of F. T. Manheim and J. R. Hein, U.S. Geological Survey; R. Stone, Minerals Management Services; and C. Morgan, Hawaii Manganese Crust EIS Project; for providing crust samples for analysis and metallurgical testing; and J. Padan, National Oceanic and Atmospheric Administration, for support of the manganese nodule studies. We also acknowledge the following Bureau of Mines personnel: J. Ritchey for providing the larger crust sample; F. E. Godoy for assistance in conducting metallurgical testing; and J. B. Zink, C. A. Hammett, D. C. Baron, and C. P. Walters for providing chemical analysis of crusts and nodules.

REFERENCES

Agarwal, J. C., Barner, H. E., Beecher, N., Davies, D. S., and Kust, R. N., 1979, Kennecott process for recovery of copper, nickel, cobalt, and molybdenum from ocean nodules. Min. Eng., v. 31 (12), pp. 1704-1709.

Allen, T., 1975, Particle size measurement. Chapman & Hall, pp. 301-312.

Alonso, A., and Daubenspek., J., 1960, Modifications in Nicaro metallurgy. Trans. Metall. Soc. of AIME, v. 217, pp. 253-257.

American Society for Testing and Materials, 1977, Annual book of standards part 19: Natural building stones; soil and rock; peats mosses, and humus. ASTM, Philadelphia, Pennsylvania, 494 p.

American Society for Testing and Materials, (1982), Standard test method for shake extraction of solid waste with water: ASTM D 3987. Annual book of standards: Part 32, Water; ASTM, Philadelphia, Pennsylvania, pp. 1423-1427.

Burns, R. G., 1976, Uptake of cobalt into ferromanganese nodules, soils, and synthetic manganese (IV) oxides. Geochim. Cosmochim. Acta, v. 40(1), pp. 95-102.

Busch, R. A., Backer, R. R., Atkins, L. A., and Kealy, C. D, 1975, Physical property data on fine coal refuse. U.S. Bur. Mines, Rept. RI 8062, 1975, 40 p.

Carlson, E. T., and Simons, C. S., 1961, Pressure leaching of nickeliferous laterites with sulfuric acid. In: Queneau, P. (ed.), Extractive metallurgy of copper, nickel, and cobalt. Interscience Publ., New York, pp. 363-397.

Caron, M. H., 1924, Process of recovering values from nickel and cobalt-nickel ores. U.S. Pat. 1,487,145, Mar. 18.

Chukhrov, F. V., Gorshkov, A. I., Sivtsov, A. V., and Berezovakaya, V. V., 1979, New mineral phases of oceanic manganese microconcretions (in Russian). Izvest. Akad. Nauk SSSR, Ser. Geol., No. 1, pp. 83-90.

Chukhrov, F. V., Gorshkov, A. I., Zvyagin, B. B., and Ermilova, L. P., 1980, Iron oxides as minerals of sedimentary environments and chemogenic eluvium. In: Varentsov, I.M. and Grasselly, Gy. (eds.), Geol. and geochem. of manganese, Akademiai Kiado, Budapest, Hungary, pp. 231-257.

Chukhrov, F. V., Zvyagin, B. B., Gorshkov, A. I., Ermilova, L. P., Korovushkin, V. V., Rudnitskaya, E. S., and Yakubovskava., N. Yu., 1976, Feroxyhyte, a new modification of FeOOH. Int. Geol. Rev., v. 19, pp. 873-890.

Dames and Moore, and E.I.C. Corporation, 1977, Description of manganese nodule processing activities for environmental studies, v. III. Processing systems technical analysis. U.S. Dept. of Commerce-NOAA, Office of Marine Minerals, Rockville, Maryland, (NTIS PB274912 set), 540 p.

Duyvesteyn, W. P. C., Wicker, G. W., and Doane, R. E., 1979, An omnivorous process for laterite deposits. Proc. Int. Laterite Symp., New Orleans, Soc. Min. Eng. AIME, New York, pp. 553-570.

Federal Register, 1980, Parts II-IX, Environmental Protection Agency; Hazardous waste and consolidated permit regulations. v. 115, no. 98, May 19, Book 2, pp. 33063-33285; 110-CFR, Parts 260-265.

Graaf, J. E, 1979, The treatment of lateritic nickel ores--a further study of the Caron process and other possible improvements--Part I. Effect of reduction conditions. Hydrometall., v. 5, pp. 47-65.

Graaf, J. E., 1980, The treatment of lateritic nickel ores--a further study of the Caron process and other possible improvements--Part II. Leaching studies. Hydrometall., v. 5, pp. 255-271.

Halbach, P., Koch, K., Renner, H-J., and Ujma, K-H., 1977, Pyrometallurgical processing of manganese nodules and lateritic nickel ores using waste materials as reducing agents. Erzmetall., v. 30, pp. 458-464.

Haynes, B. W., Law, S. L., and Barron, D. C, 1982, Mineralogical and elemental description of Pacific manganese nodules. U.S. Bur. Mines, Rept. IC 8906, 60 p.

Haynes, B. W., Barron, D. C., Kramer, G. W., and Law, S. L, 1983, Methods for characterizing manganese nodules and processing wastes. U.S. Bur. Mines, Rept. IC 8953, 10 p.

Haynes, B. W., Law, S. L., and Maeda, R., 1983a, Updated process flowsheets for manganese nodule processing. U.S. Bur. Mines, Rept. IC 8924, 100 p.

Haynes, B. W., Barron, D. C., Kramer, G. W., Maeda, R., and Magyar, M. J., 1985, Laboratory processing and characterization of waste materials from manganese nodules. U.S. Bur. Mines, Rept. RI 8938, 16 p.

Hein, J. R., Manheim, F. T., Schwab, W. C., Davis, A. S., Daniel, C. L. Bouse, R. M., Morgenson, L. A., Sliney, R. E., Clague, D. A., Tate, G. B., and Cacchione, D.A., 1985, Geological and geochemical data for seamount and associated ferromanganese crusts in and near the Hawaiian, Johnston Island, and Palmyra Island exclusive economic zones. U.S. Geol. Surv. Open File Rept. 85-292, 123 p.

Lewis, W. E., and Tandanand, S., 1974, Bureau of Mines test procedures for rocks. U.S. Bur. Mines, Rept. IC 8628, 223 p.

McKelvey, V. E., Wright, N. A., and Bowen, R. W., 1983, Analysis of the world distribution of metal-rich subsea manganese nodules. U.S. Geol. Surv. Circ. 886, 55 p.

Mero, J. L., 1965, The mineral resources of the sea. Elsevier, Amsterdam, 312 p.

Montanteme, J., Greffe, A., and Grandjacques, F., 1978, Selective reduction of nickel ore with a low nickel content. U.S. Pat. 4,073,641, Feb. 14.

Murray, J. W., and Dillard, J. G., 1979, The oxidation of cobalt (II) adsorbed on manganese dioxide. Geochim. Cosmochim. Acta, v. 43, pp. 781-787.

Murray, J., and Renard, A. F., 1891, Deep-sea deposits. Rept. on the scientific results of the voyage of HMS Challenger during the years 1873-1876; Longmans, London, 525 p.

Neuschutz, D., Scheffler, V., and Junghanss, H., 1977, Verfahren zur aufarbeitung von manganknollen durch schwefelsaure drucklaugung. (Method for the processing of manganese nodules by sulfuric acid pressure leaching.) Erzmetall., v. 30(2), pp. 61-67.

Septier, L., Dubrous, F., and Demango, M., 1979, Process for the treatment of complex metal ores containing, in particular, manganese and copper, such as oceanic nodules. U.S. Pat. 4,162,916, July 31.

Siegel, M., 1981, Studies of the mineralogy, chemical composition, textures, and distribution of manganese nodules at a site in the north equatorial Pacific ocean; Ph.D. Thesis, Harvard Univ., Cambridge, Massachusetts, 274 p.

Sridhar, R., Warner, J. S., and Bell, M. C. E., 1977, Non-ferrous metal recovery from deep sea nodules. U.S. Pat. 4,049,438, Sept. 20.

Szabo, L. J., 1976, Recovery of metal values from manganese deep sea nodules using ammoniacal cuprous leach solutions. U.S. Pat. 3,983,017, Sept. 28.

Turner, S., and Buseck, P. R., 1979, Manganese oxide tunnel structures and their intergrowths. Science, v. 203, pp. 456-458.

Turner, S., and Buseck. P. R., 1981, Todorokite: A new family of naturally occurring manganese oxides. Science, v. 212, pp. 1024-1027.

Turner, S., Siegel, M., and Buseck, P. R., 1982, Structural features of todorokite intergrowths in manganese nodules. Nature, v. 296, pp. 841-842.

U.S. Environmental Protection Agency/Corps of Engineers, 1977, Ecological evaluation of proposed discharge of dredged material into ocean waters. Environmental Effects Laboratory, U.S. Army Engineers Waterways Experiment Station, Corps of Engineers, Vicksburg, Mississippi, 103 p.

Wilder, T. C., Andreola, J. J., and Galin, W. E., 1981, Reduction processes for manganese nodules using fuel oil. Jour. Metals, v. 33, pp. 64-69.

NODULE EXPLORATION: ACCOMPLISHMENTS, NEEDS AND PROBLEMS

William D. Siapno
Deepsea Ventures, Inc.
Gloucester Point, Virginia, U.S.A.

ABSTRACT.   Nearly a century elapsed between discovery of manganese
nodules in the 1870's and the beginning of their commercial exploration
in the 1960's.   Since the latter date, however, technical progress has
been rapid, deposits of economic interest have been discovered and
mapped, and mining systems have been proven to be capable of operating
in ocean depths of 15,000 feet.

In 1984 the U.S. Government through its Department of Commerce issued
Exploration Licenses to four international consortia. This paper
presents a historic sketch of the program conducted by Ocean Mining
Associates that led to the award of one such license. The current
program is not without its problems, however, some of which are technological and some that are economic in origin.

INTRODUCTION

Early in the 1960's industry initiated investigations of the
economic potential of manganese nodules. Nodules (Fig. 1), first
discovered during the British "Challenger" Expedition of 1872-76
(Murray and Renard, 1981) had, until then, been regarded as scientific
curiosities and largely the province of academic research.
One of the early commercial entrants in deep-ocean mineral exploration was Newport News Shipbuilding and Dry Dock Company (NNS&DD Co.),
the predecessor-in-interest to Ocean Mining Associates (OMA). NNS&DD
Co. began research on nodules in 1962. The nodule exploration program
was originated from the shipyard as Deepsea Ventures, Inc., when NNS&DD
CO. was acquired by Tenneco in 1968. Deepsea Ventures is the service
contractor to OMA. OMA is currently composed of domestic subsidiaries
of ENI of Italy, Union Minière of Belgium, and U.S. Steel and the Sun
Oil Company of the United States.
In the early 1960's very little was known about the deep ocean
areas of the world. However, it had been already clearly recognized that the sea was a major storehouse of minerals. For the most
part, the continents had provided man's mineral needs. The more

247

P. G. Teleki et al. (eds.), Marine Minerals, 247–257.
© 1987 by D. Reidel Publishing Company.

Figure 1.   Manganese nodules in the Clipperton–Clarion Fracture Zone.
            Water depth is 15,000 feet.

Figure 2.   Research Vessel PROSPECTOR, the mainstay of Ocean Mining
            Associates exploration program.

easily accessible deposits, in terms of location and technology, had
largely been exploited by the mid 1900's and society, in general,
began to recognize that resources were finite, rather than limitless.
It was from this perspective that focused attention on retrieving
minerals from the sea.

ACCOMPLISHMENTS

Because there was a general paucity of data on manganese nodules,
plans to locate, delineate and evaluate these resources were necessary.
Moreover, there was a dearth of both equipment and technique to
accomplish these goals. Finding financial support for research and
development on manganese nodules was also, and continues to be a
major problem. No matter how appealing the scientific concept or how
predictable the technical potential, the economic feasibility defied
clear definition. The difficulties were many and complex. Briefly
this was the status of offshore mining when OMA's exploration program
began.

In 1964, a small cargo vessel was acquired and converted by NNS&DD
Co. for deep ocean exploration and re-named Research Vessel PROSPECTOR
(Fig. 2). Although several chartered vessels were also used later at
various times, PROSPECTOR remained the mainstay of OMA's marine
exploration surveys. Research on nodules began with a review of
pertinent literature, and with plotting the location of all reported
manganese nodule occurrences. The intent was to verify composition
and abundance at the more favorable sites and understand which salient
features of nodule deposits, such as water depth and location, and
that of the nodules themselves, such as specific gravity, shape,
size, chemical composition, were important to an exploration program.

Deepsea Ventures was an early proponent of deep-water television
(TV) (Siapno, 1975) as a survey tool (Figs. 3 and 4) and with
Hydroproducts, Inc., developed the first deep-submergence TV system
operated at the end of more than 25,000 feet of coaxial cable. The
TV required innovation in design, as did the entire handling system.
Innovation was also needed in designing surveying techniques, and in
the development and operation of oceanographic instrumentation (Fig. 5)
and sampling equipment (Fig. 6) by both the developer and the user.
Frequently, the developer and the user were one and the same. This
arrangement, however limited in some respects, assured full comprehen-
sion of the utilization of the devices. Deepsea Ventures developed a
suite of exploration equipment, including dredges, samplers, buoys,
and other tools. Wherever possible, "off-the-shelf," i.e., existing
and marketed equipment was used separately or in combination, and
modified to satisfy operational requirements.

Importantly, it was necessary to fully define what these requirements were. What needed to be measured, documented or collected?
What frequency and accuracy of measurement was needed compared to
what was rationally attainable? How necessary was each task to the
overall program? What was the expectation of successfully achieving
the desired goal? Could the data sought be best obtained by direct

or indirect measurements? How much would these cost to obtain?

All concepts and approaches had to meet the approval of the
group(s) who ultimately provided funding. Few, if any, items or
procedures were what a researcher might have chosen, had all options
been left open. Some goals of these surveys were not achieved as
effectively as desired, simply because the cost of the needed equipment
was beyond the range of financial resources available. Seagoing
personnel summed up the matter succinctly in the paraphrase: "Nodule
exploration – the last of the iron men and wooden ships." For systems
essential to the program, even when purchased directly from suppliers,
the specifications and testing were developed by Deepsea Ventures,
such as the Finite Amplitude Depth Sounder (FADS) system, InteNav
integrated satellite navigation system and Benthos Freefall Grab
Sampler (Siapno, 1976).

Despite the many technical, political and financial problems,
the "specifications" for the exploration activity were profoundly
simple. Find the highest grade deposit(s), with reserves necessary
to sustain a minimum of 20 years of mining at a rate of 3,000,000
wet tons per year, located as close as possible to the United States,
with a terrain of minimum relief, minimum exposure to climatic and
environmental hazards and located in international waters; a challenge
of great magnitude to all involved, scientists, technicians, managers
and others. From 1964 through 1972, a schedule of cruises produced
sufficient data to locate a test mining site and form the basis of
the first nodule "claim" (Kolbe and Siapno, 1974).

After a series of studies, which began in the Hydraulics Laboratory
of NNS&DD Co. and progressed to a flooded mine shaft in western
Virginia, then to shallow water tests in Chesapeake Bay, airlift
pumping of nodules was finally proven to work at a test\site in the
Atlantic Ocean on the Blake Plateau in July 1970 (Fig. 8). This
proof of concept was achieved in 2400 feet of water depth. The search
for deposits with higher contents of pay elements subsequently
proceeded in the Pacific Ocean. By mid-1972, sufficient information
had been gathered to confirm the region between the Clipperton and
Clarion fracture zones (CCFZ) as the area of major commercial interest.
Obviously, the entirety of the world's oceans was not subjected to
exhaustive search. Nonetheless, by the latter part of 1974, a careful
compilation of exploration data resulted in having Deepsea Ventures
file their first nodule claim.

Efforts now focused on upgrading designs to prove the viability
of operating at a water depth of 15,000 feet. A bulk ore carrier was
converted into the R/V DEEPSEA MINER II (Fig. 9) in 1976. Technology
proven in the Blake Plateau tests was updated (Victory et al., 1977)
incorporating various improvements to lessen operational problems in
deeper waters and to accommodate the specific characteristics of the
Pacific Ocean seafloor. In parallel with the engineering program,
the exploration group was called upon to map a test site in the CCFZ,
as well as a nearshore test site. A location approximately 13 miles
off the coast of southern California satisfied the latter requirement,
where training of personnel and testing of mining equipment was
completed. A series of deep water tests took place in 1977 and 1978,

during which all systems were used and mining feasibility was proven.
More than 500 tons of nodules were recovered during a single mining
test in the CCFZ.

Exploration continued after mining tests were completed, because
additional data were needed to support the application for an
Exploration License. In keeping with the U.S. Deep Seabed Hard
Mineral Resources Act (1980), OMA filed the application for an
Exploration License in early 1982. The license was awarded in August
1984.

Figure 3. The deep-towed platform instrumented with television
           and still-cameras in the act of deployment from the R/V
           PROSPECTOR.

PRESENT PROGRAM

Now that exploration licenses have been issued and the boundaries
of the area assigned to each U.S. license have been published, the need
for confidentiality in surveys has almost diminished. In fact, the
various consortia, who acquired U.S. licenses, resolved any differences
that may have existed and established the boundaries of areas licensed
for exploration amongst themselves. Recently, operating under the

Figure 4.   Survey Control Center on the R/V PROSPECTOR.   Instrumentation
            displayed consists of a main winch control, TV screen, tape
            recorder, Integrated Navigation System, depth recorder,
            gyro compass.

Figure 5.   Survey Data Center on R/V PROSPECTOR.   Data Recorder
            (left), Integrated Navigation System (center), Environ-
            mental Instruments, wave height analyzer and expandable
            STD (salinity-temperature-depth) recorder (right).
            Navigation plotter is in foreground.

terms of commercially negotiated Final and Supplemental Settlement
Agreements, exchanges of data became completed, whereby each group
trades exploration data collected in the other licensed areas for
similar information in its own licensed area. In some instances,
analyzed data and data products have also been exchanged. The
atmosphere, frequently marked with sharp distrust in the beginning,
has changed to one of cooperation now that operational areas are
mutually recognized.

OMA is now concentrating on compiling and analyzing data acquired
in these exchanges. The requirement is to integrate all available
information within the licensed area and carefully assess the resource
potential. Until area boundaries were legally not defined, detailed
economical evaluation studies were not feasible.

The present program also includes designing future surveys
needed to develop mining plans to meet objectives in terms of tonnage,
tenor, cost, and other factors. Similarly, plans are being formulated
to identify the equipment and ships required to perform future operations.

Figure 6.   Benthos Grab Samplers ready for launch on foredeck of
            R/V PROSPECTOR

IMPROVEMENTS

Holding an exploration license and having completed pilot tests
are not totally sufficient for implementing mining. Many improvements
remain to be made. Among these, advancements in electronic and
associated hardware development extends the promise of greatly
increased exploration capability. Electro-mechanical cables, still

Figure 7. A box core of sediment. Material in cores are analyzed
           to provide information on nodule-sediment interface
           relationships, geotechnical properties, sediment texture,
           benthic biota and microfossils.

used in conjunction with deep-towed instrumentation are known to be
severely limited in power delivery and data handling. Fiber optics,
however, permits a multifold increase in data transmission. Kevlar,
a synthetic material developed by DuPont de Neymours, Inc., is stronger
than steel; it is now being used in manufacturing cables. This material
allows larger conductors or fiber optics to be incorporated in stronger
cables of thinner diameters and; thereby also permit transmitting
more power to the instruments and equipment deployed. Integration of
these two improvements provides the opportunity to operate a greater
number of instruments simultaneously and collect more data than
possible before.
     Development of data processing onboard the towed platform, to
reduce data volume to be transmitted up the cable, is now in progress.
Conventionally, large volumes of raw data compete for space on the
carrier frequency used for transmission. Receiving processed data

Figure 8.   R/V DEEPSEA MINER on the Blake Plateau, Atlantic Ocean,
            July 1970.   The first vessel to successfully mine
            manganese nodules.

Figure 9.   R/V DEEPSEA MINER II proved nodules of commercial interest
            could be mined in 15,000 feet of water in the Pacific Ocean,
            Clipperton-Clarion fracture zone.

from the towed unit reduces both storage requirements and the demand
on shipboard computers to perform analyses. In addition to these
advancements, data compression techniques are improving the rate and
amount of survey data that are transmitted up the cable.

By the late 1980's, the satellite-based Global Positioning System
(GPS) is scheduled to provide real-time positioning, anywhere, anytime,
under any weather condition. This will resolve a myriad of age-old
problems, such as ship location, course, and heading that have haunted
the mariner since time began and has confounded the ocean explorer as
well. Positioning accuracy to $\pm$ 10 meters, will be possible with
GPS, under special circumstances. Other types of assistance from
satellites such as satellite communications extends the possibility
of creating a network of shipboard and shore-based computer systems.
Such a link should enhance possibilities to transmit dynamic data and
to evaluate near real-time natural processes such as meteorological
and oceanographic conditions and how they may affect exploration,
mining or other operations underway.

Free swimming, untethered surveying devices that do not require
large winches or sophisticated cables are coming of age. However,
total reliance on "free-swimmers" in deep-ocean surveys is not likely
in the foreseeable future. Combinations of towed and untethered
platforms still offer the most flexibility.

CONCLUSIONS

The fundamental problem of the ocean explorer is to acquire the
maximum amount of the needed data at an acceptable cost. Possibilities
for devising and employing advanced technological "wizardry" exist.
However, it is only available to those with the funds necessary to
support development, operation and maintenance of extremely sophisticated devices. There is little doubt further research and development
will provide a startling array of specialized instrumentation and
equipment for the new generation of explorers.

The immediate problem with ocean mining is symptomatic of the
generic entity of mining itself. Mining, aside a few exceptions, is
presently in economic doldrums. Ocean mining as a pioneering mining
effort has yet to find an opportunity to provide products to the
market place.

REFERENCES

Murray, J. and Renard, A., 1891, Report on deep-sea deposits. In:
    Wyville Thomson, C., (ed.) Report on the scientific results of
    the voyage of H.M.S. Challenger; Eyre and Spottiswoode, London,
    v. 5, pp. 1-525.

Siapno, W.D., 1975, TV in deep-ocean surveys, Oceanus, v. 18, no. 3,
    pp. 48-52.

Siapno, W.D., 1976, Exploration technology and ocean mining parameters;
    Mining Cong. Jour., pp. 16-22.

Kolbe, H., and Siapno, W.D., 1974, Manganese nodules: Further
    resources of nickel and copper on the deep ocean floor; Geoforum
    v. 20, pp. 63-80.

Victory, F., Siapno, W.D. and Meadows, R., 1977, Mining manganese
    from the ocean floor; Mechanical Engineering, pp. 20-25.

DEEP-OCEAN NEAR-BOTTOM SURVEYING TECHNIQUES

F. N. Spiess
Marine Physical Laboratory
Scripps Institution of Oceanography
University of California, San Diego
La Jolla, California, U.S.A.

ABSTRACT. The existence, nature and extent of seafloor mineral deposits must be determined by a sequence of operations ranging from broad area surveys to detailed on-bottom sampling and measurements. The mid-phase of the sequence would be the intensive near-bottom surveys with which this article is concerned. These can be carried out most efficiently using multiple sensor systems towed close to the seafloor. One example of such a system is the Marine Physical Laboratory's Deep Tow, whose characteristics and applications are discussed. Beyond the side-looking sonars, subbottom sounders, magnetometers, cameras, television and water-property-measuring sensors used in today's systems, however, are other types of approaches, such as acoustic backscattering measurements, and electromagnetic and chemical systems that could be used in this context. Finally, there are advanced towed vehicles under development (RUM III, Argo-Jason, PART) which can support not only the fine-scale survey function, but can combine it with some aspects of the final sampling and on-bottom measurement phase.

INTRODUCTION

Determination of the existence, nature and extent of marine mineral deposits involves a sequence of surveying and sampling operations, designed on the basis of our understanding of the settings in which economically viable assemblages may occur. The initial steps in exploration, generally, can be best carried out using geophysical techniques from ships traveling at normal speeds, complemented by seafloor sampling and photography. Once likely areas have been identified in the deep sea, however, near-bottom fine scale surveys must be made and intensive, detailed sampling carried out with a correspondingly greater investment of time at each site.

This paper focuses on the fine-scale, near-bottom survey phase, with emphasis on the concept that multiple-sensor systems are, in principle, the most efficient means for carrying out such work, and that such systems can be assembled and operated under realistic conditions. The near bottom survey data, in addition to their own direct

259

P. G. Teleki et al. (eds.), Marine Minerals, 259–275.
© 1987 by D. Reidel Publishing Company.

applicability, are essential for guiding fine-scale sampling operations, which, particularly in complex terrain, may require the use of manned submersibles.

The initial, large area surveys should naturally include the use of standard underway geophysical techniques including echo sounding, seismic reflection, magnetometry and gravity. Some of these techniques have improved substantially over the last decade, with the introduction of multi-channel seismic systems and swath-mapping, multibeam echo sounders (Renard and Allenou, 1979). Further advances can be expected as greater use is made of backscattered acoustic intensity data (de Moustier, 1985) and multiple-frequency shipboard sounding systems (Sumitomo, 1981). Advent of the NavStar or Global Positioning System (GPS) (Milliken and Zoller, 1978) will provide ocean-wide, consistent navigational control day and night to better than 100 m resolution, resulting in improved survey efficiency.

A number of near-bottom, fine-scale survey systems have been used successfully at sea. Some utilize only one or two sensing systems (e.g. Woods Hole's Angus for photography and water temperature, Deepsea Venture's closed circuit television package) while a few (the Marine Physical Laboratory's Deep Tow, the Naval Research Laboratory's sled, the Naval Oceanographic Office's Teleprobe and Lamont-Doherty Geological Observatory's SeaMARC I) have incorporated larger numbers of sensors, particularly side-looking sonars (SLS), subbottom sounders and magnetometers.

The most extensively used and diversely instrumented of these systems is the Deep Tow, and in the next section of this paper, that system is described in the resource evaluation and environmental impact contexts. In the subsequent section several potentially useful new instrumentation concepts that could be incorporated into a future system (multi-frequency acoustics, laser scanner, etc.)are discussed , and approaches treated elsewhere in this volume are mentioned. Finally, a few more complex vehicle approaches (RUM III, Argo-Jason, etc.) which can add more detailed investigative capabilities and/or carry out special types of seafloor sampling operations or analyses are presented.

DEEP TOW

Although this system has been described at various stages of its twenty-year evolution (Spiess and Mudie, 1970, Spiess and Tyce, 1973, Spiess and Lonsdale, 1982) its present status is reviewed here with three objectives in mind: to provide an example of a multifunction system, to show how such systems have been used in research related to manganese nodules, hydrothermal phenomena and seamount investigations, and to establish a starting point for visualizing other, even more powerful instrument complexes.

The present system consists of a sensor-carrying vehicle, towed 10 to 100 m off the seafloor at depths as much as 7,000 m, with a top speed of about 1 m/sec (2 knots) although 0.7 to 0.8 m/sec is more typical. The towing wire is 17.2 mm (0.68 inches) in diameter, double armored with a coaxial electrical conducting core. Cables are usually purchased in 9,000 m lengths and have breaking strength of about 15,000 kg. Onboard the towing ship there is a cable handling complex (crane, accumulator, winch), supporting electronics and data recording and display equipment.

Figure 1 is a schematic view of the vehicle itself, with its various sensor systems. These include six sonars (precision sounder, sub-bottom profiler, side-looking sonar, up-looking sounder, obstacle-avoidance sonar, and transponder interrogating and receiving system), magnetometer, cameras, television, strobe light, pressure (depth) gauge, water temperature and conductivity sensors and light transmissometer. Additional devices, including opening-closing plankton nets, remotely triggered water sampling bottles, suspended particle pump and filter systems are occasionally added for special purposes.

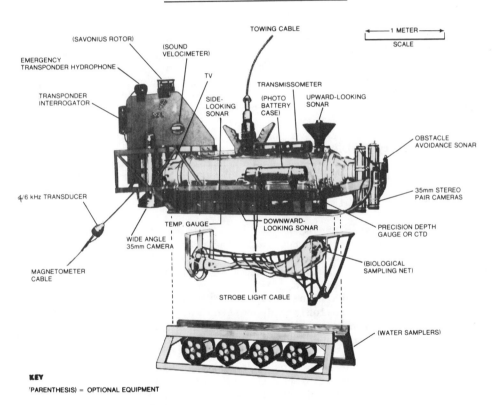

Figure 1. Deep Tow near-bottom deep sea survey vehicle (Spiess and Lonsdale, 1982).

All these systems are powered and controlled from the shipboard laboratory. Data telemetered from the various sensors are recorded in digital form and/or displayed on facsimile or chart recorders for immediate viewing and real-time interpretation.

An essential element in using systems of this kind is a towing ship with good maneuverability. The most effective ships for this purpose in the U.S. research fleet are the MELVILLE and KNORR which have cycloidal propulsion units mounted both near the bow and near the stern. Any ship with a good bow thruster (e.g. T. WASHINGTON, DE STEIGUER, NEW HORIZON) can usually be used. The U.S. Naval Research Laboratory group was successful with MIZAR, a single screw, no-bow-thruster ship, by working through a center well. The effective towing point was thus about amidships, facilitating underway maneuverability.

Navigation (usually a long-baseline near-bottom transponder system) is the key to making the most effective use of the data which these multiple sensor systems collect. Since fine-scale control of the vehicle's survey track is not feasible, with 5 to 8 km of wire out behind the towing ship, it is essential that the positions of the data points be related to one another quite accurately in order to prevent occurrence of artifacts in the overall patterns. Navigation is also required in order to be able to revisit interesting features, either with the survey vehicle itself to obtain a different, clarifying view, or with some sample device - corer, dredge, etc. The system used in most Deep Tow surveys (Spiess et al., 1966; McGehee and Boegeman, 1966; Boegeman, 1970) provides timing accuracies equivalent to one meter of range and usually involves interunit spaces of 3 to 6 km. After adjustment of transponder relative positions, using a relaxation method developed initially by Lowenstein (1966), this usually produces positions accurate to within 5 m over a major portion of the area covered by the transponder network. Sampling devices can be located in the same net by using a relay transponder (Boegeman et al., 1972) fastened to the coring wire. Typical relay transponder position uncertainty is about 10 m.

The Deep Tow system has been used in a wide variety of contexts, mostly geological/geophysical, but including seafloor search (Spiess and Sanders, 1971), marine biology (Wishner, 1979), physical oceanography (Lonsdale, 1977a) and ocean engineering (Alexander, 1981). The topics of particular interest in geology and geophysics have been spreading-center phenomena, fine scale magnetics, manganese nodule environments, sediment erosion and deposition, abyssal hills and seamounts. Work sites have mostly been in the Pacific Ocean, some in the Atlantic and one expedition each in the Red Sea and the Mediterranean (Fig. 2).

A survey made at the Galapagos Spreading Center (GSC) in 1976 serves as an initial example of how the multiple sensor concept works. For a variety of reasons based primarily on seafloor heat flow patterns and knowledge of fine-scale morphology of the central portion of the rise crest, derived in part from prior Deep Tow work on the East Pacific Rise (EPR) at 21°N (Larson, 1970) and the GSC (Klitgord and Mudie, 1974), this expedition focused on the possibilities of obtaining samples of hydrothermal fluids. Weiss et al. (1977) built a set of remotely controllable water samplers which were mounted, together with a CTD, on the deep tow fish. Initial passes through the area, emphasizing precision sounding profiles and side-looking sonar imagery, established the locations and orientations of fracturing and the existence of anomalous water temperature/conductivity patches near some fissures. Subsequent passes were then made closer to the bottom through these patches, with the sampling bottles triggered based on CTD data to collect the desired water samples. These operations obtained the first hydrothermal water samples from a spreading center environment (Weiss et al., 1977) and the simultaneous photographs revealed the first association of large clam shells with hydrothermal vents (Lonsdale, 1977b).

A second and somewhat different example occurred in the Guaymas Basin of the Gulf of California. It had been suspected that there should be hydrothermal activity there (Williams, et al., 1979), but it was not detected until a 1980 Deep Tow expedition. At that time the side-looking sonar revealed the existence of spires and rough patches on the otherwise smooth sediment cover (Fig. 3) and the up-looking sonar showed plumes of some sort rising up through the water column (Merewether et al., 1985). Temperature anomalies were then discovered and precisely positioned sediment samples were taken that showed thermal alteration of the organic-rich sediments (Simoneit and Lonsdale, 1982). Subsequently the side-looking sonar images were used to guide submersible (ALVIN) operations. A complex zone of hydrothermal activity was further photographed and sampled (Lonsdale, 1984) and is still an object of study.

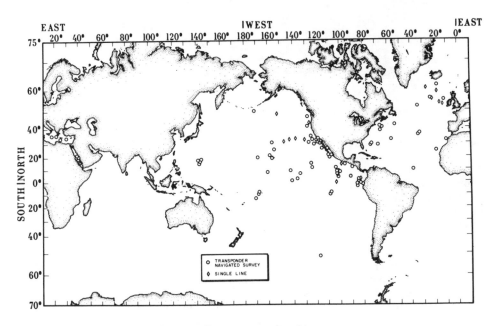

Figure 2.   Locations of Deep Tow research sites.

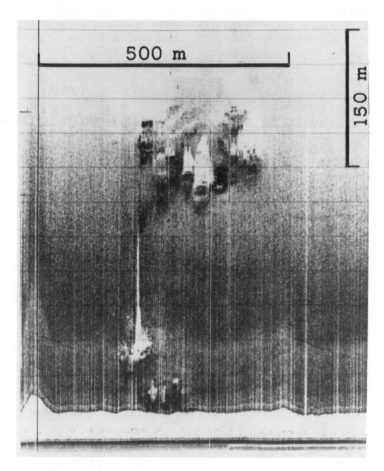

Figure 3. Side-looking sonar image of hydrothermal spire and disturbed
sediment in the Guaymas Basin. The vehicle track is along the bottom
of the figure. Dark areas are those from which acoustic backscatter is
strong. The white zone near the bottom is a measure of vehicle height
off the bottom. The long narrow white spike is the shadow cast by the
nearby spire while the mottled areas are the disturbed regions in an
otherwise smooth sediment-covered seafloor (Spiess and Lonsdale, 1982).

Sulfide mineral deposits and cobalt-rich manganese crusts usually
occur in morphologically complex environments, making it difficult to
distinguish them from their surroundings using acoustic systems.
Phosphorites and manganese nodules, however, generally are found in
simple situations and, thus, are often amenable to mapping on the basis
of their acoustic backscattering properties. Although none of the
deeply towed survey systems have been used extensively in connection

with phosphorite deposits, in one instance in the Southern California
borderland, irregular patches of these deposits were mapped with a side-
looking sonar to demonstrate how the naturally occurring patterns
could be used as a navigation base (Lowenstein, 1969; Spiess and
Lowenstein, 1971).

The Deep Tow system has been used for research in several manganese
nodule environments (Spiess and Greenslate, 1976; Spiess et al., 1978;
Karas, 1978; Spiess et al., 1984; Spiess and Weydert, 1984), although
none of these surveys involved observations as intensive as those
carried out by several of the mining consortia (e.g. Siapno, this volume). The principal systems of interest in these investigations have
been the precision echo sounder, side-looking sonar, photography, and
precisely navigated box coring. The side-looking sonar, in particular,
is useful in mapping the continuity of nodule fields, at least in
qualitative terms, as a complement to photography, with its much more
limited areal coverage. Abrupt changes from areas with no nodules to
those with sparse coverage (a few small nodules per square meter) are
observable, for example near the Line Islands (Normark and Spiess,
1976). Sharply defined bare patches in very dense nodule fields were
mapped at MANOP sites H (6° 30' N, 92° 50' W) and R (30° 20' N,
157° 50' W) (Spiess et al., 1978). While there was insufficient station
time to investigate these bare patches in detail, a box core sample
taken in one of the larger ones at site H showed no buried nodules to
a depth of nearly one meter.

A different application of this system was made during an expedition in the summer of 1983 in the vicinity of 15° N, 125° W (Spiess et
al., 1984). In this case, our group, jointly with that of R. Hessler,
also of Scripps Institution of Oceanography, conducted a study of the
environmental impact of prototype sea-floor mining operations carried
out by Deepsea Ventures, Inc. Our task was to survey the area, locate
and photograph the dredge tracks and assist Hessler's group in box
coring close to the dredge operations and in an undisturbed control
area. Hessler's goal was to determine whether, five years after the
mining operations, there was any significant effect on the populations
of small animals which live in the sediment.

The dredge tracks were clearly visible on the side-looking sonar
records (Fig. 4) and were easily mapped relative to the acoustic transponder navigation net. Photographs of the tracks (Figs. 5a and b)
showed them to be surprisingly fresh looking and with no obvious effect
on the surrounding nodule field beyond about a meter in the most extreme
case. Apparently, in the five year period since the mining operations,
there had been no current sufficient to smooth even the sharpest edges
of the furrows. The controlled box coring campaign at the work site was
successful with six cores taken all within 100 m of the dredge tracks
(Spiess et al., 1984). Preliminary analyses show no obvious differences
between the biota in the vicinity of the tracks and that in the undisturbed control area.

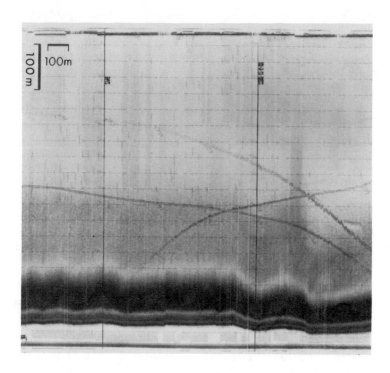

Figure 4. Side-looking sonar image of a manganese-nodule-covered region in which an experimental mining sled had been operated. The dark areas represent regions from which strong acoustic returns were received. The returns from the furrows near the edges of the sled tracks (see Fig. 5) make distinct traces through the image (Spiess et al., 1984).

QUANTITATIVE ACOUSTIC BACKSCATTER POSSIBILITIES

One potential advance in resource assessment, particularly for manganese nodules, would be in the use of statistical and frequency dependence information inherent in the intensity of backscattered acoustic energy. A research project was initiated to make acoustic backscatter measurements at a variety of frequencies under actual field conditions. Five acoustic transducers were built, operating at seven discrete frequencies (4.5, 9, 15, 28, 60, 112, and 163 kHz), each sending a fan beam aft from the Deep Tow vehicle. Knowing the topography along the track and the height of the vehicle, arrivals at any given time after transmission could be related to a particular grazing angle, while the existing Deep Tow systems could be used to obtain "ground truth" information, particularly from photography.

The principal field observations were carried out in parallel with the environmental impact investigations described above, and were complemented by sonar calibration and laboratory measurements of the acoustic reflectivity of individual nodules. Although the bulk of this work is reported in a PH.D. dissertation (Weydert, 1985b), some relevant aspects can be summarized here.

In order to determine the characteristics of individual nodules from this area, several of the nodules from Hessler's box cores were maintained in seawater at low temperature. The acoustic backscattering properties of five representative nodules were then measured in the laboratory as a function of angle at five frequencies between 45 and 160 kHz. The resulting beam patterns showed substantial variations of reflectivity as a function of angle - 10 db in 30°, for example. Average target strength in the horizontal plane increased approximately in direct proportion to 20 log D, where D is a measure of mean nodule diameter in that plane. Frequency (f) dependence was roughly proportional to 11 log f (Weydert, 1985a).

Analyses of the photographs and box cores from this expedition document substantial, large-area variability in the degree of manganese nodule coverage with superposed small scale variations in nodule sizes and abundance. The extent of areas which were statistically homogeneous with regard to size and coverage was typically of the order of a few hundred meters. The sizes of nodules in the box-core samples could best be described by Gaussian distributions, and changes in nodule coverage correlate with the thickness of the uppermost sediment layer as determined from the Deep Tow 4 kHz sounding system (Weydert, in press).

Frequency dependence of acoustic backscattering in nodule fields is different from that for nodule-free regions. Figure 6 shows the difference between normal incidence reflectivity from a bare mud location and a zone having 60–80% coverage with nodules of mean radius about 2 cm. Clearly for this nodule coverage and size, the maximum reflectivity change occurs in the vicinity of 15 kHz, not far from the frequency (12 kHz) for which the product of wave-number times radius is equal to one.

OTHER POSSIBLE MEASUREMENT SYSTEMS

Several other fine-scale survey tools could well be included in future towed instrument suites. Possibilities which would be useful, and for which various individuals have made preliminary investigations, include improved optical imaging systems, electrical methods, in-situ chemical sensors and real-time sediment composition analyses.

Present photographic and television systems suffer from a requirement to operate relatively close to the seafloor (usually 10 m or less off bottom), thus restricting the range to which a simultaneously operated side-looking sonar could provide coverage. Near-bottom operation also implies a narrow swath width for photographic data. Two approaches can be taken to relax this constraint. One is to use more sensitive receiving media (film or charge-coupled-device TV cameras) plus more powerful illumination and different light/camera geometries. Buchanan, at the Naval Research laboratories experimented in the 1960's

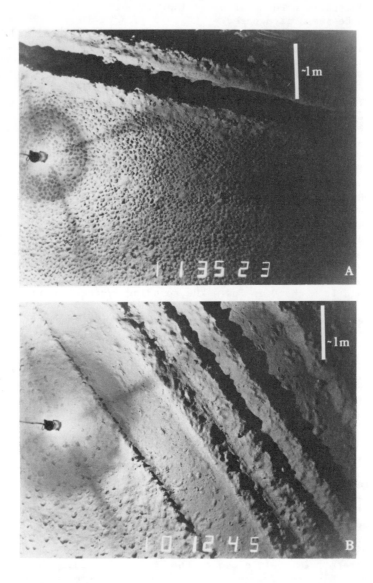

Figure 5. Photographs of edges of dredge tracks such as those shown
in Fig. 4. In A the region below the disturbed zone shows no visible
effect of the mining operations on the surroundings beyond about 10 cm
from the path of the sled runner. In B the region to the left of the
disturbed area shows nodules which are partially covered by sediment
out to about 1 m from the dredged region (Spiess et al., 1984).

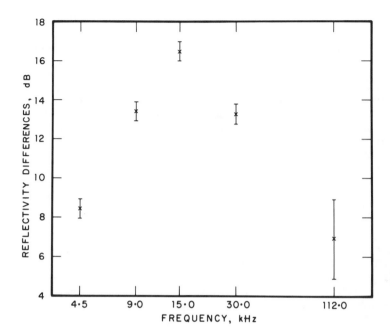

Figure 6.   Changes in normal incidence reflectivity between bare mud
and a 60% nodule-covered seafloor near lat. 15° N, long. 125° W (Spiess
et al., 1984).

with a light-behind-camera (LIBEC) system, while Ballard and Christoff
have tried use of horizontal as well as vertical separation to reduce
masking by backscattered light. The other approach (Dixon et al.,
1984) is to use a range-gated laser line scanner virtually eliminating
the backscattered interfering light field. While both approaches
would be useful in studies of manganese or phosphorite nodules or
crusts, an essential clue in examining the seafloor for exposed sulfide
deposits is the use of color differences which become more difficult to
discern at longer ranges because of the strong absorption of the red
end of the spectrum by seawater. This means that substantial additional
illumination power must be provided in the yellow portion of the spectrum for photo and TV systems while the scanning-laser types should
operate at least at two different wavelengths.
     Electrical geophysical prospecting methods have not been applied
extensively on the seafloor because of difficulties raised by the
presence of the overlying electrically conducting water. Towed direct
current versions of resistivity measuring systems, particularly if they
are operated as much as 50 m above the bottom, will require such large
amounts of power (Francis,1977) that they seem to be impractical in
deep ocean, near-bottom situations. A recent review of this problem
(Edwards, R.N., 1985) indicates that for radiated electromagnetic fields
there may be some useful configurations. Towed electrodes to detect
self-potential effects may be feasible, although there have been

difficulties with inherent electrode noise as the sensors move through the water. Interpretation of self-potential signals is complex, however, they may be useful indicators of sites at which other types of measurements, sampling or observation should be concentrated.

Particularly in prospecting for hydrothermal activity, in-situ chemical sensors may play a role. Two possibilities that would be particularly useful are the measurement of manganese concentration and detection of methane. The former would involve straightforward adaptation of existing laboratory methods. The latter would require substantial innovation, since, at present, the analyses are made by extracting the methane in gaseous form, a difficult process at the 1.5 to 3 km water depths at which one would expect ot use these techniques.

Finally, one should note the approach to continuous analysis of sediments as implemented in shallow water by Noakes (1985). In this method a device is dragged along the bottom, stirring up the sediment and pumping the resulting slurry through a series of x-ray fluorescence analyzers. In deep water one could drag a similar device and incorporate at least an analysis system for a few key components into the deeply towed vehicle and telemeter the results to the surface. Simultaneously a continuously running sample could be retained between rolled layers of filter paper in much the manner of the Longhurst-Hardy plankton recorder (Longhurst, et al., 1966).

NEW VEHICLES UNDER DEVELOPMENT

While sophisticated multi-purpose instrument survey packages will continue to grow in their capabilities and should play a major role in resource asessment, questions as to how to improve on the subsequent steps of sampling and detailed observation of complex situations remain to be answered. At least three groups in the U.S. academic research community are pursuing developments that will provide engineering and operational experience relevant to these problems. All three start with two assumptions. First, cable connected systems, although not as technologically challenging as unmanned free-swimming vehicles, have the advantages of continuous operation, availability of power, and ability to support substantial loads. Second, particularly in deep water, it is often more effective to combine the detailed survey functions discussed above with at least some of the follow-up observations or sampling capabilities in order to use station time efficiently.

The first of these developments, RUM III, is being pursued by Anderson's group at the Marine Physical laboratory of the Scripps Institution of Oceanography (Anderson and Horn, 1984). This approach builds primarily on the seafloor tractor technology which has been developed by that group over the last twenty years (Anderson and Gibson, 1974). The vehicle will have the capability of being landed on the bottom to take samples or make measurements in much the manner of RUM II (Jumars, 1978). In addition, it will operate off-bottom in a towed mode, equipped with a sensor suite modeled after that described above for the Deep Tow. RUM III is a more compact vehicle than its predecessors and will be operable from conventional research ships

equipped with 17.2 mm (0.68") well-logging cable or other similar
electromechanical cables. Its advantage over the two systems, discussed
below, is its ability to carry out operations while resting solidly on
the seafloor, with substantial electrical power available. It should
be particularly useful in suporting in-situ experiments, drilling, and
the installation and maintenance of complex long-term seafloor monitoring devices.

The second of these developments is being pursued by Ballard's
group at Woods Hole Oceanographic Institution. Called Argo-Jason, it
is essentially a marriage of the multi-sensor towed instrument package
with a deep water implementation of the conventional ROV (remotely
operated vehicle) technology which has proven so useful in the offshore
oil production context (Wernli, 1984). In a common ROV configuration a
large package ("garage") is lowered on the end of a combination strain
and electrical cable. A smaller, nearly neutrally buoyant observation
and manipulation package, having propulsion capability, is driven out
of the garage, maintaining its electrical connection through a flexible
neutrally buoyant tether cable. In the Argo-Jason context the garage
(Argo) has been given the multi-sensor survey capability, and the ROV
itself (Jason) is then available to "swim" out for close inspection,
photographic work and small object recovery activities (Ballard, 1982).

The third system, under construction by our own group in the Marine
Physical Laboratory, has as its objective the addition of a precisely
controlled instrument placement or sampling capability to the multi-
sensor survey package. This is to be achieved by the addition of
electrically powered thrusters to the instrument package itself. This
will allow the vehicle to be locally propelled to desired positions
within the transponder net while retaining the full mechanical load
handling capability of the electromechanical cable. It will thus be
able to make detailed observations in complex terrain, and by suspending
sampling devices (drills, corers, grabs), or measurement probes (e.g.
heat flow) below it, there will be a capability of taking samples or
making measurements at precisely chosen locations within the transponder
navigation net.

CONCLUSION

Building in plausible ways upon the capabilities of present
systems, it appears that efficient and effective multi-sensor near-
bottom survey packages can be designed to match the needs for assessment
of various types of mineral resources. With the addition of controlled
sampling functions such systems should be able to answer essential
questions about the extent, volume and grade of the seafloor deposits.
The improved insight derived from simultaneous collection and correlation of fine-scale data of many different types from a single vehicle
can compensate for the fact that systems, operated close to the deep sea
floor, are inherently limited in their rate of area coverage. As these
systems become more flexible in their near-bottom operations, they will
also reduce the requirement for use of manned submersibles, which
currently place further time, cost and weather constraints on follow-up

observation and sampling activities.

ACKNOWLEDGEMENTS

I am indebted to many people and agencies for support of the work reported in this paper. Our Deep Tow system was developed primarily with support of the U.S. Navy, while most of the subsequent research relevant to sulfides and manganese nodules was funded by the U.S. National Science Foundation. The studies of environmental effects of trial manganese nodule mining were supported by the U.S. National Oceanic and Atmospheric Administration. Most of the engineering and seagoing operations involved D. E. Boegeman and C. D. Lowenstein, while principal scientific collaborators have been P. F. Lonsdale, K. C. Macdonald and J. D. Mudie. Marco Weydert carried out the manganese-nodule-related analyses.

REFERENCES

Alexander, C.M., 1981, The complex vibrations and implied drag of a long oceanographic wire in cross-flow; Ocean Eng., v. 8, pp. 379-406.

Anderson, V.C. and Gibson, D.L., 1974, Seafloor soil mechanics and trafficability measurements with the tracked vehicle RUM; In: Inderbitzen, A.L. (ed.), Deep Sea Sediments: Physical and Mechanical Properties, Plenum Press, New York, pp. 347-366.

Anderson, V.C. and Horn, R.C., 1984, Remote Underwater Manipulator: RUM III; Trans. Soc. Automotive Engineers, San Diego, 7 p.

Ballard, R.D., 1982, Argo and Jason; Oceanus, v. 25, pp. 30-35.

Boegeman, D.E., 1970, High speed narrow band signal recognition circuit; U.S. Patent 3,517,214.

Boegeman, D.W., Miller, G.J., and Normark, W.R., 1972, Precise positioning for near-bottom equipment using a relay transponder; Mar. Geophys. Res., v. 1, pp. 381-396.

de Moustier C., 1985, Inference of manganese nodule coverage from Seabeam acoustic backscattering data; Geophysics, v. 50, pp. 989-1001.

Dixon, T.H., Tyce, R.C., Pivirotto, T.J. and Chapman, R.F., 1984, A range-gated laser for ocean floor imaging; Mar. Tech. Soc., v. 17, pp. 5-12.

Edwards, R.N., 1985, Electrical exploration of the seafloor, methodology: A number of concepts and a few resutls; EOS v. 66, no. 46,

876 p.

Francis, T.J.G., 1977, Electrical prospecting on the continental shelf; U.K. Institute of Geol. Science, Report No. 77/4, 41 p.

Jumars, P.A., 1978, Spatial autocorrelation with RUM (Remote Underwater Manipulator), - vertical and horizontal structure of a bathial benthic community; Deep-Sea Research, v. 24, pp. 589-604.

Karas, M.C., 1978, Studies of manganese nodules using Deep-Tow photographs and side-looking sonars; unpubl. M.S. thesis, Univ. California, San Diego, Scripps Inst. Oceanography Ref. 78-20, 38 p.

Klitgord, K.D., and Mudie, J.D., 1974, The Galapagos Spreading Centre: a near-bottom geophysical survey; Geophys. Jour. Roy. Astron. Soc., v. 38, pp. 563-586.

Larson, R.L., 1970, Near-bottom studies of the East Pacific Rise Crest and tectonics of the Gulf of California; unpubl. Ph.D. thesis, Univ. California, San Diego, Scripps Inst. Oceanography Ref. 70-22, 164 p.

Longhurst, A.R., Reith, A.D., Bower, R.E. and Seibert, D.L.R., 1966, A new system for the collection of multiple serial plankton samples; Deep-Sea Res., v. 13, pp. 213-222.

Lonsdale, P., 1977a, Inflow of bottom water to the Panama Basin; Deep-Sea Res., v. 24, pp. 1065-1101.

Lonsdale, P., 1977b, Clustering of filter-feeding macrobenthos near abyssal hydrothermal vents at oceanic spreading centre; Deep-Sea Res., v. 24, pp. 857-863.

Lonsdale, P., 1984, Hot vents and hydrocarbon seeps in the Gulf of California; Oceanus, v. 27, pp. 21-24.

Lonsdale, P., Batiza, R. and Simkin, T., 1982, Metallogenesis at seamounts on the East Pacific Rise; Mar. Technol. Soc. Jour., v. 16, pp. 54-61.

Lonsdale P.F., and Spiess, F.N., 1979, A pair of young cratered volcanoes on the East Pacific Rise; Jour. Geol., v. 87, pp. 157-173.

Lowenstein, C.D., 1966, Computations for transponder navigation; Proc. Nat. Mar. Navig. Publ. Meeting, pp. 305-311.

Lowenstein, C.D., 1969, Side looking sonar navigation; Jour. Inst. of Navig., v. 17, pp. 56-66.

McGehee, M.S., and Boegeman, D.E., 1966, MPL acoustic transponder; Rev. Sci. Instr., v. 37, pp. 1450-1455.

Merewether, R., Olsson, M. and Lonsdale, P., 1985, Acoustically detected hydrocarbon plumes rising from 2 kilometer depths in Guaymas Basin, Gulf of California; Jour. Geophys. Res., v. 90, pp. 3075-3085.

Milliken, R.J. and Zoller, C.J., 1978, Principles of operation of NAVSTAR and system characteristics; Navigation, v. 25, pp. 95-106.

Noakes, J.E., in press, A continuous seafloor sediment sampler (CS^3); a ground truth study; Jour. Geol. Soc. of London.

Normark, W. R., and Spiess, F.N., 1976, Erosion on the Line Islands archipelagic apron: effect of small-scale topographic relief; Geol. Soc. Am. Bull., v. 87, pp. 286-296.

Renard, V., and Allenou, J.P., 1979, Sea-Beam multibeam echo-sounding in "Jean Charcot",- description, evaluation and first results; Int. Hydrogr. Rev., v. 56, no. 1, pp. 35-67.

Simoneit, B.R.T., and Lonsdale, P., 1982, Hydrothermal petroleum in mineralized mounds at the seabed at Guaymas Basin; Nature, v. 295, pp. 198-202.

Spiess, F.N., Loughridge, M.S., McGhee, M.S. and Boegeman, D.E., 1966, An acoustic transponder system; Jour. Instit. Navig. v. 13, pp. 154-161.

Spiess, F.N. and Mudie, J.D., 1970, Small-scale topographic and magnetic features; In: Maxwell, A.E. (ed.), The Sea, v. 4, New concepts of sea floor evolution, Part I, Chap.7 , Wiley-Interscience, New York, pp. 205-250.

Spiess, F.N. and Sanders, S.M., 1971, Survey of chase disposal area NITNATOW; Scripps Inst. Oceanography, Ref. 71-33, 20 p.

Spiess, F.N. and Lowenstein, C.D., 1971, Seafloor pathfinding; U.S. Navy Symp. on Military Oceanography, Monterey, California, v. 1, pp. 89-102.

Spiess, F.N. and Tyce, R.C., 1973, MPL DEEP TOW instrumentation system; Scripps Inst. Oceanography, Ref. 73-4, 37 p.

Spiess, F.N. and Greenslate, A.J., 1976, Pleiades expedition, Leg 04, Preliminary cruise report; Nat. Sci. Found., IDOE Tech. Rept. no. 15, 87 p.

Spiess, F.N., Lonsdale, P., Bender, M., Kadko, D., Zampol, J., and Ford, I., 1978, MANOP cruise report, site survey areas M and H, R/V MELVILLE INDOMED: Leg 1, September - October 1977; 44 p.

Spiess, F.N. and Lonsdale, P.F., 1982, Deep tow rise-crest exploration techniques; Mar. Technol. Soc. Jour., v. 16, pp. 67-74.

Spiess, F.N. and Weydert, M., 1984, Cruise report: RAMA Leg 1 MANOP sites C&R; Scripps Inst. Oceanography, Ref. 84-8, 23 p.

Spiess, F.N., Hessler, R., Wilson, G., Weydert, M. and Rude, P., 1984, Echo I Cruise Report; Scripps Inst. Oceanography, Ref. 84-3, 22 p.

Sumitomo Metal Mining Co., 1981, Multi-frequency exploration system (MFES); Tech. info. rept. 24-8, Tokyo.

Weiss, R.F., Lonsdale, P., Lupton, J.E., Bainbridge, A.E. and Craig, H., 1977, Hydrothermal plumes in the Galapagos Rift; Nature, v. 267, pp. 600-603.

Wishner, K.F., 1979, The biomass and ecology of the deep sea bentho-pelagic (near bottom) plankton; unpubl. Ph.D. Dissert., Univ. California, San Diego, 144 p.

Wernli, R.L., (ed.), 1984, Operational guidelines for ROVs; Marine Technology Society, Washington, D.C., 226 p.

Weydert, M.M.P., 1985a, Measurements of the acoustic backscatter of manganese nodules; Jour. Acoust. Soc. Amer., v. 78, no. 6, pp. 2115-2121.

Weydert, M.M.P., 1985b, Measurements of acoustic backscatter of the deep seafloor using a deeply towed vehicle — A technique to investigate the physical and geological properties of the deep seafloor and to assess manganese nodule resources; unpubl. Ph.D. Dissert., Univ. California, San Diego.

Weydert, M.M.P., in press, Manganese nodule distributions at a site in the Eastern North Pacific; Marine Mining.

Williams, D.L., Becker, K., Lawver, L.A. and von Herzen, R.P., 1979, Heat flow at the spreading centers of the Guaymas Basin, Gulf of California; Jour. Geophys. Res., v. 84, no. B12, pp. 6757-6769.

SEAFLOOR POLYMETALLIC SULFIDES: SCIENTIFIC CURIOSITIES OR MINES
OF THE FUTURE?

Steven D. Scott
Department of Geology
University of Toronto
Toronto, Ontario
Canada M5S 1A1

ABSTRACT. The modern ocean floor is an exceptional natural laboratory
for studying processes and environments responsible for the formation of
ancient massive zinc, copper, iron, silver, gold sulfide ores that are
now being mined on land. The land deposits, in turn, provide important
insights into what can be expected to be found on the seafloor. Sulfide
deposition on modern ocean ridges and seamounts is the result of complex
interrelationships among tectonics, volcanism and heat flow, processes
that must also be important for ancient analogs. Large, ancient ore
deposits are known to have formed in back-arc basins or troughs, within
felsic volcanic rocks of submarine arcs and in sediments on thinned
continental margins. These environments have not yet been carefully
investigated for sulfide deposits on the modern ocean floor.
     The chemistry of vent fluids, as determined at several sites in
the eastern Pacific is highly variable, suggesting that chemical
processes beneath the seafloor are not everywhere the same and require
differing products of water/rock reactions at different temperatures
and/or fluid phase separation. Metal contents of end-member vent fluids
derived from similar basaltic terrains can vary widely, which also
explains the diversity in compositions of ancient ores in identical
rocks. At Guaymas Basin, reactions with sediments have stripped metals
from the fluid, lending credence to the concept that ancient massive
sulfide ores in clastic sediments may be epigenetic or diagenetic rather
than syngenetic. The efficiency of precipitation from a black smoker
is so low that this process alone cannot produce large deposits.
Efficiency can be greatly increased by having a sedimentary trap, as
exists at Guaymas Basin, or by the coalescence of large spires, as at
Explorer Ridge, that act as a "lid" to pond newly-formed plumes beneath
them.
     Seven years of seabottom exploration in the eastern Pacific has
resulted in the discovery of more than 20 sites of former or present
hydrothermal activity, and three of these sites (Galapagos Ridge, a
seamount at lat. 13°N, East Pacific Rise (EPR), and Explorer Ridge)
contain sulfide deposits of a few million tons each. These deposits are
within the size range of those being mined on land. The metalliferous
muds in the Atlantis II Deep of the Red Sea, at 94 million dry-weight

P. G. Teleki et al. (eds.), Marine Minerals, 277–300.

tons, represent a deposit as large as some of the largest mines on
land. With the exception of the Atlantis II Deep, the bulk composition
of the seafloor deposits is poorly known. Nevertheless, analyses or
random samples from the eastern Pacific Ocean, together with the sizes
of known deposits and the expectation that larger ones are yet to be
found in more favorable tectonic environments, suggest that ocean mining
may become an attractive proposition in the future. Presently, seafloor
hydrothermal deposits represent a far greater value to advancing scientific knowledge than for the metals that they contain.

INTRODUCTION

     Actively-forming and fossil seafloor sulfide deposits of zinc,
copper, iron, silver and variable gold content are known from more than
20 locations in the Pacific (Fig. 1), in several "deeps" of the Red Sea
(Degens and Ross, 1969), in the Trans-Atlantic Geotraverse (TAG) area of
the Mid-Atlantic Ridge (Rona, 1985) and on a seamount in the Tyrrhenian
Sea (Minniti and Bonavia, 1984). Additional deposits are suspected to
be present in spreading back-arc basins or troughs and in other near-
arc terrains of the southwest Pacific. Most of the occurences are

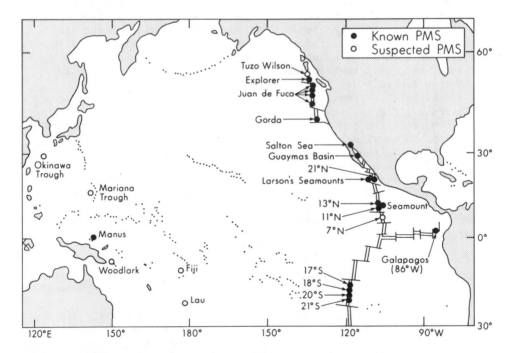

Figure 1. Distribution of known and suspected polymetallic massive
sulfide (pms) deposits and active hydrothermal vents in the Pacific.
Not all of the sites are discussed in the text. The Salton Sea is the
landward extension of the seafloor spreading in the Gulf of California.

small; nonetheless they warrant careful scientific study for the wealth
of scientific information they contain in revealing the geologic
settings and processes by which ancient massive sulfide deposits that
are now on land may have been formed and preserved. Some sites have
individual large deposits (e.g., the Atlantis II Deep of the Red Sea;
Galapagos Ridge at long. 86°W; a seamount at 13°N East Pacific Rise
(EPR); "Magic Mountain" at Explorer Ridge). Others are large by virtue
of having numerous small scattered lenses (e.g. >100 lenses at 13°N EPR
along the axis and in Guaymas Basin, Gulf of California; >60 at Explorer
Ridge).

Individual deposits and sites are described in detail by others in
this volume, and Scott (1985) recently described the processes which are
responsible for the formation and diversity of modern seafloor deposits
and their ancient analogs. In this paper, some of the scientific
advances, which are improving our understanding of sulfide-forming processes and environments of massive sulfides, are discussed from the
perspectives of both active and fossil hydrothermal systems on the
modern seafloor. These are compared to ore deposits that formed in
ancient oceans and are now part of the land mass. Such intercomparisons
will eventually improve our exploration strategies for sulfide deposits
both on land and the seafloor. Furthermore, a discourse is offered
from a strictly scientific, and not a political or economic, point of
view, whether the seafloor deposits, which are not now economical to
exploit, might become so in the future.

TECTONIC SETTINGS OF MASSIVE SULFIDES

The formation of submarine volcanogenic massive sulfide deposits,
whether ancient or modern, requires a heat source within a few km of
the seafloor to drive the circulation of seawater through the rocks,
sufficient permeability to allow deep penetration of the fluid enabling
chemical exchanges with a large volume of rock, fractures to focus
selectively the discharge of large volumes of fluid onto the seafloor,
and an efficient means of precipitating metallic sulfides from the
discharging hydrothermal fluid (Scott, 1985). All seafloor sulfide
deposits, discovered to date, are in axial or near-axial locations on
mid-ocean spreading ridges or are located on seamounts. Both types of
sites have high heat flow and a "plumbing system", but not necessarily
an efficient mechanism of precipitation. The latter point is discussed
further, below.

Investigations of sulfides on actively-spreading ridges have
provided the bulk of the present knowledge about seafloor hydrothermal
processes. However, these deposits probably do not have exact analogs
in the ancient geological record, for their fate is to be subducted
beneath continents or island arcs. An exception to this may be those
deposits that are now forming in young intercratonic ocean basins, such
as the Red Sea or the Gulf of California, which could become part of
the enclosing craton if rifting should fail.

Rifting and its attendant high heat flow has been responsible for
the formation of ancient massive sulfide deposits in several submarine

tectonic environments during the Phanerozoic and perhaps even the Precambrian. The most important are: (1) sulfides within island arcs
(e.g., the Kuroko deposits of Miocene age in Japan; the Bathurst deposits of Ordovician age in New Brunswick, Canada; the Rio Tinto deposits
of Devonian age in Spain), (2) those in back-arc basins (e.g., ophiolite-
hosted deposits of Cretaceous ·age in Cyprus and Oman, of Eocene age in
the Philippines and of Ordovician age in Newfoundland, Canada), and (3)
those in the thinned and rifted craton of a passive continental margin
(e.g., several deposits of Lower Paleozoic age in the Selwyn Basin,
Canada; Devonian-age Rammelsberg and Meggen deposits of Germany). The
island-arc and back-arc settings are predominantly volcanic and probably
experienced ore-forming processes analogous to those observed on the
seafloor today. In contrast, ancient sulfide deposits on passive continental margins occur within thick clastic sedimentary sequences that,
in some cases, are devoid of contemporaneous igneous activity; thus, a
magmatically-driven, convective, hydrothermal origin for these deposits
is not as certain. The Red Sea and the Gulf of California have some
geological features in common with these deposits, with a significant
difference in that the sediments of thinned continental margins are
underlain by a craton of granitoid composition (e.g. present-day Okinawa
Basin and Anatolian Trough, possibly) whereas those in the Red Sea and
Gulf of California are underlain by mid-ocean ridge basalts. Water/rock
reactions will differ in these two settings which, in turn, must have a
strong influence on the composition of their respective hydrothermal
fluids and the resulting deposits (Scott, 1985).
      Deposits that have formed in the arc environments (1 and 2, above)
are numerous in the ancient geological record because they have had a
high chance of preservation. All that is required is for the rifting
or extension either to fail, as was the case, for example, in the
Japanese Green Tuff Belt to preserve the Kuroko deposits (Cathles et
al., 1983), or for the tectonic regime to become compressional, as
would occur if the arc and its adjacent seafloor became accreted to a
craton. The latter is what appears to have happened in the Bathurst
district of New Brunswick, where large (>100 million tons) massive
sulfide deposits in felsic volcanic rocks and an adjacent basaltic
terrain (perhaps a contemporaneous back-arc basin) have suffered as many
as five phases of deformation while being welded onto the North American
craton. Massive sulfides on thinned passive continental margins are
rapidly buried by (or are, perhaps, deposited within) thick sediments
and, as a consequence, the probability of preservation is high regardless of their subsequent tectonic history.
      Modern examples of arc-related rifting abound in the southwest
Pacific Ocean. It is a common phenomenon for active magmatic arcs here
to split longitudinally and, if rifting continues, to produce an intra-
arc trough or back-arc basin. Vanuatu, where Pleistocene basalts occupy
a core within older volcanic rocks along the full length of the islands,
appears to be in an early stage of rifting, whereas the Ogasawara-Bonin-
Mariana arc has developed mature intra-arc troughs with true seafloor
and spreading axes like that of mid-ocean ridges. These arc-related
regions have not been as extensively studied as mid-ocean ridges, but
their spreading and related hydrothermal processes are likely the same

as for mid-ocean ridges, and Cyprus-type massive sulfide deposits are sure to be found in these settings in the future. Indeed, there is geochemical evidence (characteristic CH_4 and Mn anomalies in bottom waters; Kim, 1983) for hydrothermal venting in the Mariana Trough. Lava domes thought to be composed of rhyolite and active felsic volcanism, both associated with ancient Kuroko-type mineralization, have also been reported from the Mariana arc. Similarly, the region that extends from Papua New Guinea and New Britain around to Kermadec and North Island of New Zealand is replete with excellent examples of split arcs, intra-arc troughs and larger back-arc basins, as well as rifted cratons such as Western Woodlark Basin east of Papua New Guinea and Taupo, New Zealand.

Seamounts, particularly those with calderas, are an attractive setting for seafloor sulfides. The abundant deep fissures within calderas facilitate the convective flow of hydrothermal fluid. Replenishment of upper level magma chambers can lead to prolonged high heat flow, which is a requirement for producing large deposits. Modern examples of seafloor sulfides associated with calderas include Axial seamount on the Juan de Fuca Ridge (CASM, 1985), where active and fossil vents were found in eruptive fissures; Green Seamount west of lat. 21°N EPR (Lonsdale et al., 1982); and Palinuro Seamount in the Tyrrhenian Sea (Minniti and Bonavia, 1984). Calderas are not essential, however. A large massive sulfide deposit occurs on the block-faulted flank of a seamount east of the ridge axis at 13° N EPR (Hekinian and Fouquet, 1985). This seamount does not have a caldera. Alabaster et al. (1980) have proposed that the largest Cyprus-type massive sulfide deposits in Oman are associated with seamounts. Their evidence is that the basaltic section of the Semail ophiolite in the vicinity of the orebodies is greatly thickened and, therefore, must have been elevated. The author's observations at the Turner-Albright massive sulfide deposit in the Josephine ophiolite complex, Oregon, and at several massive sulfide deposits in the Troodos ophiolite, Cyprus, suggest the same.

Gradually, it is becoming apparent that volcanic and tectonic processes at ridge axes progress through different stages with time. These stages can be repeated as cycles. Some ridge axes, such as at the southern EPR, are but a broad swelling of the seafloor with a poorly discernable axial valley or none at all. Others such as at the southern Juan de Fuca Ridge and at 13°N EPR have a broad axial valley floored by a relatively smooth lava-lake surface. Still others, such as at Explorer Ridge or the Mid-Atlantic Ridge, have an incised axial valley with a rugged topography. These differences are probably not a simple consequence of the average spreading rate over time, because morphological features on ridge axes can be very different, notwithstanding their similar rates of spreading. Conversely, morphological features can be similar for very different spreading rates. For example, the spreading rates of the southern Juan de Fuca and Explorer Ridges are both about 6 cm/yr and, yet, they are geomorphologically very different. The Reykjanes Ridge, which is spreading at about 2 cm/yr, has the shape of the ridges of the southern EPR that are spreading at 18 cm/yr. A more likely explanation for these morphologies is that different ridges, and even different segments of the same ridge system, are at

different stages of a complex volcanic-tectonic cycle and it is the
frequency of this cycle that is proportional to spreading rate.
Volcanism must occur sometime on all of the world's 72,000 km of mid-
ocean ridges, but such activity, generally, is infrequent on slow-
spreading ridges and is more frequent on fast-spreading ridges. A key
point is that whenever volcanism is about to occur or has recently
occurred, heat flow will be high enough to produce hydrothermal activity
and sulfides. Thus, there is a likelihood of finding sulfide deposits
on the slow-spreading ridges, like the Mid-Atlantic Ridge, but the
infrequency of volcanism and the presumed shortness of the volcanic
episode means that significantly fewer deposits will be forming at any
given instant than on a fast-spreading ridge. Furthermore, deposits on
the slow-spreading ridges are likely to be smaller than those on fast-
spreading ridges, because the thermal pulse of the former is shorter
and weaker than that of the latter.

The timing of hydrothermal activity relative to the volcanic-
tectonic cycle must also be critical in determining the size of the
deposit. The largest deposits are likely to be formed just before or
just after a period of intense volcanism, that is, during the period
when the thermal pulse is either rising or decaying. At the peak of
volcanism, heat flow may be the most intense and conducive to excess
hydrothermal circulation but the time between volcanic events is not
great enough to allow a very large deposit to accumulate. Furthermore,
the thick shield of newly erupted lavas is relatively impermeable,
hence there is no plumbing system for fluid flow. However, after this
wave of volcanism has abated but while the crystallizing axial (or
seamount) magma is still hot, tectonics predominates and normal faulting
which accommodates extension of the axial region (or caldera growth)
produces the required pathways for deep circulation of seawater which
then reacts with hot rocks. Thus, hydrothermal venting is focussed and
prolonged, which satisfies two of the requirements that are essential
to the growth of large deposits. To illustrate this, the larger known
sulfide deposits in the eastern Pacific, those on the Galapagos Ridge,
on the seamount east of 13°N EPR and on Explorer Ridge, are on relatively old basalts and clearly have formed during a post-volcanic (or a
tectonic) stage. Basalts in the vicinity of the Explorer Ridge deposits
have coatings of manganese oxides, 0.1-1.0 mm thick, and are dusted by
pelagic sediments. In contrast, basalts in the vicinity of the very
small vents, found in an eruptive fissure on the floor of the Axial
Seamount caldera, are exceedingly young and glassy, and have buried a
nearby sulfide deposit (CASM, 1985). The ancient geological record
also shows that large volcanogenic massive sulfide deposits have formed
during lengthy hiatuses in volcanism. The stratigraphic horizon on
which these deposits have formed is commonly marked by a distinctive
chemical sediment ("tuffaceous exhalite" of Kalogeropoulos and Scott,
1983), whose deposition requires an extended period of volcanic
quiescence.

CHEMICAL AND PHYSICAL PROCESSES AT HYDROTHERMAL VENTS

The first analyses of high-temperature vent fluids were made on samples from 21°N EPR by Edmond et al. (1982) and Von Damm (1983). The end-member fluid (i.e., an uncontaminated hydrothermal solution) contains no sulfate or Mg and its chlorinity and salinity are about the same as seawater, in agreement with experimental studies of basalt/seawater reactions at high temperatures (Bischoff and Dickson, 1975; Mottl and Holland, 1978, 1979; Seyfried and Bischoff, 1981). The temperature of the end-member is 350°C, which was commonly regarded to be a maximum that could be achieved by a "black smoker". Lower-temperature vent fluids at 21°N are a product of simple mixing of the 350°C end-member fluid with ambient 2°C seawater, and, as a consequence, the relationships among the concentrations of dissolved consitutents and Mg or temperature are linear. A partial tabulation of the chemical composition of the 350°C end-member is given in Table I. The fluid is exceedingly enriched in most constituents relative to the concentrations found in normal seawater. Analyses of fluids from 21°N and experimental studies on mineral solubilities explain the presence of copper/zinc compositional zoning observed in chimneys (Oudin, 1983; Haymon and Kastner, 1981; Goldfarb et al., 1983; Janecky and Seyfried, 1984) and, as is discussed further below, in the interiors of mounds and in ancient ores. Copper sulfide is soluble in only the highest temperature vent fluid (>280°-300°C), whereas zinc can be transported at both high and low temperatures. Thus, zoning is a function of temperature and involves the replacement of early-formed, low-temperature, zinc-rich assemblages by high-temperature, copper-rich assemblages.

The rather simple processes operating at 21°N do not apply quantitatively to all basalt ridge axes, however. Table I also includes calculated end-member compositions of vent fluids from Axial Seamount, from two sites at 13°N EPR nearby one another and from Explorer Ridge. The calculations were made by linear extrapolation of the analytical data to Mg=0. On this basis, only 5% of the Axial Seamount fluid is end-member but 91% and 54%, respectively, at 13°N and Explorer Ridge so their extrapolations are not so uncertain. As shown in Table I, the extrapolated values for Axial Seamount are very similar in composition to the 21°N end-member fluid (except for Fe). The Explorer fluid is very different in almost all respects. The two end-member fluids from 13°N differ dramatically from one another in their Cl, Fe and Ca contents and from that of 21°N in regard to most elements. These differences must be real and not simply an artifact of contamination by ambient seawater during sampling because the extrapolation method would correct for that and, besides, many of the deviations in the extrapolated compositions of the end-member fluids from 13°N and Explorer Ridge (e.g. for CA, K and Cl) are in the wrong sense relative to that expected from seawater contamination of a 21°N-type of fluid. The reason for the major differences in the chemistries of the two fluids at 13°N, as well as among those of 13°N, 21°N and Explorer Ridge is not obvious. The abnormal composition of the fluid from Explorer Ridge may be related to its extrapolated end-member temperature of 617°C. At first glance, this temperature seems to be absurdly high, but it is

physically possible given the very high solidus temperatures (>1000°C)
of basaltic magmas. Also, the commonly accepted 350°C maximum is no
longer sacrosanct in light of the carefully documented measurement by
Delaney et al. (1984) of a 420°C vent fluid on the Endeavour segment of
the northern Juan de Fuca Ridge. If very deeply circulated water did
achieve 617°C at Explorer Ridge, confining pressure would have to be
>800 bars in order to prevent phase separaration (Sourirajan and
Kennedy, 1962). This pressure corresponds to a hydrostatic column

Table I. CALCULATED* END-MEMBER COMPOSITIONS OF EASTERN PACIFIC VENT
WATERS COMPARED TO THE COMPOSITION OF SEAWATER

	21°N	13°N (A)	13°N (B)	EXPLORER	AXIAL	GUAYMAS	SEAWATER
T, °C	350	368	–	(617)	300	315	2
pH	3.5	3.4	3.9	5.0	–	5.9	7.8
Cl (ppt)	18.26	31.83	26.15	20.36	–	21.25	19.18
ppm							
H_2S	252	157	164	60	224	162	0
SiO_2	1050	1253	1396	610	1139	735	10
Ca	647	2796	2072	1720	1152	1252	408
Ba	2	5	4	1	3	4	0
Sr	7	21	15	12	9	18	8
K	939	1290	1180	1815	–	1671	383
Mn	49	88	47	15079	30	9	0
Fe	80	663	82	15	3	4	0
Cu	2	–	–	0	–	0.04	0
Zn	5	–	–	0.05	–	0.7	0
ppb							
Ag	3	–	–	–	–	9	0
Pb	54	–	–	–	–	75	0

LOCATIONS OF VENT FIELDS

*End-member calculated for Mg=0.
Sources: 21°N--Von Damm (1983) mean values; 13°N (A) EPR--Grimaud
et al. (1984) samples 20Ti4 and 14Ti2, Chandelier site, pH measured at
20°C on 335°C fluid 14Ti2; 13°N (B) EPR--Grimaud et al. (1984) samples
24G2 and 20Til, Chainette site, temperature not measured, pH measured
at 20°C; Explorer Ridge--Tunnicliffe et al. (in press) vents 12c and
12d, pH measured at 20°C on 306°C fluid from vent 12c and 12d, pH
measured at 20°C on 306°C fluid from vent 12c; Axial Seamount--CASM
(1985) vent A, temperature from silica geothermometer; Guaymas Basin--
Von Damm (1983) mean values; Seawater data--Von Damm (1983) bottom
water at 21°N.

(i.e. ocean depth + open fracture) of 8000 m or a closed lithostatic
column of 1800 m below a seafloor with a depth of 2000 m. Some intermediate condition between hydrostatic and lithostatic is more likely.
Water/rock reaction at 617°C would, presumably, produce different
chemical compositions than those reported from the aforementioned experiments that were conducted below 500°C and from compositions measured
in lower-temperature vent systems. In such a case, fractures must open
in order for this fluid to escape to the seafloor. The sudden decompression would cause adiabatic separation of the fluid into two phases
("gas" and "liquid" of Sourirajan and Kennedy, 1962) resulting in a
loss of H_2S, precipitation of some metallic sulfide and increased
chlorinity of the residual fluid (see further, below, and Table II).
Mixing of this residuum with seawater at some depth beneath the seafloor
to produce a fluid that contained 54% residuum and 46% seawater would
add the 680 ppm Mg that was the average value analysed in the highest-
temperature vent fluids at Explorer Ridge and would cause even more
sulfides to be precipitated in fractures in the basalt. Such a process
could further account for the low concentrations of H_2S and metals that
are shown in Table I.

Vent fluids at Guaymas Basin (Table I) are also very unlike those
at 21°N, but these differences have been explained by Bowers et al.
(1985) to be the result of secondary reactions of a 21°N-type fluid,
generated in basalt, with sediment that buried the ridge axis. The sediment contains abundant carbonates, feldspar, clay minerals and organic
matter. The fluid/sediment reaction caused 80-98% of the metals to be
deposited as sulfides in the sediment, hence H_2S content is also proportionately lower in the Guaymas vent fluids. Oddly, however, Ag and Pb
are enriched in the fluids at Guaymas, presumably because there is a
source for these in the sediment. These fluids and the mounds in this
basin contain abundant hydrocarbons, which Simoneit and Lonsdale (1982)
believe to be products of rapid thermal maturation of the organic matter
in the sediment. The fluid at Guaymas is much closer in composition
(except for Mn) to those at Explorer Ridge, in contrast to that of
21°N, which gives another reason to suspect that the Explorer fluid has
been depleted as a consequence of precipitation below the seafloor.

The measured salinities and chlorinities in the Red Sea Deeps and
at a number of vent sites in the eastern Pacific fit neither the model
for 21°N nor results of basalt/seawater reaction experiments. These
experiments and the 21°N model have predicted that values should be
close to the 540 millimoles/kg chlorinity and 3.5% salinity of normal,
deep seawater. In fact, chlorinities of vent waters in the Pacific
range from 0.6 to 2 times that of seawater (Table II), and fluid
inclusions from veins in the stratigraphically lower sediments of the
Atlantis II Deep are as much as 10 times more saline than seawater (32%
equivalent NaCl; Oudin et al., 1984). Similarily, salinities of fluid
inclusions in veins and stockworks beneath several ancient, volcanogenic
massive sulfide orebodies are much higher than present-day seawater
values; e.g., 3.5 to 7% in Kuroko deposits (Pisutha-Arnond and Ohmoto,
1983; Bryndzia et al., 1983) and as high as 38% beneath some deposits
in the Precambrian of Quebec (Farr, 1984; Costa et al., 1984). At the
Atlantis II Deep, high salinities could have resulted from dissolution

of evaporites that are known to be interlayered with sediments flanking
the young ocean basin. However, this explanation clearly does not
suffice for the Pacific vents or for ancient ores in volcanic rocks.
Mixing a vent fluid of normal expected composition (e.g., 21°N) with a
very saline magmatic fluid (Sawkins and Kowalik, 1981; Bryndzia et al.,
1983; Michard et al., 1984) can explain high, but not low, salinities.

Table II. CHLORINITY AS A FRACTION OF SEAWATER (sw)
VALUE FOR SEVERAL VENT SITES IN THE EASTERN PACIFIC
(chlorinity of sw = 540 millimoles/kg)

SITES	T,°C	CHLORINITY	SOURCE
10°57'N EPR	347	0.6 x sw	1
ENDEAVOUR	380	0.8 x sw	2
21°N EPR	350	sw	3
EXPLORER	291	1.1 x sw	4
GUAYMAS	315	1.2 x sw	3
11°15'N EPR	–	1.3 x sw	1
12°50'N EPR	381	1.4 x sw	1
S. JUAN de FUCA	–	2 x sw	5

Sources: (1) Kim et al., 1984; (2) McDuff et al., 1984;
(3) Von Damm, 1983; (4) Tunnicliffe et al., in press;
(5) von Damm and Bischoff, 1985.

Phase separation or boiling is another effective means of producing
fluids with widely varying salinities (Delaney and Cosens, 1982;
Bischoff and Pitzer, 1985). The phase diagram for the binary system
$NaCl-H_2O$ (Sourirajan and Kennedy, 1962), at pressures and temperatures
of seafloor hydrothermal conditions, has an extensive solvus separating
one fluid ("liquid") with a very high salinity (as much as 50% NaCl)
from another ("gas") with a very low salinity (0.1% NaCl). Re-mixing
these two fluids in various proportions at a lower temperature could
easily produce the observed salinities of both modern and ancient
hydrothermal fluids, but would require a rather elaborate "plumbing
system" to permit the separation of the phases and their later recombin-
ation. Also, rare gases must surely undergo fractionation by the
salting-out effect (whereby the activity coefficient of the dissolved
gas increases with ionic strength of the "liquid" resulting in effer-
vescence from the gas from solution), or by virtue of the expected differences in density between the two fluids (although their densities
are not known). The consistency of rare gas compositions (Kim et al.,
1984) would seem to preclude simple phase separation. Water/rock
reactions also do not provide an adequate explanation for the observed
chlorinities/salinities, at least not at the temperatures with a 500°C
maximum, at which the experiments have been conducted. Na and Cl are

conserved in these reactions, thus low salinities cannot be generated. Loss of H_2O from the hydrothermal fluid through hydration of basalt can increase the salinity, but not by the large amounts observed at some modern vents (Table II). A 40% increase in salinity, produced by increasing the H_2O content of a normal basalt from 2% to a rather high value of 6%, requires a water/rock ratio of 0.14. A 100% increase in salinity, as was found at the Southern Juan de Fuca Ridge (Von Damm and Bischoff, 1985), requires a water/rock ratio of 0.08. These ratios are one to two orders of magnitude less than that estimated for 21°N by several researchers. If water/rock reactions can occur at much higher temperatures, as seems likely in view of arguments presented earlier in this section, NaCl may not be conserved. The possibility that oceanic crustal rocks may be a repository of NaCl under conditions, not studied experimentally to date, is suggested by the discovery of secondary Ca-amphibole enriched in Na and Cl in ferrogabbros from ridge axes and transforms (Ito and Anderson, 1983; Honnorez and Kirst, 1975; Honnorez et al., 1984). Here, then, is a mechanism that could produce both low salinities (extraction of NaCl from fluids at high temperatures) and high salinities (release of NaCl to fluids at lower temperatures) at the same site. This, indeed, has been observed in vents near 11°N EPR (Kim et al., 1984).

SIZE OF SULFIDE DEPOSITS

     With notable exception of the deposits in the Red Sea, the remaining, known seafloor sulfide deposits are modest to insignificant in size. Most consist of mounds no more than a few tens of meters in diameter, a few meters high, topped by spires as tall as 30 meters and average only about about 1,000 tons each. Many are considerably smaller. However, more than 100 of these mounds are scattered along an 18-km length of the ridge axis at 13°N EPR (Hekinian and Fouquet, 1985).
     The largest known seafloor deposit, the Atlantis II Deep in the Red Sea, has proven reserves of 94 million dry weight tons (Mustafa et al., 1984). It is only one of several deposits in the Red Sea which, unlike those in the Pacific Ocean, consist of metalliferous muds within depressions along the spreading-axis. The largest known Pacific deposits are on the Galapagos Ridge at 86°W (discovered in 1981), on a seamount at 13°N (discovered in 1981) and on Explorer Ridge (discovered in 1984). The Galapagos deposit was originally claimed to be as much as 1,000 m long, 20-200 m wide and 35 m thick and to contain several million tons of sulfides (Malahoff, 1982) but this estimate was not based on a systematic assessment. The seamount deposit discovered by the French at 13°N EPR is estimated by Hekinian and Bideau (in press) and Y. Fouquet (1985, personal communication) to contain from 2 to 4 million tons of sulfides. The areal extent of the deposit was determined from photographic and submersible observations. Thicknesses were obtained from submersible observations on fault scarps which cut the deposit and from geophysical self-potential measurements (Francis, 1985). At Explorer Ridge, deep-tow photography and submersible opera-

tions have located more than 60 deposits, ten of which are at least 150
m in diameter (the "Magic Mountain" deposit is about 300 m) and 5-10 m
thick (as much as 30 m in places) as measured on faulted sections, or
about 3-5 million tons in aggregate.

For comparison, Mavrovouni, the largest deposit in Cyprus, to which
those on the seafloor may be similar, contained about 17.5 million tons
of massive sulfides. However, the next largest, Skouriotissa, contained
only 2.5 million tons and several deposits mined on Cyprus contained no
more than a few tens of thousands of tons of massive sulfides. Furthermore, the 115 potentially mineable massive sulfide deposits discovered
in the Canadian Shield prior to the 1980's have a median content of
only 1.3 million tons and a mean of 6.2 million tons (Boldy, 1981).
Lost in these statistics, though, are the rare gigantic discoveries,
such as Kidd Creek quoted at 98 million tons but is likely much larger.
The entire Hokuroko district of Japan contained about 70 million tons
of ore, much of which has been mined. This is distributed, however,
within 12 different mines, each of which can consist of several individual deposits ranging from a little as a few tens of thousands of
tons to as much as a million tons or more. In light of these figures,
the larger seafloor sulfide deposits found to date are in the range of
many of the ancient volcanogenic massive sulfides being mined today on
land and the Atlantis II Deep is among the largest. What must be
remembered, is that all of the discoveries in the Pacific Ocean have
taken place only since 1978, and new ones are occurring at an acceler-
ating rate. Those in the Red Sea began just a decade earlier. The
recent rate of discovery of deposits on the seafloor exceeds that on
land and, together with the observation that only a miniscule fraction
of potential sites on the world's ocean floor have been closely surveyed, raises expectations ttha there are more and larger deposits yet
to be discovered.

Where will these deposits be found? Those discovered in the
eastern Pacific, including the three largest, are on mature mid-ocean
spreading ridges or on seamounts. However, the ancient geologic record
indicates that the largest volcanogenic massive sulfide deposits of
Phanerozoic age (the tectonic setting of Precambrian deposits is uncertain) are in island-arc terrains of predominantly felsic volcanic rocks
(e.g., Rio Tinto, Spain; see Monteiro and Carvalho, this volume). This
geologic setting has not yet been closely examined on the modern seafloor. Most of the world's largest Pb-Zn-Ag massive sufide deposits
occur in fine clastic sediments (shales) which were deposited on
thinned passive continental margins. This is another contemporary
geologic setting that should be examined more closely for active sulfide
formation.

The rate of accumulation of metallic sulfides at ridge-axes or
seamount sites is far less than might be expected from their flux. A
single black smoker chimney with an orifice 20 cm in diameter, that
typically vents a fluid containing 100 ppm metals at 2 in/sec (Converse
et al., 1984), produces 250 tons of metallic sulfide per year. A good-
sized field might have 100 such chimneys and a life of, say, 100 years.
During this time period, about 3 million tons of metallic sulfide
(principally FeS_2) could be produced, together with a few million tons

of non-sulfides (principally anhydrite, silica and barite) plus minerals
that are precipitated in the underlying stockworks. However, large
sulfide deposits are rare on the seafloor principally because an estimated 97% of the particles in a black smoker plume are lost into the
seawater column (Converse et al., 1984). The ultimate fate of this
considerable quantity of lost material is not known at present. This
inefficiency of precipitation is probably the most important reason why
typical ridge-axis deposits are so small. The system needs a lid to
prevent the escape of material if it is to produce a large deposit.

Campbell et al. (1985) assumed that the lid for large deposits
could be a bed of anhydrite that, owing to the mineral's retrograde
solubility, should have been precipitated when seawater was heated by
volcanic activity and later was redissolved in cold seawater, removing
any evidence of its former presence. Kuroko-type deposits commonly
have extensive gypsum (replacing anhydrite) beds associated with them
but they are below the massive sulfide ores rather than above them. If
Campbell et al. are correct in their hypothesis, one might expect to
find extensive beds of anhydrite on the modern seafloor. This is not
the case, however. Instead, as at Explorer Ridge, chimneys commonly
develop flanges and, in a forest of such chimneys, these flanges coa-
lesce to form a "roof" several meters above the sulfide mounds. When
new chimneys develop on the mounds, which is a common phenomenon, their
effluent is held under this roof long enough, perhaps, to allow the
sulfide particles in the "smoke" to settle out. Another possibility is
venting within a newly drained lava lake which has retained its solid-
ified basalt cap. Sediment is an even better trap, and this may explain
why ancient, massive sulfide deposits in sedimentary rocks are commonly
so large. In the Guaymas Basin, 80-98% of the original metal content
of the vent fluid (Table I) has been precipitated in the sediments that
bury the ridge axis (Von Damm, 1983; Bowers et al., 1985). Guaymas is
a large vent field covering 12 km^2 and could easily have more than 100
chimneys active at any given time. Using the earlier calculation of
sulfide production in a vent field and assuming that the Guaymas field
is 100 years old, a minimum of 2 to 3 million tons of metallic sulfides
may have been deposited in the sediments beneath the seafloor. Drilling by the Deep Sea Drilling Project in Guaymas Basin (holes 477 and
477A; Curray et al., 1982) had intersected some of these sulfides
within finely laminated sediments beneath a sill, which appears to have
acted as a barrier to an upward flow of hydrothermal fluid (Scott,
1985).

COMPOSITIONS OF SULFIDE DEPOSITS

Systematic, three-dimensional sampling has not been carried out
for any of the Pacific seafloor deposits. Therefore, their bulk compositions and vertical extent are very poorly known in comparison with
ancient massive sulfides. Averages of analyses for selected sites
derived from the few available data are shown in Table III. These
appear to show significant differences with respect to the type of host
rock (i.e., basalt vs. sediment) and to geographic distribution. The

low content of metals, the high ratio of Pb/(Cu+Zn) and high values
of Ba and Ca of the Guaymas samples, compared to those of other sites
listed in Table III, is explained by the differences in the compositions
of their respective vent fluids (Table I). These differences in the

TABLE III. AVERAGE CHEMICAL COMPOSITIONS OF SAMPLES FROM SELECTED
SEAFLOOR DEPOSITS

	21°N EPR	13°N EPR	EXPLORER RIDGE	AXIAL SMT	GUAYMAS BASIN	ATLANTIS II DEEP
No.	3	6	8	14	14	*
Host	B	B	B	B	S	S
Wt %						
Zn	32.3	9.3	9.0	22.2	1.0	2.07
Cu	0.8	7.7	8.1	0.4	0.2	0.45
Pb	0.3	<0.01	0.1	0.4	0.4	–
Fe	19.2	28.8	10.8	5.6	5.9	–
SiO_2	7.8	9.1	19.2	28.1	28.4	–
Ba	0.2	<0.1	7.9	9.6	14.9	–
Ca	1.3	3.1	0.2	0.2	6.7	–
ppm						
Ag	156	48	112	189	69	38.5
Au	0.2	–	0.8	4.9	0.2	0.5

* Proven reserves 94 million tons, salt-free dry weight
B   basalt   S   sediment
Sources: 21°N -- Bischoff et al. (1983); 13°N EPR -- Hekinian and
Fouquet (1985) and pers. comm. (1985); Explorer Ridge -- M.D. Hannington
and S.D. Scott, unpubl. data; Axial Seamount -- CASM (1985); Guaymas
Basin -- J.M. Peter and S.D. Scott, unpubl. data; Atlantis II--Mustafa
et al. (1984).

fluids are a consequence of different water/rock reactions in basalt
and in sediment, as discussed previously. The sediments of the Guaymas
Basin, being rich in carbonate and pelitic material, are a ready source
of the excess Pb (relative to other metals), Ba and Ca present in the
overlying deposits and act as a sink for Fe, Cu and Zn. The apparent
geographical differences in compositions in Table III are probably an
artifact of sampling. Exceptions may be Ba and Au, whose concentrations
in samples from Explorer Ridge and Axial Seamount are genuinely higher
than in those from further south on the EPR. The reason for this difference in Ba levels is not known. Hannington et al. (in press) have
assigned the very high Au values at Explorer Ridge (mean of 8 analyses
0.8 ppm; range 0.07 to 1.5 ppm) and Axial Seamount (mean of 14 analyses
4.9 ppm; range 2.9 to 6.7 ppm) to scavanging by Pb-Sb-As sulfosalt
minerals during decaying hydrothermal convection within mature mounds.

Young immature mounds with abundant black smokers, such as those at
Guaymas Basin and 21°N EPR, do not have high Au contents. Most of the
analyses which compose Table III are related to chimneys or surfaces of
mounds, as these were the most easily sampled by a submersible.
Chimneys are usually copper-rich or zinc-rich depending on their stage
of growth, and, hence, can significantly skew the distribution of
analyses. Mounds appear to be made primarily of debris from collapsed
chimneys and this debris has been recrystallized and impregnated with
minerals from circulating hydrothermal fluids. The interiors of mounds
must be sampled to obtain better estimates of their bulk composition.

Perhaps the best three-dimensional sampling was performed at
Explorer Ridge where the interior of several fossil deposits have been
exposed by faulting. Although only eight samples have been analyzed to
date (more are in progress), they show a wide compositional range (Table
IV). This further stresses the need for caution in making economic
assessments or drawing conclusions about metallogenesis based simply on
averages. The range of values in Table IV is caused primarily by a
pronounced vertical mineralogical zoning that is evident within the
incised deposits. The base and internal core of the mounds is enriched
in chalcopyrite, isocubanite, covellite and marcasite/pyrite, whereas
their tops and outer margins are enriched in sphalerite, barite and
silica. Copper-rich cores are probably common in mounds, as they are
in ancient massive sulfides, but the size of the core and degree of
segregation of zinc, silver and gold from copper depends on the stage
of maturation of the mound (Hekinian and Fouquet, 1985; Hannington et
al., in press).

The vertical zoning seen at Explorer Ridge and also at 13°N EPR
(Hekinian and Fouquet, 1985) is comparable to that in many ancient
volcanogenic massive sulfide deposits. For example, the Cu-Fe-rich
assemblage at the base of the Explorer Ridge mounds is very similar to
the "yellow ore" at the base of the Japanese Kuroko massive sulfide
lenses. Similarly, the $Zn-Ba-SiO_2$ assemblage at the top of the mounds
resembles the "black ore" at the top of Kuroko lenses. The only significant difference is that the Kuroko "black ore" contains abundant
lead. However, this may simply be a consequence of having an ample
source of lead in the Miocene felsic volcanic and Permian basement
rocks underlying the Kuroko ores. The deposits of Cyprus also display
a primary zonation of minor zinc and trace lead enrichment towards the
top of massive sulfide bodies (Constantinou, 1972) although they are
primarily copper deposits. Another feature common to these deposits is
an iron + silica layered capping on the massive sulfide lenses, such as
manifested by the "tetsusekiei" (a tuffaceous, ferruginous chert) of
Kuroko deposits (Kalogeropoulos and Scott, 1983), the ferruginous cherts
and ochres of Cyprus, and the red metalliferous sediments that mantle
modern seafloor deposits. The banded iron formations and iron + silica-
rich tuffaceous exhalites, which occur in the hangingwall of Lower
Paleozoic (e.g., Bathurst, New Brunswick) and Precambrian (e.g., Noranda)
deposits may also be analogs. Constantinou and Govett (1972) concluded
that the Cyprus ochres were produced by submarine weathering. This
process appears to be responsible for the origin of oxidized mantles on
modern seafloor deposits too, insofar that they resemble ochres. How-

ever, some of the silica-rich iron-oxide mantles on modern seafloor
deposits are of bacterial origin and bacterial filaments, similar to
those seen in seafloor deposits, have been found in samples of ferruginous chert from Cyprus, the Philippines and California (Fouquet and
Juniper, in press).

TABLE IV. MEAN AND RANGE IN CHEMICAL
COMPOSITION FOR EIGHT SAMPLES FROM
EXPLORER RIDGE

	Average	Range
Wt %		
Zn	9.0	0 - 34.3
Cu	8.1	1.0 - 29.3
Pb	0.1	0.01 - 0.7
Fe	10.8	4.1 - 20.1
S	26.8	5.2 - 43.3
SiO_2	19.2	1.3 - 31.2
ppm		
Au	0.8	0.07 - 1.5
Ag	112	0 - 640

Source: M.D. Hannington and S.D. Scott
          unpubl. data

CONCLUSIONS

     The modern seafloor offers an exceptional opportunity to study the
geologic setting and processes of massive sulfide formation and to help
solve some outstanding problems of ore genesis. In fact, because of
these studies, volcanogenic massive sulfides are among the best understood of all types of metallic ore deposits. The relationship between
different tectonic settings on the modern seafloor and the types of
deposits that these produce are certainly clearer.
     Ocean ridges display various morphologies depending on which phase
of the tectonic-volcanic evolutionary cycle they are in at any given
time. Large massive sulfide deposits, both ancient and modern, form
during the significant hiatuses in volcanism. For example, the large
deposits of Explorer Ridge overlie relatively old basalts, whereas
nearby young sheet flows host only scattered small deposits. Surfaces
of present-day seamounts would seem to offer an attractive site for
finding large sulfide deposits. The geological record indicates,
however, that the largest deposits are likely to be found on thinned or
rifted continental margins, in back-arc troughs or basins and on deep,
submarine segments of magmatic arcs.
     The simple chemical processes responsible for the composition of
the end-member vent fluids at 21°N EPR seem not to hold for fluids from

all basalt ridge crests. Temperatures higher than the 350°C measured at 21°N, once thought to be a theoretical maximum, have been found on the Juan de Fuca Ridge (420°C) and at 12°50'N EPR (381°C). Furthermore, at Explorer Ridge a value of 617°C is extrapolated for the temperature of the end-member fluid. Salinities of vent fluids are now known to range from 100% greater than that of seawater to 40% less, and metal contents of end-member fluids from different sites can vary widely. The consequence is that deposits forming on basalts should and do have a range of compositions in spite of the source rocks for the fluid constituents having more or less the same composition, i.e., that of mid-ocean ridge basalt. From this it is expected and, indeed, obser-vable in the geological record, that certain differences in compositions of deposits formed on the same rock type should exist and, where the rock type is different, the departures in compositions should be even larger. For example, Canadian Archean massive sulfide orebodies contain practically no lead, but their Zn/Cu ratios are highly variable from 0.3 to 243 (Sangster and Scott, 1976) and the ratios do not bear a simple relationship to the proportions of felsic and basaltic volcanic rocks in the footwall. On the other hand, post-Archean massive sulfide deposits within felsic volcanic rocks (e.g., in the Hokuroku district, Japan) or in pelitic sediments (e.g., at Rammelsberg, Germany) commonly contain abundant quantities of lead, whereas those in basalts (e.g., Cyprus) do not.

All of modern deposits with basalt as the host have a high ratio of metals/(sulfates + silica). At Guaymas Basin this ratio is low. Here, where the axial valley is filled with organic and carbonate-rich pelagic and pelitic sediments, hydrothermal fluids generated in basalts of the ridge axis have reacted further with the sedimetns. This has resulted in precipitation of a substantial amount of metals in the sediments and in drastically changed compositions of vented fluids and their precipitates. This observation lends credence to the argument, in the ongoing debate of syngenesis vs. epigenesis, that ancient sediment-hosted massive sulfide deposits have formed within, rather than on top of unlithified sediments.

It is clear that what is being learned about either ancient or modern massive sulfide deposits is increasing our understanding of both. Although the presently-known seafloor deposits may not be exact analogs of ancient massive sulfides, the process by which ancient massive sulfides formed can be better understood. This, in turn, leads to better appreciation of several secondary aspects of this process and may lead to more discoveries on land. For example, studies of the dispersal of plume consitutents around active vent fields will help quantify trace element distributions in cherty sediments on ancient ore horizons as a lithogeochemical guide to mineralization (Scott et al., 1983).

Are the modern seafloor deposits mere scientific curiosities, albeit informative, or do they have a real economic value? Strictly speaking, at the moment none of the seafloor deposits are ores because it has not been demonstrated that they can be recovered profitably. However, the technology is in place to extract metalliferous muds from the Atlantis II Deep when and if the market price for base metals exceeds the cost of exploitation, transportation and marketing. Considering the more solid sulfide mounds of the Pacific region, fist-sized chunks can be lifted from the relatively shallow depths of the ridge crests, thanks to devices developed by manganese nodule recovery projects (Siapno, this volume). At present, there is no means of efficiently breaking up the deposits on the seafloor to a recoverable size but, surely, the technology to do this is not insurmountable (Crawford et al., 1984). Leaving engineering aside, the likelihood of ever mining the ocean floor is more dependent on politics and the world economy than on geology. The most significant unknown about the Pacific deposits is their grade. No deposit, large or small, on the Pacific Ocean floor has yet been sampled in such a way that its bulk composition can be accurately ascertained. Random samples from various sites do contain one or more elements such as Cu, Zn, Ag and Au, whose concentrations are comparable to, or higher than, those samples taken randomly from existing massive sulfide mines on land. What is known now about the size of seafloor deposits compared to those on land is meager but deposits at three sites in the eastern Pacific (Galapagos, the seamount at 13°N and Explorer Ridge) are already within the size range of deposits that would be mined on land under favorable economic conditions. These were discovered in the course of only seven years of surveying, preponderantly aimed at purely scientific oceanographic research rather than toward commercial exploration. The geological record of ancient ore deposits indicates that there should be additional sites on the modern seafloor, such as felsic island arcs, back-arc basins and thinned continental margins, which have not been adequately explored but where large sulfide deposits are likely to exist. The discoveries to date, coupled with a reasonable expectation that larger deposits are yet to be found, may be sufficient reason to be optimistic that ocean mining will become viable in the future. For now though, massive sulfides on or below the seafloor are far more important for advancing scientific knowledge than as potential resource.

ACKNOWLEDGEMENTS

The author's education on massive sulfides has been a product of personal research on the seafloor and on land and also that of his students and colleagues, who will recognize their contributions although they may not agree with the conclusions that are drawn from them. Thanks are extended to Paul Teleki, Robby Moore, Max Dobson and Ulrich von Stackelberg who organized the NATO ARW at which this paper was presented, and IFREMER's Centre de Brest for providing the facilities for writing the manuscript during a sabbatical leave. Thanks are also due Henri Bougault, Jean-Luc Charlou, Yves Fouquet and Roger Hekinian of IFREMER and Jean Francheteau of Université de Paris for some stimulating discussions and their expert criticism of the manuscript. This continuing research on massive sulfides is supported by Operating and Strategic grants from the Natural Sciences and Engineering Research Council of Canada.

REFERENCES

Alabaster, T., Pearce, J.A., Mallic, D.I.J. and Elboushi, I.M., 1980, The volcanic stratigraphy and location of massive sulphide deposits in the Oman ophiolite, In: Panayiotou, A. (ed.), Ophiolites; Proc. Int. Ophiolite Symp., Cyprus 1979, Ministry of Agriclture and Natural Resources, Geological Survey Department, Cyprus, pp. 751–757.

Bischoff, J.L. and Dickson, F.W., 1975, Seawater–basalt interaction at 200°C and 500 bars: implications for origin of seafloor heavy-metal deposits and regulation of seawater chemistry; Earth and Planet. Sci. Letts., v. 25, pp. 385–397.

Bischoff, J.L. and Pitzer, K.S., 1985, Phase relations and adiabats in boiling seafloor geothermal systems; Earth and Planet. Sci. Letts., v. 75, pp. 327–338.

Bischoff, J.S., Rosenbauer, R.J., Aruscavage, P.J., Baedecker, P.A. and Crock, J.G., 1983, Seafloor massive sulfide deposits from 21°N, East Pacific Rise, Juan de Fuca Ridge, and Galapagos Rift: Bulk chemical composition and economic implications; Econ. Geol., v. 78, pp. 1711–1720.

Boldy, G.D.J., 1981, Prospecting for deep volcanogenic ore; Can. Inst. Mining Metall., v. 74, pp. 55–65.

Bowers, T.S., Von Damm, K.L. and Edmond, J.M., 1985, Chemical evolution of mid-ocean ridge hot springs; Geochim. et Cosmochim. Acta, v. 49, pp. 2239–2252.

Bryndzia, L.T., Scott, S.D. and Farr, J.E., 1983, Mineralogy, geochemistry and mineral chemistry of siliceous ore and altered footwall rocks in the Uwamuki 2 and 4 ore deposits, Kosaka mine, Hokuroku district, Japan; Econ. Geol., Monogr. 5, pp. 507–522.

Campbell, I.H., McDougall, T.J. and Turner, J.S., 1984, A note on fluid dynamic processes which can influence the deposition of massive sulfides; Econ. Geol., v. 79, pp. 1905–1913.

CASM (Canadian-American Seamount Expedition), 1985, Hydrothermal vents on an axis seamount of the Juan de Fuca Ridge; Nature, v. 313, pp. 212–214.

Cathles, L.M., Guber, A.T., Lenagh, T.C. and Dudas, F.O., 1983, Kuroko-type massive sulfide deposits of Japan: Products of an aborted island-arc rift? Econ. Geol., Monogr. 5, pp. 96–114.

Constantinou, G., 1972, The geology and genesis of the sulphide ores of Cyprus, unpubl. Ph.D. thesis; Imperial College, London.

Constantinou, G. and Govett, G.J.S., 1972, Genesis of sulphide deposits, ochre and umber of Cyprus; Trans. Inst. Mining and Metallurgy, v. 81, pp. B34-46.

Converse, D.R., Holland, H.D. and Edmond, J.M., 1984, Flow rates in the axial hot springs of the East Pacific Rise (21°N): Implications for the heat budget and the formation of massive sulfide deposits; Earth and Planet. Sci. Letts., v. 69, pp. 159-175.

Costa, U.R., Barnett, R.L. and Kerrich, R., 1985, The Mattagami Lake mine Archean Zn-Cu sulfide deposit, Quebec: Hydrothermal co-precipitation of talc and sulfides in a seafloor brine pool - evidence from geochemistry, $^{18}O/^{16}O$, and mineral chemistry - a reply; Econ. Geol., v. 79, pp. 1953-1955.

Crawford, A.M., Hollingshead, S.C. and Scott, S.D., 1984, Geotechnical engineering properties of deep-ocean polymetallic sulfides from 21°N, East Pacific Rise; Mar. Mining, v. 4, pp. 337-354.

Curray, J.R., Moore D.G., et al., 1982, Initial Reports DSDP 62; U.S. Govt. Printing Off., Washington, D.C.

Degens, E.T. and Ross, D.A., (eds.), 1969, Hot brines and recent heavy metal deposits in the Red Sea: a geochemical and geophysical account; Springer-Verlag, New York, 600 p.

Delaney, J.R. and Cosens, B.A., 1982, Boiling and metal deposition in submarine hydrothermal systems; Mar. Tech. Soc. Jour., v. 16, pp. 62-66.

Delaney, J.R., McDuff, R.E. and Lupton, J.E., 1984, Hydrothermal fluid temperatures of 400°C on the Endeavour Segment, northern Juan de Fuca (abs.); EOS, Trans. Amer. Geophys. Union, v. 65, p. 973.

Edmond, J.M., Von Damm, R.L., McDuff, R.E. and Measures, C.I., 1982, Chemistry of hot springs on the East Pacific Rise and their effluent dispersal; Nature, v. 297, pp. 187-191.

Farr, J.E., 1984, The geology, mineralogy and geochemistry of the 070 faults of the Corbet mine, Noranda, Quebec; unpubl. M.Sc. thesis, Univ. Toronto.

Fouquet, Y. and Juniper, S.K., in press, Formation of hydrothermal iron-silica deposits within microbial mats in the deep sea; Nature.

Francis, T.J.G., 1985, Resistivity measurements of an ocean sulfide deposit from the submersible Cyana; Mar. Geophys. Researches, v. 7, pp. 419-438.

Goldfarb, M.S., Converse, D.R., Holland, H.D. and Edmond, J.M., 1983, The genesis of hot spring deposits on the East Pacific Rise, 21°N;

Econ. Geol., Monogr. 5, pp. 184–197.

Grimaud, D., Michard, A. and Michard, G., 1984, Composition chemique et
    composition isotopique du strontium dans les eaux hydrothermales
    sous-marins de la dorsale Est Pacifique a 13° Nord; Compt. Rend.
    Acad. Sci., v. II, no. 13, pp. 865–870.

Hannington, M.D., Peter, J.M. and Scott, S.D., in press, Gold in sea-
    floor polymetallic sulfide deposits; Econ. Geol.

Haymon, R.M. and Kastner, M., 1981, Hot spring deposits on the East
    Pacific Rise at 21°N: Preliminary description of mineralogy and
    genesis; Earth and Planet. Sci. Letts., v. 53, pp. 363–381.

Hekinian, R. and Bideau, D., in press, Volcanism and mineralization of
    the ocean crust on the East Pacific Rise; Metallogeny of Basic and
    Ultrabasic Rocks Symposium, 1985, Edinburgh.

Hekinian, R. and Fouquet, Y., 1985, Volcanism and metallogenesis of
    axial and off-axial structures on the East Pacific Rise near 13°N;
    Econ. Geol., v. 80, pp. 221–249.

Honnorez, J. and Kirst, P., 1975, Petrology of rodingites from the
    equatorial Mid-Atlantic fracture zones and their geotectonic
    significance; Contr. to Miner. and Petrol., v. 49, pp. 233–257.

Honnorez, J., Mevel, C. and Montigny, R., 1984, Occurence and signifi-
    cance of gneissic amphibolites in the Vema Fracture zone, equatorial
    Mid-Atlantic Ridge; Geol. Soc. London, Spec. Publ., no. 13, pp.
    121–130.

Ito, E. and Anderson, A.T., Jr., 1983, Submarine metamorphism of gabbros
    from the Mid-Cayman Rise: Petrographic and mineralogic constraints
    on hydrothermal processes at slow-spreading ridges; Contr. to Miner.
    and Petrol., v. 82, pp. 371–388.

Janecky, D.R. and Seyfried, W.E., Jr., 1984, Formation of massive
    sulfide deposits on oceanic ridge crests: Incremental reaction
    models for mixing between hydrothermal solutions and seawater;
    Geochim. et Cosmochim. Acta, v. 48, pp. 2723–2738.

Kalogeropoulos, S.I. and Scott, S.D., 1983, Mineralogy and geochemistry
    of tuffaceous exhalites (tetsusekiei) of the Fukazawa mine, Hokuroku
    district, Japan; Econ. Geol., Monogr. 5, pp. 412–432.

Kim, K.-R., 1983, Methane and radioactive isotopes in submarine hydro-
    thermal systems; unpubl. Ph.D. thesis, University of California-
    San Diego.

Kim, K.-R., Welhan, J.A. and Craig, H., 1984, The hydrothermal vent fields at 13°N and 11°N on the East Pacific Rise: ALVIN 1984 results (abs.); EOS, Trans. Amer. Geophys. Union, v. 65, p. 973.

Lonsdale, P.F., Batiza, R. and Simkin, T., 1982, Metallogenesis at seamounts on the East Pacific Rise; Mar. Technol. Soc. Jour., v. 16, pp. 54-61.

Malahoff, A., 1982, A comparison of the massive submarine polymetallic sulfides of the Galapagos rift with some continental deposits; Mar. Technol. Soc. Jour., v. 16, pp. 39-45.

McDuff, R.E., Lupton, J.E., Kadko, D. and Lilley, M.D., 1984, Chemistry of hydrothermal fluids, Endeavour Ridge, Northeast Pacific (abs.); EOS, Trans. Amer. Geophys. Union, v. 65, p. 975.

Michard, G., Albarede, F., Michard, A., Minster, J.-F., Charlou, J.-L. and Tan, N., 1984, Chemistry of solutions from the 13°N East Pacific Rise hydrothermal site; Earth and Planet. Sci. Letts., v. 67, pp. 297-307.

Minniti, M. and Bonavia, F.F., 1984, Copper-ore grade hydrothermal mineralization discovered in a seamount in the Tyrrhenian Sea (Mediterranean): Is the mineralization related to porphyry-coppers or to base metal lodes? Mar. Geol., v. 59, pp. 271-282.

Monteiro, J.H. and Carvalho, D., 1986, Submarine explosive volcanism and sulfide deposition in ancient active margins: The case of the Iberian Pyrite Belt; this volume, pp.368-380.

Mottl, M.J. and Holland, H.D., 1978, Chemical exchange during hydrothermal alteration of basalt by seawater, I. Experimental results for major and minor components of seawater; Geochim. et Cosmochim. Acta, v. 42, pp. 1103-1115.

Mottl, M.J. and Holland, H.D., 1979, Chemical exchange during hydrothermal alteration of basalt by seawater, II. Experimental results for Fe, Mn and sulfur species; Geochim. et Cosmochim. Acta, v. 43, pp. 869-884.

Mustafa, H.E.Z., Nawab, Z., Horn, R. and Le Lann, F., 1984, Economic interest of hydrothermal deposits. The Atlantis II Project; Proc. Offshore Mineral Resources, Second Int. Sem., Brest, France, pp. 509-539.

Oudin, E., 1983, Hydrothermal sulfide deposits of the East Pacific Rise (21°N). Part I: Descriptive mineralogy; Mar. Mining, v. 4, pp. 39-72.

Oudin, E., Thisse, Y. and Ramboz, C., 1984, Fluid inclusion and mineralogical evidence for high-temperature saline hydrothermal circulation in the Red Sea metalliferous sediments: Preliminary results; Mar. Mining, v. 5, pp. 3-31.

Pisutha-Arnond, V. and Ohmoto, H., 1983, Thermal history and chemical and isotopic compositions of the ore-forming fluids responsible for the Kuroko deposits in the Hokuroku district of Japan; Econ. Geol., Monogr. 5, pp. 523-558.

Rona, P.A., 1985, Black smokers and massive sulfides at the TAG hydrothermal field, Mid-Atlantic Ridge 26°N (abs.); EOS, Trans. Amer. Geophys. Union, v. 66, p. 936.

Sangster, D.F. and Scott, S.D., 1976, Precambrian, strata-bound, massive Cu-Zn-Pb sulfide ores of North America; In: Wolf, K.H. (ed.), Handbook of stratabound and stratiform ore deposits; Elsevier, v. 6, pp. 129-222.

Sawkins, F.J. and Kowalic, J., 1981, The source of ore metals at Buchans: Magmatic versus leaching models; Geol. Assoc. of Canada, Spec. Paper 22, pp. 255-267.

Scott, S.D., 1985, Seafloor polymetallic sulfide deposits: Modern and ancient; Mar. Mining, v. 5, pp. 191-212.

Scott, S.D., Kalogeropoulos, S.I., Shegelski, R.J. and Siriunas, J.M., 1983, Tuffaceous exhalites as exploration guides for volcanogenic massive sulphide deposits; Jour. Geochem. Expl., v. 19, pp. 500-502.

Seyfried, W.E. and Bischoff, J.L., 1981, Experimental seawater-basalt interaction at 300°C, 500 bars, chemical exchange, secondary mineral formation and implications for the transport of heavy metals; Geochim. et Cosmochim. Acta, v. 45, pp. 135-147.

Siapno, W.D., 1986, Nodule exploration: Accomplishments, needs and problems; this volume, pp.244-254.

Simoneit, B.R.T. and Lonsdale, P.F., 1982, Hydrothermal petroleum in mineralized mounds at the seabed of Guaymas Basin; Nature, v. 295, pp. 198-202.

Sourirajan, S. and Kennedy, G.C., 1962, The system $H_2O-NaCl$ at elevated temperatures and pressures, Amer. Jour. Sci., v. 260, pp. 115-141.

Tunnicliffe, V., Botros, M., de Burgh, M.E., Dinet, A., Johnson, H.P., Juniper, S.K. and McDuff, R.E., in press, Hydrothermal vents of Explorer Ridge, northeast Pacific; Deep Sea Research.

Von Damm, K.L., 1983, Chemistry of submarine hydrothermal solutions at 21° North, East Pacific Rise and Guaymas Basin, Gulf of California; unpubl. Ph.D. thesis, Mass. Inst. Technol. - Woods Hole Ocean. Inst., Rept. WHOI-84-3.

Von Damm, K.L. and Bischoff, J.L., 1985, Southern Juan de Fuca Ridge
    hot spring chemistry (abs.); EOS, Trans. Amer. Geophys. Union, v.
    66, p. 926.

SULFIDE DEPOSITS ON THE SEA FLOOR: GEOLOGICAL MODELS AND RESOURCE
PERSPECTIVES BASED ON STUDIES IN OPHIOLITE SEQUENCES

Randolph A. Koski
U.S. Geological Survey
345 Middlefield Road
Menlo Park, California 94025 U.S.A.

ABSTRACT. Fossil hydrothermal systems exposed in ophiolites include
subsea-floor feeder-zone mineralization in volcanic rocks, diabase, and
high-level gabbro, massive sulfide deposited by fluids discharging at
the sea-floor/seawater interface, and peripheral Fe- and Mn-oxide facies
deposited on the surrounding sea floor. The stratigraphic and structural
position of Cyprus-type massive sulfide deposits in ophiolitic volcanic
sequences indicate that hydrothermal activity was focused along ridge-
parallel faults either in the axial graben or on the ridge flanks, and
in seamounts on or near the paleoridge axis. Paleotransform faults were
less favorable tectonic settings for sulfide mineralization. Recent
geochemical and stratigraphic studies indicate that most Tethyan and
Cordilleran ophiolites formed in marginal basins or in suprasubduction
settings; relatively few well-studied ophiolites appear to represent
oceanic crust generated at mid-ocean spreading axes. The size,
composition, and distribution of ophiolite-hosted deposits suggest that
pyritic sulfide masses of >1 million tons averaging >1% Cu should be
present in contemporary ridge settings. By analogy with ophiolites, rift
zones behind and between ensimatic arcs (e.g., Lau Basin, Mariana
Trough) and above subduction zones (e.g., Andaman Sea) have considerable
potential for additional discoveries of massive sulfide in the deep
ocean.

INTRODUCTION

A fundamental tenet of the plate-tectonic paradigm poses that new
oceanic crust is continuously manufactured along the earth-encircling
system of oceanic ridges where sea-floor spreading occurs. Much of our
current understanding of the magmatic and volcanic processes operative
at these accretionary plate boundaries is derived from studies of
ophiolites on land (e.g., Moores and Vine, 1971; Gass and Smewing, 1973;
Coleman, 1977). Similarly, the recognition that massive sulfide deposits
within the volcanic stratigraphy of ophiolite sequences represent
mineralization on the ancient sea floor has led to an increased
understanding of hydrothermal processes in oceanic rift zones and fueled

301

P. G. Teleki et al. (eds.), Marine Minerals, 301–316.
© *1987 by D. Reidel Publishing Company.*

speculation that hydrothermal circulation and sulfide deposits should
occur at modern sites of sea-floor spreading (e.g., Sillitoe, 1972;
Constantinou and Govett, 1973; Spooner and Fyfe, 1973).

The remarkable discoveries of active hot springs, sulfide deposits,
and exotic benthic organisms at several localities along the East
Pacific Rise invite comparison with fossil hydrothermal systems and
deposits in ophiolites on land. Although many gross geological and
compositional similarities exist (Oudin et al., 1981), it is apparent
that massive sulfide deposits of the size and composition of ophiolite-
hosted deposits have not yet been found in contemporary sea-floor
settings. This paper attempts to review current concepts of ophiolite
formation and the results of detailed studies of associated sulfide
occurrences; elucidate the spatial and genetic relationship of sea-floor
hydrothermal deposits preserved in ophiolites to their paleogeologic
setting; and evaluate the potential for mineral deposits occurring in
contemporary oceanic rift zones that have not yet been studied in
detail.

OPHIOLITE AS OCEANIC CRUST

The modern definition, developed by the Geological Society of America
Penrose Field Conference (1972), states that ophiolite is an association
of mafic and ultramafic rocks that includes from bottom to top, when
complete, a basal ultramafic complex consisting of harzburgite,
lherzolite, dunite, and chromitite, usually with a pervasive tectonite
fabric; a cumulate section of peridotite, pyroxenite, and gabbro; a
sheeted diabase dike complex; and a basaltic volcanic sequence, largely
pillowed (Figure 1). Intrusions of gabbro and plagiogranite occur at
intermediate levels below the dike complex. The volcanic sequence may be
overlain by pelagic and metalliferous sedimentary rocks (radiolarian
chert, shale, and limestone) or volcaniclastic debris shed from volcanic
centers in nearby island arcs. Rocks structurally underlying the
ophiolite may include crystalline basement and shallow-water sedimentary
facies (continental margin assemblage), a tectonic melange that contains
continental margin, rise, and abyssal sedimentary units, and,
immediately below the ultramafic section, a basal plate of high-grade
metamorphic rocks, especially amphibolite (Moores, 1982).

Ophiolite complexes are interpreted to be allocthonous fragments of
the oceanic lithosphere and the upper mantle that were generated during
sea-floor spreading. Subsequently, these fragments were uplifted and
overthrust onto (i.e., obducted) and incorporated into the orogenic
belts of continental margins during convergent plate interaction.
Complete ophiolite sections have dimensional, geophysical, and
stratigraphic characteristics that are similar to layers 1, 2, and 3 of
the oceanic crust and the depleted mantle immediately below the Moho
(Figure 1; Coleman, 1977).

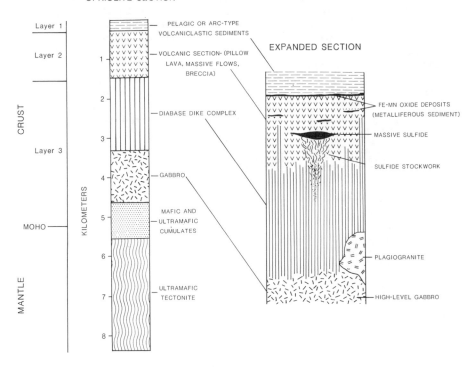

Figure 1. Left, a generalized cross section of a complete ophiolite
sequence and the comparable crustal seismic layers. Right, an expanded
section showing the position of massive sulfide, the underlying
stockwork, and Fe-Mn oxide deposits in the ophiolite sequence.

SULFIDE DEPOSITS IN OPHIOLITES

Volcanic strata in the upper part of ophiolite sequences host a
distinctive type of Fe-Cu-Zn sulfide mineralization, commonly referred
to as "Cyprus-type" deposits. Where these massive sulfide accumulations
are undeformed, they occur as tabular, lenticular, saucer-shaped, or
wedge-shaped bodies within pillow lava and pillow-lava breccia (Figure
1); the lateral margins of the sulfide masses are often bounded by
faults. Vein mineralization in pipe-, funnel-, or keel-shaped stockwork
systems generally extends downward, together with decreasing vein
density, into the sheeted dike complex below the lavas.

   Large deposits (e.g., Skouriotissa and Mavrovouni, Cyprus, and
Lasail, Oman) are as much as 500 to 700 m long, 150 to 200 m wide, and
50 m thick, and range in size from a few thousand to nearly 20 million
metric tons. However, deposits over 5 million tons are uncommon. Massive
sulfide in ophiolites tends to occur in clusters of closely spaced

deposits. For example, numerous deposits located along the northern
flank of the Troodos massif are clustered together in groups spaced 7.4
to 12.5 km apart (Solomon, 1976). Within groups, deposits are spaced 2.0
to 4.5 km apart. Adamides (1984) concluded that deposits clustered
within districts in Cyprus represent contemporaneous and genetically
related events on the sea floor.

The tonnages and percentages of Cu and Zn in selected deposits from
ophiolites in Cyprus, Oman, and Newfoundland are presented in Table I.

Table I. Size and grade for selected Cyprus-type massive
sulfide deposits in Cyprus, Oman, and Newfoundland.

Deposit	Tonnage	Cu(%)	Zn(%)
Cyprus			
Skouriotissa	6,000,000	2.2	–
Mavrovouni	15,000,000	4.0	0.5
North Mathiati	2,800,000[a]	0.2	–
Apliki	1,650,000	1.8	–
Kalavasos A	1,000,000	0.25–2.5	0.7
Sha	350,000	0.5–1.2	–
Klirou East	420,000	1.1	1.4
Kinousa underground	300,000	2.4	3.4
Limni	16,000,000[b]	1.0	
Oman			
Lasail	8,000,000	2.3	0.2
Bayda	750,000	2.7	1.3
Aarja	3,170,000	1.7	0.8
Newfoundland			
York Harbor	332,000	1.9	4.7
Tilt Cove	9,000,000[c]	1–12	–
Betts Cove	130,682	2–10	–
Little Bay	3,400,000[d]	0.8–2.0	–
Miles Cove	200,000[e]	1.45	–
Whalesback	4,181,708	0.85–1.1	–

(a) Mined for pyrite. (b) From stockwork zone. (c) Includes
1,319,500 grams Au. (d) Includes 180,000 grams Au. (e)
Contains 0.3 grams Au/ton, 11 grams Ag/ton. Data for Cyrpus
from Bear (1963) and Adamides (1980). Data for Oman from
Oman Mining Company, LLC. Data for Newfoundland from Swinden
and Kean (1984).

For a massive-sulfide type ore, the average base-metal content is low;
with a few exceptions, copper ranges from 0.5 to 5 percent and zinc from
0.1 to 2 percent. Cobalt is locally concentrated to 0.35 percent, but is

generally near 0.01 percent (Constantinou, 1980). Silver and gold
average from 3 to 30 grams/ton and 0.1 to 3 grams/ton, respectively
(Hutchinson, 1973). Despite low precious metal contents, gold has been
recovered from several deposits. For example, 9 million tons of copper
ore yielded about 1,320,000 grams of gold from the Tilt Cove massive
sulfide deposit in the Betts Cove ophiolite, Newfoundland (Table I;
Swinden and Kean, 1984).

The sulfide assemblage in Cyprus-type deposits is characteristi-
cally simple: abundant pyrite or, less commonly, pyrrhotite with minor
amounts of marcasite, chalcopyrite, and sphalerite. Chalcopyrite is
usually ubiquitous, and may increase in and toward the stockwork zone
and along controlling structures. Sphalerite is generally subordinate to
chalcopyrite, although it is erratically distributed and may be locally
abundant. In a few deposits (e.g., York Harbor, Newfoundland; Turner-
Albright, Oregon), sphalerite is a major constituent of the ore and
occurs with Fe sulfide in a variety of banded and breccia textures.
Sphalerite-rich zones of the Bayda deposit in the Samail ophiolite,
Oman, contain the fossilized impressions of hydrothermal vent worms
(Haymon et al., 1984b). Zn- and Cu-sulfide-rich chimney fragments and
traces of vent fauna have also been described from deposits in Cyprus
(Oudin and Constantinou, 1984). Minor sulfide minerals in Cyprus-type
deposits include galena, jordanite, bravoite, bornite, mackinawite,
isometric cubanite, idaite, tennantite, and wurtzite (Oudin et al.,
1981; Ixer et al., 1984).

Gangue minerals that typically constitute a very small percentage
of the massive sulfide ore are quartz, chlorite, calcite, and gypsum. In
several deposits, a zone with abundant secondary silica is present
beneath the massive sulfide lens (Constantinou and Govett, 1973;
Constantinou, 1980). In this zone, pyrite is mixed with quartz, red
jasper, or massive cherty material.

Below massive sulfide or the silica-rich zone, stockwork veins
consist of Fe, Cu, and Zn sulfides, quartz, and chlorite. The amount of
sulfide in these underlying vein networks can be significant: the Limni
deposit, the largest in Cyprus, produced 16 million tons of ore with a
grade of Cu near 1%, all from stockwork mineralization (Adamides, 1975,
1984). The basaltic wallrock in the stockwork zone has been converted to
a dense, fine-grained assemblage of quartz, one or more phyllosilicates
(chlorite, illite, smectite), and pyrite. The amount of Mg, Fe, S, Cu,
Zn, and Co have been increased in the basalt of the stockwork zone
(Constantinou, 1980). Alteration has also been extensive in the sheeted
dike complex where Mg, Na, and Sr show a net enrichment and Si, Ca, Zn,
and Ba show a net depletion (Adamides, 1984).

Many of the sulfide deposits in Cyprus, and a few elsewhere, are
overlain by Mn-poor, Fe-oxide-rich sedimentary rocks. The lateral extent
of this ferruginous material generally coincides with the limits of the
massive sulfide lens, indicating that these deposits form by in situ sea
floor oxidation of sulfide (Constantinou and Govett, 1973). However, in
the Lasail deposit of Oman, laminated metalliferous sediment composed
largely of magnetite, hematite, and quartz extends laterally for tens of
meters beyond the massive sulfide. In ophiolitic rocks of the British
Solomon Islands, massive sulfide is overlain by a layer of laminated Fe-

oxide-rich sediment and a compact sinter containing anhydrite, barite, opaline silica, and pyrite (Taylor, 1974, 1983). This sinter is, in turn, overlain by Fe-rich chert and manganiferous wad. These successions appear to represent the zonation from sulfide to oxide facies deposition on the sea floor. Amorphous Mn-Fe-rich deposits (umbers) and manganiferous cherts are common features in the upper part of ophiolites (Figure 1). The presence of vein networks below some umbers, their stratigraphic occurrence, and the chemical and mineralogical similarity between umbers and metalliferous sediments on the flanks of modern-day ridges indicate that these oxide deposits formed during hydrothermal discharge in topographically elevated areas of ocean ridges, but away from their axes of spreading (Robertson, 1975).

ENVIRONMENT OF SULFIDE DEPOSITION

Cyprus-type deposits apparently represent hydrothermal mineralization at ancient oceanic spreading axes, and are a composite of stockwork or feeder-zone deposition in volcanic rocks, diabase, and gabbro in the subseafloor and massive sulfide deposition at the sea-floor/seawater interface. The stratigraphic position at which sulfide mineralization occurs within the volcanic sequence is an indicator of proximity to the paleospreading axis. For example, the location of the Betts Cove massive sulfide deposit, near the contact between basal volcanic rocks and sheeted dike complex in the Bay of Islands ophiolite, Newfoundland (Upadhyay and Strong, 1973), suggests that mineralization occurred on the axis of spreading. Similarly, most deposits in the Troodos Complex occur at the contact between the Lower and Upper Pillow Lavas (Constantinou and Govett, 1973; Constantinou, 1980), and thus are also spatially associated with volcanism at the ridge axis. However, Adamides (1984) has determined that the period of principal hydrothermal activity is coincident with the extrusion of the earliest Upper Pillow Lavas in an off-axis setting, presumably on the flanks of the ridge. Boyle and Robertson (1984) have studied the stratigraphy in the Mathiati area of Cyprus, and concluded that a continuous metallogenic episode began with sulfide deposition along the walls of a central graben and continued with deposition of Fe-Mn oxide sediments in off-axis depressions of the ridge flanks.

The volcanic and tectonic setting of sulfide deposition has been deduced from recent, detailed field studies in ophiolites. According to Alabaster et al. (1980) and Alabaster and Pearce (1985), massive sulfide deposits in the Sohar district of the Samail ophiolite in Oman occur along the stratigraphic interface between lavas that erupted at the spreading axis and lavas that formed seamounts away from the axis; the ore-forming hydrothermal episode is inferred to have been generated by heat from off-axis magma chambers associated with seamount volcanism. Earlier, Smewing et al. (1977) proposed that the sulfide deposits near Sohar formed along a 250-m-high escarpment parallel to the axis of spreading. Similarly, the spatial relations at the Bayda deposit in this district suggest that deposition of massive sulfide occurred along steep, closely-spaced fault scarps subparallel to the paleoridge axis

(Haymon et al., 1984a).

In the northern part of the Troodos Complex in Cyprus, Varga and Moores (1985) have recognized a series of structural grabens that resemble axial valleys of modern, slow-spreading ridges. Axis-parallel normal faults within these grabens may represent subsea-floor channelways that controlled the flow of hydrothermal fluid and the distribution of several massive sulfide deposits (e.g., Skouriotissa and Mavrovouni) in the region. In a detailed study of the Kalavasos district of Cyprus, Adamides (1980, 1984) determined that hydrothermal fluids were chanelled along normal fault structures parallel to the spreading axis (Figure 2) near the intersection of the ridge and a paleotransform fault system, known as the Arakapas fault belt (Simonian and Gass, 1978). Massive sulfide deposits formed in the hanging wall of the normal faults, but the Arakapas fault belt appears to be nearly devoid of sulfide mineralization (Robertson, 1978). Another recognized fracture zone in an ophiolite, the Coastal Complex of western Newfoundland, contains quartz-carbonate veins and sulfide mineralization in zones of tectonic brecciation (Karson, 1984), but no massive sulfide deposits have been found.

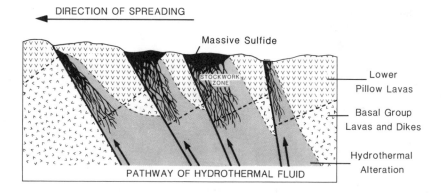

Figure 2. Diagram showing relationship of massive sulfide deposits in the Kalavasos district, Cyprus, to fault structures in the Troodos ophiolite. The faults are parallel to the paleoridge axis and served as pathways for ascending hydrothermal fluids in the ridge-crest hydro-thermal system. The massive sulfide bodies represent deposition at the sea-floor/seawater interface that overlies stockwork or feeder-zone sulfide mineralization in subsea-floor volcanic and dike rocks. Modified from Adamides (1980).

ORIGIN OF OPHIOLITE COMPLEXES

Recent studies indicate that Tethyan, Appalachian, and circum-Pacific ophiolites may have formed in a variety of sea-floor spreading environments (Figure 3; Coleman, 1984). The hypothesis that ophiolites

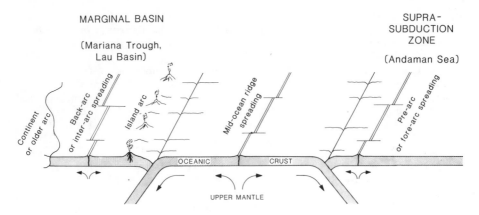

Figure 3. Diagrammatic representation of sea-floor spreading in mid-ocean ridge and marginal basin environments. Most Tethyan and Cordilleran ophiolites may have formed in backarc, interarc, or suprasubduction settings. Suprasubduction settings include forearc regions and rift zones that preceded the development of a mature arc. Spreading axes in the Lau Basin, Mariana Trough, and Andaman Sea appear to be tectonic analogs in the present ocean.

are oceanic lithosphere formed at mid-ocean ridges has been supported in a few cases. According to Laurent (1980), the composition of pillow lavas and the presence of pelagic sediments indicate that Quebec Appalachian ophiolites are fragments of oceanic crust derived from a wide ocean. In another example, a variety of geochemical data (pyroxene compositions, major and trace element contents, Sr isotope ratios) for gabbro and basalt from the Dun Mountain ophiolite, New Zealand, have been cited as evidence for a mid-ocean ridge origin (Davis et al., 1980).

The trace-element composition of ophiolitic lavas has become a widely used criterion for determining the tectonic origin of ophiolites. This approach is based on the geochemical differences of volcanic rocks in modern oceanic-ridge and island-arc environments, particularly the ratios of Ti, Zr, Y, Cr, Nb, and the rare earth elements that are considered to be immobile during metamorphism, alteration, and weathering (Pearce and Cann, 1971, 1973; Pearce, 1975; Sun and Nesbitt, 1978). Discriminant graphical analysis of the stable trace elements has convinced several researchers that ophiolites represent oceanic crust generated by sea-floor spreading in marginal basins behind or between ensimatic island arcs (Figure 3; Pearce, 1975; Smewing et al., 1975; Upadhyay and Neale, 1979; Coish et al., 1982; Shervais and Kimbrough, 1985). In these studies, ophiolitic volcanic rocks have compositions that are similar to mid-ocean ridge basalt and island-arc basalt or have compositions that are transitional between these end-member types. The concept of a near-arc origin for several Cordilleran ophiolite complexes in the western United States is consistent with the close temporal and spatial association of volcanic and sedimentary rocks with island-arc affinities (Evarts, 1977; Xenophontos and Bond, 1978; Harper, 1984).

Additional stratigraphic, chemical, and petrographic data on Tethyan and Appalachian ophiolites have led to the broader concept that oceanic crust has been generated at a variety of sites above subduction zones and at various stages during the evolution of marginal basins (Figure 3). Thus, ophiolites are thought to have formed by sea-floor spreading in a suprasubduction setting, and the construction of a mature island arc is not essential (Pearce et al., 1984; Moores et al., 1984; Adamides, 1984). For example, the composition of volcanic rocks in the Troodos Complex, Cyprus, and the absence of arc-type lavas and volcaniclastic material suggest that this ophiolite formed above a subduction zone prior to the formation of an island arc (Pearce et al., 1984; Robinson et al., 1983; Schmincke et al., 1983). Furthermore, the succession of lava types in the Samail ophiolite, Oman, are thought to represent a period of volcanism that continued during a tectonic transition from a spreading axis to an embryonic arc environment (Pearce et al., 1981; Alabaster et al., 1982). Cameron et al. (1979, 1980) and Crawford et al. (1981) have suggested another model in which lava sequences in ophiolites that contain boninite (mafic andesite with high Mg), such as Troodos, Cyprus, and Betts Cove, Newfoundland, are the result of volcanism in a forearc setting. Finally, Robertson and Dixon (1984) have proposed that the island arc-like geochemical signatures of volcanic rocks and the development of basal amphibolite zones in Tethyan ophiolites are related to a change from extensional to compressional tectonics at an oceanic spreading axis. As regional plate motions cause spreading to become asymmetrical, one of the plates becomes stationary, subsides, reverses direction of movement, and ultimately plunges below the opposite plate. Volcanic rocks formed in this suprasubduction setting will have the anomalous boninitic compositions found in several ophiolites.

In summary, it appears that ophiolites may have formed at mid-ocean ridges, in marginal basins behind or between island arcs, or in suprasubduction zone settings (Figure 3). However, few of the well-studied Tethyan or Cordilleran ophiolite complexes are now inferred to have formed at a large-ocean spreading axis.

CONCLUSIONS AND IMPLICATIONS FOR SEAFLOOR MINERAL RESOURCES

Several conclusions derived from this review are relevant to current and future mineral resource investigations of oceanic ridges:

1. Cyprus-type sulfide deposits in ophiolites represent hydrothermal mineralization at ancient oceanic spreading axes. Massive sulfide formed at the sea-floor/seawater interface; stockwork or feeder-zone mineralization formed in the volcanic rocks, sheeted diabase, and gabbro below the sea floor.

2. The massive sulfide generally represents only a small part of the fossil geothermal system. Large amounts of metal can also be deposited in the underlying feeder system or be mobilized into the water column and precipitated in Fe- and Mn-oxide deposits beyond the limit of sulfide.

3. Massive sulfide deposits in ophiolites tend to occur in clusters which implies that sea-floor metallogenesis has been episodic and dependent on favorable tectonic and volcanic conditions.

4. The location of massive sulfide deposits within the volcanic stratigraphy of ophiolites suggests that mineralization occurred in various settings both on and near the ridge axis. Normal faults parallel to the ridge axis and near-axis seamounts have been structurally favorable environments for metallogenesis. Paleotransform faults have been less favorable, though deposits have formed along axial structures near the intersection of major transform faults.

5. The geochemistry and stratigraphy of volcanic rocks indicate that many Tethyan and Cordilleran ophiolites are fragments of oceanic crust formed by sea-floor spreading in marginal basins behind or between island arcs or in suprasubduction settings. Relatively few well-studied ophiolites are now thought to have formed at mature ocean ridges.

New concepts of ophiolite genesis and the tectono-stratigraphic setting of Cyprus-type massive sulfide deposits can be a valuable guide in the search for and evaluation of hydrothermal mineral deposits in modern ocean basins. Land-based ophiolites provide an opportunity to observe the third (vertical) dimension of geothermal systems that occur in oceanic rift zones. Characteristics such as the depth of the magma chamber (the high-level gabbro in ophiolites), the porosity of the crust, the type of fluid pathways between the heat source and the sea floor, and the nature and distribution of rock alteration can add to conceptual models of fluid convection at oceanic ridges. Furthermore, much of the chemical modeling of metal-forming geothermal systems in contemporary sea-floor settings is supported by studies of rock alteration and stockwork mineralization below volcanogenic massive sulfide deposits on land (e.g., Spooner, 1980).

The characteristics of Cyprus-type deposits suggest that pyritic massive sulfide masses of >1 million metric tons, averaging >1% Cu, should occur in modern ridge settings. Malahoff (1982) has described a pyritic and Cu-rich deposit on the Galapagos Rift with dimensions 1000 m long, 20 to 200 m wide, and 35 m thick that are comparable to those of onland deposits. Most deposits discovered on the East Pacific Rise, however, are comparatively small (a few hundred to thousands of tons of massive sulfide), have mound and chimney structures, and have Zn-sulfide-rich compositions (Spiess et al., 1980; Hekinian et al., 1980; Haymon and Kastner, 1981; Styrt et al., 1981; Koski et al. 1984). Although the compositions of known deposits on the present sea floor do not appear to be strictly comparable to Cyprus-type deposits, their genesis appears to involve similar volcanogenic and hydrothermal processes.

Perhaps the most useful concepts derived from recent research on
ophiolites pertain to the variety of tectonic settings in which oceanic
crust containing massive sulfide deposits has formed previously.
Although numerous sites of active hydrothermal discharge and sulfide
deposition have been located along the medium- and fast-rate spreading
axes of the eastern Pacific, studies of both Tethyan and Cordilleran
ophiolites suggest that active rift environments behind and between
ensimatic island arcs (e.g., Lau Basin and the Mariana Trough in the
western Pacific Ocean) and above subduction zones (e.g., the Andaman Sea
in the eastern Indian Ocean) have considerable potential for additional
discoveries (Figure 3). Furthermore, ridge-parallel faults along the
spreading axes and seamounts on or near the axes are favorable sites for
the formation of large sulfide deposits. Conversely, fracture zones
appear to be environments that are less promising for large-scale
hydrothermal activity, but may be important as recharge zones in fluid
convection cells.

ACKNOWLEDGMENTS

I thank J.P. Albers and R.C. Evarts for their careful reviews of this
manuscript and helpful suggestions for its improvement. L.M. Benninger
drafted the illustrations.

REFERENCES

Adamides, N.G., 1975, Geological history of the Limni concession,
    Cyprus, in the light of the plate tectonics hypothesis; Inst. Mining
    and Metallurgical Trans., Sec. B., v. 84, pp. B17–B23.

Adamides, N.G., 1980, The form and environment of formation of the
    Kalavasos ore deposits – Cyprus; In: Panayiotou, A. (ed.),
    Ophiolites; Proc. Int. Ophiolite Symp., 1979, Cyprus Geol. Surv.
    Dept., pp. 117–128.

Adamides, N.G., 1984, Cyprus volcanogenic sulphide deposits in relation
    to their environment of formation; unpubl. Ph.D. thesis, University
    of Leicester, 383 p.

Alabaster, T., and Pearce, J.A., 1985, The interrelationship between
    magmatic and ore-forming processes in the Oman ophiolite; Economic
    Geology, v. 80, pp. 1–16.

Alabaster, T., Pearce, J.A., Mallick, D.I.J., and Elboushi, I.M., 1980,
    The volcanic stratigraphy and location of massive sulfide deposits
    in the Oman ophiolite; in Paniyotou, A. (ed.), Ophiolites; Proc. of
    the Int. Ophiolite Symp., 1979, Cyprus Geol. Surv. Dept., pp. 751–
    757.

Alabaster, T., Pearce, J.A., and Malpas, J., 1982, The volcanic
    stratigraphy and petrogenesis of the Oman ophiolite complex;
    Contrib. Mineral. and Petrol., v. 81, pp. 168-183.

Bear, L.M., 1963, The mineral resources and mining industry of Cyprus;
    Cyprus Geol. Surv. Depart. Bull. no. 1, 208 p.

Boyle, J.F., and Robertson, A.H.F., 1984, Evolving metallogenesis at the
    Troodos spreading axis; In: Gass, I.G., Lippard, S.J., and Shelton,
    A.W. (eds.), Ophiolites and oceanic lithosphere; Geol. Soc. of
    London Spec. Pub. 13, pp. 169-181.

Cameron, W.E., Nisbet, E.G., and Dietrich, V.J., 1980, Petrographic
    dissimilarities between ophiolitic and ocean-floor basalts; In:
    Panayiotou, A. (ed.), Ophiolites; Proc. Int. Ophiolite Symp., 1979;
    Cyprus Geol. Surv. Dept., pp. 182-192.

Cameron, W.E., Nisbet, E.G., and Dietrich, V.J., 1979, Boninites,
    komatiites, and ophiolitic basalts; Nature, v. 280, pp. 550-553.

Coish, R.A., Hickey, R., and Frey, F.A., 1982, Rare earth element
    geochemistry of the Betts Cove ophiolite, Newfoundland: complexities
    in ophiolite formation; Geochimica et Cosmochimica Acta, v. 46, pp.
    2117-2134.

Coleman, R.G., 1977, Ophiolites; Springer-Verlag, New York, 229 p.

Coleman, R.G., 1984, The diversity of ophiolites; Geologie en Mijnbouw,
    v. 63, pp. 141-150.

Constantinou, G., 1980, Metallogenesis associated with the Trooodos
    ophiolite; In: Panayiotou, A., (ed.), Ophiolites; Proc. Int.
    Ophiolite Symp., 1979, Cyprus Geol. Surv. Dept., pp. 663-674.

Constantinou, G., and Govett, G.J.S., 1973, Geology, geochemistry, and
    genesis of Cyprus sulfide deposits: Economic Geology, v. 68, pp.
    843-858.

Crawford, A.J., Beccaluva, L., and Serri, J., 1981, Tectono-magmatic
    evolution of the west Philippine-Mariana region and the origin of
    boninites; Earth and Plan. Sci. Letts., v. 54, pp. 346-356.

Davis, T.E., Johnston, M.R., Rankin, P.C., and Stull, R.G., 1980, The
    Dun Mountain ophiolite belt in East Nelson, New Zealand; In:
    Panayiotou, A. (ed.), Ophiolites; Proc. Int. Ophiolite Symp., 1979,
    Cyprus Geol. Surv. Dept., pp. 480-496.

Evarts, R.C., 1977, The geology and petrology of the Del Puerto
    ophiolite, Diablo Range, central California Coast Ranges; In:
    Coleman, R.G., and Irwin, W.P. (eds.), North American ophiolites;
    Oregon Dept. of Geology and Mineral Industries, Bull. no. 95, pp.

121-139.

Gass, I.G., and Smewing, J.D., 1973, Intrusion, extrusion and metamorphism at constructive margins: evidence from the Troodos massif, Cyprus; Nature, v. 242, pp. 26-29.

Geological Society of America Penrose Field Conference, 1972, Ophiolites; Geotimes, v. 17, pp. 24-25.

Harper, G.D., 1984, The Josephine ophiolite, northwestern California; Geol. Soc. America Bull., v. 95, pp. 1009-1026.

Haymon, R.M., and Kastner, M., 1981, Hot spring deposits on the East Pacific Rise at 21 N: preliminary descriptions of mineralogy and genesis; Earth and Planet. Sci. Letts., v. 53, pp. 363-381.

Haymon, R.M., Koski, R.A., and Collinson, T.B., 1984a, The geology, geochemistry, and tectonic setting of the Bayda sulfide deposit in the Samail ophiolite, Oman; abs., Geol. Soc. of America, Ann. Mtg., v. 16, p. 534.

Haymon, R.M., Koski, R.A., and Sinclair, C., 1984b, Fossils of hydrothermal vent worms from Cretaceous sulfide ores of the Samail ophiolite, Oman; Science, v. 223, pp. 1407-1409.

Hekinian, R. Fevrier, M., Bischoff, J.L., Picot, P., and Shanks, W.C., 1980, Sulfide deposits from the East Pacific Rise near 21 N; Science, v. 207, pp. 1433-1444.

Hutchinson, R.W., 1973, Volcanogenic sulfide deposits and their metallogenic significance; Economic Geology, v. 68, pp. 1223-1246.

Ixer, R.A., Alabaster, T., and Pearce, J.A., 1984, Ore petrography and geochemistry of massive sulphide deposits within the Semail ophiolite, Oman: Trans. Inst. Mining and Metallurgy, v. 93, Sect. B, pp. B114-B124.

Karson, J.A., 1984, Variations in structure and petrology in the Coastal Complex, Newfoundland: anatomy of an oceanic fracture zone; In: Gass, I.G., Lippard, S.J., and Shelton, A.W., (eds.), Ophiolites and oceanic lithosphere; Geol. Soc. of London Spec. Publ. 13, pp. 131-144.

Koski, R.A., Clague, D.A., and Oudin, Elisabeth, 1984, Mineralogy and chemistry of massive sulfide deposits from the Juan de Fuca Ridge; Geol. Soc. America Bull., v. 95, pp. 930-945.

Laurent, R., 1980, Environment of formation, evolution and emplacement of the Appalachian ophiolites from Quebec; In: Panayiotou, A. (ed.), Ophiolites; Proc. Int. Ophiolite Symp., 1979, Cyprus Geol. Surv. Dept., pp. 628-636.

Malahoff, Alexander, 1982, A comparison of the massive submarine
    polymetallic sulfides of the Galapagos Rift with some continental
    deposits; Mar. Technol. Soc. Jour., v. 16, pp. 39-45.

Moores, E.M., 1982, Origin and emplacement of ophiolites; Rev. Geophys.
    and Space Phys., v. 20, pp. 735-760.

Moores, E.M., Robinson, P.T., Malpas, J., and Xenophontos, C., 1984,
    Model for the origin of the Troodos massif, Cyprus, and other
    mideast ophiolites; Geology, v. 12, pp. 500-503.

Moores, E.M., and Vine, F.J., 1971, Troodos Massif, Cyprus and other
    ophiolites as oceanic crust: evaluation and implications; Royal Soc.
    Phil. Trans., v. 268, pp. 443-466.

Oudin, Elisabeth, and Constantinou, Georges, 1984, Black smoker chimney
    fragments in Cyprus sulphide deposits; Nature, v. 308, pp. 349-353.

Oudin, E., Picot, P., and Pouit, G., 1981, Comparison of sulphide
    deposits from the East Pacific Rise and Cyprus; Nature, v. 291, pp.
    404-407.

Pearce, J.A., 1975, Basalt geochemistry used to investigate past
    tectonic environments on Cyprus; Tectonophysics, v. 25, pp. 41- 67.

Pearce, J.A., Alabaster, T., Shelton, A.W., and Searle, M.P., 1981, The
    Oman ophiolite as an arc-basin complex: evidence and implications;
    Royal Soc., Phil. Trans., v. A300, pp. 299-317.

Pearce, J.A., and Cann, J.R., 1971, Ophiolite origin investigated by
    discriminant analysis using Ti, Zr, and Y; Earth and Planet. Sci.
    Letts., v. 12, pp. 339-349.

Pearce, J.A., and Cann, J.R., 1973, Tectonic setting of basic volcanic
    rocks determined using trace element analyses; Earth and Planet.
    Sci. Letts., v. 19, pp. 290-300.

Pearce, J.A., Lippard, S.J., and Roberts, S., 1984, Characteristics and
    tectonic significance of suprasubduction zone ophiolites; In:
    Kokelaar, B.P., and Howells, M.F. (eds.), Marginal basin geology;
    Geol. Soc. of London, Spec. Publ. 16, pp. 77-94.

Robertson, A.H.F., 1975, Cyprus umbers: basal-sediment relationships on
    a Mesozoic ocean ridge; Jour. Geol. Soc. of London, v. 131, pp. 511-
    531.

Robertson, A.H.F., 1978, Metallogenesis along a fossil oceanic
    fracture: Arakapas fault belt, Troodos Massif, Cyprus; Earth and
    Planet. Sci. Letts., v. 41, pp. 317-329.

Robertson, A.H.F., and Dixon, J.E., 1984, Introduction: aspects of the geological evolution of the eastern Mediterranean; In Dixon, J.E., and Robertson, A.H.F., (eds.), The geological evolution of the eastern Mediterranean; Geol. Soc. of London Spec. Publ. 17, pp. 1–74.

Robinson, P.T., Melson, W.G., O'Hearn, T., and Schmincke, H.-U., 1983, Volcanic glass compositions of the Troodos ophiolite, Cyprus; Geology, v. 11, pp. 400–404.

Schmincke, H.-U., Rautenschlein, M., Robinson, P.T., and Mehegan, J.M., 1983, Troodos extrusive series of Cyprus: a comparison with oceanic crust; Geology, v. 11, pp. 405–409.

Shervais, J.W., and Kimbrough, D.L., 1985, Geochemical evidence for the tectonic setting of the Coast Range ophiolite: a composite island arc–oceanic crust terrane in western California; Geology, v. 13, pp. 35–38.

Sillitoe, R.H., 1972, Formation of certain massive sulphide deposits at sites of sea-floor spreading; Trans. Inst. Mining and Metallurgy, v. 81, Sect. B, pp. 141–148.

Simonian, K.O., and Gass, I.G., 1978, Arakapas fault belt, Cyprus: a fossil transform fault; Geol. Soc. America, Bull., v. 89, pp. 1220–1230.

Smewing, J.D., Simonian, K.O., Elboushi, I.M., and Gass, I.G., 1977, Mineralized fault zone parallel to the Oman ophiolite spreading axis; Geology, v. 5, pp. 534–538.

Smewing, J.D., Simonian, K.O., and Gass, I.G., 1975, Metabasalts from the Troodos Massif, Cyprus: genetic implication deduced from petrography and trace element geochemistry; Contrib. Mineral. and Petrol., v. 51, pp. 49–64.

Solomon, M., 1976, Volcanic massive sulphide deposits and their host rocks--a review and explanation; In: Wolf, K.H. (ed.), Handbook of stratabound and stratiform ore deposits; v. 6; Elsevier, Amsterdam, pp. 20–54.

Spiess, F.N., Macdonald, K.C., Atwater, T., Ballard, R., Carranza, D., Cordoba, D., Cox, C., Diaz-Garcia, V.M., Francheteau, J., Guerrero, J., Hawkins, R., Haymon, R., Hessler, R., Juteau, T., Kastner, M., Larson, R., Luyendyk, B., Macdougall, J.D., Miller, S., Normark, W., Orcutt, J., and Rangin, C., 1980, East Pacific Rise: hot springs and geophysical experiments; Science, v. 207, pp. 1421–1433.

Spooner, E.T.C., 1980, Cu-pyrite mineralization and seawater convection in oceanic crust - The ophiolitic ore deposits of Cyprus; In: Strangway, D.W. (ed.), The Continental crust and its mineral deposits; Geol. Assoc. Canada Spec. Paper 20, pp. 685–704.

Spooner, E.T.C., and Fyfe, W.S., 1973, Sub-sea-floor metamorphism, heat and mass transfer; Contrib. Mineral. and Petrol., v. 42, pp. 287-304.

Styrt, M.M., Brackmann, A.J., Holland, H.D., Clark, B.C., Pisutha-Arnond, V., Eldridge, C.S., and Ohmoto, H., 1981, The mineralogy and the isotopic composition of sulfur in hydrothermal sulfide/sulfate deposits on the East Pacific Rise, 21 N latitude; Earth and Planet. Sci. Letts., v. 53, pp. 382-390.

Sun, Shen-su, and Nesbitt, R.W., 1978, Geochemical regularities and genetic significance of ophiolitic basalts; Geology, v. 6, pp. 689-693.

Swinden, H.S., and Kean, B., 1984, Volcanogenic sulfide mineralization in the Newfoundland Central Mobile Belt; In: Mineral deposits of Newfoundland--a 1984 perspective; Newfoundland Depart. Mines and Energy, Rept. no. 84-3, pp. 55-77.

Taylor, G.R., 1974, Volcanogenic mineralization in the islands of the Florida Group, B.S.I.P.; Trans. Inst. Mining and Metallurgy , v. 83, Section B, pp. B120-B130.

Taylor, G.R., 1983, Manganese oxide mineralogy in an exhalative environment from the Solomon Islands; Mineralium Deposita, v. 18, pp. 113-125.

Upadhyay, H.D., and Neale, E.R.W., 1979, On the tectonic regimes of ophiolite genesis; Earth and Planet. Sci. Letts., v. 43, pp. 93-102.

Upadhyay, H.D., and Strong, D.F., 1973, Geologic setting of the Betts Cove copper deposits, Newfoundland: an example of ophiolite mineralization; Economic Geology, v. 68, pp. 161-167.

Varga, R.J., and Moores, E.M., 1985, Spreading structure of the Troodos ophiolite, Cyprus; Geology, v. 13. pp. 846-850.

Xenophontos, Costas, and Bond, G.C., 1978, Petrology, sedimentation, and paleogeography of the Smartville terrane (Jurassic) - bearing on the genesis of the Smartville ophiolite; In: Howell, D.G., and McDougall, K.A. (eds.), Mesozoic paleogeography of the western United States; Pacific Coast Paleogeography Symp. 2; Soc. Econ. Paleont. and Miner., pp. 291-302.

RECENT HYDROTHERMAL METAL ACCUMULATION,
PRODUCTS AND CONDITIONS OF FORMATION

Harald Bäcker and Joachim Lange
PREUSSAG, Marine Technology Department
Arndtstrasse 1
D-3000 Hannover

ABSTRACT. Hydrothermal activity leading to the concentration and
accumulation of metals on and beneath the ocean floor can be observed
in various present marine settings which are influenced by tensional
tectonic forces and by volcanism. Typical end products are metalliferous
sediments, ferromanganese crusts, massive sulfides and stockwork
mineralizations. They are formed by chemical precipitation from heated
seawater which received its metal load through interaction with the hot
volcanic rock. The type of product deposited at a specific site from
ore solutions, which are relatively uniform, depends mainly on exit
conditions, and mixing with ambient seawater before, during and after
discharge at the ocean floor. The distribution, in time and space, of
hydrothermal metal formation is bound to important regional structural
units, but there is also an intimate relationship with the cyclic
development of tectonism and volcanism within the seafloor spreading
framework.

INTRODUCTION

Metal enrichments related to recent seafloor hydrothermal activity
have become a favorite objective of marine research during the last
twenty years, following the discovery of hot brines and associated
metalliferous sediments in the Red Sea.
Although the character of the ore fluids and the principal metals
involved appear to be surprisingly similar at various locations, a
variety of products precipitate from the fluids before and after discharging at the ocean floor. Few of them show sufficiently high metal
contents and quantities to be of any economic interest, at present. A
few occurrences warrant detailed exploration work and might represent
metal accumulations competitive with land-bound deposits, after development of adequate mining methods has taken place. There is now a fairly
good agreement about the character and mode of formation of hydrothermal
precipitates, but due to the limited size of areas covered by detailed
surveys, little is known about the distribution of ore forming hydrothermal activity in time and space.

P. G. Teleki et al. (eds.), Marine Minerals, 317–337.

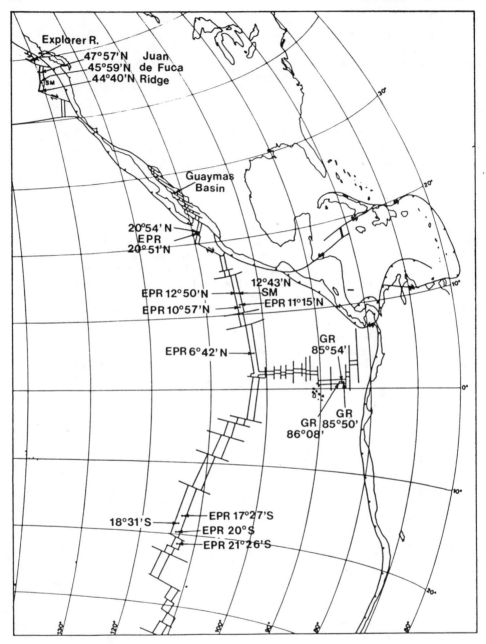

Figure 1. Known sites of hydrothermal discharges and sulfide formation
in the East Pacific Ocean.

A number of factors, however, can be identified as those principally responsible for the occurrence and character of hydrothermal activity and metal formation on the ocean floor, and which are useful as guides for further exploration.

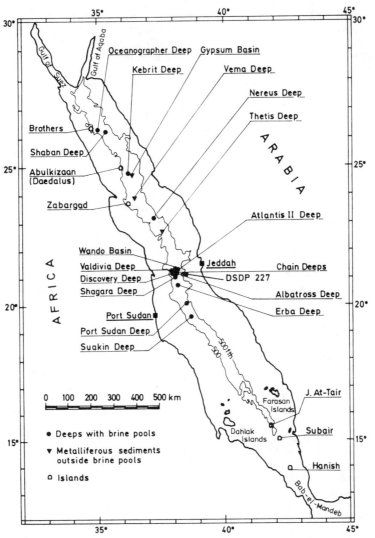

Figure 2. The Red Sea. Location of known occurrences of brines and of metalliferous sediments.

HYDROTHERMAL PRODUCTS

There are various types of material on the ocean floor and in the

subsurface that are entirely or partly derived from hydrothermal
activity: metalliferous sediments, ferro-manganese crusts, massive
sulfides, and disseminated sulfides in sediments and rocks. Typically,
iron, manganese, sulfide sulfur, zinc and copper are enriched with
respect to normal seafloor sediments and to the suspected basaltic
source rock (Table I). Locally, there are also important accumulations
of silicates, predominantly iron smectites, and sulfates (anhydrite).

Rarely do these materials represent the total yield of submarine
hydrothermal sources concerning those components which are insoluble
in seawater or pore water. Instead, they are end products of differentiation processes, the efficiency of which determines the metal grades
and, hence, the economic value of such deposits.

Metalliferous sediments are defined as "unconsolidated accumulations of variable proportions of hydrothermal, detrital, hydrogenous,
and biogenous material" (Heath and Dymond, 1977; Meylan et al., 1981).
The hydrothermal component is very fine-grained. Metalliferous sediments enriched in Fe, Mn, Cu, Cr, Ni and Pb have been described by
Bostrom and Peterson (1966) and subsequently by many other authors from
the flanks of the East Pacific Rise (EPR), and from elsewhere in the
ocean where submarine volcanism is known or can be suspected to occur.
This material, which contains up to about 15% iron, but less than 0.5%
base metals, is economically uninteresting, but may serve as an exploration guide. This type of metalliferous sediments appears to be very
uniform, exhibits rather constant Fe/Mn ratios, can be linked to widespread hydrothermal activity and/or has been subjected to transport by
ocean currents over long distances. Occurrences with rapidly changing
Fe/Mn ratios, on the other hand, indicate proximity of local sources.
Metalliferous sediments, deposited from a semi-stable brine pool,
exhibit a variety of facies, some of which show economically interesting
base metal grades (Table I and Bäcker, 1976). The largest deposit
known so far, the Atlantis II Deep in the Red Sea, covers about 60 km^2
and contains about 30 million tons of iron, 2 million tons of zinc and
500,000 tons of copper.

Ferromanganese crusts on hard-rock outcrops, which locally attain
a thickness of as much as 17 cm, are usually of hydrogenous origin, but
certain geochemical indicators point to important hydrothermal influences: Fe/Mn ratios considerably outside the range of 0.1 - 2, high
Zn/Co ratios, and low Ni and Cu contents.

Massive sulfides occurs as more or less porous hard-rocks in the
form of mounds, chimneys and masses of blocks. In unaltered state they
are mainly composed of well-crystallized sulfides of iron, zinc and
copper, with not much else admixed with them. The base metal grades
are highly variable and locally very high, however, average values on a
deposit scale are not yet available (Haymon and Kastner, 1981; Lafitte
et al., 1984; Oudin, 1983).

Sediment and rock hosted sulfides, well-known from fossil deposits
and first noted in young sediments at the island of Volcano (Honnorez,
1969), are being found in various tectonic settings, such as incipient
and advanced spreading centers, transform faults, and other shear zones.
Strongly transformed and mineralized breccias from faults associated
with surficial massive sulfides were recently recovered from the

Table I. Selected, typical hydrothermal-sedimentary products

	% Fe	% Mn	% Cu	% Zn	ppm Co	ppm Ni
Mn-rich crusts, average of 6 samples, Galapagos Rift (GR) 85°50'W	2,5	46,3	0,038	0,097	30	350
Fe-rich crusts, East Pacific Rise (EPR) lat. 13°N	22,2	2,8	0,010	0,038	100	100
Metalliferous sediments, limonite mud, average of 100 samples, Atlantis II Deep	51,3	1,1	0,187	0,226		
Metalliferous sediments, manganite mud, Chain Deep, average of a 620 cm long core	13,3	30,5	0,104	0,948		
Metalliferous sediments, sulfide mud, average of 11 samples, Atlantis II Deep	21,6	0,16	1,4	16,8		
Metalliferous sediments, carbonate-free portion of EPR axial sediment lat. 19° - 21°S	28,1	9,1	0,128	0,039	110	550
Massive sulfides, EPR lat. 21,5°S	35,8	0,006	6,8	9,1	1400	62
Massive sulfides, EPR lat. 12°50'N	27,0	0,052	1,0	17,8	430	31
Massive sulfides, GR long. 85°50'W	36,3	0,032	4,1	1,4	460	5
Oxidic remnants of sulfides, average of 13 stations, GR 85°50'W	42,1	0,5	0,800	0,400	90	40

Figure 3. Mineralized basalt breccias from the Galapagos Rift,
hydrothemal site near long. 86°, in 2,600 m water depth.

Galapagos Rift (GR) at long. 85°50' (Fig. 3) and the East Pacific Rise
(EPR) at lat. 18.5° and 21.5°S (Bäcker et al., 1985). Hydrothermal
solutions had penetrated through the sediments in the Guaymas Basin,
Gulf of California (Lonsdale et al., 1980), south of the GR at long.
86°10' and near the Kebrit Deep in the Red Sea, and have produced some
hydrothermal products at the sedimentary surface, mainly silicates and
sulfides, but major metal accumulations are expected to be present at
depth.

DISTRIBUTION OF THE MINERAL OCCURRENCES

Mineralizations resulting from submarine hydrothermal activity are
mainly known from volcanic terrain, although a few recent discoveries
appear to have formed along faults, such as in the northern Red Sea
(Bonatti et al., 1985). Most of the known metal occurrences align along
the divergent plate boundaries. Intra-plate volcanism yielded little
major occurrences, so far. Rona (1984) mentions about 55 sites from
seafloor spreading centers, but their locations reflect more the limited
data base than the true global metal distribution pattern. All major
findings made to date are from the East Pacific (Fig. 1) and the Red
Sea (Fig. 2). In the East Pacific, sulfide occurrences were found
along the southern and northern branches of the EPR, at the Galapagos
Rift, in the Gulf of California, at the Juan de Fuca, Gorda and the
Explorer Ridges. In the Red Sea, the main sites with active metallogenesis are located in the central to northern portions within pull-
apart basins and small isolated spreading centers. The spreading rate
is often regarded as a major factor for governing the distribution and
quantity of the mineral occurrences, but the known sites cover the
whole range from "slow" (Red Sea at 0.5 - 3 cm/year) to "ultrafast"
spreading centers (EPR lat. 17 - 22°S at 16 cm/year). Along a longer
section of the EPR, the presently visible metallogenesis is not uniform
and not randomly distributed. Francheteau and Ballard (1983) observed
highest activities near topographic highs of the rift axis, between two
transform faults, while the rift portions near the fracture zones
appeared to be too "cool" for the maintenance of major hydrothermal
circulation. The occurrence of important hydrothermal fields in axial
graben structures of the EPR at lat. 21.5°S and 18.5°S (Bäcker et al.,
1985), 13°N (Hekinian et al., 1984) and of the Galapagos Rift (Malahoff
et al., 1983) point to the importance of tectonic movements for the
initiation and maintenance of ore-forming hydrothermal activity.

TECTONICS

The graben structures are the result of uplift and tensional forces
active along the spreading centers. The most common tectonic features
are normal faults dipping in the direction of the rift axis and arranged
en echelon, are striking in rift direction. Within the neo-volcanic
zone in prevailing ponded lava terrain collapse pits frequently align
along incipient faults (Fig. 4, Fig. 6., 1. and 2.). Further rifting

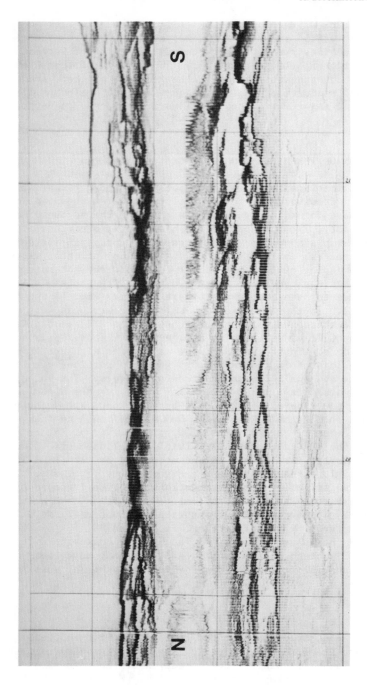

Figure 4. An example of prevailing tectonic and collapse features around hydro—
thermal sites (EPR, Lat. 12°50' N) shown in uncorrected side-scan sonar records,
700 m in length, 200 m in swath-width on both sides. Walls of collapse pits are
strongly influenced by tectonic displacements which are arranged "en echelon."

leads to the development of major marginal faults, usually termed the
"walls" (Fig. 6, pt. 3). Within a hydrothermal field most sources are
related to faults within the graben area, or are located at the base of
the "walls." The importance of these faults as channel-ways for the hydrothermal solutions are underlined by the presence of strongly mineralized breccias (Fig. 3). Transcurrent directions are less frequent and
generally not well-defined by morphological features. Their importance
for the location of ore bodies seems to be overrated by many geologists.
Certain sections of the spreading centers are clearly characterized by
the preponderance of destructive tectonic features accompanied by the
formation of talus piles and collapse pits (Fig. 4; Fig. 7, Section 3
to 5). On bathymetric maps parallel lineaments prevail.

     This setting is locally replaced by constructive non-linear fea-
tures produced by lava flows.

VOLCANISM

     Magmatism seems to be the most important factor in the evolution
of hydrothermal sedimentary ore formation, first, as the necessary heat
source to generate and maintain a hydrothermal circulation system, and
secondly, as the source for the metals involved. It is, therefore, of
paramount importance, for the understanding of the formation and distribution of ore bodies, to investigate the character and evolution of
volcanic events and products on the ocean floor.

     Along the spreading centers, the source of the erupted lavas is
the mantle, which undergoes decompression melting when rising to fill
the gap between diverging plates. The rise of the magma into the crust
results from its density being lower than that of the surrounding mantle
rocks (Whitehead et al., 1984). According to Whitehead et al. (op.
cit.), this rise should take place in regularly spaced protrusions,
which form magma chambers preferentially at the mid-points of individual
ridge segments. This view is supported by the observed rise of the EPR
axis and increased volcanic and hydrothermal activity about midway
between transform faults (Francheteau and Ballard, 1983). In Iceland,
the lateral influence of individual magma chambers along the rift can
be traced as far as 70 km distant (Sigurdsson and Sparks, 1978). At
Kilauea, Hawaii, both rifting and eruptions are episodic, and the
episodes migrate along the rift zones. However, the frequency and
volume of eruptions remain roughly constant (Holcomb and Clague, 1983).
Thus, the magmatic plumbing systems are not stationary, they reach
distant places, providing for local heat transfer and possibly also for
hydrothermal pathways by back-drainage of lava.

     On the deep ocean floor two groups of basalt flows can be observed:
pillow lavas and several types of ponded or sheeted flows. Both appear
to have the same composition, but the crystalline fabric of the sheet
flows is poorly developed (Hekinian, 1984). In intermediate- to fast-
spreading centers sheet flows represent early, brief, but voluminous
eruptions (Ballard et al., 1979), while smaller persistent eruptions
produce well channeled pillow flows. Pillows, however, also occur at
the margins of sheet flows. Major sheet flows accumulate in lava lakes,

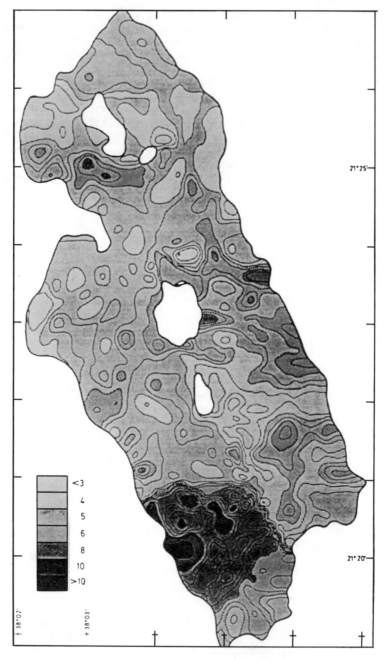

Figure 5. Distribution of Zn/Cu ratio in the sediments of the
Atlantis II Deep, Red Sea, reflecting sources and metal dispersion
(data base: 600 cores).

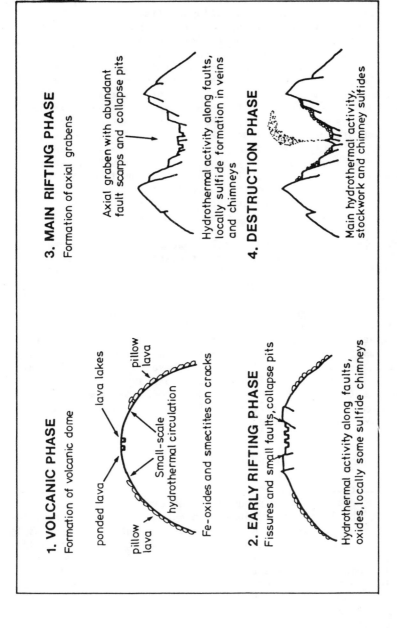

Figure 6. Schematic presentation of phases of activity along the axis of intermediate- to fast-spreading divergent plate boundaries.

which show abundant indications of lava level fluctuations, such as
"bath tub rings", lava cellars and collapse pits.

Most sulfide occurrences found so far are somehow related to ponded
lava flows. Higher heat production and/or clogging of large volumes of
water could be the reasons for this. Ponded lavas are not so well
documented from slow spreading centers, but this observation does
not exclude the presence of major metal deposits, as documented by the
Atlantis II Deep in the Red Sea. While most of the known sulfide
occurrences have been found in rifted portions of elongated axial
shield volcanoes, some are in axial and off-ridge caldera volcanoes (EPR
lat. 13°N: Hekinian et al., 1983; EPR lat. 21°N: Lonsdale et al.,
1982; Juan de Fuca Ridge lat. 46°N: Canadian-American Seamount Expedition, 1985). Such volcanoes are probably the result of excess magma
production and, hence, the possible locus of extended hydrothermal
activity and metallogenesis. Morphological data from a seamount chain
at the western flank of the EPR at lat. 10°N (Fornari et al., 1984)
show a change from conical domes to large truncated cones as the edifice
moves away from the axis.

Recent observations in the Semail ophiolite of Oman (Alabaster and
Pearce, 1985) have shown that the largest massive sulfide deposit
(Lasail, 8 million tons) formed in the interim between a spreading axis
lava eruption and an off-axis seamount eruption, and the latter drove
the hydrothermal circulation.

Little is known about the influence that magma composition has on
the type and extent of metallogenesis. Although most known marine
massive sulfides are linked to the eruption of quartz-normative theo-
leiitic basalts (MORB), that described by Malahoff et al. (1983) from
the GR at long. 85°50' belongs to the FeTi basalts of the HAM (high
amplitude magnetic anomaly) section of the Galapagos Rift. While the
initial samples from this deposit were described as copper-rich with
only minor amounts of zinc, recently collected samples from further west
more closely resemble EPR sulfides.

HYDROTHERMAL CIRCULATION AND THE ORIGIN OF THE METALS

The combination of tensional faulting and volcanism in a submarine
environment is well suited to produce hydrothermal circulation. Little
is known about the lateral and subsurface extent of such a circulation.
Frequent surficial circulation during magmatic events is evidenced by
the wide-spread occurrence of iron oxides and smectites on the walls of
shrinkage cracks in fresh lava flows. Significantly large metal deposits, however, require persistent deep circulation. The lowermost limit
is defined by the magma chamber, the top of which is suspected to be at
a depth of 2 km. On the other hand, circulation into the range of
influence of the magma chamber is required to furnish the heat necessary
for the deposition of major sulfide bodies. Hydrothermal circulation
models (Strans and Cann, 1982) have shown that conductive heatflow into
a single 1,000-m deep fracture could support only isolated hot vents.

Taking the absence of sedimentary sequences along most spreading
centers and the relatively uniform composition of the known ore fluids

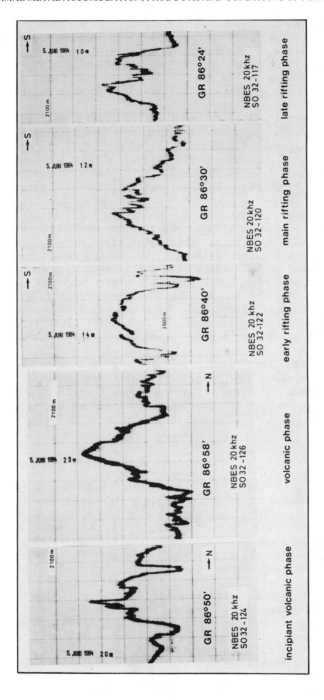

Figure 7. Bathymetric sections across the Galapagos Rift, at long. 86–95° W (narrow-beam echosounder). Development of the axial part from the upbuilding of an axial volcano to faulting and to subsidence.

into account, it can be assumed that the ore-forming "basaltophilic
elements" Fe, Mn, Cu and Zn (Smirnov, 1967) are leached from the basalt.
This view is supported by laboratory tests (Bischoff and Dickson, 1975,
Mottl and Holland, 1978). The basalt used by Bischoff and Dickson (op.
cit.) contained 132 ppm Zn, 22 ppm Cu and 18 ppm Pb. Galapagos Rift
basalts analyzed by Clague et al. (1981) showed 100 ppm Zn, 60 ppm Cu
and 0.027 ppm Ag.

While the general availability of the transition metals from
basaltic rocks and the leaching process appear to be well demonstrated,
there is some disagreement about the size of deposits, the amount of
available source rock and the observed degree of rock alteration. Most
rock samples obtained from hydrothermal sites show little or no high-
temperature alteration, which implies that most leaching takes place at
depth.

To generate the 2 million-ton Zn-deposit in the Atlantis II Deep
from a basaltic source rock with a concentration of 100 ppm Zn, 67 km^3
of rock is required, assuming a 10% transfer efficiency and negligible
zinc dispersion from the deposit area. The leached basalt body can
have an areal extent of 4.5 x 15 km and a thickness of about 1 km, with
its uppermost surface about 100 m below the seafloor.

A similar ore body on the EPR would require a much larger volume
of source rocks, because the majority of the metal load of the sources
has been dispersed. Additional source rock quantities are required to
furnish the stockwork mineralization suspected to exist below the
hydrothermal vents.

Information from land-based metallogenic provinces indicate that
entire deposits or parts of them may originate from remobilized older
occurrences. Such a stockpile effect could use the energy and metal
input of several magmatic-tectonic cycles to build up one major surficial deposit.

FRACTIONATION PROCESSES

There is evidence that ore fluids, derived from basalt-seawater
interaction at elevated temperatures, have a relatively uniform composition and temperature before they give rise to the formation of a variety
of deposits (Edmond, 1982; Edmond et al., 1982). Where such fluids
could be observed and sampled (e.g., EPR lat. 21° and 13°N) they were
found to be transparent and homogeneous and showed temperatures around
300–350° C. Sulfate and magnesium were depleted in respect to seawater.
Deviation from the physical and chemical character of this "standard
high-temperature basalt/seawater interaction product" indicates external
influences and fractionation processes.

One such process has been noted at the low-temperature vents of the
Galapagos Rift (long. 86° W), where sulfate and magnesium decrease uniformly with temperature. This trend for SO_4 and Mg can be extrapolated
to zero at 380° and 344° C, respectively (Edmond et al., 1979; Edmond
et al., 1982). If the temperature decrease and the sulfate + magnesium
increase are caused by mixing of the original mineralizing fluid with
unaltered seawater, and takes place within the rocks at shallow depth,

it can be assumed that part of the metal content is incorporated within
a stockwork mineralization below the seafloor.

Important fractionation processes can also be expected when the
ore fluids have to pass through sediments before reaching the seafloor.
In the hydrothermal mounds south of the Galapagos Rift axis at long.
86° W, pelagic sediments have been replaced by an Fe-rich smectite and
manganese has precipitated at the sediment/seawater interface (Moorby
and Cronan, 1983).

In the Guaymas Basin (Gulf of California), there is evidence for
the presence of high-temperature hydrothermal solutions in sediments in
the dissolution of calcium carbonate and siliceous organisms, in re-
crystallization of clays, and in the precipitation of a variety of new
mineral phases, among them sulfides and barite (Lesmes et al., 1984).
Organic material in the sediments has been transformed into petroleum
(Simoneit and Lonsdale, 1982). Similar products have recently been
found near the Kebrit Deep, Northern Red Sea, which, in contrast to the
occurrences in the Gulf of California, contain massive zinc sulfides
(Puchelt and Laschek, 1984).

In the Red Sea the density of the mineralizing fluids, increased
by the uptake of NaCl from Miocene evaporites, has caused the formation
of submarine brine pools with well-defined and relatively stable brine/
seawater interfaces. Within such restricted environments, physico-
chemical characteristics, such as pH and Eh, can stay constant for long
periods of time. As a result, chemical reactions and the character and
distribution of the precipitated sediments are controlled by the location and activity of the sources and mixing at the brine/seawater in-
terface(s). In the Atlantis II Deep, the character and changes in such
a semiclosed system, such as temperature, dissolved metals and brine
levels, could be observed over a period of about 20 years (Hartmann,
1985). A major feature of such a system is the continuous removal of
most manganese from the brine basin, which migrates into marginal
basins (e.g. the Chain Deeps near Atlantis II Deep) or forms a halo
around the deep (Bäcker, 1976).

Base metal sulfides precipitate preferentially around the hydrothermal discharge vents, but even within this group a certain fractionation takes place, which is mainly governed by the distance from the
sources and morphological features (Fig. 5).

Some of the brine pools in the Red Sea have not shown any significant change of brine level, temperature or composition over periods of
several years (Albatross, Kebrit, Shaban). In contrast, the stratigraphic record of the 25,000 year old Atlantis II Deep indicates that
two major stable brine periods existed, during which sulfides and
silicates were preferentially deposited. In the interim, the stable
brine-pool broke down because of external events, such as earthquakes
and lava flows, and highly disturbed oxidic sediments were deposited,
including manganese (the "Central Oxidic Zone"). However, lamination
in the two sulfidic zones in the Atlantis II Deep also indicates varying
conditions, such as fluctuating hydrogen sulfide supply or changing pH
or temperature.

At open-ocean hydrothermal sites, such as those along the East
Pacific Rise, sulfide deposition is restricted to the vents. Obviously,

the major part of the metal content is ejected from black smokers and dissipated by the mid-depth oceanic circulation over a distance of hundreds of kilometers. Sediments derived from the hydrothermal effluents show high iron and manganese contents, and elevated values of Zn, Cu, Ag, Sb and silicate (Edmond et al., 1982; Marchig et al, 1982; Kunzendorf et al., 1984). This portion may exceed 90% of the total metal content, but there is no quantitative data available, at present. The quantity and character of the massive sulfide deposits around the single vents seem to be largely the result of self-clogging of the hydrothermal plumbing system.

POSTGENETIC PROCESSES

Following the self-closure of a black smoker, clear hydrothermal effluents continue to leak from the chimney through small fissures and pores. Apparently the dissolved metal load has been precipitated before, filling voids in the chimney structure. One mode of chimney growth includes the initial precipitation of anhydrite from heated seawater, to form a protective wall or cap with subsequent replacement by sulfide phases (Haymon, 1983; Campbell et al., 1984). In the Atlantis II Deep, anhydrite forms major accumulations locally, probably as a result of short intervals of increased inflow of hot brines.

Post-depositional transformations are already common in very young metalliferous sediments. These include the transformation of geothite to hematite and magnetite (Bischoff, 1969) and an increase in crystallinity (Weber-Diefenbach, 1977). Volcanic and tectonic events cause the destruction and eventually the resedimentation of deposits. Open-ocean massive sulfides are subjected to oxidation, if not covered in time by lava flows. Most of the active sulfide chimneys are accompanied by ocre-colored mounds of decayed sulfides.

SPATIAL AND TEMPORAL RELATIONSHIPS

The time-space distribution of the hydrothermal activity is especially important in respect to economically interesting sulfide deposits. If the volcanic and hydrothermal activity is mainly restricted to regional highs between transform faults (Francheteau and Ballard, 1983), potential mining sites would be limited to small portions of the rift system, but they would be comparatively easy to find.

On the other hand, there are important indications of migrating and fluctuating activity of both volcanism and hydrothermalism. The lateral transformation, perpendicular to the rift axis, of a rift valley into an axial high has been described from the Galapagos Rift at long. 86° W (Van Andel and Ballard, 1979). For the EPR near the mouth of the Gulf of California, Lewis (1979) concludes from the morphological development of volcano-tectonic features, that rift valleys are not steady-state features, but the result of variations in magma supply rate. Recent work at the EPR between lat. 17° and 22° S (Bäcker et al., 1985; Renard et al., 1985) suggests the presence of cycles that start with a constructive magmatic phase and end with the dismantling of a morphological high by faulting and collapse. In this model each phase has its own specific style of hydrothermalism (Fig. 6). In the Galapagos Rift, the development of the crustal zone of the ridge can be recognized well on echosounder profiles (Fig. 7). Major sulfide accumulations are expected to form, when metal and heat supply from two or more volcanic phases, or from extended activity through surplus magma production concentrate at one location, and the resulting deposits do not become covered by new lava flows.

CONCLUSIONS

Hydrothermal activity on the ocean floor is a wide-spread phenomenon in areas which are characterized by submarine volcanism and faulting. This activity is episodic rather than continuous and occupies its place within a cyclic but complicated tectonic-magmatic development. Though shallow hydrothermal circulation during effusive phases leads to a number of oxide and silicate products, economically significant base metal accumulations depend on deep hydrothermal circulation within the influence of major heat sources. High-temperature hydrothermal fluids, which received their metal load through basalt leaching, show relatively uniform composition. Precipitates derived from these fluids, however, are highly variable, which is due to a number of fractionation processes, active before and after discharge on the ocean floor. Economically important hydrothermal-sedimentary ore deposits can be expected only under environmental conditions which lead to the concentration of the valuable metals copper, zinc and silver at the expense of iron, manganese and silica.

REFERENCES

Alabaster, T. and Pearce, J.A., 1985, The interrelationship between magmatic and ore-forming hydrothermal processes in the Oman ophiolite; Econ. Geol., v. 80, no. 1, pp. 1-16.

Bäcker, H., 1976, Fazies und chemische Zusammensetzung rezenter Ausfallungen aus Mineralquellen im Roten Meer; Geol. Jahrb. D, v. 17, pp. 151-172.

Bäcker, H., Lange, J. and Marchig, V., 1985, Hydrothermal activity and sulphide formation in axial valleys of the East Pacific Rise crest between 18° and 22°S; Earth Planet. Sci. Lett., v. 72, no. 1, pp. 9-22.

Ballard, R.D., Holcomb., R.T. and Van Andel, T.H., 1979, The Galapagos Rift at 86° W: 3. Sheet flows, collapse pits, and lava lakes of the rift valley; Jour. Geophys. Res., v. 84, no. B-10, pp. 5407-5422.

Bischoff, J.L., 1969, Goethite-hematite stability relations with
    relevance to sea water and the Red Sea brine system; In: Degens,
    E.T. and Ross, D.A. (eds.), Hot brines and recent heavy metal
    deposits in the Red Sea; Springer-Verlag, New York, pp. 402-406.

Bischoff, J.L. and Dickson, F.W., 1975, Seawater-basalt interaction at
    200° C and 500 bars: Implications for origin of sea-floor heavy-
    metal deposits and regulation of seawater chemistry; Earth Planet.
    Sci. Lett., v. 25, pp. 385-397.

Bonatti, E., Colantoni, P., Rossi, P.L. and Taviani, M., in press, Ore
    formation in the Red Sea: occurrence of metalliferous ores outside
    the axial valley;   Mem. Soc. Geol. Ital., v. 27.

Bostrom, K. and Peterson, M.N.A., 1966, Precipitates from hydrothermal
    exhalations on the East Pacific Rise; Econ. Geol., v. 61, pp. 1258-
    1265.

Campbell, I.H., McDougall, T.J. and Turner, J.S., 1984, A note on fluid
    dynamic processes which can influence the deposition of massive
    sulfides; Econ. Geol., v. 79, no. 8, pp. 1905-1913.

Canadian-American Seamount Expedition, 1985, Hydrothermal vents on an
    axis seamount of the Juan de Fuca ridge; Nature, v. 313, pp. 212-214.

Clague, D.A., Frey, F.A., Thompson, G. and Rindge, S., 1981, Minor and
    trace element geochemistry of volcanic rocks dredged from the
    Galapagos spreading center; role of crystal fractionation and mantle
    heterogeneity; Jour. Geophys. Res., v. 86, pp. 9469-9482.

Edmond, J., 1982, The chemistry of ridge crest hot springs; Marine
    Technol. Soc. Jour., v. 16, pp. 23-25.

Edmond, J.M., von Damm, K.L., McDuff, R.E. and Measures, C.I., 1982,
    Chemistry of hot springs on the East Pacific Rise and their effluent
    dispersal; Nature, v. 297, no. 5863, pp. 187-191.

Edmond, J.M., Measures, C., McDuff, R.E., Chan, L.H., Collier, R.,
    Grant, B., Gordon, L.I. and Corliss, J.V., 1979, Ridge crest hydro-
    thermal activity and the balances of the major and minor elements
    in the ocean: the Galapagos data; Earth Planet. Sci. Lett., v. 46,
    no. 1, pp. 1-18.

Fornari, D.J., Ryan, W.B.F. and Fox, P.J., 1984, The evolution of craters
    and calderas on young seamounts: insights from SeaMARC I and Sea-
    Beam sonar surveys of a small seamount group near the axis of the
    East Pacific Rise at 10° N; Jour. Geophys. Res., v. 89, no. B-13,
    pp. 11069-11083.

Francheteau, J. and Ballard, R.D., 1983, The East Pacific Rise near
    21° N, 13° N and 20° S: inferences for along-strike variability of
    axial processes of the Mid-Ocean Ridge; Earth Planet. Sci. Lett.,
    v. 64, no. 1, pp. 93-116.

Hartmann, M., 1985, Atlantis II Deep geothermal brine system; chemical
    processes between hydrothermal brines and Red Sea deep water; Marine
    Geol., v. 64, no. 1/2, pp. 157-177.

Haymon, R.M., 1983, Growth history of hydrothermal black smoker chim-
    neys; Nature, v. 301, no 5902, pp. 695-698.

Haymon, R.M. and Kastner, M., 1981, Hot spring deposits on the East
    Pacific Rise at 21° N: preliminary description of mineralogy and
    genesis; Earth Planet. Sci. Lett., v. 53, pp. 363-381.

Heath, G.R. and Dymond, J., 1977, Genesis and transformation of metal-
    liferous sediments from the East Pacific Rise, Bauer Deep, and
    Central Basin, northwest Nazca plate; Geol. Soc. Am. Bull., v. 88,
    no. 5, pp. 723-733.

Hekinian, R., 1984, Undersea volcanoes; Sci. American, v. 251, no. 1,
    pp. 34-43.

Hekinian, R., Fevrier, M., Avedik, F., Cambon, P., Charlou, J.L.,
    Needham, H.D., Raillard, J., Boulegue, J., Merlivat, L., Moinet, A.,
    Manganini, S. and Lange, J., 1983, East Pacific Rise near 13° N:
    geology of new hydrothermal fields; Science, v. 219, no 4590, pp.
    1321-1324.

Hekinian, R., Renard, V. and Cheminee, J.L., 1984, Hydrothermal deposits
    on the East Pacific Rise near 13° N: geological setting and
    distribution of active sulfide chimneys; In: Rona, P.A., Bostrom,
    K., Laubier, L. and Smith, K.L. (eds.), Hydrothermal processes at
    seafloor spreading centers; Plenum, New York, pp. 571-594.

Holcomb, R.T. and Clague, D.A., 1983, Volcanic eruption patterns along
    submarine rift zones; Proc. OCEANS '83, San Francisco, v. 2, pp.
    787-790.

Honnorez, J., 1969, La formation actuelle d'une gisement sous-marine de
    sulfures fumerolliens a Vulcano (Mer tyrrhenienne); Partie I.: Les
    mineraux sulfures des tufs immerges a faible profondeur; Mineralia
    Deposita, v. 4, pp. 114-131.

Kunzendorf, H., Walter, P., Stoffers, P. and Gwozdz, R., 1984, Metal
    variations in divergent plate-boundary sediments from the Pacific;
    Chem. Geol., v. 47, no. 1/2, pp. 113-133.

Lafitte, M., Maury, R. and Perseil, E.A., 1984, Analyse mineralogique de
    cheminees a sulfures de la dorsale Est Pacifique (13° N); Mineralia

Deposita, v. 19, no. 4, pp. 274-282.

Lesmes, D., Gieskes, J.M., Campbell, A.C., Stout, P.M. and Brumsak,
  H.J., 1984, Geochemistry of hydrothermally altered surface sediments
  in the Guaymas Basin, Gulf of California; EOS, Trans. Am. Geophys.
  Union, v. 65, no. 45, p. 974.

Lewis, B.T.R., 1979, Periodicities in volcanism and longitudinal magma
  flow on the East Pacific Rise at 23° N; Geophys. Res. Lett., v. 6,
  no. 10, pp. 753-756.

Lonsdale. P., Batiza, R. and Simkin, T., 1982, Metallogenesis at seamounts of the East Pacific Rise; Mar. Technol. Soc. Jour., v. 16,
  no. 3, pp. 54-61.

Lonsdale, P.F., Bischoff, J.L., Burns, V.M., Kastner, M. and Sweeny,
  R.E., 1980, A high-temperature hydrothermal deposit on the seabed
  at a Gulf of California spreading center; Earth Planet. Sci. Lett.,
  v. 49, no. 1, pp. 8-20.

Malahoff, A., Embley, R.W., Cronan, D.S. and Skirrow, R., 1983, The
  geological setting and chemistry of hydrothermal sulfides and
  associated deposits from the Galapagos Rift at 86° W; Marine Mining,
  v. 4, no. 1, pp. 123-137.

Marchig, V., Gundlach, H., Moller, P. and Schley, F., 1982, Some geochemical indicators for discrimination between diagenetic and
  hydrothermal metalliferous sediments; Marine Geol., v. 50, no. 3,
  pp. 241-256.

Meylan, M.A., Glasby, G.P., Knedler, H.E. and Johnston, J.H., 1981,
  Metalliferous deep-sea sediments; In: Wolf, H.H. (ed.), Handbook
  of stratabound and stratiform ore deposits; Elsevier, Amsterdam,
  pp. 77-178.

Moorby, S.A. and Cronan, D.S., 1983, The geochemistry of hydrothermal
  and pelagic sediments from the Galapagos Hydrothermal Mounds Field,
  D.S.D.P. Leg 70; Mineral. Mag., v. 47, no. 3, pp. 291-300.

Mottl, M.J. and Holland, H.D., 1978, Chemical exchange during hydrothermal alteration of basalt by seawater; I. Experimental results for
  major and minor components of seawater; Geochim. Cosmochim. Acta,
  v. 42, pp. 1103-1115.

Oudin, E., 1983, Hydrothermal sulfide deposits of the East Pacific
  Rise (21° N); Part 1, Descriptive mineralogy; Marine Mining, v. 4,
  no. 1, pp. 39-72.

Puchelt, J. and Laschek, D., 1984, Marine Erzvorkommen im Roten Meer;
  Fridericiana, Zeitschr. Univ. Karlsruhe, v. 34, pp. 3-17.

Renard, V., Hekinian, R., Francheteau, J., Ballard, R.D. and Backer, H., 1985, Submersible observations at the axis of the ultrafast-spreading East Pacific Rise (17°30' to 21°30' S); Earth Planet. Sci. Lett., v. 75, pp. 339-353.

Rona, P.A., 1984, Hydrothermal mineralization at seafloor spreading centers; Earth-Science Rev., v. 20, pp. 1-104.

Sigurdsson, H. and Sparks, S.R.J., 1978, Lateral magma flow within rifted Icelandic crust; Nature, v. 274, no. 5667, pp. 126-130.

Simoneit, B.R.T. and Lonsdale, P.F., 1982, Hydrothermal petroleum in mineralized mounds at the seabed of Guaymas Basin; Nature, v. 295, no 5846, pp. 198-202.

Smirnov, V.I., 1967, Sources of ore material; Inst. Mining Metal. Trans., Sect. B., v. 76, p. 229.

Strans, M.R. and Cann, J.R., 1982, A model of hydrothermal circulation in fault zones at mid-ocean ridge crests; Geophys. Jour. Roy. Astr. Soc., v. 71, pp. 225-240.

Van Andel, T.H. and Ballard, R.D., 1979, The Galapagos Rift at 86° W: 2. Volcanism, structure and evolution of the rift valley; Jour. Geophys. Res., v. 84, no. B-10, pp. 5390-5406.

Weber-Diefenbach, K., 1977, Geochemistry and diagenesis of recent heavy metals ore deposits at the Atlantis II Deep (Red Sea); In: Klemm, D.D. and Schneider, J.H., (eds.), Time- and stratabound ore deposits; Springer-Verlag, Berlin, pp. 419-436.

Whitehead, J.A., Dick, H.J.B. and Schouten, H., 1984, A mechanism for magmatic accretion under spreading centers; Nature, v. 312, no. 5990, pp. 146-148.

THE CHEMISTRY OF SUBMARINE ORE-FORMING SOLUTIONS

J.M. Edmond, *K.L. Von Damm and T.S. Bowers
Department of Earth, Atmospheric and
Planetary Sciences
Massachusetts Institute of Technology
Cambridge, MA   01239,   USA

ABSTRACT. Hydrothermal solutions have been sampled from two important types of submarine hot springs; from the sediment-starved crest of the East Pacific Rise at lat. 21°N and between 10° and 13°N from the sediment-buried axis in the Guaymas Basin in the central part of the Gulf of California. The former solutions are acid with relatively high concentrations of metals and sulfide. The latter are "spent" of their ore metals through reaction with sedimentary carbonate and organic carbon. The reaction with carbon releases large quantities of ammonia. In the Guaymas Basin, therefore, a sediment-hosted ore body may be in the process of formation. At 21°N, the major proportion of the transported ore is dispersed into the water column as black, fine-grained sulfide "smoke". Mechanisms for ore retention at such systems are as of now unknown.

INTRODUCTION

The combined results of terrestrial and oceanographic exploration for massive sulfide ore bodies suggest three distinct regimes for their emplacement.

Mature, open ocean, sediment-starved ridge axes.

This regime hosts simple Cyprus or ophiolite type massive sulfides (Franklin et al., 1981). On land, these are seen to be lenticular Fe-Cu-Zn sulfide bodies deposited on pillowed or sheet-flow terrain. They usually fall in the size range 1 - 10 million tons (Mt) with grades of Cu and Zn ranging up to about 10%. They are underlain by mineralized stringer or stockwork zones, the fossil feeder channels for the ore-forming solutions. They are overlain by an oxidized gossan or "ochre", formed during post-depositional exposure to oxygenated abyssal waters. Above this are either further lava flows or a sedimentary section

* Present address: Environmental Sciences Division, Oak Ridge
  National Laboratory, P. O. Box X, Oak Ridge, Tennessee 37831

P. G. Teleki et al. (eds.), Marine Minerals, 339–347.
© 1987 by D. Reidel Publishing Company.

grading from oxidized and silicified metalliferous sediments, or
"umbers", into abyssal claystones or volcanoclastic, usually andesitic,
detritus. The latter is indicative of formation of the deposit at a
spreading center close to an active margin or in a back-arc extensional
regime. In either case the analogy with open-ocean spreading centers is
reasonably good. The active systems discovered over the past 6 years on
the East Pacific Rise (EPR) and Juan de Fuca Ridge are small and
dominated by constructional features, the famous "black smoker"
chimneys, and the talus they produce. The fluids generally have
temperatures in the 350 ± 5°C range, with two examples (of about 12)
occuring at 380°C and 405°C. The analyzed fluid compositions are
clearly adequate to form ore bodies, given sufficient flux. However, in
the systems found to date, the overwhelming proportion of transported
ore metals is vented to the free-water column, where it precipitates as
an extremely fine-grained "smoke" of sulfide minerals. This is not
observed to settle out around the vents, but is initially dispersed in
the regional oceanic circulation, then oxidized and sedimented out to
form the metalliferous sediments or umbers. Mechanisms for more
efficient ore retention or localization are discussed below.

Rifted margins, intracratonic rifted basins and extensional back arc
regimes.

At these sites and those close to active margins the spreading axis
can be buried by sediments if the spreading rate is slow relative to the
sedimentary accumulation rate (Gustafson and Williams, 1981). The
composition of the sediment can be extremely diverse. At rifted margins
the ridge is usually buried by fluvial debris, e.g., from the Colorado
in the Salton Trough and the Yangtze and Yellow rivers in the Okinawa
Trough. Under certain oceanographic conditions, the narrow seaways thus
formed can be sites of very high biological productivity in the surface
waters overlying the ridge axis, e.g., in Gulf of California, such that
the sediment blanket can contain large amounts of organic carbon
associated with opaline and calcareous tests. Depending on local
climate, intracratonic rifts can accumulate continental playa-type salt
deposits; examples are the Eastern Rift zone in East Africa, and the
Afar Triangle. If the rifts are intermittently open to the sea, thick
sections of marine evaporites can accumulate, as in the Red Sea and in
the proto-South Atlantic. Where the ridge axis is buried, the volcanic
activity, associated with spreading, forms dykes and sills that intrude
into the unconsolidated fill, instead of surficial pillow lavas and
sheet flows. In addition to the basalts, the entire sediment column
participates in the hydrothermal reactions, whereby the sediments are an
additional source of ore metals. Where calcareous or organic rich
horizons are present, they increase the pH of the ascending fluids, by
providing carbonate or ammonia, thus stripping the sedimentary layers of
ore-forming metals. The results are stratiform, sediment-hosted,
volcanogenic, massive sulfides. Because ore precipitation is contained
within the sediment column, it is highly efficient, losses in solution
being limited by sulfide solubility at intermediate to high pH levels.
This type of ore body is generally much larger than the Cyprus type,

with several tens of millions of tons of ore present. Depending on the
host sediment type, very high ore grades can be reached. One can
imagine turbidites composed of immature feldspathic sand as a fertile
source of lead, for instance.

Containment of hydrothermal solutions by evaporite dissolution to
form brines appears to be unique to the Red Sea at the present time.
This is the only case where cooling by conductive heat loss (double
diffusion at the brine interface) has been demonstrated as an important
ore localization mechanism (McDougall, 1984).

Seamounts

With increasing utilization of sonar swath-mapping (SEABEAM) and
acoustic side-scan imaging (SEAMARC, GLORIA) techniques, young seamounts
are being found with increasing frequency. These commonly display a
summit morphology similar to that of sub-aerial basaltic volcanoes with
calderas and secondary collapse pits and cones. A growing body of
opinion identifies calderas as the depositional environment of the
Japanese Kuroko-type deposits (Ohmoto and Takahashi, 1983) and of some
of the large Canadian massive sulfide deposits of Proterozoic age.
Hydrothermal mineral deposits have, so far, been observed on four
contemporary seamounts. At EPR 21°N, two off-axis seamounts were
explored. One had a peculiar atacamite-barite association in its summit
caldera. The other had sulfides and low-temperature (10°C) activity
with fluids that precipitated gel-like iron hydroxides as they entrained
the ambient 2.2°C seawater (Lonsdale et al., 1982). At 13°N on the EPR,
a large flank sulfide deposit has been found (Hekinian et al., 1983).
On Juan de Fuca, extinct EPR-type chimneys and low-temperature
Galapagos-type activity have been observed inside the summit caldera.
Much more exploration is obviously needed in this type of terrain to
broaden this catalog of features and processes.

CHEMISTRY

Hydrothermal fluids from active vents on sediment-starved ridge
axes have been sampled at 21°N and 10-13°N on the EPR (Table I, Von Damm
et al., 1985(a)). The solutions are acid, with pH readings between 3.0
and 3.5, and contain millimolar quantities of H_2S, Fe and Mn. They are
completely depleted in Mg and sulfate, are variably enriched in K and Ca
relative to ambient seawater, and have sharply elevated levels of Li, Rb
and Be. Ionic strengths range from 60% to 130% of the seawater value.
The striking impression is one of large compositional variability,
unexpected for such uniform primary reactants as tholeiitic basalt and
seawater. It is difficult to identify the responsible mechanisms as
inter-element correlations are not well developed. An exception is the
relationship between Ca and Sr, where a linearly increasing trend is
observed over a concentration range of about a factor of five (Fig. 1).
The trend passes through the origin, indicating a solid-solution control
with a distribution coefficient of about unity. This is consistent with
the non-radiogenic isotopic composition of the hydrothermally produced
Sr. Large amounts of Ca and Sr must be mobilized in the primary

Table I. Chemistry of hot springs on the East Pacific Rise, at 21°N, and in the Guaymas Basin, Gulf of California. Notation: M=moles/kg, m=millimoles/kg, μ=micromoles/kg, n=nannomoles/kg, NGS=National Geographic Smoker, OBS=Ocean Bottom Seismometer, SW=Southwest, HG=Hanging Gardens.

		21°N				Guaymas Basin					Seawater
		NGS	OBS	SW	HG	1	2	4	5	7	
T°C		237	350	355	351	291	291	315	287	300	2
Li	μ	1,033	891	899	1,322	1,054	954	873	933	1,076	.26
Na	m	510	432	439	443	489	478	485	488	490	464
K	m	25.8	23.2	23.2	23.9	48.5	46.3	40.1	43.1	49.2	9.8
Rb	μ	31	28	27	33	85	77	66	74	86	1.3
NH_4	m	<0.01	<0.01	<0.01	<0.01	15.6	15.3	12.9	14.5	15.2	<0.01
Be	n	37	15	10	13	12	18	29	29	17	0.02
Mg	m	0	0	0	0	0	0	0	0	0	52.7
Ca	m	20.8	15.6	16.6	11.7	29.0	28.7	34.0	30.9	29.5	10.2
Sr	μ	97	81	83	65	202	184	226	211	212	87
$^{87}Sr/^{86}Sr$		0.70302	0.70317	0.70335	0.70303	-	-	0.70518	-	-	0.70905
pH		3.8	3.4	3.6	3.3	5.9	5.9	5.9	5.9	5.9	7.8
Cl	m	579	489	496	496	601	589	599	599	606	541
Si	m	19.5	17.6	17.3	15.6	12.9	12.5	13.8	12.4	12.8	0.16
Al	μ	4.0	5.2	4.7	4.5	0.9	1.2	3.7	3.0	1.0	0.005
Mn	μ	1,002	960	699	878	139	222	139	128	139	<0.001
Fe	μ	871	1,664	750	2,429	56	49	77	33	37	<0.001
Co	n	22	213	66	227	<5	<5	<5	<5	<5	0.03
Cu	μ	<0.02	35	9.7	44	<0.02	<0.02	1.1	0.1	0.007	0.007
Zn	μ	40	106	89	104	4.2	1.8	19	2.2	2.2	0.01
Ag	n	<1	38	26	37	230	<1	2	<1	<1	0.02
Cd	n	<1	155	144	180	<10	<10	46	27	<10	1
Pb	n	183	308	184	359	265	304	230	<20	<20	0.01
H_2S	m	6.57	7.30	7.45	8.37	5.82	3.95	4.79	4.11	5.98	0
As	n	<30	247	214	452	283	732	1,074	516	711	27
Se	n	0.6	72	70	60	82	87	-	-	92	2.5

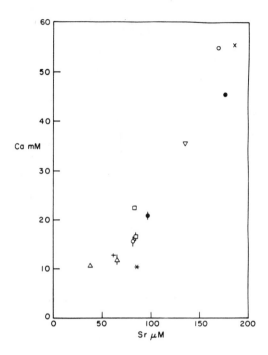

Figure 1. The relationship between Ca and Sr for hydrothermal fluids from 21°N and 10–13°N, EPR.

seawater-basalt reaction and then reprecipitated, probably as epidote, in the conduits. The processes controlling the extent of precipitation, i.e., the final concentrations of Ca and Sr, are not known, although the Ca-Sr trend is so regular that it must approximate equilibrium. Thermodynamic calculations using the computer code EQ3/6 (Bowers et al., 1985) show that the concentrations of the ore-forming elements at 21°N are not controlled by solubility equilibrium of their sulfides (Fig. 2). In fact, the solutions must be cooled by more than 50°C before saturation is reached. As the first phase to separate upon cooling or seawater entrainment is a model "Fe-talc", it is tempting to speculate that the ore-metal mobility is controlled by solid solution equilibria with secondary alumino-silicate phases present.

These modelling results shed light on one of the more peculiar features of the stockwork zones, the general high degree of chloritization. Because the observed solutions are free of Mg, the process measures a source of Mg in the conduits, a source which can only be seawater leakage. This, if pervasive, could lead to formation of "smoke" in the conduits and growth of particles by progressive precipitation and agglutination to a size capable of settling out of the

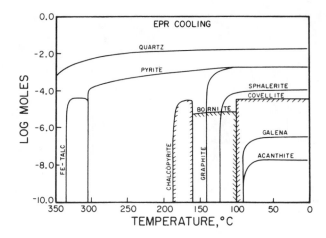

Figure 2. Identity and amount of minerals calculated to precipitate from 1 kg of fluid as a result of cooling the East Pacific Rise (21°N) OBS hydrothermal solution at 500 bars.

exit plumes. This would greatly enhance the efficiency of ore localization. Since leakage is self-sealing, continuing tectonic activity will be required to maintain it. In this model, the extreme undiluted fluids found thus far are incapable of forming large accumulations of ore because of the unconfined nature of the sulfide precipitation.

The sulfide samples recovered from the EPR are texturally quite unlike those described in ore deposits. This is because the initial precipitate, as shown by the EPR chimneys, is recrystallized and chemically annealed by the primary fluid streaming through it after burial. Analogously to a reflux system, the gangue and minor metals will be progressively concentrated in the zone of active primary precipitation. This explains the coarse grain size of the sulfides observed in Cyprus-type deposits, and the elevated concentrations of minor elements, such as Ag, reported in the ochre and uppermost few meters of the sulfide bodies. It also cautions against extrapolating estimates of ore grades, based on surficial samples, to the entire deposit.

At the sediment-covered axes where hot springs have been sampled (Guaymas Basin, Gulf of California and, indirectly, the Red Sea brines), the presence of high concentrations of mantle-derived isotope ^{3}He in solution (Lupton, 1979; Lupton et al., 1977) indicates that the fluids circulated through and reacted with the oceanic crust, rather than having been derived from conductively heated formation waters. The solution compositions at Guaymas (Von Damm et al., 1985(b)) can be modelled successfully as the result of the reaction of EPR-type solutions with the sediment fill at elevated temperatures (315°C). The

fluids in the Guaymas Basin are remarkable for their near-neutral pH
level, high alkalinity and low metal contents (Table I). Most unusual
is that NH_4^+ is a major ion (15mM) in these waters. These conditions
result from a reaction with the biogenic carbonate in the sediments and
from the thermocatalytic breakdown of immature planktonic carbon (C:N
of about 6). Both reactions increase the pH and, hence, both act to
localize ore from the ascending fluids. It is likely that organic
carbon (i.e. black shale) alone can act as a sedimentary pH "front",
rather than as the commonly inferred redox boundary. As expected, the
thermodynamic calculations show all ore metal concentrations to be
controlled by solubility (Fig. 3), incidentally validating the
experimental constants in the data base used in the model. The Pb and
Sr isotopes occur in the island-arc fields, a strong indication, in the
case of Pb, that the detrital material derived from the Tertiary
volcanic province of the adjacent Mexican mainland has reacted
extensively with the ore fluids (Chen et al., 1983).

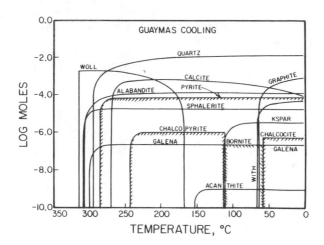

Figure 3. Identity and amount of minerals calculated to
precipitate from 1 kg of fluid as the result of Guaymas (#4)
hydrothermal solution at 500 bars.

From all these features it can be concluded that the hydrothermal
waters escaping from the seafloor in the Gulf of California are "spent",
or devoid of their metal constituents and that a localized or
disseminated ore deposit is forming within the sediment column.
   Data for the chemical composition of hydrothermal waters at
seamounts are not yet available.

CONCLUSIONS

The formation of ore-bearing fluids requires heat and soluble
anions. Seawater, evolved continental brines or, in the case of hydrous
fractionated magmas, "magmatic water" supply the latter. The heat
source is igneous and, hence, is predominantly located at accreting and
consuming plate boundaries. The types of ore bodies formed are
characteristic of the various stages of evolution of the oceanic plates.
The active examples of ore forming systems discovered in submarine
exploration, so far, appear to be reasonable analogs of major deposits
found on land, but formed in the equivalent oceanic-tectonic
environments. However, studies of contemporary deposits provide us
information about the processes of reintrusion, remetamorphism and
tectonic dismemberment associated with obduction. These processes
greatly complicate the study of ore deposits and often render their
original form unrecognizable.

Many problems remain to be solved. The reaction mechanisms
responsible for the observed wide range in compositions of the ridge
crest hydrothermal fluids remain to be identified. The mechanism of ore
localization of Cyprus-type deposits has yet to be established.
Seamount calderas require further intensive study. Above all, active
deposits of commercial scale have to be found for the various
speculations and inferences, presented above, to be tested.

REFERENCES

Bowers, T., Von Damm, K.L., and Edmond J.M., 1985, Chemical evolution of
    mid-ocean ridge hot springs; Geochim. Cosmochim. Acta, v. 49, pp.
    2239-2252.

Chen, J.H., Wasserburg, G.J., Von Damm, K.L., and Edmond J.M., 1983, Pb,
    U and Th in hot springs on the East Pacific Rise at 21°N and Guaymas
    Basin, Gulf of California; EOS, v. 64, p. 724.

Franklin, J.M., Sangster, D.M., and Lydon, J.W., 1981, Volcanic-
    associated massive sulfide deposits; In: B.J. Skinner (ed.), Econ.
    Geol. 75th Anniv. Vol., pp. 485-627.

Gustafson, L.B. and Williams, N., 1981, Sediment hosted stratiform
    deposits of Copper, Lead and Zinc; In: B.J. Skinner (ed.), Econ.
    Geol. 75th Anniv. Vol., pp. 139-178.

Hekinian, R., Fevnier, M., Avedik, F., Cambon, P., Charlon, J.L.,
    Needham, H.D., Raillard, J., Boulegne, J., Merlivat, L., Moinet, A.,
    Manganini S., and Lange, J., 1983, East Pacific Rise near 13°N:
    geology of new hydrothermal fields; Science, v. 219, pp. 1321-1324.

Lonsdale, P., Batiza, R., and Simkin, T., 1982, Metallogenesis at
    seamounts on the East Pacific Rise; Mar. Tech. Jour., v. 16, pp.
    54-60.

Lupton, J.E., 1979, Helium-3 in the Guaymas Basin: evidence for injection of mantle volatiles in the Gulf of California; Jour. Geophys. Res., v. 84, pp. 7446-7452.

Lupton, J.E., Weiss, R.F., and Craig, H., 1977, Mantle helium in the Red Sea brines; Nature, v. 266, pp. 244-246.

McDougall, T.J., 1984, Fluid dynamic implications for massive sulphide deposits of hot saline fluid flowing into a submarine depression from below; Deep-Sea Res., v. 31, pp. 145-170.

Ohmoto, H. and Takahashi, T., 1982, Geologic setting of the Kuroko deposits, Japan: Part III, Submarine calderas and Kuroko genesis; In: Ohmoto, H. and Skinner, B.J. (eds.), The Kuroko and related Volcanogenic Massive Sulphide Deposits; Econ. Geol. Monog., v. 5, pp. 39-54.

Von Damm, K.L., Edmond, J.M., Measures, C.I., Grant, B.C., Walden, B., and Weiss, R.F, 1985(a), Chemistry of submarine hydrothermal solutions at 21°N, East Pacific Rise; Geochim. Cosmochim. Acta, v 49, pp 2197-2220.

Von Damm, K.L., Edmond, J.M., Measures C.I., and Grant, B.C., 1985(b), Chemistry of submarine hydrothermal solutions at Guaymas Basin, Gulf of California; Geochim. Cosmochim. Acta, v. 49, pp 2221-2237.

TRACE ELEMENT AND PRECIOUS METAL CONCENTRATIONS IN EAST PACIFIC RISE,
CYPRUS AND RED SEA SUBMARINE SULFIDE DEPOSITS

Elisabeth Oudin
Bureau de Recherches Géologiques et Minières
B.P. 6009
45060 Orléans Cédex, France

ABSTRACT. Trace elements, including Se, Ge, Ga, Hg, Cd, Ni, Co, As, Sb,
Ag and Au have been analyzed in submarine sulfide-rich samples from the
East Pacific Rise, Cyprus and the Red Sea's Atlantis II Deep. In the East
Pacific Rise samples, fractionation results from evolution of the fluid
composition and temperature as it mixes with seawater. Se, Co, Ni become
enriched in high-temperature Cu-dominated samples, while Cd, Ag, Ge, Ga,
As and Sb become enriched in Zn-dominated samples. Similar trends are ob-
served in samples from Cyprus except for selenium which was not detected.
Differences are observed in the enrichment factors between EPR and Red
Sea samples, when compared with mid-ocean ridge basalts, probably reflecting differences in the source-rocks and fluid composition. In the Red
Sea samples, gold shows a good correlation ($\sim$ 0.7) with base metal and
sulfur as well as with Cd and As except in a zone where episodic boiling
has occurred and where Au, Ag, Sb and Cd can be considerably enriched.
The Au/Ag ratio is fairly constant (0.02). The trace elements either form
rare specific minerals or occur in trace quantities in common minerals.

INTRODUCTION

     Samples from three different hydrothermal fields on the East Pacific
Rise and from the Red Sea metalliferous sediments collected in the hydrothermally active SW basin of Atlantis II Deep have been analyzed for Se,
Ge, Ga, Hg, Cd, Ni, Co, As, Sb, Ag and Au by atomic absorption or spark
source mass-spectrometry.
     The analyzed elements occur in trace quantities in the samples where
sulfides and/or sulfates (East Pacific Rise) and/or oxides and/or silicates (Red Sea) predominate. The samples from the East Pacific Rise (EPR)
were collected during American cruises in the hydrothermally active field
near lat. 21°N (Spiess et al., 1980), and in the inactive fields on the
Juan de Fuca Ridge (Normark et al., 1982), and the Galapagos Ridge
(Malahoff, 1982). The samples from Atlantis II were collected during the
R/V Valdivia cruise Va 29, in 1980. A few samples from the massive sulfide deposits in the Troodos ophiolite complex of Cyprus were also analyzed for comparison with their modern analogs from the EPR.

349

P. G. Teleki et al. (eds.), Marine Minerals, 349–362.
© 1987 by D. Reidel Publishing Company.

Table I. Trace elements in East Pacific Rise and Red Sea samples. Semi-quantitative analyses by Spark emission spectroscopy (CEA) (in ppm). Detection limits for Se 20, Co 8, Ge 2, Ag 0.2, Cd 1, Ni 5, As 20, Ga 2, Sb 20, Ag 0.06-0.2. Au (detection limit 10 ppm) was also analyzed but not detected. na : not analyzed ; nd : not detected. I and II samples are from 21°N on the East Pacific Rise. I : Chalcopyrite-anhydrite from black smoker chimney walls. IIA : Zinc sulfide (wurtzite) from an inactive chimney (1157-2-1A) or associated with anhydrite (Cont. next page).

	Sample Number	Se	Ge	Ga	Hg	Cd	Ni	Co	Ag	As	Sb
I	1149-1-5	490	3	nd	nd	119	55	75	23	nd	nd
	1150-1-2B	510	5	20	nd	740	5	100	60	40	nd
	1149-1-1	510	5	5	nd	550	22	100	69	nd	nd
	1149-2-15	580	3	nd	nd	9	70	80	35	nd	nd
	1156-3-2	590	3	nd	nd	4	7	75	6	nd	nd
	1149-2-18	600	3	20	nd	10	30	60	32	nd	nd
	1158-1-1	730	3	nd	nd	3	15	70	4	nd	nd
	1158-1-2	810	3	5	nd	5	25	100	4	nd	nd
	982-R-5	810	3	nd	nd	5	20	100	9	nd	nd
IIA	1157-2-1A	340	30	70	nd	1980	5	10	81	40	40
	1159-1-6	220	5	70	0.5	2380	nd	10	17	nd	nd
	1159-1-1A	150	4	125	nd	1460	5	10	30	40	nd
IIB	1159-1-1	70	90	40	nd	255	10	10	464	510	130
	1159-1-4	50	70	25	nd	740	20	10	150	310	90
	1159-1-3	40	28	25	1.2	830	5	10	90	260	70
	1159-1-2	20	45	25	1.9	720	7	10	90	330	60
III	22D2A	na	85	25	na	1380	4	50	70	na	na
	22D2B	na	180	10	na	635	2	25	215	na	na
	22D2(3)	na	90	2	na	610	3	10	47	na	na
IV		nd	4	5	nd	8	48	125	5	180	nd
V A	34526	nd	3	10	nd	16	25	10	6	720	nd
	36549	nd	12	15	nd	9	20	15	40	570	20
	34525	nd	25	30	na	96	45	10	50	430	60
V B	CH18A	nd	45	180	11	1810	8	7	132	190	330
	CH18B	nd	15	90	0.6	1260	5	nd	82	nd	60
V C	34527	nd	3	nd	na	nd	5	10	0.2	nd	nd
	34521	nd	3	5	0.06	2	20	10	1	240	nd
VI		nd	7.3	36.7	7.2	295	46	140	180	345	62

Table I (continued). IIB: Zinc sulfide associated with pyrite and opal from mounds. III: Samples from Juan de Fuca inactive field; 22D2A is dominated by hexagonal wurtzite associated with minor pyrrhotite, isocubanite and chalcopyrite; 22D2B is dominated by spherolitic collomorphous sphalerite and minor wurtzite associated with baryte, pyrite, galena, jordanite and opal: 22D2(3) is dominated by collomorphous zinc sulfide. IV: Sample of massive collomorphous pyrite from the Galapagos Ridge. V: Samples from Cyprus; 34526, 34549 and 34525 are fossil black smoker chimney fragments of chalcopyrite included in a matrix of pyrite and opal collected at Peristerka; CH18A and B are zinc-sulfide samples from a fossil mound at Peristerka; 34527 and 34521 are massive pyrite samples collected in Kambia and Mathiati. VI: Average content in Red Sea sediments (mean value of 27 analyses for Se, Ge, Ga, Hg, as well as analyses reported in Table I for Cd, Ni, Co, Ag, As, Sb).

EAST PACIFIC RISE SAMPLES

The mineralogy of samples from 21°N has been described in detail (Oudin, 1981; Haymon and Kastner, 1981; Styrt et al., 1981; Zierenberg et al., 1984). Samples from Juan de Fuca have been described by Koski et al. (1984) and samples from the Galapagos by Oudin (1982) and Skirrow and Coleman (1982).

The anhydrite and opal present in some samples were removed and concentrates of predominantly Cu-Fe-Zn sulfides were analyzed. The results are presented in Table I.

Compared to zinc-rich samples (IIA and IIB) from 21°N and from Juan de Fuca (III), the copper-rich high-temperature (350°C) "black smoker" chimney walls (I) from 21°N are enriched in selenium, cobalt and nickel, and depleted in cadmium, silver, germanium, gallium, arsenic, antimony and mercury. High levels of cadmium, silver, germanium gallium, arsenic and antimony are also reported by Zierenberg et al. (1984) in zinc-dominated samples from 21°N. High cadmium, silver and cobalt content are mentioned by Hekinian et al. (1980) in zinc- and pyrite-rich samples from the inactive zone at 21°N. According to Bischoff et al. (1983), both Juan de Fuca and EPR 21°N zinc-rich samples have similar high arsenic content, but Juan de Fuca samples are depleted in antimony (19 and 34 ppm analyzed in two samples). The massive pyrite from the Galapagos (IV) is enriched in arsenic, nickel and cobalt, and depleted in the other elements.

The copper-iron sulfides from the copper-rich "black smoker" chimney walls at 21°N, analyzed with an electron microprobe, contain as much as 0.1% selenium (Table II) which has probably replaced sulfur.

Nickel and cobalt contents are too low to be detected with the microprobe. Cadmium, germanium, gallium and some of the silver are probably present in the zinc sulfides (Zierenberg et al., 1984). Arsenic is a constituent of jordanite ($Pb_{14}As_6S_{23}$) and Ag-bearing tennantite, and both are rarely observed associated with zinc sulfides and opal in samples from 21°N (Oudin, 1981).

Table II. Selenium content in copper-iron sulfides from the East Pacific Rise at lat. 21°N and the Red Sea (Atlantis II Deep). The mineral compositions of samples from the two sites were determined with a CAMEBAX microprobe (Analysts : C. Gilles and G. Vard).

	Mineral	Selenium content (%)	(Number of analyses)
EPR	isocubanite	0.00	(19)
	chalcopyrite	0.04 - 0.10	(8)
	bornite	0.00 - 0.12	(15)
	idaïte	0.07 - 0.08	(3)
Red Sea	isocubanite	0.00 - 0.04	(5)
	chalcopyrite	0.00 - 0.06	(27)

Jordanite has also been noted in samples from Juan de Fuca where it is more abundant than at 21°N, and is associated with arseniferous galena (Oudin, 1982 ; Koski et al., 1984).

The IIA-type zinc sulfides of table I are either from active (when associated with anhydrite as in samples 1159-1-6 and 1159-1-1A) or extinct (1157-2-1A) high-temperature chimneys. Zinc sulfides of IIB group are associated with pyrite and opal and represent lower temperature samples usually collected at the surface of the hydrothermal mounds (Oudin, 1981). High-temperature zinc sulfides are enriched in cadmium and gallium. The relatively high selenium content probably results from the presence of chalcopyrite, a minor constituent. Low-temperature zinc sulfides are, in the contrast, found to be enriched in germanium, silver, arsenic and antimony. As- or Ag-bearing minerals mentioned above have been identified to occur in this low-temperature facies.

A similar pattern is observed in a single sample (III/22D2A and B) from Juan de Fuca. The observed mineralogical zonation corresponding to the temperature gradient (Oudin, 1982 ; Koski et al., 1984) is : pyrite and opal (low-temperature assemblage, in contact with seawater) ; followed by collomorphous sphalerite (minor wurtzite, pyrite, opal, As-bearing, baryte, galena and tennantite) in sample 22D2B ; succeeded by a high-temperature association of hexagonal wurtzite with minor pyrrhotite, isocubanite ($CuFe_2S_3$, cubic), and chalcopyrite in sample 22D2A.

SAMPLES FROM CYPRUS

Three types of samples from the massive sulfide deposits of Cyprus have been analyzed : in samples VA in Table I collected in Peristerka mine, chalcopyrite is abundant or predominant. In these samples, chalco-

pyrite forms fossil black smoker chimney fragments that are partly replaced later by pyrite and silica (Oudin and Constantinou, 1984). Samples
denoted VB were also collected in Peristerka ; sphalerite and wurtzite
are predominant and are associated with accessory chalcopyrite, pyrite
and minor galena. These samples are likely from a fossil mound near the
black smokers similar to that observed at the EPR. Samples denoted VC are
dominated by pyrite ; sample 34527 is from the Kambia mine and sample
34521 from the Mathiati deposit.

The Cyprus and EPR samples have similar ranges in germanium, gallium,
mercury, cadmium, nickel, silver and antimony. However, selenium was not
detected in samples from Cyprus, the cobalt content is also much lower,
and they are enriched in arsenic. Those differences are either primary or
secondary, i.e., due to post-depositional transformation such as recrys-
tallization during diagenesis or low-grade metamorphism. Copper-rich samples are slightly enriched in nickel, cobalt and arsenic and depleted in
germanium, gallium, mercury, cadmium, silver and antimony when compared
with Zn-rich samples.

As a consequence, except for selenium and to a lesser extent arsenic, similar variations in trace element content are observed in Cu-rich
and Zn-rich samples from both Cyprus and EPR. In Cyprus the higher arsenic content, especially in Cu-rich samples may be caused by the high content of silica ; in EPR samples, arsenic-bearing minerals are associated
with the later forming low-temperature opal (Oudin, 1981). A pyrite sample (denoted IV in table 1) collected on the Galapagos Ridge is also very similar to pyrite-dominated samples of Cyprus, except for its cobalt
content, which is distinctly lower in the latter. However, local cobalt-
enrichment is known in Cyprus ore (Constantinou and Govett, 1973).

RED SEA SAMPLES

More than one hundred samples, from 6 cores collected in the SW basin in Atlantis II Deep, have been analyzed for Au, Ag, As, Sb, Ge, Ga,
Hg, Cd, Co and Ni (table III). The samples are from the sulfide-rich
units (SU1, SU2), the sulfide and amorphous phase rich unit (SAM), the
sulfide, oxide and anhydrite-rich unit (SOAN), the oxide-rich unit (CO),
the oxide- and anhydrite-rich unit (OAN) as well as from the "normal"
sediments rich in detrital material associated with oxides and pyrite
(DOP) that underlie the hydrothermal deposits. The samples were analyzed
on a dry salt-free basis. The samples are described in table III according to their major constituents, namely SiO_2, Fe_2O_3, base metal content
(Zn, Pb, Cu), sulfur and trace elements.

In Red Sea sediments, the high contents of SiO_2 and Fe_2O_3 preclude
a direct comparison with EPR samples. Samples were selected to study
trace elements in sulfides and are representative in their average base
metal and sulfur contents, of the base metal sulfide-rich fraction of
the sediments when compared with published analyses (Bischoff, 1969 ;
Hendricks et al., 1969). In samples of table III, the zinc content varies
from 0.01 to 33.8 %, averaging 6.5 %. The copper content varies from
35 ppm to 2.8 % (average 1 %). The lead content is even lower : 10 to
4125 ppm with an average of approximately 1000 ppm. The sulfur content
(both reduced and oxidized) varied from 0.6 % to 21.7 % (average 10.6 %).

Table III. Chemical analyses of 111 samples from the SW basin of the Atlantis II Deep. SiO$_2$, Fe$_2$O$_3$, Zn, S are expressed in wt% ; Cu, Pb, Ag, As, Sb, Ge, Ga, Hg, Cd, Co, Ni are in ppm ; Au is in ppb. All samples have been analyzed on a dry salt-free basis. SiO$_2$ analyzed by colorimetry, S by titration. Au, Fe$_2$O$_3$, Zn, Cu, Pb, Ag, As, Sb, Cd, Co, Ni analyzed by atomic absorption ; detection limits are : Ag 1, Co 10, Ni 10, Cd 2, Sb 20, As 50. + have been analyzed for SiO$_2$ and Fe$_2$O$_3$, Zn, Cu, Pb, S by Thisse (1982). * Ge, Ga, Hg have been analyzed by semi-quantitative spark source mass-spectrometry (CEA). For detection limits of these elements, see Table I (except for Sb : 10). See text for units nomenclature.

Core N°	Unit	Sample depth (m)	SiO$_2$	Fe$_2$O$_3$	Zn	Cu	Pb	S	Au	Ag	As	Sb	Ge	Ga	Hg	Cd	Co	Ni
274	SU2	105.0	32.5	20.7	3.94	8040	1784			100	240	<20			.	135	257	32
274	SU2	155.0+	27.2	14.3	23.28	16630	3300	20.40	4220	214*	1070*	70*	15	100	5.8	480*	153*	50*
274	CO	295.0	16.9	38.6	0.18	2600	368			9	200	<20				< 2	32	<10
274	CO	392.0	8.4	37.4	0.11	235	16			< 1	100	<20				< 2	27	<10
274	CO	412.0	10.3	50.0	0.15	1100	136			4	310	<20				< 2	<10	<10
274	CO	515.0	22.7	36.3	0.20	5220	392			30	390	<20				< 2	58	<10
274	CO	668.0	8.4	54.3	0.07	3140	80			15	140	<20				< 2	<10	<10
274	CO	732.0	12.3	39.2	0.09	1290	-176			2	380	<20				< 2	<10	<10
274	SU1	770.0+	37.4	20.1	5.12	23460	1100	5.84	1950	405*	300*	120*	5	45		980*	109*	90*
274	SU1	813.0	38.3	29.2	0.05	3340	240			16	290	30				3	70	<10
274	SU1	918.0	24.2	30.0	0.20	1830	232			10	< 50	<20				< 2	30	15
274	SU1	946.0	23.3	24.6	0.09	13000	1560			118	720	70				3	150	78
274	SU1	997.0	18.4	7.0	17.00	20100	1624		2370	104	790	110				357	122	132
274	SU1	1003.0	12.9	22.1	8.24	14500	1528			62	600	60				226	117	98
336	SAM	55.0+	17.7	35.3	9.53	11050	1200	10.40		90*	150*	<10*	2	8	3.5	180*	133*	20*
336	SAM	97.0	14.9	26.9	1.96	6800	1040			72	300	<20				90	215	12
336	SAM	143.0	25.5	27.0	3.83	11800	2664			147	320	75				197	268	38
336	SAM	233.0	24.2	26.2	4.00	8020	632		660	79	120	<20				204	148	22
336	SAM	312.0+	17.3	20.2	8.17	8150	1300	15.30	2300	103*	120*	10*	6	25	6.6	190*	137*	70*
336	SAM	380.0	31.5	20.5	4.66	7480	1352		1880	78	160	<20				110	270	34
336	SAM	414.0	16.2	16.0	7.55	9500	1616		1960	103	570					232	255	58
336	OAN	433.0+	4.7	77.8	0.15	4445	45	0.56	270									
336	SOAN	474.5	29.9	27.2	0.60	3960	312			18	120	<20				18	48	44
336	SOAN	520.0+	14.4	13.1	18.64	1400	2200	21.70	2990	137*	820*	70*	15	100	6.0	360*	105*	30*
336	SOAN	540.0	10.7	6.8	9.71	9200	1176		1870	65	480					248	162	36
336	SOAN	594.0	19.7	17.3	4.47	11100	744		1120	58	170	<20				204	146	59
336	CO	662.0	12.6	35.8	0.09	6220	280			18	250	<20				< 2	27	10
336	CO	738.0+	4.7	82.7	0.05	4600	16	1.20	400									
338	SAM	3.5	26.5	27.2	4.79	7880	720			71	180	<20				178	140	40
338	SAM	40.0+	19.3	50.8	3.53	9830	1900	7.50		188*	510*	70*	7	20	14.2	340*	102*	10*
338	SAM	109.0	20.2	32.5	4.56	11600	1352			112	290	<20				275	152	13
338	SAM	145.0+	22.2	40.3	9.37	12710	1400	5.98		40*	200*	20*	5	12		290*	120*	20*
338	SAM	211.0	26.3	30.4	3.58	5850	1168			72	280	<20				175	190	14
338	SAM	254.0+	23.1	37.4	11.63	10730	2100	8.86	2200	129*	310*	15*	5	12		270*	144*	25*
338	SAM	311.0	42.7	16.6	6.55	6440	880			66	180	<10				227	138	19
338	SAM	363.0+	21.5	37.1	14.70	9570	2400	11.60	2500	123*	320*	10*	3	15		260*	227*	40*
338	SAM	415.0	22.2	26.4	6.93	7430	1376			72	330	<20				213	252	25
338	SAM	463.0+	20.5	28.6	17.08	10580	2200	14.80	2800	135*	290*	10*	2	10		340	176	95
338	SAM	535.0+	25.5	42.9	4.99	9650	1800	8.04		78*	400*	10*	7	20		133*	36*	20*
338	SAM	568.0	28.3	24.6	1.71	12500	2392			90	840	30				55	185	85
338	SOAN	607.0	0.0	0.0					2980									
338	SOAN	610.0	17.1	24.7	4.720	27000	784			176	60	30				362	162	16
338	SOAN	650.0	27.1	26.0	0.380	8020	136			22	60					3	87	42
338	SOAN	686.0	16.4	25.5	0.225	5650	248			46	< 50					2	135	34
338	SOAN	710.0							940									
338	SOAN	713.0	45.1	6.9	2.200	6740	576			48	130					116	116	68
338	SOAN	732.0+	23.3	27.2	7.540	15000	1700		1140	98*	570*	20*	4	35	4.6	190*	81*	110*
338	SOAN	767.0+	26.0	28.7	3.580	8740	900	12.00		64*	420*	20*	2	45	2.5	127*	75*	60*
338	SOAN	784.0	18.0	22.5	3.040	15100	1360	9.00		87	470					172	155	92

Table III. continued.

Core N°	Unit	Sample depth (m)	SiO2	Fe2O3	Zn	Cu	Pb	S	Au	Ag	As	Sb	Ge	Ga	Hg	Cd	Co	Ni
198	AM	29.0+	13.8	40.8	2.83	6640	1220	9.50	1370	135*	190*	50*	3	20	nd	168*	73*	20*
198	SU2	89.0	9.3	29.4	0.41	2450	128			4	130	20				2	10	10
198	SU1	158.5	28.2	20.8	0.18	5330	64			16	50	20				3	20	12
198	SU1	194.5	28.0	24.4	0.25	6400	352			40	70	20				3	130	45
198	SU1	202.0+	28.1	21.4	9.89	28350	2117	12.90	3840	270*	620*	110*	6	100	5.6	460*	211*	180*
198	SU1	248.0+	27.8	19.6	15.34	19300	1720	14.20	2400	128*	300*	50*	8	60	4.6	360*	84*	100*
198	DOP	395.5	3.0	49.5	0.05	1090	32			< 1	< 50	20				2	55	10
198	DOP	497.0	3.5	53.5	0.17	140	16		<100	< 1	< 50	20				2	14	10
198	DOP	620.0	17.0	10.3	0.05	370	24			< 1	< 50	20				3	16	45
198	DOP	747.5	16.1	4.0	0.03	35	10			< 1	< 50	20				5	10	45
264	SAM	72.0+	20.4	35.2	8.71	6850	816	8.36	1800									
264	SU2	410.0+	22.1	33.7	15.36	9550	1500	10.70	1670									
264	SU2	492.0+	31.5	27.3	9.15	10600	1800	9.90	2600									
264	DOP	992.0+	32.4	22.1	0.37	234	50	15.80	<100									
268	SAM	50.0+	22.5	43.3	4.89	7400	1100	5.14	1480	103*	140*	20*	5	25	6.8	210*	127*	50*
268	SAM	103.0+	20.3	39.5	12.27	10150	1150	8.50	2100	105*	140*	20*	4	15	4.0	320*	125*	20*
268	SAM	160.0+	18.2	51.9	4.58	6840	1780	6.92	2200	94*	460*	10*	4	10		54*	102*	30*
268	SAM	190.0		28.3	6.00	8150	1505		2000	95	250					214	242	20
268	SAM	220.0+	31.9	34.4	4.94	6840	760	3.64	1900	93*	180*	90*	12	25		47*	40*	40*
268	OAN	257.0		20.7	2.84	6900	750		2300	119	320	30				157	192	14
268	OAN	285.0+	22.9	46.0	3.19	8520	800	8.05	1900	133*	380*	30*	10	35	6.2	145*	107*	40*
268	OAN	341.0		35.1	5.60	11600	265		1700	103	80					288	183	18
268	OAN	395.0+	27.0	38.3	5.85	7500	1000	5.55	1870	92*	220*	30*	9	25		180*	137*	20*
268	SOAN	409.0		23.6	9.90	8650	1900		2000	114	350					292	313	40
268	SOAN	456.0+	20.8	40.4	6.26	7700	1400	7.80	1530	103*	310*	20*	10	50		121*	131*	40*
268	SOAN	495.0							2000									
268	SOAN	526.0							1700									
268	SOAN	555.0+	21.7	31.0	13.51	11230	2200	13.80	2200	134*	430*	20*	10	50		214*	145*	80*
268	SOAN	587.0		27.3	5.60	9900	1730		2100	90	1300					166	245	42
268	SOAN	633.5			16.36	16200												
268	SOAN	636.0+	24.8	25.9	16.62	15780	2300	18.30	3250	120*	910*	30*	12	70		310*	98*	60*
268	SOAN	637.0			16.20	15900												
268	SOAN	649.5		13.7	22.80	19100	2780			172	1280	80				575	274	82
268	SOAN	662.0		11.2	26.40	21000	2560		4200	232	140	110				603	365	80
268	SOAN	672.5		11.7	29.00	21000	2070			170	1200	120				682	380	25
268	SOAN	683.5		17.6	14.20	21800	2080			83	680	40				325	347	74
268	SOAN	689.0							4600									
268	SOAN	694.5		16.5	15.50	17100	1195			97	540					380	302	64
268	SOAN	714.0		21.7	4.12	24000	1760		3780	120	820					129	295	115
268	SOAN	742.5			1.72	25100												
268	SOAN	745.0+	26.9	28.6	5.58	25120	800	15.40	4020	138*	890*	20*	10	60		350*	101*	70*
268	SOAN	745.5			1.58	22800												
268	SOAN	754.5		27.5	0.96	20900	790			80	150					37	258	98
268	SOAN	766.5		26.3	3.00	24600	680			116	120					130	252	115
268	SOAN	777.0		23.2	0.41	15500	405			48	120					13	185	34
268	SOAN	789.0							480									
268	SOAN	791.5		5.7	0.03	3900	145			10	200						68	
268	CO	835.0+	9.5	49.1	0.02	4200	30	9.40	390									
268	CO	931.5		36.8	0.05	17200				48							110	20
268	CO	936.0		35.2	0.04	24800				113							200	20
268	CO	942.0							1080									
268	CO	969.5		41.0	0.28	24000	30			265	150					10	245	28
268	CO	981.5		11.4	7.60	28000	550			800		60				645	49	63
268	CO	985.0+	2.0	7.9	16.33	4030	750	19.50	21300	3140*	60*	260*	8	5	16.4	1280*	6*	5*
268	CO	985.5		7.2	0.17	8420	970			1710	60	150				1430	31	25
268	CO	991.0		13.8	23.00	8080	2800			1090	200	230				1550	60	42
268	CO	994.5		7.4	33.80	7250	2275			1270	130	370				1850	21	
268	CO	998.5		43.0	6.60	9450	4125			250	650	50				425	186	130
268	CO	1001.5		29.5	0.96	7300	180			34	130					54	52	
268	CO	1008.0		34.2	0.98	11900	165			36	60						64	
268	CO	1011.0		32.4	0.12	11500	125		560	23	320					470	58	16
268	CO	1027.0+	7.9	43.0	0.04	4500	30	20.30	1860									
268	CO	1040.0		31.8	0.02	560	95			3							52	
268	CO	1078.0		55.0	4.00	4250												
268	CO	1097.0+	8.9	77.2	0.01	3630	10	2.88	420									
268	CO	1108.0		38.8	11.00	9220	430		2270	23	410					470	58	

The average Zn/Cu ratio is 7.6 and the average Zn/Pb ratio 54. The mean
values for Se, Ge, Ga, Hg, Cd, Ni, Co, Ag, As, Sb are presented in
Table I (VI). In the EPR the oceanic basalts are a likely source for metal and sulfur (Bischoff et al., 1983). Enrichment factors for Red Sea
samples with respect to mid-ocean ridge basalts (see Bischoff et al.,
1983) are presented in Figure 1 and compared with enrichment factors for
Zn-rich EPR samples.

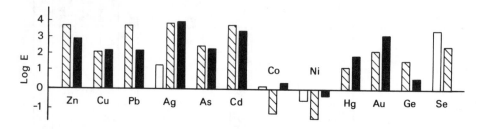

☐ Cu-rich samples (Black Smokers)
☒ Basal Zn-rich mounds (Bishoff,1983)
■ Red Sea samples

Figure 1. Ratio of the elements in sulfide samples from the EPR and Red
Sea and in mid-ocean ridge basalt ("enrichment factor" of Bischoff et
al., 1983).

Cu, Ag, As and Cd are enriched at the same degree in EPR and Red Sea samples. Hg and Au are more strongly enriched in Red Sea samples than in EPR
samples. Nickel is depleted, compared to mid-ocean ridge basalts in EPR
samples and slightly less so in Red Sea samples. Cobalt is clearly depleted in Zn-rich EPR samples but is moderately enriched in Red Sea samples.
The variations observed in the enrichment factors of the elements in EPR
samples have already been discussed by Bischoff et al. (1983) as having
been caused by several factors, including the degree of leaching from the
source rock, and fractionation from the fluid during deposition. Another
reason may be that the elements remain in solution and are dispersed in
seawater. Temperature fractionation in Zn- and Cu-rich samples from the
EPR has been discussed earlier in this paper. For the EPR, Cu-rich sam-
ple-enrichment factors, represented in Figure 1, show that cobalt in
these higher temperature facies is slightly enriched compared to that of
mid-ocean ridge basalts and massive pyrite deposits from the Galapagos
Ridge (Table I). Selenium was analyzed in some of the Red Sea copper iron
sulfides (Table II) but because of the low copper content, it was not detected in the whole sample.
      The differences observed in the enrichment factors between EPR and
Red Sea samples probably also reflect differences in the sources of the
metals, the hydrothermal fluid composition and the process of deposition.
In the Red Sea, the hydrothermal fluids are highly saline, as indicated
by the hot brines predominantly hydrothermal in origin that overlie the
sediments, and by fluid inclusion data (Oudin et al., 1984). The salinity
of EPR hydrothermal fluids is approximately that of seawater. The salini-

ty of the fluid increases probably because it circulates in the evaporitic sequence bordering the metalliferous deeps and the black shales that are intercalated with these evaporites. These evaporites are an additional possible source of metals (Manheim, 1974). It is believed that the influence of salinity on the leaching and precipitation of the various elements is not negligible.

In the Red Sea, metals can also become fractionated to form sulfides, silicates and oxides. This appears to be indicated by the lack of correlation among certain elements such as Ge, Ga, Co, Ni with sulfur and base metals. As anticipated, zinc, copper and lead correlate well with sulfur and zinc also correlates with cadmium and silver (Thisse, 1982). Cadmium-zinc and silver-lead correlations in these samples are presented in Figure 2a and 2b. Gold correlates with sulfur (r = 0.6 ; Figure 2c), zinc (r = 0.71 ; Figure 2d), copper (r= 0.69) and lead (r = 0.76) also with arsenic and cadmium.

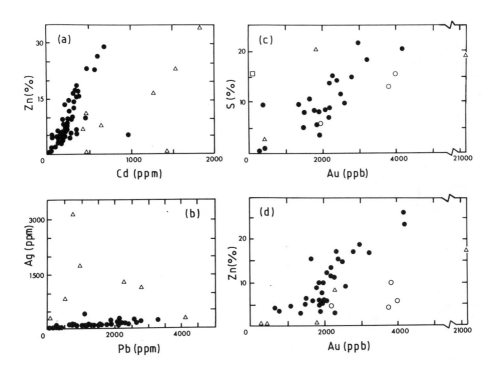

Figure 2. Correlation diagrams for metal contents in Red Sea sediments.

△ Boiling zone

○ Cu-rich sample with moderate zinc content

□ Pyrite dominated sample from DOP

Exceptions to these trends mostly belong to the top of a distinct zone at the bottom of one core (268). In this zone, episodic boiling at 390°C degrees occurs as shown by fluid inclusions in the sulfates. The sulfates associated with sulfides are deposited by epigenetic hydrothermal fluids circulating in hematitic sediment (Oudin et al., 1984). Samples from this zone (hereafter termed "the boiling zone") are distinguished on Figure 2. Another exception (Figure 2c) includes a sulfur-rich base-metal-poor sample of the DOP zone below the hydrothermal sediments where sulfur is probably formed by bacterial reduction of seawater sulfate (Shanks and Bischoff, 1980). A third group of exceptions to the gold correlations pertains to a few copper-rich samples that contain moderate amounts of zinc (Figure 2c and 2d). The data in the cadmium-versus-zinc diagram (Figure 2a) refers to two groups of sphalerite : cadmium-poor (low-temperature ?) sphalerite is best represented compared to cadmium-rich (high-temperature ?) sphalerite. High-temperature sphalerite from the boiling zone is enriched in cadmium but two samples from this zone are enriched in cadmium with a very low zinc content (< 0.2 % Zn). This suggests that cadmium is present in other minerals too, as indicated by microprobe analyses of silver bearing minerals. The silver content of the samples varies from 10 to 3140 ppm with an average of 180 ppm. The gold content varies from less than 100 ppb up to 4600 ppb for most samples. The average gold content is approximately 2000 ppb. Hendricks et al. (1969) report an average gold content of 480 ppb in 110 samples from Atlantis II, Discovery and Chain Deeps. However, the sulfur and base metal content of their samples were extremely low compared with samples described herein. It is a well established fact that in the Atlantis II Deep, the SW basin is considerably enriched in base metal and sulfur compared to other parts of the Deep and other Deeps. One sample collected at the top of the boiling zone exhibits extremely high silver ($\sim$ 3000 ppm) and gold (21300 ppb) concentrations and contains cadmium and antimony as well. Silver and gold enrichment related to boiling fluids are well documented in the literature, especially in porphyry copper deposits and epithermal deposits ; their possible concentration in present-day submarine sulfide deposits has been recently discussed by Delaney and Cosens (1982). In the Red Sea Au-Ag-rich samples described in this paper, silver-bearing sulfides and sulfosalts (Oudin et al., 1984) were detected by microscope, with argentite (Ag_2S) as the predominant mineral. Electrum (Au-Ag alloy) occurs in the same samples either as minute inclusions in the anhydrite or in the sulfides. The percentage of silver in electrum varies from grain to grain, and is reflected in the color variation of the mineral. This is the first occurrence of a gold mineral in a submarine sulfide deposit. Argentite (Ag_2S) has also been found by Weber-Diefenbach (1977) in a core from Atlantis II NW basin. The average Au/Ag ratio in the Red Sea samples described in this paper is fairly constant (approximately 0.02) and within the range that is typical of massive sulfide deposits (Boyle, 1979).

CONCLUSIONS

The mineralogical and geochemical studies of samples from EPR, the Red Sea and Cyprus lead to the following conclusions :

Distributions of minor elements such as Se, Ge, Ga, Hg, Cd, Ni, Co, As, Sb and Ag in East Pacific Rise samples indicate that geochemical fractionation occurs in the hydrothermal fluid as a consequence of its mixing with seawater on the ocean-floor. Se, Ni, Co concentrate in the high-temperature chalcopyrite-dominated "black smokers". High-temperature zinc sulfides contain high amounts of Cd and Ga. Lower temperature zinc sulfides are enriched in Ge, Ag, Sb and As where Ag, Sb and As also form minerals such as tennantite and jordanite.

Samples from Cyprus are fossil equivalents of EPR samples. The mineralogy and composition of samples from Cyprus are similar to those from EPR except for the absence of selenium. Selenium-bearing copper-iron sulfides were either not deposited in Cyprus or selenium was removed by leaching during aging of the deposit.

The enrichment factors of Se, Ge, Ga, Hg, Cd, Ni, Co, As, Sb and Ag (i.e. the ratio between the concentration of these elements in submarine sulfide samples and mid-ocean ridge basalts) are different in EPR and Red Sea samples. These differences probably reflect their distinct geological settings and processes of formation. EPR deposits are formed in the sea-floor as the result of seawater-basalt interaction near the spreading axis. Red Sea deposits are formed in large deeps at the axial rift as a result of seawater-sediments (mainly evaporites)-basalt interaction.

Boiling of the hydrothermal fluid in the Red Sea Atlantis II Deep SW basin is a localized phenomenon. It is however an efficient concentrating process for gold and silver with a distinct geochemical signature. Other cores analyzed by Preussag are locally enriched in gold and silver (Bäcker, personal communication, 1985). The distribution of these Au-Ag-rich samples could possibly give an indication of the extension in time and space of the high-temperature event in the SW basin.

ACKNOWLEDGMENTS

We thank H.E. Z. Mustafa and Z.A. Nawab for allowing us to analyze core samples from Atlantis II and for their general support and assistance. Samples from EPR were kindly provided by : J. Edmond and H.D. Holland (21°N), J. Bischoff and R. Koski (Juan de Fuca) and A. Malahoff (Galapagos). We are also grateful to F. Le Lann who is the BRGM representative consultant to the Saudi-Sudanese Red Sea Commission and we would like to thank Y. Thisse for his assistance and S. Delaroque for typing the manuscript. An anonymous reviewer considerably improved this manuscript and is gratefully acknowledged.

REFERENCES

Bäcker, H., and Richter, H., 1973, Die Rezente Hydrothermal-Sedimentäre Lägerstätte Atlantis II - Tief im Roten Meer ; Geol. Rundsch., v. 3, pp. 687-741.

Bischoff, J.L., 1969, Red Sea geothermal brine deposits : their mineralogy, chemistry, and genesis; In: Degens, E.T. and Ross, D.A. (eds.), Hot brines and recent heavy metal deposits in the Red Sea; Springer Verlag, Berlin, pp. 368-401.

Bischoff, J.L., Rosenbauer, R.J., Aruscavage, P.J., Baedecker, P. A., and Crock, J.G., 1983, Sea-floor massive sulfide deposits from 21° N, East Pacific Rise, Juan de Fica Ridge and Galapagos Rift: Bulk chemical composition and economic implications; Econ. Geol. v. 78, no. 8, pp. 1711-1720.

Boyle, R.W., 1979, The geochemistry of gold and its deposits; Geol. Surv. Canada, Bull. 280, 584 p.

Constantinou, G. and Govett, G.J.S., 1973, Geology, geochemistry and genesis of Cyprus sulfide deposits; Econ. Geol., v. 68, pp. 843-858.

Delaney, J.R. and Cosens, B.A., 1982, Boiling and metal deposition in submarine hydrothermal systems; Mar. Technol. Soc. Jour., v. 16, no. 3, pp. 62-66.

Haymon, R. and Kastner, M., 1981, Hot springs deposits on the East Pacific Rise at 21° N: Preliminary description of mineralogy and genesis; Earth and Planet. Sci. Lett.,v. 53, pp. 363-381.

Hekinian, R., Fevrier, M., Bischoff, J.L., Picot, P. and Shanks, W.C., 1980, Sulfide deposits from the East Pacific Rise near 21° N; Science, v. 207, pp. 1433-1444.

Hendricks, R.L., Reisbick, F.B., Mahaffer, E.J., Roberts, D.B. and Peterson, M.N.A., 1969, Chemical composition of sediments and interstitial brines from Atlantis II, Discovery and Chain Deeps; In: Degens, E.T. and Ross, D.A. (eds.), Hot brines and recent heavy metal deposits in the Red Sea; Springer Verlag, Berlin, pp. 407-440.

Koski, R.A., Clague, D.A. and Oudin, E.,1984, Mineralogy and chemistry of massive sulphide from the Juan de Fuca Ridge; Geol. Soc. America Bull., v. 95, pp. 930-945.

Malahoff, A., 1982, Comparison of the massive submarine polymetallic sulfides of the Galapagos Ridge with some continental deposits; Mar. Technol. Soc. Jour., v. 16, no. 3, pp. 39-45.

Manheim, F.T., 1974, Red Sea geochemistry; In: Initial Reports of the Deep Sea Drilling Project, National Science Foundation, Washington, D.C., v. 23, pp. 975-998.

Normark, W.R., Lupton, J.E., Murray, J.W., Delaney, J.R., Johnson, H.P., Koski, R.A., Clague, D.A. and Morton, J.L., 1982, Poly-metallic sulfide deposits and water-column of active hydrothermal vents on the Southern Juan de Fuca Ridge; Mar. Technol. Soc. Jour., v. 16, no. 3, pp. 46-53.

Oudin, E., 1981, Etude minéralogique et géochemique des dépôts sulfurés sous-marins actuels de la Ride Est Pacifique (21°N); Bur. Rech. Géol. et Min., Orléans, France, Rept. no. 25, 241 p.

Oudin, E., 1982, Etude minéralogique de trois échantillons de sulfures sous-marins prélevés sur la Ride de Juan de Fuca (Pacifique NE); Bur. Rech. Géol. et Min., Orléans, France, Rept. no. 82 SGN 688 MGA.

Oudin, E., 1982, Mineralogical study of hydrothermal sulphides collected on the Galapagos Rift at 0°45'N: A comparison with other rift sulphides; Bur. Rech. Geol. et Min., Orleans, France, Rept. no. 82 SGN 841 MGA.

Oudin, E. and Constantinou, G., 1984, Black smoker chimney fragments in Cyprus sulphide deposits; Nature, v. 308, no. 5957, pp. 349-353.

Oudin, E., Thisse, Y. and Ramboz, C., 1984, Fluid inclusion and mineralogical evidence for high-temperature saline hydrothermal circulation in the Red Sea metalliferous sediments: Preliminary results; Marine Mining, v. 5, no. 1, pp. 3-31.

Shanks, W.C. and Bischoff, J.L., 1980, Geochemistry, sulfur isotope composition and accumulation rates of Red Sea geothermal deposits; Econ. Geol., v. 75, pp. 445-459.

Skirrow, R. and Coleman, M., 1982, Origin of sulphur and geothermometry of hydrothermal sulphides from the Galapagos Rift, 86°N; Nature, v. 229, pp. 142-144.

Spiess, F.N., MacDonald, K., Atwater, T., Ballard, R., Carranza, A., Cordoba, D., Fox, G., Diaz Garcia, V.M., Francheteau, J., Guerrero, J.T., Hawkins, J., Haymon, R., Hessler, R., Juteau, T., Kastner, M., Larson, R., Luyendyk, B., MacDougall, J.D., Miller, S., Normark, W., Orcutt, J. and Rangin, C., 1980, Hot springs and geophysical experiments; Science, v.207, pp.1421-1432.

Styrt, M.M., Barckmann, A.J., Holland, H.D., Clark, B.C., Pisutha-Arnaud, V., Elridge, C.S. and Ohmoto, H., 1981, The mineralogy and the isotopic composition of sulfur in hydrothermal sulfide/sulfate deposits on the East Pacific Rise, 21°N latitude; Earth and Planet. Sci. Lett., v. 53, pp. 382-390.

Thisse, Y., 1982, Sédiments métallifères de la fosse Atlantis II (mer Rouge): Contribution à l'étude de leur contexte morpho-structural et de leurs caractéristiques minéralogiques et géochemiques; Thèse de 3ème cycle, unpubl., Université d'Orléans, France, 155 p.

Weber-Diefenback, K., 1977, Geochemistry and diagenesis of recent
    heavy metal deposits at the Atlantis II Deep (Red Sea); In:
    Klemm, D.D. and Schneider, H.J. (eds.), Time- and strata-bound
    ore deposits; Springer Verlag, Berlin, pp. 419-436.

Zierenberg, R.A. and Shanks, W.C., 1983, Mineralogy and geochemistry
    of epigenetic features in metalliferous sediments, Atlantis II
    Deep, Red Sea; Econ. Geol., v. 78, pp. 57-72.

POSSIBILITY OF MINERAL ENRICHMENT IN THE BLACK SEA

Erol IZDAR and Mustafa ERGÜN
Dokuz Eylül University, Institute of Marine Sciences
    and Technology
Izmir, Turkey

ABSTRACT. The Black Sea is the largest anoxic water body in the world,
where the oxygen-hydrogen sulfide interface exists between 100 and 250
m water depth. The thickness of sediments in the Black Sea basin is in
excess of 15 km, and contains high concentrations of metallic salts.
The results of the sampling project carried out from the R/V K. PIRI
REIS indicate the presence of elemental concentrations of Co: 25 ppm,
Cr: 25 ppm, Cu: 34 ppm, Mo: 105 ppm, Ni: 128 ppm, Vn: 167 ppm, U:
13 ppm, Sr: 78 ppm and Mn: 500-1000 ppm, in the upper 60 cm of the
sediment layer in certain horizons of the marginal areas. Magnetic
data strongly suggest regional subsidence along the marginal areas. It
is therefore very important to investigate the marginal areas extending
from Istanbul to Sinop, in connection with anaerobic conditions of the
Black Sea.

INTRODUCTION

The Black Sea has always held much interest to oceanographers and
geologists because of its quite unique hydrologic, and a result,
sedimentologic regime. It is a large ($423,000$ km^2), deep ($2,000$ m)
semi-isolated marine basin situated between two Alpine mountain ranges,
the Caucasus Mountains to the north and east, and the North Anatolian
Mountains to the south. The maximum water depth is 2124 m and it has a
water volume of $534,000$ km^3.

The Black Sea is the largest anoxic water body in world ocean.
Dissolved molecular oxygen is present only in the uppermost surface
layers of the Black Sea. This layer varies in thickness between 100
and 250 m. Below this layer, and throughout the entire depth of the
Black Sea, hydrogen sulfide supplants oxygen. This pronounced strati-
fication delineates two entirely different environments within the
Black Sea with respect to their chemistry and biology. Fluctuation in
the oxygen-hydrogen sulfide interface is found to lead to the formation
or dissolution of various mineral phases ranging from oxides to sulfides
(Degens and Ross, 1974).

From another perspective the marginal area of southern Black Sea
from Istanbul to Sinop is also very interesting given the high magnetic

P. G. Teleki et al. (eds.), Marine Minerals, 363–374.

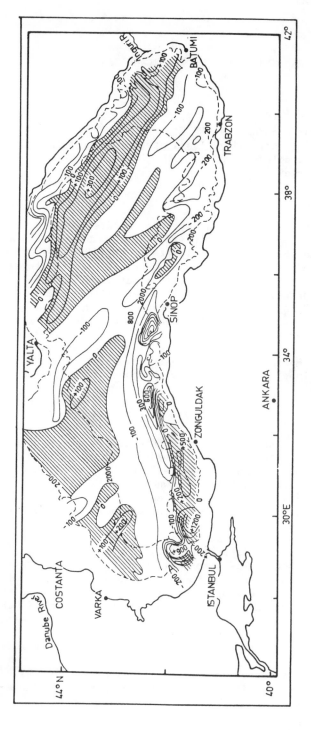

Fig. 1. Magnetic anomaly map of Black Sea (redrawn from Ross, et al., 1974). Contour interval is 100 gammas. Positive anomaly areas are shaded. (Reprinted with permission from A.A.P.G.)

anomalies and strong tectonism. The expected high geothermal gradient should cause various chemical and biological changes to occur under anoxic conditions. Therefore, the Black Sea continues to require investigation of its water and sediment chemistry, biological activity and geological framework and processes.

GENERAL FRAMEWORK OF BLACK SEA

Based on bathymetry the Black Sea can be divided into four principal physiographic provinces: shelf, basin slope, basin apron and abyssal plain. Continental slopes are steep. At present, half of Europe and part of Asia drain into the Black Sea. The total drainage area is about 2,290,200 km^2.

The Black Sea basin, whose origin probably dates back to the Middle or Late Mesozoic is thought to be a remnant of the ancient Tethys Sea. The sediment thickness in the basin exceeds 15 km. Geophysical studies have shown that the central Black Sea basin has a total crustal thickness of 18 to 24 km, including a sedimentary layer 8 to 16 km thick (Neprochnov et al., 1974). The Bouguer gravity anomalies of the Black Sea tend to follow the general topography with the strongest anomalies found in the central part. The Black Sea may be a relict of an ocean crust where a "granite-free" gabbroic layer has been intruded in place. It is also interesting that the marginal areas of the Black Sea show a "granitic" seismic velocity underlying the marginal part of the basin. Magnetic data suggest regional subsidence along the marginal areas. These magnetic anomalies tend to mirror that of the surrounding Caucasus and North Anatolian mountains, which suggest that the trend of these features may extend into the present Black Sea (Fig. 1).

The Black Sea may, in the plate-tectonic context of the Tethyian geology, be considered as a marginal back-arc basin which developed behind the subduction zone that dipped beneath the Rhodope-North Anatolian fragment.

In general, the heat flow values are relatively low (0.92+0.23 $\mu cal/cm^2/sec$) determined by Erickson and Simmons (1974). However, they concluded that after correcting the effects of rapid sedimentation the actual heat flow may on the order of 2.2 $\mu cal/cm^2/sec$.

WATER AND SEDIMENT CHEMISTRY

The hydrologic balance for the Black Sea is shown in Table I. Shimkus and Trimonis (1974) present data on the sediment and salt load which is annually carried by the major rivers into the Black Sea basin. The total load, divided by the size of individual drainage area, gives the total amount of yearly denudation in tons per square kilometer (Table II). The average denudation rate for the entire source area is 0.063 mm/year or about 100 tons/km^2/year (Degens et al., 1976). Close examination of the basinal sediment record reveals that annual layering (varves) can be traced as far back as the last climatic optimum (Atlantic), and that thousands of microlayers often measuring less than

Table I. Hydrologic balance for the Black Sea

GAINS	AMOUNT m^3/sec	LOSSES	AMOUNT m^3/sec
Inflow through the Bosphorus	6,100	Outflow through the Bosphorus	12,600
Precipitation	7,600	Evaporation	11,500
Runoff	10,400		
	24,100		24,100

1 mm can be followed across the entire abyssal plain. Today's total
suspended load is in the order of 200 x 10^6 tons/year. Since that load
will produce annually a blanket 0.3 mm thick, all the observed changes
and rates of sedimentation recorded in the sediments deposited over the
past 20,000 years simply reflect changes in the amount of river runoff
and the land surface characteristics in terms of vegetation, soil and
climate.

The Black Sea is the largest anoxic basin in the world. A well
developed oxygen-hydrogen sulfide interface occurs at about 100 to 250
m. This interface is lowered down along the coasts because of fresh
water inflows and it upwells to 100 m in the middle of the basin.
Salinities are 17.5 to 19% above and about 22% below the oxygen-hydrogen
sulfide interface. Hydrogen sulfide values are 0.09 cm^3/liter above
the interface, 0.47 cm^3/liter at 200 m, 2.35 cm^3/liter at 400 m and 6.6
cm^3/liter at 2100 m.

The Black Sea has existed as a catch basin since Cretaceous time
(Brinkmann, 1974). In taking the whole sedimentary column on top of
the basement as a measure of all material that has been deposited since
then, a mean sedimentation rate (with respect to the compacted sediments)
of 10 cm/1000 years can be calculated. An annual amount of salt of
about 87.5 x 10^6 tons is being carried to the Black Sea from the
catchment area. If only 2.5% of this is metalliferous, then 218,700
tons of metallic material is accumulating annually in the sediments of
the basin. At this rate, the accumulated amount of metalliferous
material could have reached 109,350 x 10^6 tons during the last 500,000
years. As these are deposited in a sulfidizing environment there will
be a weight increase of 10-25%. Therefore, 10,000 x 10^6 tons of sulfide
could also have been deposited during the last 500,000 years.

MAGNETIC ANOMALIES OF THE BLACK SEA

Hydrothermal activity leading to the concentration, accumulation
of metals on and beneath the ocean floor can be observed in various
present marine settings that are influenced by tensional tectonic forces
and by volcanism. Magmatism is one of the important factors in the
evolution of hydrothermal sedimentary ore formation; first, as the
necessary heat source to generate and maintain a hydrothermal circulation
system, and secondly, as the source for the metals involved. It is,
therefore, of paramount importance for the understanding of the formation

Table II. Denudation in the source area of the Black Sea Basin. Tons in metric units.

River(s)	Detritus (a) (10^6 ton/yr)	Salts (a) (10^6 ton/yr)	Total Load (b) (10^6 ton/yr)	Size of Area (km^2)	Amount of Weight (ton/km^2yr)	Denudation Volume (m^3)	Denudation Rate (mm/yr)
Danube	83.00	52.51	135.51	681,000(b)	199.0	124.4	0.125
Dnestr	2.50	2.79	5.29	61,900	85.5	53.5	0.054
Bug	0.53	1.35	1.88	34,000	55.5	34.6	0.035
Dnepr	2.12	10.79	12.91	383,500(b)	24.0	15.0	0.015
Don	6.40	8.43	14.83	446,500	33.2	20.8	0.021
Kuban	8.40	1.95	10.35	63,500	163.0	102.0	0.102
Caucasian rivers	6.79		7.30	24,100	303.0	189.5	0.190
Rioni	7.08	2.16	7.60	15,800	481.0	301.0	0.301
Coruh	15.13		16.20	16,700	971.0	607.0	0.607
Turkish coast rivers	17.00	6.70	23.70	231,500	102.4	64.0	0.064
Bulgarian coast rivers	0.50	0.80	1.30	22,200	58.5	36.6	0.037

(a) Data from Shimkus and Trimonis, 1974.
(b) Reduced area.

and distribution of ore bodies, to investigate the character and
evolution of volcanic events and products on the ocean floor.

The residual magnetic field along the margin area tends to mirror
that of the surrounding Caucasus and North Anatolian mountains, which
suggests that these features may continue their trend into the present
Black Sea (Fig. 1). Magnetic data strongly suggest regional subsidence
along the marginal area.

The volcanism occurred mainly during the Campanian and Maestrichtian ages and continued into the Eocene in this region. The volcanics
have been described as andesitic, but also include basaltic tuffs,
flows, pillows, agglomerates, conglomerates and tuffaceous sandstones,
associated with flysch. Along the Turkish margin of the Black Sea, the
basement seen on reflection seismic profiles corresponds most probably
to these volcanic bands. Magnetic anomalies are associated with volcanics and granitic intrusives in the Northern Anatolia. The bathymetric
profiles with the magnetic anomalies are shown in Fig. 2 from Istanbul
to Sinop. These were derived from surveys of the Department of Navigation
- Hydrography and Oceanography (1965). This zone is the area of interest
because there must be a high geothermal gradient to reactivate metalliferous particles within the sediments. It is a known fact that there
has been high influx of these particles in the Black Sea. These reactivated particles may find further enrichment at the sediment-seawater
boundary (Izdar, 1984). But ore deposit quality enrichment at depths
will only take place within certain narrow zones, as a result of element
migration and geothermal activities.

DISCUSSION OF MINERAL ENRICHMENT IN THE BLACK SEA

In the regions of the ocean where the consumption of oxidants
exceeds the supply of oxygen from the atmosphere, the water chemistry
is radically changed. Important marine geochemical processes such as
deep-sea interactions between pore water and sea water, recycling of
sea water through hydrothermal vents and metalliferous sediment deposition
all involve, at some stage, changes in the redox state of seawater.
Specific examples of these changes are the aerobic and anaerobic
degradation of organic matter, the autotrophic oxidation of sulfides,
and the oxidation of dissolved manganese. Copper and cadmium are
depleted in the anoxic portion of the water column, nickel remains
unchanged, and iron and manganese become enriched. Iron and manganese
binding is rapid and the reaction is bacterially catalyzed. The
distribution of the isotope radium-226 is closely associated with that
of dissolved manganese under these conditions.

Three stages of element enrichment can be expected in the Black
Sea (See schematic section, Fig. 3):

(1) The first stage of enrichment is caused by the biological processes under stagnant conditions at the interface of waters having
different densities. Bacterially decomposed particles are accumulated
as thin coatings around cellular materials.

(2) The second stage of enrichment occurs at the seawater-sediment
boundary as sediment accumulates. Mineral enrichments take place as

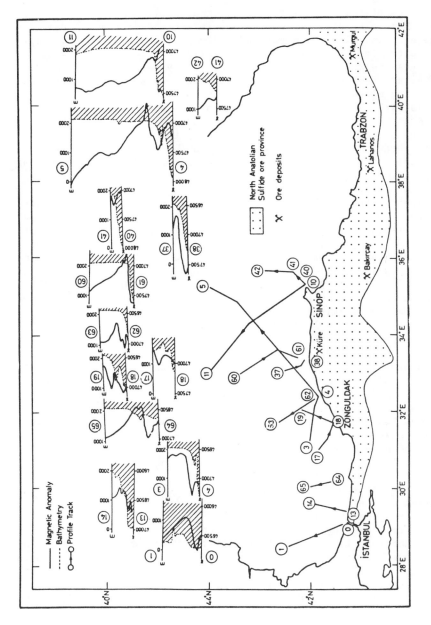

Fig. 2. Magnetic anomaly (in gammas) and bathymetry (in meters) for the region from Istanbul to Sinop. The North Anatolian sulfide ore province is shown with the major ore deposit areas.

surficial cementations, crusts and impregnations.

(3) The third stage of enrichment is activated by geothermal flux in certain zones. The rapid rate of sedimentation in the Black Sea, combined with high concentration of metallic constituents, is well known. These metallic components can be easily reactivated with the aid of marine geochemical processes, such as deep-sea pore water - sea water passing through the hydrothermal vents.

Gells rich in iron and manganese float in water and are usually deposited at the oxygen-hydrogen sulfide interface at depths of 170 to 250 m along the coasts. These may represent the first stage of enrichment. Uranium enrichment does occur in the abyssal plain, which is an example of secondary mineralization (Institute of Mineral Research and Exploration, 1980). The map of the distribution of U_3O_8 for the upper 60 cm of bottom sediments is shown in Fig. 4. Uranium has been brought in from the catchment area by rivers. However, these deposits do not have any economic importance. Several other elements are also present, as shown by the results of the sampling project, carried out with the R/V K. PIRI REIS since 1982. Based on 50 gravity-corer and boomerang samples collected, the element concentrations were determined to be: Co: 25 ppm, Cr: 25 ppm, Cu: 34 ppm, Mo: 105 ppm, Ni: 128 ppm, Vn: 167 ppm, U: 13 ppm, Sr: 78 ppm and Mn: 500-1000 ppm for the upper 60 cm of sediments. There may be even higher concentrations in areas where geochemical and tectonic settings are conducive to deposition of these metals (Izdar, 1984).

The deep-water and the nearshore areas of the Black Sea basin have been subject to similar geological and structural developments as the North Anatolian belt. This region has been described as a metallogenic province extending from Persia to Yugoslavia, that contains volcano-sedimentary hydrothermal sulfide deposits formed since the Later Cretaceous (Fig. 2). Geochemical processes that likely have produced the ancient hydrothermal sulfide ore bodies on land, may also be active at present in the terrigenous sediments of the Black Sea basin, now under marine conditions. On land, the sulfide ore bodies are closely associated with volcanic rocks (Fig. 3), where magnetic anomalies are traceable into the southern Black Sea. Therefore, the southern part of the Black Sea can be described as an area of rapid, contemporary mineralization, infused with andesitic volcanism.

A project investigating the sediment flux in the Black Sea, using a sediment trap made by the Woods Hole Oceanographic Institute, is presently underway in collaboration with the University of Hamburg and the Institute of Marine Sciences and Technology, Ismir. It is expected that this study will clarify certain aspects of element distributions and reactions taking place within the sediment flux under anoxic conditions (Izdar et al., 1984). The collected data have been analyzed and results will be published soon.

CONCLUSIONS

There are three kinds of mineral enrichment processes in the Black Sea basin: (1) gells rich in iron and manganese at the oxygen-hydrogen

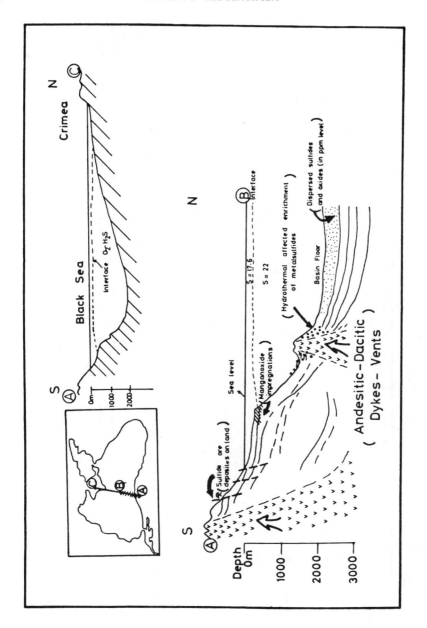

Fig. 3. Schematic section for the stages of element enrichment in the Black Sea.

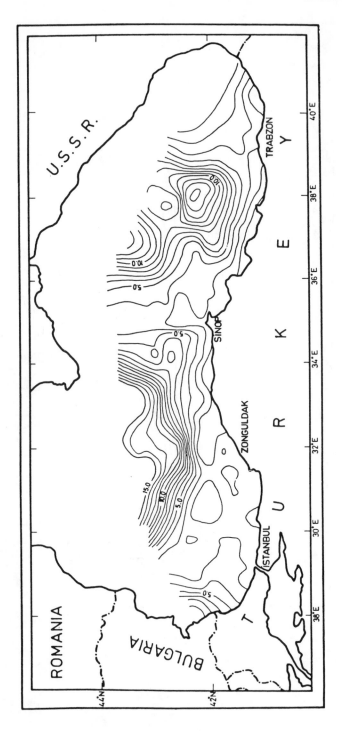

Fig. 4. The distribution of U_3O_8 (in ppm) in the upper 60 cm of bottom sediments (Institute of Mineral Research and Exploration, 1980).

sulfide interface, (2) uranium enrichment in the abyssal plain, and (3) hydrothermal sulfide accumulation within the terrigenous sediments. The Black Sea basin is thought to be a remnant of the ancient Tethys. It has served as a catch basin since Cretaceous time, in which the rate of sedimentation, containing high concentrations of elements, is rapid. The deep and nearshore areas of the Black Sea basin have been undergoing the same geological development as the North Anatolian belt that contains volcano-sedimentary hydrothermal sulfide ore bodies.

However, many unsolved problems remain. The southern margin of the Black Sea along the Turkish coasts should be more thoroughly investigated with respect to its water and sediment chemistry, and its structural setting. Particular attention should be given to bottom morphology and magnetic anomalies between the coast and the abyssal plain. The metal-rich zones could be investigated further with the use of electrical methods to determine their location and extent.

In addition, a few very important processes resulting from the redox conditions in seawater should merit more attention. These are (1) the bacterial reactions at the oxygen hydrogen-sulfide interface, (2) the uptake or release of trace elements, Mn, Fe, Cu, Ni, Cd, Co and the uranium isotope series, as well as thorium and radium in anoxic systems, and (3) the bacterial catalysis of manganese oxidation. The mechanisms which control these reactions are both chemical and biological in nature, and to be thoroughly understood, they must be the subject of further interdisciplinary investigations.

REFERENCES

Brinkmann, R., 1974, Geologic relationship between Black Sea and Anatolia; In: Degens, E.T. and Ross, D.A. (eds.), the Black Sea - geology, chemistry and biology; Am. Assoc. Petrol. Geol., Memoir 20, pp. 63-76.

Degens, E.T., Polska, A. and Erickson, E., 1976, Rate of soil erosion; In: Svensson, B.G. and Soderland, R. (eds.), SCOPE Report 7, Ecol. Bull. Stockholm, v. 22, pp. 185-191.

Degens, E.T., and Ross, D.A. (eds.), 1974, The Black Sea - geology, chemistry and biology; Am. Assoc. Petrol. Geol., Memoir 20, 633 p.

Department of Navigation - Hydrography and Oceanography (Turkey), 1965, Geophysical investigations in the Black Sea; Istanbul, Rept. D515-G/F5, 20 p.

Erickson, A. and Simmons, G., 1974, Environmental and geophysical interpretation of heat flow measurements in Black Sea; In: Degens, E.T. and Ross, D.A. (eds.), the Black Sea - geology, chemistry and biology; Am. Assoc. Petrol. Geol., Memoir 20, pp. 249-278.

Institute of Mineral Research and Exploration (Turkey), 1980, A preliminary project report for the investigation of Black Sea bottom sediments, Ankara, 20 p.

Izdar, E., 1984, Elementanreicherungen in jungeren marinen Sedimenten
    im Schwarzen Meer; Lagerstatten Colloquium, Institut fur Angewandte
    Geologie, Berlin.

Izdar, E., Konuk, T., Honjo, S., Asper, V., Manganini, S., Degens,
    E.T., Ittekot, V. and Kempe, S., 1984, First data on sediment trap
    experiment in Black Sea deep water; Naturwissenschaften, v. 71,
    pp. 478-479.

Neprochnov, Y.P., Neprochnova, A.F. and Mirlin, Y.G., 1974, Deep struc-
    ture of Black Sea basin; In: Degens, E.T. and Ross, D.A. (eds.),
    The Black Sea - geology, chemistry and biology; Am. Assoc. Petrol.
    Geol., Memoir 20, pp. 35-49.

Ross, D.A., Uchupi, E. and Bowin, C.O., 1974, Shallow structure of the
    Black Sea; In: Degens, E.T. and Ross D.A. (eds.), The Black Sea -
    geology, chemistry and biology; Am. Assoc. Petrol. Geol., Memoir
    20, pp. 11-34.

Shimkus, K.M. and Trimonis, E.S., 1974, Modern sedimentation in Black
    Sea; In: Degens, E.T. and Ross, D.A. (eds.), The Black Sea -
    geology, chemistry and biology; Am. Assoc. Petrol. Geol., Memoir 20,
    pp. 249-278.

SEAFLOOR VOLCANISM AND POLYMETALLIC SULFIDE DEPOSITS IN ANCIENT ACTIVE
MARGINS: THE CASE OF THE IBERIAN PYRITE BELT

J.H. Monteiro and D. Carvalho
Servicos Geologicos de Portugal
R. Academia das Ciencias, 19-2
1294 LISBON, PORTUGAL

ABSTRACT. Massive polymetallic sulfide deposits occur throughout the
Iberian Pyrite Belt. The mineralization is related to submarine felsic
volcanism that took place during the Upper Devonian-Lower Carboniferous
evolution of the South Portuguese Zone. The geotectonic setting of the
IPB is reexamined taking into account the results from active margin
drilling under Deep Sea Drilling Program/IPOD. Changes of the mode of
subduction from M-type to C-type, appears to be a reasonable model for
the geological conditions prevailing at the IPB. It is suggested that
the link between field geology and geochemical studies carried in
ancient subduction complexes coupled with the studies of modern analogs,
allows a better understanding of convergent margins and their ore
deposits.

INTRODUCTION

The Iberian Pyrite Belt (IPB), famous for its massive sulfide
deposits has been explored for more than 3000 years. Large concentrations of base metal sulfides occur throughout the Belt, forming several
lenses or bodies of massive sulfides. Total sulfide content of the Belt
is estimated to be in excess of 1000 million tons. Over 20% has been
mined out, and perhaps 10-15% eroded away. Known reserves exceed 650
million tons.

Mineralization in the Belt is related to felsic submarine volcanism
that took place during the Upper Devonian-Lower Carboniferous evolution
of the South Portuguese zone of the Iberian Peninsula. This volcanism
was explosive, generating a wide variety of tuffs, accompanied by a few
lavas, breccias and aglomerates. Hercynian magmatic activity was wide
spread in the IPB.

Plate tectonic reconstructions of the Hercynian Orogeny by several
authors; (Bard, 1971, Carvalho, 1972, Bard et al., 1973, Dewey, 1973,
Ribeiro, 1974, Hurley et al., 1974) suggest the South Portuguese zone
was an active margin during the Upper Paleozoic, including the
metallogenic province known as the IPB.

Carvalho (1972) pointed out that the ore deposits were related to

375

P. G. Teleki et al. (eds.), Marine Minerals, 375–387.
© *1987 by D. Reidel Publishing Company.*

metallogenic processes of plate tectonics involving a subduction zone, dipping to the NW. Soler (1973) considered the geotectonic setting of the IPB as a Cordilleran-type mountain belt originating in a south-dipping subduction zone and somewhat arbitrarily, places the continental crust where the Pyrite Belt would be located. Schermerhorn (1975) raised objections to the plate tectonic models based on the apparent lack of andesitic volcanism, and high-pressure metamorphism.

More recently, several authors reexamined the geotectonic models for the Iberian Pyrite Belt. Among them Munha (1983) studied the Hercynian magmatism in the IPB and derived geotectonic implications based on petrographic and geochemical data. In this study, Munha interprets the dominant bimodal volcanism to reflect the inhomogenous geochemical nature of the mantle under a former active continental margin, combined with complex melting relationships attending the initial stages of extensional back-arc spreading. Ribeiro & Silva (1983) suggest subduction toward the SW (using present coordinates) during the Middle Devonian, the South Portuguese Zone being a back-arc basin related to this subduction as suggested by the petrological studies of Munha (1981). The same authors indicate that after Visean stage the back-arc closed by means of subduction directed toward the NW and accretion of a prism of sediments escaping subduction took place in the South.

Studies resulting from active margin drilling under the Deep Sea Drilling Program / International Phase of Ocean Drilling, suggest that there may be a systematic relationship between subduction and plate motion geometry, back-arc spreading, magma types, vertical tectonics and the occurrence of fore-arc accretion versus tectonic erosion. Uyeda and Kanomori (1979) and Uyeda (1982) recognized that, to solve the problems related to extensional tectonics and volcanism in active convergent margins, it is necessary to consider the existence of two basically different modes of subduction; C-Type (Chilean Type) and M-Type(Mariana Type).

The purpose of this paper is to stress the relationship between massive sulfide ore deposits and submarine volcanism and to reexamine the geotectonic setting of the IPB. Study of fossil, convergent margins can give insight into active ones. Results of these studies should also improve the exploration methods used in metallogenetic provinces such as the IPB.

THE SULFIDE ORE BODIES

Sulfide ore bodies occur at different levels in the volcanic - siliceous complex (VS), and are associated with submarine felsic tuffs of quartz-keratophyric to rhyolitic composition of the Upper Devonian-Lower Carboniferous "Zona Sul Portuguesa". This volcanic complex over-lies a sequence of phyllites and lenticular quartzites of Late Devon-ian age and is overlain, with local disconformities, by flysch depos-its ranging from the Visean to the Namurian - Lower Westphalian stage (Fig. 1). The entire sequence was folded and slightly metamorphosed during the Hercynian orogeny.

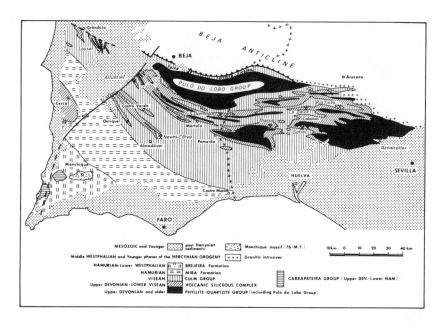

Fig. 1. General geology of the Iberian Pyrite Belt (adapted from
Carvalho et al., 1976; and Oliveira, 1979).

The volcanogenic ore bodies are stratiform polymetallic sulfides
reaching dimensions of 4.5 km by 1.5 km with a thickness of 80 m. Tonnages range from a few hundred thousand tons to some tens of millions
of tons (Mt) of massive sulfides. Barriga and Carvalho (1983), following Alvarez (1974), list the size and ore content distributions:

Very large deposits	(>20 Mt)	- 75%
Large deposits	(5-20 Mt)	- 17%
Medium deposits	(1-5 Mt)	- 8%
Small deposits	(< 1 Mt)	- 1%

Pyrite bodies are generally enveloped by halos of disseminated
pyrite and, in many cases, contain stockworks with considerable amounts
of Cu and chloritic alteration on the footwall. The metal content
varies widely, with average values of:

S	44-48%
Fe	39-44%
Cu+Pb+Zn	2-7 %
Au	0.1-1.5 g/t
Ag	5-40 g/t

Other metals, such as Co and Sn are present in significant amount.
Both the disseminated mineralizations and the stockworks generally have
a Cu content between 0.5 and 1.5%. The Neves Corvo recent discovery is
unusually rich with ore bodies, containing some tens of Mt, Cu grades
exceeding 7%.

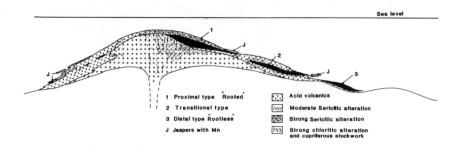

Fig.2 Schematic cross section of the different types of massive sulfide
ore bodies in the Iberian Pyrite Belt (adapted from Carvalho, 1979).

        Carvalho (1979) has distinguished three main groups of deposits
(Fig. 2): the rooted or proximal type, the transition type and the
rootless or distal type. The former is distinguished by location near
the volcanic centers, and the occurrence of a footwall stockwork with
quartz sericitic and chloritic hydrothermal alterations. The distal type
corresponds to deposits with no direct connections to volcanic centers.
In certain cases these may be completely interstratified with
sedimentary rocks, penecontemporaneous in age. They do not exhibit
typical stockworks; hydrothermal alterations may be missing and
sedimentary structures in the ore are frequent. Between these two
extreme groups, all intermediate situations are possible, and are
included in the transitional type.
        Chemically derived sediments, usually called jaspers and sometimes
cherts occur closely associated with the mineralization. Certain jaspers
are very often accompanied by primary rhodocrosite and rhodonite,
sometimes in such concentrations, that the derived oxides could be
exploited as manganese ores.

VOLCANISM

        Volcanic rocks occur in the volcanic - sedimentary complex (VS),
which is a heterogenous rock group with varying thickness and rapidly
changing volcano-sedimentary facies. Thickness ranges from a few tens of
meters to about 800 m; this unit is the only one with abundant bimodal

volcanics and is of particular interest, because it is the exclusive host for the stratabound sulfide and manganese mineralizations. Sulfides are associated with felsic tuffs and manganese with jaspers.

The volcanic rocks, well described in Munha (1983), are predominantly felsic ($\approx$ 80% of the outcropping volcanic areas), mainly meta-rhyolites (quartz keratophyres), meta-basalts and meta-dolerites (spilites, albite-diabases) with less abundant intermediate rocks. Felsic volcanics occur as pyroclastics, volcanic breccias, agglomerates, lava flows and occasionally volcanic chimneys. The volcanics are interfingered with shales and radiolarian cherts indicating submarine deposition. The occurrence of felsic lavas with columnar jointing and ignimbritic flows indicate transient subaerial or extremely shallow submarine volcanic activity.

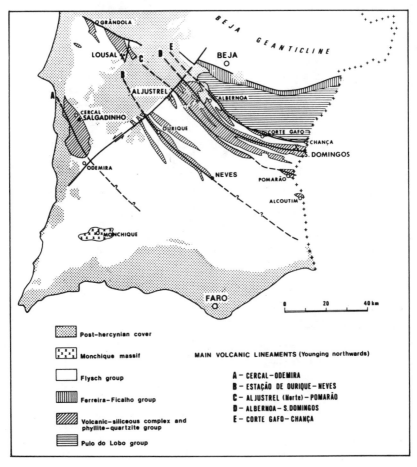

Fig.3 Volcanic lineaments in Iberian Pyrite Belt (adapted from Carvalho, 1976).

Mafic volcanics occur both as lava flows and intrusives. They are present at various levels of the VS complex but predominantly occur at or near the base, or at the top of the complex. Radiolarian chert filling the interstices in pillow lavas indicates that volcanism took place in a submarine environment.

A reduction of clastic input appears to occur as the volcanic activity became more important (Oliveira, 1983). The same author also noted that although not yet proved, the existence of gravity quartzite slumps cannot be excluded.

Based on age determinations, Carvalho (1976) concluded that the lower boundary of the VS is diachronous. Volcanic activity migrated towards the north, probably, linked to an active northward-dipping subduction zone. The major volcanic activities lineaments are shown in Fig. 3 and a schematic representation of the space and time distribution in the IPB are illustrated in Fig. 4.

Volcanic rocks of andesitic type also occur north of the IPB in the Evora-Beja massif associated with quartz porphyries. This massif was uplifted and was the source area for the flysch deposits. MacGuillavry (1961) pointed out that the greywackes contain fragments of preexisting volcanic rocks. This indicates that some parts of the area were covered with volcanic rocks, and were also uplifted during volcanism.

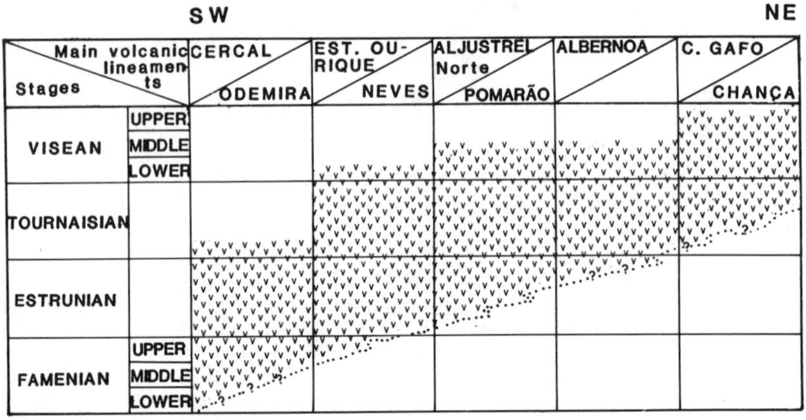

S W						NE
Main volcanic lineaments / Stages	CERCAL / ODEMIRA	EST. OU-RIQUE / NEVES	ALJUSTREL Norte / POMARÃO	ALBERNOA	C. GAFO / CHANÇA	
VISEAN — UPPER						vvvv
VISEAN — MIDDLE			vvvvv	vvvv	vvvv	vvvv
VISEAN — LOWER			vvvvvv	vvvv	vvvv	vvvv
TOURNAISIAN		vvvvvv	vvvvvv	vvvvv	vvvv	...?
ESTRUNIAN	vvvvvv	vvvvvv	vvvvvv	vvvv	..?..?.	
FAMENIAN — UPPER	vvvvvv	vvvv				
FAMENIAN — MIDDLE	vvvv					
FAMENIAN — LOWER	..?					

Fig.4 Space and time distribution of volcanism in the Iberian Pyrite Belt (adapted from Carvalho, 1976).

HYDROTHERMAL ACTIVITY

Carvalho (1976) and Barriga and Carvalho (1983) pointed out that ore genesis in the IPB is a consequence of hydrothermal activity that is penecontemporaneous with volcanism, evident in the sulfide ores themselves, the metalliferous and siliceous sediments, and hydrothermal wall-rock alterations around the ore zones. A pervasive hydrothermal metamorphism affects most volcanic rocks of the IPB (Munha and Kerrich, 1980; Munha et al., 1980).

For the sulfide ores, oxygen isotope data (Barriga and Kerrich, 1981) indicate formation temperatures of $215°$ to $270°C$ for the stockworks of Feitais-Estacao, Chanca and Cerro Colorado (Rio Tinto) and $340°C$ for the Salgadinho. For the Feitais-Estacao ore are $150°C$ and for the overlying jaspers $110°$ C. This steep thermal gradients at time of ore formation and compositional variation revealed by microprobe data were probably caused by mixing of the mineralizing fluid with seawater.

Barriga and Carvalho (1983) have suggested that the deposits formed as a consequence of hydrothermal metamorphism of volcanic rocks. Heat dissipation from the volcanic piles, and possibly from underlying magma chambers, may have driven seawater convection through the highly permeable volcanic rocks, extracting metal from them and generating the massive sulfide deposits. However, Barriga and Carvalho (op. cit.) stated that a magmatic fluid contribution to the ore cannot be ruled out. The association of acidic volcanics with hydrothermal alterations and ore bodies is well known. Fig. 5 is a map of the spatial distribution of the hydrothermal alteration, pyrite "gossan" and volcanic agglomerate at Chanca deposit. The importance of explosive phases of volcanism with ore genesis is not yet well understood but the generation of voids is a significant factor for the circulation of mineralized fluids.

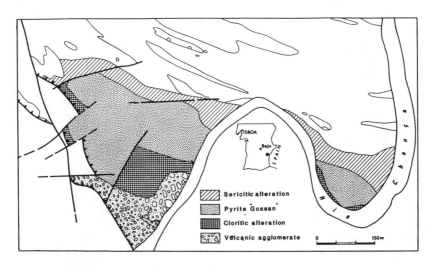

Fig.5 Map of hydrothermal alterations, pyrite "gossans" and volcanic agglomerate in the Mina da Chanca province (Carvalho, D. and Inverno, C., 1976).

DISCUSSION

Uyeda and Kanamory (1979) considered two different modes of subduction, the so-called Chilean type (C-type) and Mariana type (M-type). The

fundamental differences are related to the degree of coupling between the
upper and lower plate. The two modes have the following characteristics:

Chilean type   Mariana type
(high stress regime)   (low stress regime)

- Compressive tectonic   - Tensional tectonic
  stress   stress
- Dip of Bennioff zone   - Steep dip of the
  less steep   Benioff zone
- Porphyry coopper ores   - Massive sulfide ores
- Andesitic volcanism   - Bimodal volcanism
- Uplift 10xM-type   - Less uplift (subsi-
                                       dence in the fore arc)

An extremely significant result of recent DSDP active margins
drilling program is that it has established the history of vertical
movements, sediment subduction and extensional tectonics. For the
Pacific active margins (Japan, Mariana and Middle America Trench) the
occurrence of widespread volcanism in fore arcs (Mariana fore-arc) is
also an important discovery (Hussong and Uyeda, 1981).

Tectonic erosion of the leading edge of the overriding
plate(Hussong et al., 1976; von Huene et al., 1980) and cooling of the
upper mantle by the subduction of a cold plate both play a role in
fore-arc subsidence in M-type subduction.

Studies of the Marianas and Japan arcs have shown that uplift of
arcs may be related to a change in the mode of subduction from M-type to
C-type. The stronger coupling of C-type subduction should favor
offscraping of trench sediments or even ocean crust whereas weaker
coupling of M-type may allow easier subduction of sediments and crustal
fragments. Sediment supply is an important factor, but in general M-type
arcs are more distant from continents and, hence, have a poor supply of
sediments. Graben structures developed due to bending of the subducting
plate. These are more frequent on M-type subduction because of the
presence of steeper Benioff zones. The grabens could play an important
role in the subduction of sediments (Hilde and Sharman, 1978).

Inferred models of the geodynamic development of the IPB can be
reexamined in light of recent research results from the study of active
arcs.

It is proposed here that a change from M-type to C-type subduction
took place before the emplacement of the Baixo Alentejo flysch group.
This change may be related to a change in the rate of thrusting of the
overriding plate which is the driving mechanism-mentioned by Ribeiro &
Silva (1983). Dewey (1980) postulated that arcs could change from
extensional to compressional depending on the relative motion of the
overriding plate. Fig.6 adapted from Carvalho (1972) indicate in a
schematic way the evolution of the Paleozoic continental margin of
Southern Portugal.

Based on these studies a extensional fore-arc or inter-arc location
appears to be a probable site for the volcanism related to the ore
bodies in the IPB. This volcanism in a M-type mode of subduction is a more
simpler model,given the geology and position of the IPB than a back-arc

setting with a SW-dipping plate. Geochemical data presented by Munha (1983) on Ca-Mg-Fe compositional variations of clinopyroxenes for the andesitic rocks and the TiO2/Zr ratio for the intermediate rocks are similar to rocks of the boninite type series compatible with the petrology of volcanic rocks from the fore-arc sites (Meijer et al., 1981). However detailed sampling on comprehensive geochemical studies are necessary in order to support a final conclusion. Rocks that would be expected to occur in ensialic back arc environments such as evaporites, red beds, peralkaline rocks, and limestones are absent giving, further support to the argument for a fore-arc location. Subsidence of the fore-arc of an M-type subduction zone (Hussong and Uyeda, 1981) may have depressed the area below the Carbonate Compensation Depth (CCD) and this can be a reason for the general absence of calcareous sediment in the IPB.

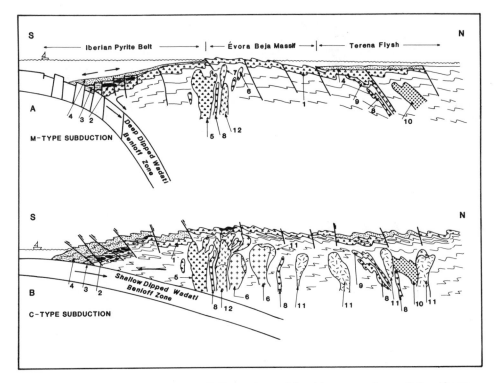

Fig.6 Evolution of the Upper Paleozoic subduction complex of Southern Portugal. A-stage before the second phase of folding (pre-Westphalian-D). B-pattern after the 2nd phase of folding. 1-Silurian-Devonian; 2-volcanic rocks of the Pyrite Belt; 3-volcanic siliceous complex (VS); 4-flysch sequences; 5-basic to ultrabasic complex; 6-diorites to calc-alkaline granites; 7-quartz porphyries; 8-basic and ultrabasic rocks related with deep fractures; 9-hyperalkaline rocks; 10-older granite; 11-alkaline granites; 12-quartz diorites.

Migration of volcanic fronts northward in the IPB can be explained
by tectonic erosion by the subducting plate associated with roll back of
the trench, similar to the model of Hussong & Uyeda (1981) for the
Marianas arc. The lack of clastic component and even possible collapse
of the margin front (Oliveira, 1983) can also be related to the tectonic
erosion by the subducting plate caused by the horst and graben
structures developed in the M-type subduction according the models of
Hilde (1983) and Uyeda (1984).

The back-arc environment related to M-type subduction was
completely closed after collision between Ossa-Morena and Centro-
Iberian zones during early Carboniferous (Ribeiro and Silva, 1983). If a
C-type mode followed then it was responsible for the uplift of the arc
and the production of the sediments that were emplaced as accretionary
prism. Strong coupling with the overriding plate perhaps leads to
emplacement of the Baixo Alentejo flysch group by offscraping of the
down going slab.

Andesitic quartz porphyries and associated volcanics of ·the Evora-
Beja Massif and diorites to calc-alkaline granites to the North are
compatible with inner arc magmatism of Chilean-type subduction.

The occurrence of sulfide ores in the IPB originated by hydothermal
activity in close association with felsic volcanics, with the right
thermal and chemical gradients can explain the fact that sometimes
several stacked cycles of felsic volcanism are present, and only one
(the better developed) contains significant sulfide mineralization,
despite the presence of abundant metalliferous sediments in other
cycles. Lowell and Rona (1984) studied transport models based on
hydrothermal convection systems in the oceanic crust, and arrived to the
conclusion that given the chemical and thermal conditions of the "black
smokers" from active rifts, these models cannot account for ore deposits
of more than 3 Mt. It is perhaps in fore-arc or inter-arc settings that
the right set of conditions occur for the generation of large to very
large massive sulfide orebodies, such as the huge deposits of the IPB.
These are difficult sites to investigate   at active margins, because of
sediment cover and subsidence.

Sawkins (1984) considered metal deposits that are related to fore-
arc felsic magmatism and the possibility of occurrence of granitic
intrusives between the magmatic arc and trench. The IPB map (Fig. 1)
shows outcropping of granitic rocks (Campo Frio) NW of Rio Tinto. These
rocks (granites, granodiorites, quartz-diorites and diorites) are
perhaps the deeper equivalents of the magmatism, at the IPB.

CONCLUDING REMARKS

These studies lead to the opinion that the Southern Portuguese
zone is an ideal location to study the evolution of the convergent
margins. The link between field geology and geochemical studies carried
in this old subduction complex coupled with the study of modern analogs
allows a better understanding of convergent margins and their ore
deposits. Exposed on land, subduction complexes allow the studies to
be made of deeper levels. They can serve as a kind of "ground truth"
to the geophysically dominated observations of modern active margins.

REFERENCES

Alvarez, 1974, Los yacimientos de sulfuros polimetálicos del SO ibérico
    y sus métodos de prospección; unpubl. Ph.D thesis, Univ. Salamanca.

Bard, J.P., 1971, Sur l'alternance des zones métamorphiques dans le
    segment hercynien sud-ibérique; comparaison de la variabilité des
    caractères géotectoniques de ces zones avec les orogènes
    "orthotectoniques".Bull. Geol. y Min., v. 82, no. 3-4, pp.324-345.

Bard, J.P., Capdevilla, R., Matte, P., and Ribeiro, A., 1973,
    Geotectonic model for the Iberian Variscan Orogen; Nature, v. 241,
    107, pp., 50-52.

Barriga, F.J.A.S., 1983, Hydrothermal metamorphism and ore genesis at
    Aljustrel, Iberian Pyrite Belt; unpubl. Ph.D thesis, Univ. Western
    Ontario, p. 388

Barriga, F.J.A.S. and Kerrich, R., 1981, High O fluids, circulation
    regimes and mineralization at Aljustrel, Iberian Pyrite Belt; Geol.
    Soc. Amer., Abstr. Pap. (Boulder, Colo.), v.13, no. 7, pp. 403-404.

Barriga, F.J.A.S., and Carvalho, D., 1983, Carboniferous volcanogenic
    sulfide mineralizations in South Portugal Iberian Pyrite Belt ;
    Serv. Geol. Portugal, Mem., no. 29, pp. 99-113.

Carvalho, D., 1972, The metallogenetic consequences of plate tectonics
    and the Upper Paleozoic evolution of Southern Portugal; Estudos
    Notas e Trabalhos do Servico Fomento Mineiro, v.20, pp. 297-320.

Carvalho, D., 1976, Consideracoes sobre o vulcanismo da regiao de
    Cercal-Odemira. Suas relacoes com a faixa piritosa; Comunic. Serv.
    Geol. Portugal,v.60, pp. 215-238.

Carvalho, D., 1979, Geologia, metalogenia e metodologia da
    investigaçao de sulfuretos polimetalicos do Sul de Portugal;
    Comunic.Serv. Geol. Portugal, v.65, pp. 169-191.

Carvalho, D., Conde, L., Enrille, J., Oliveira, V.,Schermerhorn,
    L.J.G., 1976, Livro-guia das excursoes geologicas na faixa piritosa
    iberica; Comunic.Serv.Geol.Portugal, v.60, pp. 271-315.

Carvalho, D., Correia, H., Inverno, C., 1976, Contribuiçao para
    o conhecimento geologico do Grupo Ferreira-Ficalho. Suas relaçoes
    com a Faixa Piritosa e Grupo do Pulo do Lobo; Mem. e Noticias Univ.
    Coimbra, v.22, pp.145-169.

Dewey, J.F., 1973, Tibetan, Variscan, and Precambrian basement
    reactivation: products of continental collision. Journ.Geology, v.
    81, pp. 683-692.

Dewey, J.F., 1980, Epidosites, sequence and style at convergent
    plate boundaries. In: Strongway, D W (ed), The Continental crust and
    its mineral deposits; Geol. Assoc. Canada, special Paper no. 20, pp.
    553-573.

Hilde, T.W.C., 1983, Sediment subduction versus accretion around the
    Pacific; Tectonophysics, v.99, pp. 381-397.

Hilde, T.W.C., and Sharman, G.F., 1978, Fault structure of the
    descending plate and its influence of the subduction process; EOS,
    v. 59, pp. 1182.

Hurley, P.M., Boudda, A., Kanes, W.H., and Nairn, A.E.M., 1974,
    A plate-tectonic origin for Late Precambrian- Paleozoic orogenic
    belt in Morocco; Geology, v.2, no. 7, pp. 343-344.

Hussong, D.M., Edwards, P.B,, Johnson, S.H., Campbell, J.F., Sutton,
    G.H., 1976, Crustal structure of the Peru-Chile Trench; 8-12 S
    latitude, In: Sutton, G.H., Manghnani, M.H., Moberly, R.; and
    McAfee, E.U. (eds.), The Geophysics of the Pacific Ocean Basin and
    Its Margin; Geophys. Monogr., Am. Geophys. Union 19, pp. 17-85.

Hussong, D.M., and Uyeda, S., 1981, Tectonic processes and the
    history of the Mariana arc: A synthesis of the results of deep sea
    drilling project leg no.60; In: Initial Reports of the Deep Sea
    Drilling Project, v. 60, National Science Foundation, Washington,
    D.C., pp. 909-929.

Lowell and Rona, 1984, Hydrothermal models for the generation of
    massive sulfide ore deposits; Journ. Geophys. Research, B, Paper
    4B5005.

MacGillavry, H.J., 1961, The Upper Paleozoic of Baixo Alentejo,
    Southern Portugal; Econ. Geol., v. 72, pp. 527-548.

Meijer, A., Anthony, E., and Reagan, M., 1981, Petrology of volcanic
    rocks from the fore-arc sites. In: Initial Reports of the Deep Sea
    Drilling Project; v. 60, National Science Foundation, Washington,
    D.C., pp. 709-729

Munha, J. 1981, Igneous and metamorphic petrology of the Iberian
    Pyrite Belt volcanic rocks; unpubl. Ph. D. Thesis, Univ. of Western
    Ontario, p.711.

Munha, J., 1983, Hercynian magmatism in the Iberian Pyrite Belt;
    Serv. Geol. Portugal, Mem., no. 29, pp. 39-81.

Munha, J., Fyfe, W.S., and Kerrich,R., 1980, Adularia, the
    characteristic mineral of felsic spilites; Berlin, v. 75, no. 1, pp.
    15-19.

Munha, J., and Kerrich, R., 1980, Seawater-basalt interaction
    in spilites from the Iberian Pyrite Belt; Berlin, v. 73, no. 2, pp.
    191-200.

Oliveira, J.T., Horn, M., Paproth, E., 1979, Preliminary note on the
    stratigraphy of the Baixo Alentejo Flysch Group, Carboniferous of
    Portugal and on the paleo-geographic development compared to
    corresponding units in Northwest Germany; Comunic. Serv. Geol.
    Portugal, v.65, pp.151-168.

Oliveira, J.T., 1983, The Marine Carboniferous of South Portugal:
    A stratigraphic and sedimentological approach; Serv. Geol. Portugal,
    Mem.no.29, pp.3-37.

Ribeiro, A., 1974, Contribution a l´etude tectonique de
    Tras-os-Montes oriental; Serv.Geol.Portugal.Mem. no. 24, p. 168.

Ribeiro, A., and Silva, J.B., 1983, Structure of the South portuguese
    zone. Serv.Geol.Portugal, Mem. n.29, pp. 83-89.

Sawkins, F.J., 1984, Metal deposits in relation to plate tectonics,
    Springer-Verlag, Berlin, 325 p.

Schermerhorn, L.J.G., 1975, Spilites, regional metamorphism and
    subduction in the Iberian Pyrite Belt: Some Comments; Geologie en
    Mijnbouw, v. 54, no. 1, pp. 23-35.

Soler, E., 1973, L´association spilites-quartz keratophyres du
    Sud-Ouest de la peninsule ibérique. Geologie en Mijnbouw, v. 52, pp.
    277-287.

Uyeda, S., 1982, Subduction zones: an introduction to comparative
    subductology; Tectonophysics, v. 81, pp. 133-150.

Uyeda, S., 1984, Subduction zones: their diversity, mechanism and
    human impacts; GeoJournal, v.8, no. 4, pp. 381-406.

Uyeda, S., and Kanamori, H., 1979, Back-arc opening and the mode of
    subduction; Jour. Geophys. Res., v. 84, pp. 1049-1061.

von Huene, R., Langseth, M., Nasu, N., Okada, H., 1980, Summary,
    Japan Trench Transect. Init. Rep. DSDP LVI-LVII, pt.1, U.S. Govern.
    Print off, Washington, D.C., pp. 473-488.

SCIENTIFIC RATIONALE FOR ESTABLISHING LONG-TERM OCEAN BOTTOM OBSERVATORY/LABORATORY SYSTEMS

John R. Delaney[1], Fred N. Spiess[2], Sean C. Solomon[3], Robert Hessler[2], Jill L. Karsten[1], John A. Baross[1], Robin T. Holcomb[4], Denis Norton[5], Russell E. McDuff[1], Fred Sayles[6], John Whitehead, Jr.[6], Dallas Abbott[7], and LeRoy Olsen[8]

[1]School of Oceanography, Univ. of Washington, Seattle, Washington, U.S.A.; [2]Scripps Institution of Oceanography, U.C.S.D., La Jolla, California, U.S.A.; [3]Dept. of Earth and Planetary Sciences, M.I.T., Cambridge, Massachusetts, U.S.A.; [4]U.S. Geological Survey, Vancouver, Washington, U.S.A.; [5]Dept. of Geology, Univ. of Arizona, Tucson, Arizona, U.S.A.; [6]Woods Hole Oceanographic Inst., Woods Hole, Massachusetts, U.S.A.; [7]School of Oceanography, Oregon State Univ., Corvallis, Oregon, U.S.A.; [8]Applied Physics Laboratory, Univ. of Washington, Seattle, Washington, U.S.A.

ABSTRACT. The oceanographic community is in a position scientifically and technologically to initiate programs leading to the installation of one or more permanently instrumented observatory/laboratory complexes on submarine spreading centers. The dynamic nature of these systems is well established. Yet, there has been no long term, inter-disciplinary effort focused on specific sites to document rates of change in system components, nor the interactions linking the physical, chemical, and biological processes involved. The ultimate goal of this natural laboratory approach would be to establish, then model, the temporal, and the spatial, co-variation among the active processes involved in generation and aging of 60 percent of the planetary surface. The technological and intellectual stimulation involved in successful implementation of natural seafloor laboratories will provide a new generation of dynamically-based, quantitatively testable models of ocean lithosphere genesis and of the biological and chemical consequences of its formation.

The complex and interrelated magmatic, deformational, hydrothermal and biological processes operating at ridge crests span a broad range of time and space scales. Consequently, a wide variety of coordinated and synchronized measurements will be essential to permit integrated interpretation of important cause-and-effect relationships. A number of seafloor-, borehole-, and water column-mounted instrument arrays currently exist or may be readily adapted for use. Power requirements, data acquisition, and sensor development are among the components of

P. G. Teleki et al. (eds.), Marine Minerals, 389–411.

system architecture which must be developed to provide maximum
flexibility to individual investigators and optimal coordination with
other participants. Ideally, intensive characterization efforts will be
focused on the unit element of accretion, or the ridge segment scale
(50-100 km), although a number of specific sub-systems may be studied in
greater detail at smaller scales. In addition, integration of the time-
series data into evolving numerical simulations of spreading center sub-
systems will be a powerful feedback component in the evolution of the
field program.

INTRODUCTION

The modern exploratory phase of oceanography began with the voyage of
the HMS CHALLENGER in 1872 and its systematic surveys of the shape,
composition and biota of the ocean floor. Since then, a significant
fraction of the oceanographic research effort has been directed toward
spatial resolution and broad scale sampling of geological and
geophysical features throughout the ocean basins. Since the end of
World War II, exploration and mapping of the ocean basins has
intensified and has become increasingly sophisticated. The rapidly
developing technologies of swath mapping, seismic profiling, and global
positioning systems will soon provide an unparalleled capacity for
spatial resolution of features on and beneath the ocean floor over a
broad spectrum of scales.
     One of the principal intellectual developments of this exploratory
stage of oceanography, the theory of plate tectonics, focused attention
on the 60,000 km length of geologically active mid-ocean ridge as the
zone along which much of the earth's crust is generated from mantle
material. Oceanic crust has been produced along these linear volcanic
features over the past 200 million years, and probably throughout much
of geologic time. This global rift system is a sharply focused, high
energy zone which is the most extensive and active surface feature on
Earth. The broad range of activities involved may be viewed as a single
system, in which the transfer of energy from the mantle is focused
beneath spreading centers and dispersed through the lithosphere,
hydrosphere, and biosphere.
     Recent discoveries indicate that many fundamental geological,
chemical and biological processes are concentrated within a relatively
narrow band centered on these zones of crustal divergence. For example,
high-temperature seawater circulation at spreading centers is a major
component of terrestrial heat and mass transfer. Biological communities
associated with hydrothermal vents flourish in a highly unstable,
nutrient-rich environment, in sharp contradiction to the well-
established precept that deep-sea faunal communities are the product of
stable, nutrient-poor environments. Such discoveries underscore the
value of the exploration and mapping phase to our understanding of
global processes.
     However, a number of basic questions about the mechanics and
consequences of oceanic crustal generation and movement will never be
adequately addressed by simply improving the resolution of mapping

techniques or by continued exploration of less well-known portions of the ocean basins. Our knowledge of the time-dependent processes which form and modify the ocean crust lags far behind our ability to locate the products of those processes in space. The oceanographic community must expand beyond the exploration phase and establish an integrated scientific inquiry focused on spatial and temporal resolution of the processes which generate oceanic crust.

With a philosophy akin to that which underlay the establishment of magnetic, astronomical and volcano observatories, it is feasible for the oceanographic community to initate a major international program leading to the installation of one or more permanent 'observatories' on the floor of the ocean. The goal of this program would be to document and model changes in related processes occurring at an active ridge crest over a time span during which those processes show significant variation. The multidisciplinary interaction resulting from the successful implementation of such a Long-term Ocean Bottom Observatory (LOBO) should provide the stimulus for a new generation of insights into the thermal, mechanical, chemical, and biological interactions involved in production and early evolution of 60 percent of the earth's surface.

KEY QUESTIONS

The principal goal of these ocean bottom observatories is to make concurrent and continuous measurements of interrelated magmatic, tectonic and hydrothermal activities. The measurements must be made over time periods long enough and areas large enough to permit scientists to address, in a quantitative fashion, a number of fundamental problems.

o What are the rates of arrival of basaltic magma at the base of the oceanic crust? Do the partial melts from the mantle enter a subaxial magma chamber episodically or continuously?

o Is plate divergence episodic or continuous? On what time scale? On what spatial scale?

o Under what conditions and at what rates does the zone of brittle failure (cracking front) encroach on a solidifying magma chamber? How does the heat source geometry evolve?

o What is the duration, extent and character of hydrothermal activity at any given site?

o On what time scales and for what reasons do the vent-related biological communities grow? Recruit? Reproduce? Die?

o What proportion of the global organic carbon flux does the chemosynthetic productivity at vents represent?

o How does the process of oceanic crustal accretion affect the overlying water column in the short term? In the long term?

o What are the episodicities of, and feedback mechanisms linking, magmatic, tectonic, hydrothermal and biological activity at a mid-ocean ridge?

o Of what physical and chemical significance to the heat and mass transport in submarine hydrothermal systems is the fact that the critical point for seawater occurs at a hydrostatic pressure of 310 bars or a depth of 3100 meters?

The core of these problems consists of identifying and quantifying, on a variety of scales, the causes of, and mechanisms linking, heat and mass transfer from the mantle to the crust and between the crust and the ocean.

THE SYSTEM

Spatial Scales

A 'unit element of accretion' along the global system of spreading centers has recently been identified as the 'ridge segment' (Francheteau and Ballard, 1983; Macdonald and Fox, 1983; Schouten et al., 1985). A ridge segment is bounded along strike by any one of the following: a major or minor transform fault, a zero-age offset/overlapping rift zone (Rea, 1978; Schouten and Klitgord, 1982; Macdonald and Fox, 1983; Lonsdale, 1983), or a propagating transform zone (Hey et al., 1980). The segments are approximately 50 to 100 km long and commonly display a single longitudinal high located at some distance from the bounding faults. On the East Pacific Rise (EPR), these shallow regions have been found to be sites of active hydrothermal discharge, greater abundance of fluid lavas relative to pillow lavas, and fewer faults and fissures (Francheteau and Ballard, 1983). A transition to increasing axial depth and greater fissure and pillow lava abundances is noted toward the distal ends of the segment, suggesting that magma migrates along axis away from a localized magmatic center situated beneath the longitudinal summit of the ridge (Ballard et al., 1982; Francheteau and Ballard, 1983). The common occurrence of these features in the form of the ridge segment suggests that it is the minimum scale at which an essential range of crustal generation processes must be investigated.

Temporal Scales

The complex and diverse processes, which compose the magmatic, tectonic and hydrothermal systems acting at a spreading center, occur over a range of time scales which spans many orders of magnitude. Examples in Table I show this span, which ranges from stabilities in plate motion vectors on the order of 1-10 million years to doubling rates for hydrothermal vent bacterial populations which are on the order of a few hours. It is evident that a number of processes, particularly those associated with the hydrothermal systems and earthquake activity, show significant variation over time spans which are readily monitored in a few years. Although the dynamic nature of these ridge crest systems has been established, there has been no long term, multi-disciplinary effort focused on one or more sites to document the nature of their evolution, the rates of change of features in the system, nor the relationships linking the physical, chemical and biological processes.

Table I. Activity time scales at ridge crests.

YEARS

10^{-8} (0.3 sec): Period of hydrothermally-related tremors (Riedesal et al., 1982)

10^{-6} (30 sec): Duration of rise crest earthquake swarms (Riedesal et al., 1982)

10^{-5} (5 min): Mixing of vent water with seawater

10^{-4} (50 min): Transit time through 5 km of crust by 350°C hydrothermal fluid
             (using a velocity of 1 m/s) (Converse et al., 1982)

             Frequency of rise axis microearthquakes (in 3 week period) (Riedesal
             et al., 1982)

10^{-3} (8 hr): Doubling rate for thermophilic bacteria (Baross and Deming, 1983;
             Karl et al., 1984)
             Half-life of hydrogen sulfide in a low temperature vent (Cline and
             Richards, 1969)
             Duration of brief volcanic eruption (single pulse) (Holcomb, 1980)
             Duration of single volcanic earthquake swarm (Newhall et al., 1984)

10^{-2} (4 dy): Growth of a 100-cm long black smoker chimney (Goldfarb et al., 1983)

10^{-1} (1 mo): Tidal cycle influences on sediment/heat/plume dispersal in water
             column
             Duration of brief volcanic eruption (multiple pulses) (Holcomb,
             1980)
             Measureable ground deformation precursory to eruption (Newhall et
             al., 1984)

10^{0}: Fluctuations in the growth rate of hydrothermal clams (Fatton and
             Roux, 1981)
             Duration of sustained eruption (tube-fed flows) (Holcomb, 1980)
             Increased heat flow precursory to volcanic eruption (Newhall et al.,
             1984)

10^{1}: Lifetime of individual hydrothermal vent smoker (Macdonald et al.,
             1980)
             Time required for a steady-state, medium rate ridge to spread 1
             meter
             Lifetime of a single hydrothermal clam (Lutz et al., 1985)

10^{2}: Eruption cycle of a fast spreading ridge (Macdonald, 1983)

10^{3}: Eruption cycle of a slow spreading ridge (Crane and Ballard, 1981)

10^{4}: Lifetime of a magma chamber (Lister, 1983)
             Reversal of magnetic field polarity (Cox et al., 1964)
             Lifetime of a hydrothermal vent field (Macdonald et al., 1980)

10^{5}: Stability of magnetic field polarity (Cox et al., 1964)
             Episodicity in maximum magmatic output on the East Pacific Rise
             (Ballard and van Andel, 1977; Lichtman and Eissen, 1983)
             Instability of non-transform offsets (Schouten et al., 1985)

10^{6}: Changes in absolute plate motions (based on width of Hawaiian-
             Emperor bend)

10^{7}: Episodes of seafloor spreading (Ewing and Ewing, 1967)

Estimates regarding the scale of magmatic and major tectonic processes acting at a spreading center are much more tenuous than those of hydrothermal processes, and it is clear that those occurring at the highest end of the range cannot be addressed by a decade scale monitoring program. The intermediate scale phenomena, such as tilt, strain, and magmatic activity, are reasonable candidates for time-series studies. Although complete cycles of magmatic and tectonic processes will not be documented within ten to twenty years, important information on the rates of change can be expected. For example, with the resolution of present seafloor geodetic technology, observations made over a period of only ten years should definitively establish whether spreading occurs continuously or episodically. Volcano observatories on Iceland and Hawaii have already demonstrated the utility of this approach in understanding the dynamics of magma injection, inflation, fissuring, and eruption (e.g., Hauksson, 1983; Marquart and Jacoby, 1985).

System Components

The lithosphere generating system at mid-ocean ridges comprises at least seven subsystems or components, depicted in Figure 1. These components are: 1) the upper mantle magma source, which undergoes partial melting as it ascends beneath a spreading center; 2) the magma chamber, in which partial melts accumulate and undergo modification; 3) the brittle 'lid' of oceanic crust, which formed by solidification of magma through extrusion, intrusion, or crystallization within the chamber; 4) interstitial seawater, which saturates the brittle portion of the crust and circulates in response to the thermal gradients imposed by magma accumulation; 5) the mass of seawater overlying the ridge crest; 6) organisms which inhabit both the benthic and sub-sea floor water masses; and, 7) particulate materials generated in the water column.

At least three dynamic interfaces moderate the exchange of energy and mass among these components: 1) the brittle-ductile transition between the mantle-magma chamber system and the overlying crust; 2) the water-rock boundaries, including surfaces exposed in cracks, on grain boundaries, and at the seafloor; and, 3) the cell walls of all living organisms.

System Interactions

Activity along segments of submarine spreading centers commonly involves interplay among two or more subsystems, each of which may be described on a variety of scales.

Magma Supply/Spreading Rate Balance   The process of spreading must be continuous on a global scale, or the major plates would experience remarkable acceleration. This assumption does not require that spreading be continuous on the time and space scales involved in decade-long studies of ridge segments. Discrete extensional events are expected based on experience from Iceland (e.g. Hauksson, 1983). Without debating the causes of spreading, it is safe to conclude that magma arrives at the lower crust on a regular basis from the mantle beneath, although it is not clear whether its injection from below is continuous or intermittent (cf. Whitehead et al., 1985). Specific linkages between magma arrival at the base of the crust and the triggering of surface extension are not known for ridge crests, but have been postulated in Iceland (Sigurdsson and Sparks, 1978).

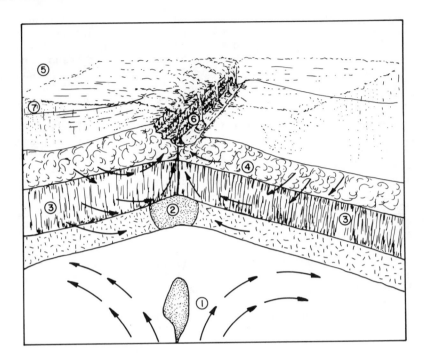

Figure 1. Generation of oceanic lithosphere at ridge crests involves at least seven sub-systems or components. In the broadest view, these components comprise a single system in which energy is focused beneath ridge crests by processes in the upper mantle, and dissipated in the lithosphere, hydrosphere and biosphere during the formation of oceanic crust. Ultimately, it is the interactions in this overall system which Long-term Ocean Bottom Observatories can help to define and quantify. The sub-systems include: 1) sub-axial mantle material undergoing decompression and partial melting; 2) a magma chamber in which partial melts accumulate and undergo modification; 3) the portion of the oceanic crust and lithosphere which can sustain brittle fracturing; 4) the interstitial water derived from seawater which saturates the brittle portion of the lithosphere; 5) the mass of seawater overlying the seafloor; 6) organisms which inhabit near bottom and sub-seafloor water masses; and, 7) sediments generated in the overlying seawater.

Magma Chamber Dynamics   Activity in a magma chamber includes input of
molten material from below, loss of heat and mass by crystallization of
material on the ceiling, walls, and floor of the chamber, and, injection
of magma into the overlying crust and seawater. The heat transferred
outward from the magma chamber drives circulation of modified seawater
through the cracked portion of the crust and the seawater, in turn,
cools the magma chamber. The rates at which cooling solidifies magma in
the chamber, allowing a 'cracking front' to encroach on the residual
magma, have been estimated, but are not known (Lister, 1977; 1983). A
large but unknown fraction of volatile fluxes from the mantle to the
surface reservoir are channeled through magma chambers at ridge crests
(Javoy et al., 1982; Des Marais and Moore, 1984). Some of these
volatiles are the raw materials for chemosynthesis.

Fluid Circulation   Circulating fluid reacts with the crystalline
portion of crust as it is heated. The fluid eventually becomes too
buoyant to remain in the vicinity of the heat source. If fluid
temperature is in the range 350 to 450°C, its circulation may be
strongly influenced by the distinctly non-linear behavior of water
properties in the vicinity of the critical point of seawater, which is
near 405°C and 310 bars (Bischoff and Rosenbauer, 1984). Continued
decompression results in mixing, venting, and precipitation of the
dissolved components. The rates at which hydrothermal discharge change
locally and regionally are largely unknown, although there has been much
indirect inference to indicate changes must occur on relatively short
time scales (e.g., Macdonald et al., 1980; Lutz et al., 1985).

Phase Precipitation   Decompression of the rising hydrothermal fluid and
mixing with seawater result in the deposition of mineral phases within
the cracked crust and at the exit site. Edifices constructed from the
precipitates are spatially discontinuous. The controls on their spacing
and on vent spacing are unknown but are thought to be related to the
size of the local convecting circulation cell. The duration of venting
at a particular location is also unknown. It may be as short as a year
(Francheteau and Ballard, 1983), although fields of vents may persist as
long as a thousand years (Macdonald et al., 1980). The mineral
precipitates begin to equilibrate with seawater soon after deposition,
but the controls on these reactions and the role bacteria may play in
the reaction rates are poorly understood.

Hydrothermal Plume Formation   Vented water mixes with overlying
seawater, creating a classic plume and flat-topped cap (Lupton et al.,
1985), which represents the integrated thermal output from the
hydrothermal system. Vent-generated particulates entrained in this
plume eventually redissolve, become oxidized, and/or settle onto the
seafloor as identifiable hydrothermal products (e.g., Heath and Dymond,
1977; Dymond, 1981; Bonatti, 1983). This plume represents an unknown
factor in deep ocean circulation and may be an important dispersal
mechanism for planktonic larvae of certain vent animals.

Biological Systems   Following initiation of hydrothermal output at any particular site, vents are colonized by a wide range of chemosynthetic bacteria, which flourish on nutrients within the venting fluid and provide the basis of the food chain (Corliss et al., 1979; Jannasch and Wirsen, 1979). Large animals colonize the vent fields, grow, and must reproduce before venting ceases. The paths by which larvae and bacteria are dispersed are poorly understood.

A TIME-SERIES APPROACH TO MAJOR PROCESSES

Mechanisms active at a spreading center may be classified as either tectonic, magmatic, hydrothermal, or biological, although this artificial delineation obscures the natural linkages and interactions among these four types of processes. Major questions about each of these processes could be investigated using time-series observations. Some questions are best addressed with continuous monitoring over longer periods of time, while others can be examined with discrete episodes of sampling or by mapping of a specific site. The configuration of a system which will allow meaningful interfacing among measurements of diverse type and duration is likely to be complex and may change dramatically in the course of conducting experiments and making observations. Therefore, the system must have built-in flexibility. Schematic examples of the types of instruments and communication links involved in such a study of a ridge crest are presented in Figures 2 and 3. In the following sections, approaches and instrumentation required to address representative problems are described.

Tectonic Processes

Background Interplate motion studies, the lineation of magnetic anomalies, and seismic observations all imply that, on the scale of hundreds of kilometers, the various plates are in motion relative to one another (e.g., Minster and Jordan, 1978) and that they behave, to a reasonable approximation, as if they are rigid. The variability of appearance of fracturing and lava flow types at spreading centers (e.g., Ballard et al., 1979), as well as the apparent variability of hydrothermal activity, imply, however, that at some small spatial scale the relative motion of the plates is episodic (Goguel, 1983). Even in the simplest steady-state model, there must be a zone in which the newly-formed crustal material is, in some sense, accelerated from zero velocity to that characteristic of the relative motion of the plates.

Observations of surface morphology at intermediate-rate spreading centers seem to indicate that a large fraction of the acceleration takes place in a zone which is at most a few kilometers wide (e.g., Normark, 1980). Moreover, the existence of intense fissuring at some of the spreading centers, where volcanic activity is not particularly fresh (e.g., the northern portion of the area studied by several researchers at 21°N EPR), implies that spreading may be nearly continuous while volcanism is not (Lichtman et al., 1984).

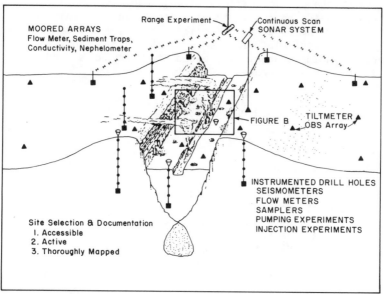

Figure 2. A schematic overview of a natural seafloor laboratory, or Long-term Ocean Bottom Observatory (LOBO), including arrays of instruments which simultaneously monitor temporal variations in a wide variety of processes operating along a ridge crest. A broad spectrum of active experiments can also be conducted at LOBO sites.

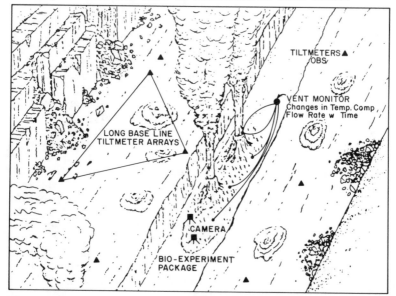

Figure 3. An enlargement of a portion of Figure 2 indicating the schematic details of a focused LOBO study designed to document variations in activity at a specific hydrothermal vent site.

Geodetic Approach   The combination of these factors indicates that a series of strain observations during the course of a few years, within areas several kilometers wide and on the order of ten kilometers along strike, at intermediate-rate ridges should provide useful results (assuming measurement accuracies of a few centimeters). The time-series and spatial patterns of increasing strain within the acceleration zone, if measured over decades, should provide practical constraints for models on crustal accretion, fissuring as related to (or even controlling) hydrothermal activity, and insight into plate edge seismic processes.

Limitations on possible strain measurement techniques have been considered (National Research Council, Committee on Geodesy, 1983; Spiess, 1985, 1986) and these imply that intermediate-rate spreading centers, with their narrow zone of strain build-up would be fruitful for initial investigations. Longer observation times would be required for displacements to build up to measurable values at slow spreading rate ridges. Fast spreading sites are also of interest, although acceleration zones may be wider in some cases. If so, these would require a longer spread of observation points along a rise crest section, together with greater along-strike coverage.

In addition to strain measurements, measurements of variations in elevation and tilt will indicate processes such as normal faulting, which play a critical role in the translation of newly-formed crust from the rise axis to the flanking abyssal hills.

Seismological Studies   Earthquake activity at a ridge crest is one of the primary manifestations of the spreading process. An integral part of any LOBO system at a ridge crest site, should be a network of bottom or sub-bottom seismic stations sufficient in number and areal density to permit locaton of the epicenters and depths of nearby earthquakes and to allow estimations of the character of faulting associated with the larger events. With a LOBO system on a slow-spreading ridge, we could study the earthquake cycle for M_s = 5.5-6 events, including recurrence times, time and space distribution of activity, and phenomena precursory to the largest earthquakes. With an observatory on an intermediate- or fast-spreading ridge, the thickness of the brittle layer and its evolution off-axis (now known only approximately from estimates of flexural rigidity) could be studied, as could the magnitudes and mechanisms of smaller on-axis earthquakes (not observable teleseismically). For either type of ridge system, the earthquake locations and fault-plane solutions could be compared to the distribution of faults and fissures mapped by photographic and side-scan sonar techniques, to the distribution and activity levels of hydrothermal vents, and to the measurements of deformation (strain, tilt, vertical motion). Important new insights into the development of rigidity in the young crust could also be provided by relating pre- and post-seismic displacements to actual release of seismic energy, as a function of both time and space.

Magmatic and Volcanic Processes

As the mantle ascends beneath spreading centers, it is subjected to
decompressional partial melting. Recent models of melt rising through
the parent rock indicate that the melt should easily rise by means of
compaction at concentrations between 0.5-5% melt (Richter and McKenzie,
1984). In this manner, partial melts may either accumulate in a magma-
rich body near or within the base of the crust or at some unknown depth
below the crust. The large amount of crust-free peridotite and thin or
absent crust found at transform fracture zones (Whitehead et al., 1985)
argue against the former hypothesis. When a magma-rich zone forms at
depth, the zone may generate buoyancy-driven diapirs, whose spacing may
be a function of the segmentation of spreading centers. The depths at
which such events would occur are estimated to be 10-50 km below
spreading centers, and the presence of these features may be detectable
with appropriately scaled deformation studies. Time scales associated
with the deep events are unknown but may be long. For processes
occurring within the individual magma chambers, which may be detected by
seismic, tilt, bottom depth, or other studies, the time and length
scales may approximate those of land-based volcanic eruptions
(instantaneous to years).

Hypotheses   The apparent similarity between segments of mid-ocean
ridges and the rift zones of large shield volcanoes suggests that,
although the volcanic processes may be complex, testing of fairly
sophisticated, specific hypotheses based on subaerial analogues may
proceed. For example, on most shield volcanoes, there is a restricted
center of magmatic upwelling from which magma migrates laterally at
shallow levels along the rift. Further, the plumbing system stays in
quasi-equilibrium over short time intervals (years to decades), such
that eruptions or changes at one location cause readjustments at another
location within the edifice up to tens of kilometers away.

Geodetic Studies   Based on knowledge of the behavior of subaerial
volcanoes, one can assume that surface distortions associated with and
indicative of space and time variations in the magma chamber and its
associated 'plumbing' will be present. Over a period of 5 to 10 years,
measurements of tilt, elevation change, and dilation extending several
kilometers off the rise crest and tens of kilometers along strike, if
made in conjunction with seismic/acoustic observations, would allow
models of crestal magma chamber behavior to be tested. In this regard,
site selection criteria become very important; thorough seismic,
bathymetric side scan sonar, photographic, geochemical and biological
surveys of candidate ridge crest segments is essential to wise selection
of one or more locations that exhibit optimal levels of vigorous
activity.

Seismological Constraints   Several aspects of magmatic activity can
potentially be addressed by seismological measurements (e.g. Reid et
al., 1977; East Pacific Rise Group, 1980). The three-dimensional
seismic velocity structure of the crust and upper mantle, in the

vicinity of a LOBO site should be determined by a combination of seismic tomography (including both shots and natural sources) and multi-channel seismic reflection records. Hypothetically, any large magma bodies should be resolved as limited regions of low seismic velocity and high attenuation. The geometry of the heat source and the permeable portions of the system are fundamental to developing high quality numerical models of either the magmatic evolution of the magma chamber or the changes in circulation of the hydrothermal system.

Seafloor gravity and electromagnetic measurements can provide important independent information required for modeling such systems. If selected portions of the system are observed using some type of seismic imaging or by other methods, then these studies should be repeated at regular intervals to test for changes in the depth or configuration of the postulated magma body.

Hydrothermal Processes

Many lines of evidence demonstrate that hydrothermal systems are dynamic. Therefore, it is reasonable to expect that their characteristics, such as temperature, flow rates, and composition, will change with time. The highly interactive nature of the fluid, solid, and biological products of these systems necessitates that the time evolution of each of these parameters be evaluated in order to determine their actual rates, changes, and the feedback loops linking important variables. Much of the technological capability to carry out most of the time-series monitoring of vent systems is available today. Although some new development of more sophisticated, high temperature sensors is required, most of the basic instrumentation can be built now.

Fluid Physics   Vigorous near-surface hydrothermal activity is known to be the source of acoustic and seismic energy (RISE team, 1980; Riedesel et al., 1982). This fact suggests that the level and distribution of hydrothermal flow can be monitored by networks of bottom hydrophone and ocean bottom seismometer stations. Records from such instruments would provide important complementary data to measurements of the flux of fluid, heat, and chemical species at known vent areas. The quantitative relationship between acoustic and seismic signals, on one hand, and the characteristics of hydrothermal activity (and distinguishing such signals from those of magmatic origin), on the other, are poorly known at present. One of the challenges of the LOBO measurement strategy will be to define such relationships to sufficient level that a specific seismic signal will be recognizable as a particular phenomenon. This approach will involve measurement of acoustic and seismic spectra as functions of distance from known hydrothermal sources, the search for causes of excitation in "organ pipe" modes, and the identification and hypocentral location of isolated "bursts" of hydrothermally generated signals.

Fluid Chemistry   Hydrothermal circulation of varying intensity is characteristic of young oceanic crust. The best studied hydrothermal discharge systems lie on the ridge axis (Edmond et al., 1979; Von Damm et al., 1984; Von Damm, 1983). The chemistry of the fluids exiting from these axial vents results from the interaction of a variety of processes: 1) low temperature percolation of seawater into a relatively large volume of crust, accompanied by reaction with basalt at progressively higher temperatures; 2) heating, coupled with more extensive water-rock interaction, followed by ascent of hydrothermal fluid of a somewhat uniform character in relatively confined upflow zones; 3) mixing of hydrothermal fluid with seawater during its ascent; this mixing may occur either in the subsurface or at the seafloor; 4) further reactions experienced by the ascending fluid, including homogeneous precipitation, exchange with basalts and/or sediments, and microbially-mediated redox transformations. The extent to which each process leaves its imprint on an exiting fluid is, in large part, determined by the fracture geometry and its evolution, and by fluid flow rates. Thus, it is tied to intensity changes in earthquake activity, to fracture abundance and spacing, to distance from a major heat source, and to the amount of rock with which the fluid has previously interacted.

Development of chemical instrumentation for a seafloor observatory would most likely involve adaptation of present laboratory capabilities to high pressures and, in some cases high temperatures, found in the ridge crest environment. Several types of measurements, useful for defining the chemistry of hydrothermal solutions, could utilize existing electrochemical sensors. With limited development, the following components could be measured using available or slightly modified electrodes: Ca^{2+}, K+ (Na+), NH^{4+}, pH, H_2S, Cl-, Eh, pCO_2, O_2, SiO_2. A wide variety of other elements and constituents of interest, such as Fe^{2+}/Fe^{3+}, Mn^{2+}/Mn^{4+}, as well as trace elements, may require modification of techniques such as flow injection analysis and ion chromatography (Johnson et al., 1986).

Microbiological Studies   A variety of biologically important energy sources are generated when hot water reacts with basalt. These energy sources include H_2S, reduced metals, CH_4, H_2, CO, and NH_3. Vent bacteria that are able to utilize these elements and compounds have been described (e.g., Corliss et al., 1979; Jannasch and Wirsen, 1979). These bacteria are the primary biological product in hydrothermal vent environments and provide the organic carbon and nitrogen-based nutrients, either directly or indirectly (as endosymbionts, for example), to the extensive and diverse animal communities. Little is known about the complete range of vent environments where such microorganisms grow, or their productivity over time. Uncertainties also exist regarding variability in species composition over time as a function of changing chemistry of the hydrothermal fluids. The same is true of biological responses to marked variability in other factors such as temperature, pH, or solute concentration.

Two general classes of experiments could be conducted using a seafloor observatory to measure variability in microbial biomass, growth rates, specific activities, species composition, and related geochemical characteristics. One type of experiment would include semi-continuous measurements for long and short time periods; the other class of study would include long-term (over 1 year), infrequently sampled experiments.

The short-term experiments would involve the use of multiple in situ devices that could sample venting water at different time intervals and inject specific chemicals into different sampling flasks. Examples include: poisons for microbial biomass determinations, gas and other chemical analyses; radioisotopes for measuring specific kinds of microbial activities, including growth rates, rates of sulfur oxidation and fixation of CO_2; and, chemicals for extracting specific classes of organic compounds such as ATP, RNA, and DNA.

The long term, infrequently sampled experiments would include: colonization experiments, microbial leaching of metals (precious and rare earths), microbial decomposition, modification and utilization of industrially important compounds, hazardous compounds and other recalcitrant organic compounds having either a positive or negative environmental significance, and experiments related to the microbial production of industrially important chemicals, including petrochemical compounds, and high temperature active enzymes.

The colonization experiments would involve following the ecological successions of species on implanted clean surfaces of a variety of chemical compositions. Experiments would also determine the role of bacteria in mediating inorganic alteration of vent precipitates and in 'preparing surfaces' for local settlement of various species, the significance of protozoa in the food chain, larval settlement, benthic community development and succession, and in situ growth rates of micro- and macro-organisms. These colonization experiments must involve time-lapse photography, periodic retrieval, and laboratory examination of sub-sections of the implanted surface material.

Macrobiological Studies   Hydrothermal vent environments are spatially discontinuous and offer a highly localized, abundant bacterial food source for vent animals. Many chemical and physical factors, including chemical composition of hydrothermal fluids, flow rates of vent water, temperature, and availability of oxygen, influence species composition, abundance, and growth rates. Species that live near vents must be able to grow to reproductive maturity while conditions are favorable, compete for space around vent openings, and have the ability to disperse to and colonize new vents. A useful way to study these processes and activities, is to conduct time-series observations and analyses of established and newly-formed vent environments on a time scale of a year to a decade.   At present, little is known about the variability of animal communities with relation to changes in the chemistry and flow rate of vent water, nor about the development and demise of the community during the course of vent life. The few opportunities in which the same vent sites have been visited more than once revealed distinct changes in the community during the course of a few years and certainly within the time scale proposed for use of a bottom observatory (Laubier and Desbruyeres, 1984; Hessler, et al., 1985).

Community distribution, species interactions and community response to environmental changes can be determined through collections and photographic surveys at different scales in conjunction with related chemical, microbiological and physical measurements. Supplemental

plankton/larvae collections, manipulative field experiments (tagging,
settling plates, cleared bottom recolonization, transplants of animals
from different sites, etc.) and laboratory analyses of periodically
sampled, representative species for determination of physiological
condition, fecundity and gut contents, will provide additional
information on general health, and physiology of vent communities.
Time-lapse photography with concurrent environmental assays will reveal
other facets of species activity and interactions.

Inorganic Products   Problems associated with inorganic products of
hydrothermal systems range from the complex interplay of circulating
fluids with silicate walls of the deep conduits, to reactions with and
through sulfide structures at the ocean floor, to mechanisms of
precipitation, dispersal, degradation, and accumulation of vent products
in the water and sediment columns.   Investigation of these processes
requires a time-series approach, using a variety of techniques such as
deep, zero-age drilling, water-column sediment traps, repetitive coring,
and repetitive sampling of sulfide structures in varying stages of
decay.   Arrays of sediment traps may be used to document flux-vs-
distance relationships for a suite of water-borne hydrothermal
precipitates, as well as to examine the bio-geochemical parameters that
influence the evolution of these particles.

Water Column Plume   Lupton et al. (1985) and Massoth et al. (1984) have
established that the flat-topped 'crown' to the plume of mixed
hydrothermal fluid (and vent particulates), which overlies a vent field,
may be mapped in the water column using CTD and/or nephelometer readings
as far as 10 km from the source. This discovery is particularly
important, because the conservative parameters of salinity and
temperature not only define the plume, but also may be used to model its
behavior. A large percentage of the thermal output of the system can be
monitored using moored CTD/current meter/sediment trap arrays which
encompass this plume layer. New instruments may have to be developed,
but a time-series strategy of sampling near- and far-field plumes,
combined with a full scale, conventional physical oceanographic mooring
program, will allow quantitative definition of the vent field thermal
output and certain chemical fluxes. If this monitoring approach proves
feasible, the results will provide extremely powerful constraints on the
overall thermo-chemical modeling efforts for the system, in that much of
the output can be quantified through time.

Theoretical Modeling

The upper mantle-crustal-deep ocean system represented by a ridge crest
segment comprises a complex heat and mass transfer system which has
played a dominant role in earth history. Yet, beyond certain
generalizations, we have little actual insight into the specific rate
limiting processes of this primary planetary process. A powerful
approach to understanding the system is to make use of the natural
feedback and refinement capability offered by iteration between

quantified observations and theoretical predictions. To date, modeling
and/or simulation efforts have been directed at individual sub-systems
within the vertically stacked upwelling regimes of mantle and magma,
hydrothermal fluid, and water column plume. Obviously, each of these
domains requires much wider system definition than is represented by the
actively upwelling portion. Furthermore, the numerical models which
presently exist for these sub-systems (e.g., subcrustal mantle behavior
- Whitehead et al., 1984, Richter and McKenzie, 1984; magma chamber
dynamics - Huppert and Sparks, 1984; crustal fracturing and fluid
circulation near plutons - Norton, 1984; plume modeling - Turner and
Gustafson, 1978) do not focus on the scale of change which the LOBO
program could be documenting. Therefore, a program of the magnitude
suggested for a LOBO installation will require a major theoretical
modeling effort. If this is pursued while the field program is being
developed, and the models are designed to be easily interfaced with
newly obtained real-time observations, it could provide invaluable
guidance to the continued development of the field program.

CONCLUSIONS

The broad range of activities involved in generating oceanic crust may
be viewed as a single, complex system, in which the transfer of energy
from the mantle is focused beneath spreading centers and dispersed
through the lithosphere, hydrosphere, and biosphere. Actual time scales
on which the feedback mechanisms associated with this energy transfer,
impact the mechanical, chemical, and biological fluxes are virtually
unknown. By treating the system as a whole and focusing on the time and
space variations within a combined observational and theoretical
framework, oceanographers, as a group, have the potential within several
decades of gaining new understanding of the processes by which a large
fraction of the planetary surface is generated and modified through
geologic time.

ACKNOWLEDGMENTS

This document is the product of many discussions with a wide variety of
colleagues over the past 2.5 years, in particular, D. Bibee, R. Batiza,
H. Curl, Jr., D.A. Clague, E.E. Davis, A. Dobson, M. Dobson, J. Dymond,
F. Duennebier, R. Embley, D. Forsythe, S.R. Hammond, G.R. Heath, H.P.
Johnson, M. Leinen, B.T.R. Lewis, C.R.B. Lister, J.E. Lupton, K.C.
Macdonald, G. Massoth, J. Morton, W.R. Normark, J. Orcutt, G.M. Purdy,
and V. Tunnicliffe. Most recently, a small workshop was convened near
Seattle, supported by the University of Washington College of Ocean and
Fisheries Sciences for the purpose of exploring the scientific arguments
for development of long-term, in situ studies on the ocean floor. This
paper is a direct result of that workshop.

REFERENCES

Ballard, R.D., Holcomb, R.T., and van Andel, Tj.H., 1979, The
    Galapagos Rift at 86°W: sheet flows, collapse pits, and lava
    lakes of the rift valley; Jour. Geophys. Res., v. 84, pp. 5407-
    5422.

Ballard, R.D. and van Andel, Tj.H., 1977, Morphology and tectonics
    of the inner rift valley at lat. 36°50'N on the Mid-Atlantic
    Ridge; Geol. Soc. Am. Bull., v. 88, pp. 507-530.

Ballard, R.D., van Andel, Tj.H., and Holcomb, R.T., 1982, The
    Galapagos Rift at 86°W: 5. variations in volcanism, structure,
    and hydrothermal activity along a 30-km segment of the rift
    valley; Jour. Geophys. Res., v. 87, pp. 1149-1161.

Baross, J.A. and Deming, J.W., 1983, Growth of black smoker
    bacteria at temperatures of at least 250°C; Nature, v. 303, no.
    5916, pp. 423-426.

Bischoff, J.L. and Rosenbauer, R.J., 1984, The critical point and
    two-phase boundary of seawater, 200-500°C; Earth Planet. Sci.
    Letts., v. 68, pp. 172-180.

Bonatti, E., 1983, Hydrothermal metal deposits from the oceanic
    rifts: a classification. In: Rona, P.A., Bostrom, K., Laubier,
    L., and Smith, Jr, K.L. (eds.), Hydrothermal processes at
    seafloor spreading centers, Nato Conf. Ser. in Mar. Sci.,
    Plenum Press, New York, v. 12, pp. 491-502.

Cline, J.D. and Richards, F.A., 1969, Oxygenation of hydrogen
    sulfide in seawater at constant salinity, temperature, and pH;
    Environ. Sci. Tech., v. 3, pp. 838-843.

Converse, D.R., Holland, H.D., and Edmond, J.M., 1982, Hydrothermal
    flow rates at 21°N; EOS, Trans. Amer. Geophys. Union, v. 63,
    pp. 472.

Corliss, J.B., Dymond, J., Gordon, L.I., Edmond, J.M., Von Herzen,
    R.P., Ballard, R.D., Green, K., Williams, D., Bainbridge, A.,
    Crane, K., and van Andel, Tj.H., 1979, Submarine thermal
    springs on the Galapagos Rift; Science, v. 203, pp. 1073-1083.

Cox, A., Doell, R.R., and Dalrymple, G.B., 1964, Reversals of the
    Earth's magnetic field; Science, v. 144, pp. 1537-1543.

Crane, K. and Ballard, R.D., 1981, Volcanics and structure of the
    FAMOUS Narrowgate rift: evidence for cyclic evolution: AMAR;
    Jour. Geophys. Res., v. 86, no. B6, pp. 5112-5124.

Des Marais, D.J. and Moore, J.G., 1984, Carbon and its isotopes in mid-oceanic basaltic glasses; Earth Planet. Sci. Letts., v. 69, pp. 43-57.

Dymond, J., 1981, Geochemistry of Nazca Plate surface sediments: an evaluation of hydrothermal, biogenic, detrital, and hydrogenous sources; In: Kulm, L.D., Dymond, J., Dasch, E.J., and Hussong, D.M. (eds.), Nazca Plate: Crustal formation and Andean convergence, Geol. Soc. Am. Mem., no. 154, pp. 133-174.

East Pacific Rise Study Group, 1980, Crustal processes of the mid-ocean ridge; Science, v. 213, pp. 31-40.

Edmond, J.M., Measures, C., McDuff, R.E., Chan, L.H., Collier, R., Grant, B., Gordon, L.I., and Corliss, J.B., 1979, Ridge crest hydrothermal activity and the balances of the major and minor elements in the ocean: the Galapagos data; Earth Planet. Sci. Letts., v. 46, pp. 1-18.

Ewing, J. and Ewing, M., 1967, Sediment distribution on the mid-ocean ridges with respect to spreading of the seafloor; Science, v. 156, pp. 1590-1592.

Fatton, E. and Roux, M., 1981, Etapes de l'organisation microstructurale chez Calyptogena magnifica Boss et Turner, bivalve à croissance rapide des sources hydrothermales oceaniques. Compte Rendu, Acad. Sci. Franc., v. 293, pp. 63-68.

Francheteau, J. and Ballard, R.D., 1983, The East Pacific Rise near 20°N, 13°N, and 20°S: inferences for along-strike variability of axial processes of the mid-ocean ridge; Earth Planet. Sci. Letts., v. 64, pp. 93-116.

Goguel, J., 1983, A short note on continuity or discontinuity in the global tectonic plate velocities; Tectonophys., v. 100, pp. 1-4.

Goldfarb, M.S., Converse, D.R., Holland, H.D., and Edmond, J.M., 1983, The genesis of hot spring deposits on the East Pacific Rise, 21°N; Econ. Geol. Monogr., v. 5, pp. 184-197.

Hauksson, E, 1983, Episodic rifting and volcanism at Krafla in North Iceland: growth of large ground fissures along the plate boundary; Jour. Geophys. Res., v. 88, pp. 625-636.

Heath, G.R. and Dymond, J., 1977, Genesis and diagenesis of metalliferous sediments from the East Pacific Rise, Bauer Deep and Central Basin, Northwest Nazca Plate; Geol. Soc. Am. Bull., v. 88, pp. 723-733.

Hessler, R.R., Smithey, Jr., W.M., and Keller, C.H., 1985, Spatial and temporal variation of giant clams, tube worms, and mussels at deep-sea hydrothermal vents; In: Jones, M.L. (ed.), Hydrothermal vents of the Eastern Pacific: an overview, Bull. Biol. Soc. Washington, no. 6, pp. 411-428.

Hey, R., Duennebier, F.K., and Morgan, W.J., 1980, Propagating rifts on mid-ocean ridges; Jour. Geophys. Res., v. 85, pp. 3647-3658.

Holcomb, R.T., 1980, Kilauea volcano Hawaii: chronology and morphology of the surficial lava flows; Unpubl. Ph.D. Thesis, Stanford University, California, 321 p.

Huppert, H.E. and Sparks, R.S.J., 1984, Double-diffusive convection due to crystallization in magmas; Ann. Rev. Earth Planet. Sci., v. 12, pp. 11-37.

Jannasch, H. and Wirsen, C.O., 1979, Chemosynthetic primary production of East Pacific seafloor spreading centers; Bioscience, v. 29, pp. 592-598.

Javoy, M., Pineau, F., and Allegre, C.J., 1982, Carbon geodynamic cycle; Nature, v. 300, pp. 171-173.

Johnson, K.S., Beehler, C.L., Sakamoto-Arnold, C.M., and Childress, J.J., 1986, In situ measurements of chemical distributions in a deep-sea hydrothermal vent field; Science, in press.

Jones, W.J., Leigh, J.A., Mayer, F., Woese, C.R. and Wolfe, R.S., 1983, Methanococcus jannaschi sp. nov., an extremely thermophilic methanogen from a submarine hydrothermal vent; Archiv. fur microbiologie, v. 136, pp. 254-261.

Karl, D.M., Burns, D.J., Orrett, K., and Jannasch, H.W., 1984, Thermophilic microbiolactivity in samples from deep sea hydrothermal vents; Mar. Biol. Letts., v. 5, pp. 227-231.

Laubier, L., and Desbruyeres, D., 1984, Les oasis du fond des océans; La Recherche, v. 15, pp. 1506-1517.

Lewis, B.T.R., 1979, Periodicities in volcanism and longitudinal flow on the East Pacific Rise at 23°N; Geophys. Res. Letts., v. 6, pp. 753-756.

Lichtman, G.S. and Eissen, J.-P., 1983, Time constraints on the evolution of medium-rate spreading centers; Geology, v. 11, pp. 592-595.

Lichtman, G.S., Normark, W.R., and Spiess, F.N., 1984, Photogeologic study of a segment of the East Pacific Rise axis near 21°N latitude; Geol. Soc. Am. Bull., v. 95, pp. 743-752.

Lister, C.R.B., 1977, Qualitative models of spreading center processes, including hydrothermal penetration; Tectonophysics, v. 37, pp. 203-218.

Lister, C.R.B., 1983, On the intermittency and crystallization mechanisms of sub-seafloor magma chambers; Geophys. J. R. Astron. Soc., v. 73, pp. 351-366.

Lonsdale, P., 1983, Overlapping rift zones at the 5.5°S offset of the East Pacific Rise; Jour. Geophys. Res., v. 88, pp. 9393-9406.

Lupton, J.E., Delaney, J.R., Johnson, H.P., and Tivey, M.K., 1985, Entrainment and vertical transport of deep ocean water by buoyant hydrothermal plumes; Nature, v. 316, pp. 621-623.

Lutz, R.A., Fritz, L.W., and Rhoads, D.C., 1985, Molluscan growth at deep-sea hydrothermal vents; In: Jones, M.L. (ed.), Hydrothermal vents of the Eastern Pacific: an overview, Bull. Biol. Soc. Washington, no. 6, pp. 199-210.

Macdonald, K.C., 1983, Crustal processes at spreading centers; Rev. Geophys. Space Physics, v. 21, no. 6, pp. 1441-1454.

Macdonald, K.C. and Fox, P.J., 1983, Overlapping spreading centers: new accretion geometry on the East Pacific Rise; Nature, v. 302, pp. 55-58.

Macdonald, K.C., Becker, K., Spiess, F.N., and Ballard, R.D., 1980, Hydrothermal heat flux of the "black smoker" vents on the East Pacific Rise; Earth Planet. Sci. Letts., v. 48, pp. 1-7.

Marquart, G. and Jacoby, W.R., 1985, On the mechanism of magma injection and plate divergence during the Krafla rifting episode in NE-Iceland; Jour. Geophys. Res., v. 90, no. 12, pp. 10,178-10,193.

Massoth, G.J., Baker, E.T., Feely, R.A., and Curl, Jr., H.C., 1984, Hydrothermal signals away from the Southern Juan de Fuca Ridge; EOS, Trans. Amer. Geophys. Union, v. 65, pp. 1112.

Minster, J.B., and Jordan, T.H., 1978, Present-day plate motions; Jour. Geophys. Res., v. 83, pp. 5331-5354.

National Research Council, 1983, Seafloor referenced positioning: needs and opportunities; Rept. of the Committee on Geodesy, Panel on Ocean Bottom Positioning, Natl. Academy Press, Washington, D.C., 53 p.

Newhall, C.G., Dzurisin, D., and Mullineux, L.S., 1984, Historical unrest at large quaternary calderas of the world, with special reference to Long Valley, California;   U.S. Geol. Surv. Open-File Report 84-939, pp. 714-742.

Normark, W.R., 1980, Delineation of the main extrusion zone of the East Pacific Rise at lat. 21°N; Geology, v. 4, pp. 681-685.

Norton, D., 1984, Theory of hydrothermal systems; Ann. Rev. Earth Planet. Sci., 12, pp. 155-177.

Rea, D.K., 1978, Asymmetric sea-floor spreading and a non-transform axis offset: the East Pacific Rise 20°S survey area; Geol. Soc. Am. Bull., v.89, pp. 838-844.

Reid, I.D., Orcutt, J.A., and Prothero, W.A., 1977, Seismic evidence for a narrow zone of partial melting underlying the East Pacific Rise at 21°N; Geol. Soc. Am. Bull., v. 88, pp. 678-682.

Richter, F.M. and Mckenzie, D., 1984, Dynamical models for melt segregation from a deformable matrix; Jour. Geol., v. 92, pp. 729-740.

Riedesal, M., Orcutt, J.A., Macdonald, K.C., and McClain, J.S., 1982, Microearthquakes in the black smoker hydrothermal field, East Pacific Rise at 21°N; J. Geophys. Res., v. 87, pp. 10613-10623.

RISE Team, 1980, East Pacific Rise: hot springs and geophysical experiments; Science, v. 207, pp. 1421-1433.

Schouten, H., Klitgord, K.D., and Whitehead, J.A., 1985, Segmentation of mid-ocean ridges; Nature, v. 317, pp. 225-229.

Schouten, H. and Klitgord, K.D., 1982, The memory of the accreting plate boundary and the continuity of fracture zones; Earth Planet. Sci. Letts., v. 59, pp. 255-266.

Sigurdsson, H. and Sparks, S.R.J., 1978, Lateral magma flow within rifted Icelandic crust; Nature, v. 274, pp. 126-130.

Spiess, F.N., 1985, Sub-oceanic geodetic measurements; Inst. Electric and Electronic Eng., Trans. Geosci. Rem. Sens., v. GE-23, pp. 502-510.

Spiess, F.N., 1986, Deep ocean near-bottom surveying and sampling techniques; Trans. Geosci. Rem. Sens., v. GE-23, pp.

Turner, J.S. and Gustafson, L.B., 1978, The flow of hot saline solutions from vents in the seafloor: some implications for exhalative massive sulfide and other ore deposits; Econ. Geol., v. 73, pp. 1082-1100.

Von Damm, K.L., 1983, Chemistry of submarine hydrothermal solutions at 21°N, East Pacific Rise and Guaymas Basin, Gulf of California; Unpubl. Ph.D. thesis, Woods Hole Ocean. Inst.-Mass. Inst. Tech., Rept. WHOI-84-3, 240 p.

Von Damm, K.L., Grant, B., and Edmond, J.M., 1984, Preliminary report on the chemistry of hydrothermal solutions at 21°N, East Pacific Rise; In: Rona, P.A., Bostrom, K., Laubier, L.; and Smith, Jr., K.L. (eds.), Hydrothermal Processes at Seafloor Spreading Centers, Plenum Press, New York, pp. 369-389.

Whitehead, J.A., Dick, H.J.B., and Scouten, H., 1985, A mechanism for magmatic accretion under spreading centers; Nature, v. 312, pp. 146-148.

ELECTRICAL METHODS IN THE EXPLORATION OF SEAFLOOR MINERAL DEPOSITS

T.J.G. Francis
Institute of Oceanographic Sciences
Wormley
Godalming
Surrey, GU8 5UB U.K.

ABSTRACT. The range of electrical experiments which have been carried out at sea in order to learn about the resistivity (or conductivity) structure of the seabed is reviewed. Some of these experiments were directly related to mineral exploration. Others could be modified to make them more suitable for mineral exploration and appraisal. Electrical methods are particularly relevant to the study of sulfide ore deposits.

INTRODUCTION

A very important element in the future progress of mineral exploration in the oceans will be the development of geophysical techniques which will allow mineralized bodies buried in the sediments and hard rocks of the ocean floor to be detected. On land, various electrical techniques have proved the most suitable for detecting metalliferous ore bodies, because the minerals they contain are often good conductors of electricity. But the applicaton of these techniques to the seabed, even in the shallow waters of the continental shelves, has been severly restricted by the high conductivity of sea water. Nevertheless, a small number of electrical experiments have now been tried in the oceans.

REVIEW OF THE USE OF ELECTRICAL METHODS AT SEA

The first attempts to use electrical methods at sea in order to learn something about the seabed were described by Schlumberger et al. (1934). They laid cables along the seabed out to a distance of 1 km from the shore and made resistivity measurements in order to measure the thickness of the unconsolidated sediment in water depths of up to 38 m. Such measurements would be done considerably more precisely and cheaply today by seismic reflection profiling.

In the 1960's, scientists at Imperial College, studying the feasibility of geophysical techniques for offshore mineral exploration, attempted to make small scale resistivity measurements in St. Ives

413

P. G. Teleki et al. (eds.), Marine Minerals, 413–419.
© *1987 by D. Reidel Publishing Company.*

Bay on the north Cornish coast (Bruckshaw and Taylor Smith, 1963;
Marke, 1965). The county of Cornwall in southwest England has a
history of copper and tin mining extending back many centuries. The
objective of the Imperial College group's work was to determine
whether geophysical techniques could be used to map mineralized areas
offshore. However, their electrical measurements were unconvincing,
largely because of the scale on which they were conceived. Nevertheless,
they reached the conclusion that to get sufficient penetration of
current into the seabed the electrodes should be towed along the
seafloor rather than at the sea surface.

    Some ten years later, electrical measurements on a much larger
scale were carried out off the north Cornish coast by Francis (1977).
A naval minesweeper, chosen because of its large generating capacity,
was used to tow an array of electrodes over half a kilometer long.
By towing the array along the sea surface it was possible to maintain
constant spacing between the electrodes and at the same time to cover
the ground rapidly. Direct currents of up to 2000A were passed
between current electrodes approximately 300 m apart and the potential
field thus generated measured between pairs of silver-silver chloride
non-polarizing electrodes. Measurements of the resistivity of the
seabed were made in this manner along 2040 km of ship's track. In
one area off the Cornish coast a low resistivity seabed was discovered
which was clearly associated with the presence of sulfide mineralization.
However, because the electrode spacing needs to be much larger than
the water depth, it was concluded that, whilst the method can give
useful measurements of seabed resistivity in continental shelf
depths, it would be impracticable in ocean depths.

    At much the same time that the above work was going on in England,
Corwin (1976) was exploring the use of the self-potential method in
offshore areas of the United States. Silver-silver chloride non-
polarizing electrodes were towed behind a small boat on field trials
in Penobscot Bay, Maine, an area where sulfide bodies found on land
near the shore are known to produce self-potentials (SPs). Self-
potentials as high as 300 mV were measured in water depths of about
12 m. In one case they exceeded the adjacent SPs measured on land,
indicating that the deposit might be more heavily mineralized offshore.
However, it must be noted that these measurements were made in shallow
water in a narrow inlet. With increasing water depth in open water,
SPs at the sea surface would become smaller and smaller and eventually
be lost in other natural sources of potential field. Perhaps Corwin's
most important contribution was to point out that the background
noise level of potential measurements made with silver-silver chloride
non-polarizing electrodes in sea water (typically a few tenths of a
millivolt) are generally much less than the noise level observed in
self-potential data on land (ranging up to $\pm$ 10 mV). Thus the higher
sensitivity of the SP method at sea may counteract to some extent the
attenuation of SP anomalies caused by the high conductivity of sea water.

    Recently, Wynn and Grosz (1986) have reported experiments on the
Atlantic Continental Shelf of the United States off the coast of
Virginia using an induced polarization (IP) streamer towed close to
the seafloor. Measurements were made along a line 8 km long in water

depths averaging 7 m and IP anomalies were detected which are
interpreted to be due to concentrations of ilmenite in the seabed
sediments. Wynn and Grosz believe that IP surveys could be used to
define areas of the continental shelf where titanium-bearing
concentrations of heavy-minerals occur.

A number of electrical experiments have been carried out on the
sea floor, both on the continental shelf and in the deep ocean, in
order to ascertain the geotechnical properties of the sediments.
Kermabon et al. (1969) developed a probe which was used to measure in-
situ vertical profiles of resistivity at a number of stations on the
Tyrrhenian abyssal plain to sub-bottom depths of 7.5 m. Hulbert et
al. (1982) developed another probe, designed to be driven into coarse-
grained sediments with a vibracorer, which was used to study carbonate
sediments in shallow water off Florida to sub-bottom depths f 5 m.
Jackson (1975) developed a focussed resistivity pad, similar in
concept to the focussed resistivity tools used in well logging, to
lay on the sea floor and measure the resistivity of the underlying
sediment. While all of these instruments have operated successfully,
none is suitable for mineral exploration purposes because all were
designed essentially for making spot measurements in soft sediments.
However, the technique developed by Jackson might be developed for
deep ocean operation and applied to the appraisal of sulfide deposits
already known to outcrop on the sea floor.

Measurements to determine the conductivity structure below the
ocean floor by magnetotelluric sounding have been carried out since
1965 (Filloux, 1977). This technique involves the observation of the
horizontal components of the magnetic and electric fields of naturally
occurring electromagnetic signals. Since the conductivity of a layer
of sea water several kilometers in thickness eliminates frequencies
above about 10 c.p.h. (0.003 Hz), the method is only suitable for
studying the conductivity of the mantle on a scale of hundreds of
kilometers and is, thus, insensitive to the conductivity structure of
the crust and uppermost mantle. Thus, magnetotelluric methods cannot
be used for mineral exploration purposes on the ocean floor as they
can be on land (Telford et al., 1976).

In the last decade, various active-source electromagnetic sounding
experiments have been conducted on the ocean floor. Young and Cox
(1981) lowered an 800 m long transmitting antenna to the seafloor
close to the axis of the East Pacific Rise at 21°N. Signals from
this antenna in the frequency range 0.25-2.25 Hz were detected 19 km
away by observing the electric field at two 9-m long crossed antennas
lying on the ocean floor. The observations were modelled to yield a
conductivity structure of the crust and upper mantle to a depth of 30
km. As it stands now, this is not a viable exploration technique,
but the success of this experiment suggests that active-source
electromagnetic sounding, possibly using different frequencies and
different source-receiver configurations, could be developed into one.

Very recently Edwards et al. (1985) reported the first results
of the MOSES experiment, carried out to determine the electrical
conductivity and thickness of the sediment in Bute Inlet, British
Columbia, where the water depth is 640 m. In this experiment a

vertical antenna was used with electrodes near the sea surface and on
the seabed. The transmitted power was 1.25 kW at a frequency of
about 0.125 Hz. The field thus generated was detected by an ocean
bottom magnetometer. The transmitter-receiver separation was varied
between 150 and 2000 m. The authors of this experiment are already
considering its application to studying the conductivity structure in
the vicinity of sulfide deposits on mid-oceanic ridges.

The first electrical measurements to be made on an ocean floor
sulfide deposit have been reported by Francis (1985). Simple DC
resistivity measurements were made with an Wenner array of 10 m spacing
between electrodes deployed along the sea floor by the submersible
CYANA. In addition, self-potentials were observed. The results of
this work, carried out on a seamount close to the axis of the East
Pacific Rise at 13°N, are presented in Figure 1. Pillow basalts near
the base of the seamount were found to be about forty times as resistive
at the seawater, in good agreement with downhole logging measurements
in Deep Sea Drilling Project (DSDP) drill holes. The resistivity of
the sea bottom where sulfides are exposed was found to be one to two
orders of magnitude less than that of the pillow basalts. At one
site the seabed was almost twice as conductive as the overlying sea
water. Furthermore, self-potentials were found to be associated with
the sulfide, up to 10 mV between a pair of electrodes 10 m apart, but
were undectable on the pillow basalts. It was concluded that SP
measurements, not necessarily made from a submersible, might be the
easiest electrical method for mapping the areal extent of an ocean
floor sulfide deposit. Active electrical methods would be required
to define its internal variability and three dimensional extent.

BOREHOLE GEOPHYSICS

Appraisal of any substantial deposit will require drilling.
The means to carry this out have already been developed for the Deep
Sea Drilling Project and the Ocean Drilling Program (Serocki and
McLerran, 1984). The capability of spudding-in on bare rock, first
achieved on Leg 106 of the Ocean Drilling Program in 1985, is
particularly relevant to the appraisal of ocean floor mineral deposits.
But, because holes drilled in oceanic depths will always be much more
expensive than comparable holes on land, it will be necessary to conduct
appraisal operations with a minimum number of holes. Here, the use
of borehole geophysical techniques will be important, particularly
resistivity and electromagnetic methods which allow the properties of
the rock well away from the hole (100+ meters) to be studied. Both
single hole and hole-to-hole techniques will be useful. Considerable
progress has been made in recent years in the application of borehole
geophysical techniques for mineral exploration on land (Daniels and
Dyke, 1984). Some of these techniques should be directly applicable
beneath the ocean floor. Large-scale resistivity measurements,
sensitive to the rock properties as far as about 100 m from the hole,
have already been made in a number of DSDP drill holes (Francis, 1981;
Becker et al. 1982).

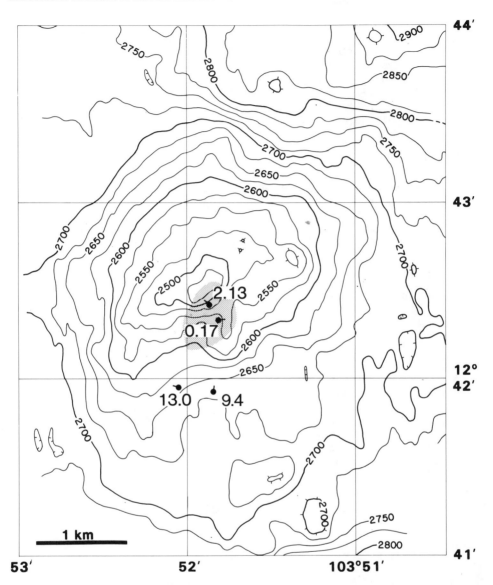

Figure 1. Resistivity measurements of the seabed plotted on a bathymetric map of a seamount close to the axis of the East Pacific Rise. Resistivity values in ohm-m. The solid circles mark the position of the submersible during the measurements, the 'tails' the direction of the array. Contour interval is 25 m. The resistivity of the sea water immediately above the bottom was 0.28 ohm-m. The hatched area indicates the approximate area of outcrop of the sulfide deposit (from Francis, 1985).

CONCLUSIONS

Electrical experiments already made on the ocean floor and borehole geophysical techniques already tried on land and down a few oceanic drill holes provide a basis for developing viable electrical methods for the exploration appraisal of ocean floor mineral deposits. These methods will be especially useful for the exploration and appraisal of sulfide ore deposits.

REFERENCES

Becker, K., Von Herzen, R.P., Francis. T.J.G., Anderson, R.N., Honnorez, J., Adamson, A.C., Alt, J.C., Emmermann, R., Kempton, P.D., Kinoshita, H., Laverne, C., Mottl, M.J. and Newmark, R.L., 1982, In situ electrical resistivity and bulk porosity of the oceanic crust Costa Rica Rift; Nature, v. 300, pp. 594-598.

Bruckshaw, J.M. and Taylor Smith, D., 1963, The feasibility of geophysical methods of offshore mineral exploration; Imperial College, London, Geophysics Report 63-4.

Corwin, R.F., 1976, Offshore use of the self-potential method; Geophys. Prosp., v. 24, pp. 79-90.

Daniels, J.J. and Dyke, A.V., 1984, Borehole resistivity and electromagnetic methods applied to mineral exploration; IEEE Trans. on Geoscience and Remote Sensing, v. FE-22, pp. 80-87.

Edwards, R.N., Law, L.K., Wolfgram, P.A., Nobes, D.C., Bone, M.N., Trigg, D.F. and Delaurier, M., 1985, First result of the MOSES experiment: Sediment conductivity and thickness determination, Bute Inlet, British Columbia, by Magnetometric Off-Shore Electrical Sounding; Geophysics, v. 50, pp. 153-160.

Filloux, J.H., 1977, Ocean floor magnetotelluric sounding over North Central Pacific; Nature, v. 269, pp. 297-301.

Francis, T.J.G., 1977, Electrical prospecting on the continental shelf; Inst. Geol. Sci. Rept., no. 77/4, 41 p.

Francis, T.J.G., 1981, Large scale resistivity experiment at Deep Sea Drilling Project Hole 459B; Init. Repts., Deep Sea Drilling Project, v. 60, Washington (U.S. Govt. Printing Office), pp. 841-852.

Francis, T.J.G., 1985, Resistivity measurements of an ocean floor sulfide mineral deposit from the submersible CYANA; Mar. Geophys. Res., v. 7, pp. 419-438.

Hulbert, M.H., Bennett, R.H. and Lambert, D.N., 1982, Seabed geotechnical parameters from electrical conductivity measurements; Geo-Marine

Lett., v. 2, pp. 219-222.

Jackson, P.D., 1975, An electrical resistivity method for evaluating
    the in-situ porosity of clean marine sands; Mar. Geotechnol., v. 2,
    pp. 91-115.

Kermabon, A., Gehin, C. and Blavier, P., 1969, A deep-sea electrical
    resistivity probe for measuring porosity and density of consolidated
    sediments; Geophysics, v. 34, pp. 554-571.

Marke, P.A.B., 1965, The development and use of offshore mineral
    exploration techniques; unpubl. Ph.D. thesis, Imperial College,
    London.

Schlumberger, C., Schlumberger, M. and Leonardon, E.G., 1934, Electrical
    exploration of water-covered areas; Trans. Amer. Inst. Min. Metall.
    Engrs., Contribut. 71, pp. 122-134.

Serocki, S.T. and McLerran, A.R., 1984, The Ocean Drilling Program, a
    technical overview; World Oil, v. 199, no. 2, pp. 52-58.

Telford, W.M., Geldart, L.P., Sheriff, R.E. and Keys, D.A., 1976.
    Applied Geophysics; Cambridge University Press, 860 p.

Wynn, J.C. and Grosz, A.E., 1986. Application of induced polarization
    method to offshore placer resource evaluation, 18th Annual Offshore
    Technology Conference, Houston, Texas (in press).

Young, P.D. and Cox, C.S., 1981. Electromagnetic active source sounding
    near the East Pacific Rise, Geophys. Res. Lett., v. 8, pp. 1043-1046.

SOURCES OF CONFUSION: WHAT ARE MARINE MINERAL RESOURCES?

A. A. Archer
4 Brook Rise
Chigwell
Essex IG7 6AP
United Kingdom

ABSTRACT. Earth science literature is littered with specialists'
terms, some necessary, others jargon, which can give rise to
serious misunderstandings between earth scientists, as well as
between them and laymen. The terminology applied to mineral
resources is a notable example. Relatively recent attempts at
clarification are summarized and by emphasizing that the essential
characteristic of any mineral resource is that it must satisfy
certain economic criteria, another attempt is made to reduce the
present confusion.

THE COMPLAINT

    Most encouragingly, the practical benefits that follow from
seeking advice from geologists and those who work with them are
being ever more widely recognized.
    Much less fortunately, the earth scientists' vocabulary is
becoming increasingly specialized as their knowledge increases.
Indeed, several vocabularies have been spawned within the earth
sciences, so that now it is not only laymen who are in difficulty.
Many of those who have spent their careers in one field find
themselves increasingly isolated by a language barrier unless armed
with (and prepared constantly to use) the appropriate lexicon.
    Specialized words may be essential for communication between
specialists, or greatly facilitate it. But their use may also be
simply an attempt to impress outsiders, thus becoming jargon.
Guilt is not, of course, restricted to earth scientists:

    "Although solitary under normal prevailing circumstances,
racoons may congregate simultaneously in certain situations of
artificially enhanced nutrient availability'. . . . I have an
uncharitable suspicion that the sentence means no more than that
racoons live alone, but gather at bait" (Howard, 1984).

421

P. G. Teleki et al. (eds.), Marine Minerals, 421–432.
© 1987 by D. Reidel Publishing Company.

Abundant plain words to replace jargon ("gibberish") are available
to English speakers in the Concise English Dictionary. The
campaign is greatly complicated when the plain words employed have
been given specialized meanings, thus completely confounding those
innocent of such usage. The relevant plain English definitions of
examples that have the potential to be major obstacles to
understanding are:

> Resource: "Means of supplying what is needed, stock that can
> be drawn on, *available assets; country's collective means for
> support and defence."

> Reserve: "Thing reserved for future use, extra stock or amount.
> . . State of being kept unused but available."

To these might be added:

> Deposit: "Thing stored or entrusted for safe keeping:. . .
> layer of precipitated matter, natural accumulation."

> Occur (hence occurrence): Be met with, be found, exist, in
> some place or conditions."

Thus a well-educated layman draws little if any distinction
between "resources" and "reserves" and should be excused if
confused by specialized meanings. Note also that there is an
overlap with one meaning of "deposits". The asterisk (*) denotes
usage that is "chiefly U.S. (often also Canadian, Australian,
etc.)."

Examples of confusion

A 'stock' of anything is commonly understood to be finite,
readily counted or measured. It is therefore unsurprising that
those unfamiliar with the subject have great difficulty in
accepting that the stock of a mineral in the ground, to be drawn
upon when needed, cannot be given as a finite quantity (see, for
example, House of Lords, 1982). Moreover, such a number is often
sought so that by dividing it by present world consumption, the
period after which supplies will be exhausted becomes clear! This
fanciful conception (Zwartendyk, 1974), the Static Life Index, can
also put in its deceiving appearance in computer modeling of
possible futures, in which future demand is varied, and given the
grandiose label 'Dynamic Life Index' if the 'stock' remaining to be
discovered is added.

The similarity between the ordinary meanings of "reserves" and
"resources" may, perhaps, explain the National Coal Board's
estimate of the recoverable "reserves" of coal in Great Britain.
For many years these were said to be 45 billion tons, providing a
Static Life Index of 300 years; this quantity included coal that

could not be won, even at exorbitant cost, with the methods now
available.

The establishment of the United Nations Ad Hoc Sea-Bed Committee,
the Sea-Bed Committee, and the Third Law of the Sea Conference and
the adoption by it of the United Nations Convention on the Law of
the Sea on April 30, 1982, followed an initiative by Ambassador
Pardo, Malta, in August, 1967, when the General Assembly adopted a
"Declaration and treaty concerning the sea-bed and ocean floor,
underlying the seas beyond the limits of present national
jurisdiction, and the use of the resources in the interests of
mankind" as an item on its agenda. The mineral resources of the
continental shelf having become subject to national jurisdiction by
the Convention on the Continental Shelf, 1958, the resources to
which the item referred were almost all on the very deep sea-bed
and far from land. Nevertheless, it was clear from statements made
in the Committees and the Conference that many shared the view
that, as soon as their legal status had been agreed, these
resources would be exploited, generating such wealth that it would
make an effective contribution to the economies of the developing
countries. This misunderstanding of the concept of "resources,"
which became reflected also in the opposite view that their
development must be strictly limited to avoid adverse consequences
for existing land-based producers, was a serious obstacle to the
negotiations.

The 1958 Convention had created a legal definition of the
continental shelf that is quite distinct from any recognized by
marine scientists, thus providing a brand new ambiguity that is
only slightly reduced by Article 76 of the 1982 Convention, Article
133 of which includes these novel definitions:

"'resources' means all solid, liquid or gaseous mineral
resources in situ in the Area ...." (beyond the limits of
national jurisdiction)

"resources, when recovered from the Area, are referred to as
'minerals'"!

CLARIFICATION

History

The need for rigid definition of the terms "reserves" and
"ore", when applied to a particular property by a mining company,
has long been recognized; the scope for misrepresentation in, for
example, a prospectus, is obvious. Definitions may be enshrined in
national legislation or laid down by professional bodies
(Institution of Mining and Metallurgy, 1984). The terminology used
to describe individual properties will not be further pursued in
this contribution. The need to codify the terminology used in
regional, national and international contexts was recognized by the

United States Geological Survey (USGS) and United States Bureau of
Mines (USBM) almost 40 years ago (1948). Some 8 years later,
"Mineral Reserves and Mineral Resources" were discussed in a very
important paper by Blondel and Lasky (1956) on behalf of a committee
of the Society of Economic Geologists with international
membership. The scope of this work is indicated in its opening
sentences:

> "What exactly is meant by the question: 'What are the iron
> ore reserves of such and such a country?' And along with
> this, 'What rules must be observed in estimating these
> reserves?' Such questions have been the subject of numerous
> discussions. The answer to the first has depended in the past
> on who asked the question and on who furnished the answer; the
> answer to the second has depended on who made the estimate.
> Yet clearly it is desirable that there be a single answer to
> each..... The crux of the confusion lies in the fact that the
> basic notions, and the terms used, do not have the same
> meaning for everyone, and these variations in sense arise
> largely from the circumstances of their usage ... that of the
> miner and .... that of the economist."

Today, 30 years since this was published, the crux of the
problem remains that the idea that economists or economics could
have anything to do with the subject is still rejected as a heresy
by many geologists (Zwartendyk, 1981) as well as miners and
others. "Let us proceed on the facts rather than become involved
with the arcane, woolly, 'art' of the economists" is not an
unfamiliar cry! But it reflects a fundamentally absurd attitude.
Rather the "two approaches - that of the miner and that of the
economist - are conceptually different, but they are not
necessarily in conflict. In reality they are supplemental. The
first has limited application, and the second depends on the
first. Each is based on its own set of considerations" (Blondel and
Lasky, 1956, p. 688). These definitions were, therefore, suggested
(op.cit., p.691):

> Reserves: "mineral material considered as being exploitable
> under existing conditions, including cost, price, technology
> and special local circumstances."

> Resources: the reserves and "that mineral material which, to
> be exploited, demands conditions more favorable than those
> currently existing, as well as further exploration to bring
> into consideration undiscovered deposits ..."

McKelvey and his diagram

Resistance to the lucidly expressed principles enunciated by
the Society of Economic Geologists' expert committee proved to be
firm and the cudgel was taken up by McKelvey (1972, 1974). The

diagram that he devised (Fig. 1), more than any words, however
lucid, makes the essential principles abundantly clear, that is
that the different categories are dependent upon both physical and
economic criteria. The degrees of both geological assurance and
economic feasibility <u>must</u> be taken into account.

Following discussions between the USGS and the USBM, a
classification that embraces the McKelvey diagram was adopted by
the United States Department of the Interior in 1976 (U.S. Bureau
of Mines and U.S. Geological Survey, 1976). This includes these
definitions:

> Resource: "a concentration of naturally occurring solid,
> liquid, or gaseous materials in or on the earth's crust in
> such form that economic extraction of a commodity is currently
> or potentially feasible."

> Reserve: "that portion of the identified resource from which
> a usable mineral and (sic) energy commodity can be
> economically and legally extracted at the time of
> determination. The term <u>ore</u> is used for reserves of some
> minerals."

Revised USGS and USBM definitions

The definition of a resource adopted in 1976, like that
published by Blondel and Lasky, has a serious flaw. It includes a
sub-economic category, the material in which requires "a
substantially higher price or a major cost-reducing advance in
technology to be economically viable" (McKelvey, 1974, p. 77).
Thus one limit of the resources of any mineral is left to
subjective judgement, that is the bottom line of McKelvey's diagram
is left undefined (Brobst, 1979). It is a brave (or foolish) man
who is prepared to predict the lower costs that will follow the
introduction of uninvented technology, or the price that Man will
be prepared to pay at some undefined time in the future. Resources
estimated in accordance with these definitions also include
material that has not been found, involving a complementary form of
clairvoyance!

A system that is "more workable in practice" (U.S. Bureau of
Mines and U.S. Geological Survey, 1980) was, therefore, devised.
Sub-economic resources were restricted explicitly to material that
has been identified and the new classification incorporates a tactful
(or tactical ?) retreat from economics by creating a new category:

> Reserve base: "That part of an identified resource that meets
> specified minimum physical and chemical criteria related to
> current mining and production practices, including those for
> grade, quality, thickness, and depth."

But the definition continues, to include:

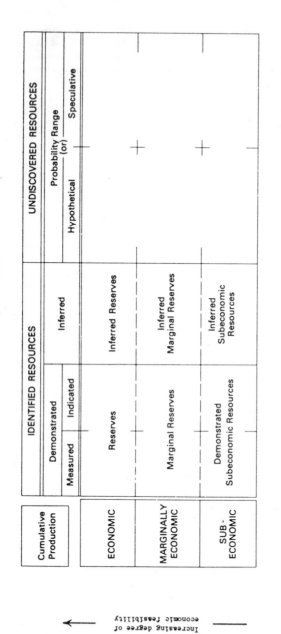

Figure 1. Major elements of USBM and USGS mineral-resource classification (modified from U.S. Bureau of Mines and U.S. Geological Survey, 1980, p. 5).

"It may encompass those parts of the resources that have a
reasonable potential for becoming economically available
within planning horizons beyond those that assume proven
technology and current economics."

It is pointed out that this 'reserve base' category has been
called the 'geologic reserve' by others. This very confusing
amendment necessitated revision of the definition of 'reserves':

Reserves: "That part of the reserve base which could be
economically extracted or produced at the time of
determination...... Reserves include only recoverable
materials: thus, terms such as "extractable reserves" and
"recoverable reserves" are redundant....."

The clear identification of quantity estimated to be the
reserves as the part that can be recovered, rather than the in situ
quantity in the previous system, is helpful in principle, but
depends on it being remembered that the definition has been
changed.

United Nations definitions

To be accepted internationally, the terms used must be
unequivocally compatible with different economic systems. Although
it does not appear in any of the official United States
definitions, the word 'profitable' can be found in the explanatory
texts (for example, U.S. Bureau of Mines and U.S. Geological Survey,
1980), a word that in theory forms no part of a Centrally Planned
Economy. And the states with Centrally Planned Economies, that is
those in Eastern Europe and elsewhere, were not represented on the
committee of the Society of Economic Geologists. Any system to be
used for reporting internationally must be readily translatable,
notably in the official languages of the United Nations (English,
French, Spanish, Russian, Chinese and Arabic). 'Reserves' and
'resources' present a problem as there are no corresponding words
for both in some languages.
    In 1975, the Committee on Natural Resources recommended that
its parent organization, the United Nations Economic and Social
Council, should adopt a Canadian proposal that the terminology and
classification of mineral resources should be studied, with a view
to adopting a system that would facilitate the compilation of
international statistics.
    As a result, an international group of experts met in New York
early in 1979 and within a week had agreed on a report (United
Nations, 1979) that was adopted by the Committee on Natural
Resources and by the Economic and Social Council later that year.
The most radical proposal was, perhaps, that other than for
reporting on individual properties, the term "reserves" should be
abandoned. The group stressed the dynamic nature of the concept of
resources, so that estimates apply only at the time they are

produced and must be updated at intervals. Estimates should not be
restricted only to those deposits that are workable at the time,
but the subjective element inevitably associated with the future
should be limited by defining the future period involved, as first
proposed by Zwartendyk (1972), thus minimizing the inherent
weakness of the concept of resources. The utility of attempting to
estimate "the total amount of mineral commodities that will
ultimately become available from the earth's crust for mankind's
use" was questioned.

Considerable stress was laid on the need clearly to distinguish
between in-situ and recoverable quantities. The need for great
caution and professional guidance in the compilation of the
statistics was also emphasized: "the adoption of a satisfactory
classification system for international use cannot, by itself, solve
all problems of compilation and evaluation." In order to avoid
misunderstandings arising from the different meanings attached to
such words as "reserves" and "resources", the group adopted an
orderly arrangement of letters and numbers to identify different
categories. This controversial idea was selected deliberately to
ensure that users read the definitions, rather than assume that
words mean what they think they mean!

Mineral resources, representing all of the quantities that "might
be of economic interest over the foreseeable period of the next few
decades" (normally restricted to the next two or three decades),
are subdivided into three basic categories, R-1, R-2 and R-3 if the
estimates are of in-situ quantities, and r-1, r-2 and r-3 if
recoverable, depending on the degree of assurance attached to the
estimates. The R-1 (r-1) and perhaps R-2 (r-2) categories are
described as R-1-E or R-2-E (etc.) if they "are considered to be
exploitable in a particular country or region under the prevailing
socio-economic conditions with available technology," while the
balance is described as R-1-S or R-2-S (etc.). The suffix "M" is
available as optional extra to identify an intermediate, marginal,
category, if this is felt necessary in some countries.

Material with a lower economic potential than is required for
it to qualify as part of the resources should be described simply
as an 'occurrence.'

British Geological Survey definitions

The so-called alpha-numeric system adopted by the United
Nations is described above as controversial. This indeed proved to
be the case when the Institute of Geological Sciences (now the
British Geological Survey, BGS) came to standardize its terminology
in the light of the U.N. agreement. Notwithstanding a certain
natural conservatism, such a system was adopted, consisting of a
simplification of that adopted by the U.N.

Thus R-1-E and R-2-E became E1 and E2 and r-1-E and r-2-E
became pE1 and pE2 ('p' for 'part') (Fig. 2).

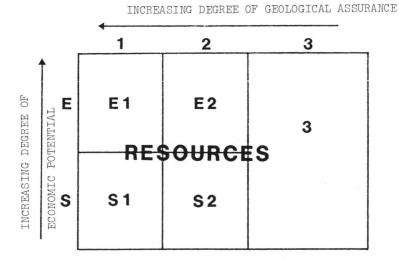

Figure 2.   The BGS classification of in-situ mineral resources (the prefix 'p' is used to indicate the recoverable part).

Recognizing that this notation is not appropriate for the text of all its publications, the BGS classification (House of Lords, 1982, pp. 298-306) includes this definition:

> Mineral resource: "a mineral occurrence which is economically workable now or is likely to become economically workable, normally within the next 20 or 30 years. For this purpose, occurrences include material already mined or partially processed, e.g. mine tips. In reporting resources a distinction must be made between the quantities present in situ and those that are recoverable."

The resources that are currently workable (in category El, for example) may, if necessary, be described as "economic resources", rather than as "reserves", which in common with the U.N. system should be abandoned except in the narrow sense that the term is employed by the mining industry.

THE CONFUSION ELIMINATED

Such confusion as now remains will be entirely eliminated if the following are kept in mind:

1. The cost of recovering a mineral commodity and the price which it commands are dependent upon
     (a) Such physical factors as geographical location, grade, size and mineralogical complexity of the deposits; and

    (b) Such economic factors as strength of demand,
        interest rates and fiscal regimes.

2. The definition of a mineral resource and a classification
   system must, therefore, have two main elements

    (a) A geological component; and
    (b) An economic component

3. As one of the main objects of assessing the resources of a
   mineral is related to questions of future supply,
   consideration should not be limited only to those deposits
   currently thought to be workable economically, but a line must
   be drawn somewhere above estimates based on crustal
   abundance.

4. A reasonable place to draw this arbitrary line is below
   deposits that, at the time of estimation, seem likely to be
   workable in the light of plausible economic and technological
   developments in the next 20 or 30 years.

5. A mineral resource is, therefore, defined as a mineral
   occurrence which is economically workable now or is likely to
   become economically workable, normally within the next 20 or
   30 years.

6. It is essential that it is perfectly clear whether an estimate
   refers to in-situ or to recoverable quantities.

7. Simple book profitability is not the determinant of "economically
   workable". Such factors as social costs and benefits, trade
   policies and strategic stocks may be taken into account in
   countries with free market economies, as well as those with
   Centrally Planned Economies.

Exceptions

    At the risk of re-introducing confusion, there are special
circumstances when, for practical reasons, the economic component
must be set aside. An example is the national sand and gravel
resource assessment programme undertaken by the British Geological
Survey. The economics of the winning of these and other low ex-
quarry price minerals is strongly influenced by transport costs
(among other factors). As it is impossible to predict what
relatively nearby demand may be created even within the near
future, it was found necessary to rely on physical criteria alone
(Archer, 1969).
    It must be emphasized that this was forced by the particular
circumstances of the case and that the criteria used, such as
thickness, depth and overburden ratio, are surrogates for the
normally essential economic factors.

MARINE MINERAL RESOURCES

Concealment by seawater does not call for the use of special criteria to decide whether an occurrence of a mineral should be described as a mineral resource. The same definition applies wherever minerals are found.

Unfortunately, this has not always been followed. No one has benefited from the misuse of the term and exaggerated estimates of the "resources" beyond the continental shelf given publicity shortly before Ambassador Pardo's initiative in 1967, which were largely responsible for serious misunderstandings at the U.N. Law of the Sea Conference.

What are marine mineral resources? They are mineral occurrences below the sea which are either economically workable now or are likely to become economically workable within the next 20 or 30 years.

It is quite simple, really, and important, notwithstanding which it would be unsafe to assume that abuse of the term will not continue.

REFERENCES

Archer, A. A., 1969, Background and problems of an assessment of sand and gravel resources in the United Kingdom; Proc. 9th Commonwealth Mining and Metallur. Cong., Inst. Mining and Metallurgy, London, v. 2, pp.495-508.

Blondel, F. and Lasky, S. G., 1956, Mineral reserves and mineral resources; Economic Geology, v. 51, no. 7, pp. 686-697.

Brobst, D. A., 1979, Fundamental concepts for the analysis of resource availability; In: Smith V. K. (ed.) Scarcity and growth reconsidered; Resour. Future, Johns Hopkins Univ. Press, Baltimore, pp. 106-142.

Institution of Mining and Metallurgy, 1984, Royal Charter and Bye-laws; 31 p.

House of Lords, 1982, Strategic minerals; Select Commm. on the European Communities, Session 1981-82, 20th rept., Her Majesty's Stat.Off., London, pp. 17, 31 et seq., 295.

Howard, P., 1984, The state of the language; Hamish Hamilton, London, 44 p.

McKelvey, V. E., 1972, Mineral resource estimates and public policy; Amer. Scientist, v. 60, pp. 32-40.

McKelvey, V. E., 1974, Potential mineral reserves; Resources Policy, v. 1, no. 2, pp. 75-81.

United Nations, 1979, The international classification of mineral
    resources; Econ. and Social Coun., Comm. Nat. Resour., 6th
    Session, Rept. E/C. 7/104, pp. 1-13. (see also: Schanz, J.
    J., 1980, The United Nations endeavor to standardize mineral
    resource classification; Nat. Resour. Forum, v. 4, no. 3, pp.
    307-313.

U.S. Bureau of Mines and U.S. Geological Survey, 1948, Mineral
    resources of the United States; Public Affairs Press,
    Washington, D.C., 212 p. (see also: App. to the Hearings
    before a Subcomm. of the Comm. on Public Lands, U.S. Senate, 80th
    Cong. Washington, 1947).

U.S. Bureau of Mines and U.S. Geological Survey, 1976, Principles of
    mineral resources classification system of the U.S. Bureau of
    Mines and the U.S. Geological Survey; U.S. Geol. Surv. Bull.
    1450-A, 5 p.

U.S. Bureau of Mines and U.S. Geological Survey, 1980, Principles of
    resource/reserve classification for minerals; U.S. Geol. Surv.
    Circ. 831, 5 p.

Zwartendyk, J., 1972, What is 'mineral endowment' and how should we
    measure it?, Dept. Energy, Mines and Resources, Ottawa, Rept.
    MR-126.

Zwartendyk, J., 1974, The life index of mineral resources: a
    statistical mirage; Can. Inst. Mining Bull., v. 76, no. 750,
    pp.67-70.

Zwartendyk, J., 1981, Economic issues in mineral resource adequacy
    and in the long-term supply of minerals; Economic Geology, v.
    76, no.5, pp.999-1005.

THOUGHTS ON APPRAISING MARINE MINERAL RESOURCES

DeVerle P. Harris
Department of Mining and Geological Engineering
The University of Arizona
Tucson, Arizona 85721   USA

ABSTRACT. This paper shows that basic principles developed for
estimating mineral endowment and potential mineral supply of a land
region apply also to such estimation for a marine region. However,
estimation methods for marine minerals may differ from those for land
minerals as a consequence of differences in modes of occurrence and in
available information. Because many mineral endowment measures, e.g.,
thickness of ferromanganese crust and concentration of cobalt, are
correlated, as also are bathymetric and geologic measurements, the
multivariate statistical methods of factor, regression, and correspondence analyses may be especially useful in the estimation of the metal
endowment of a marine region. This paper examines the use of these
methods, with cobalt in ferromanganese crusts serving as a reference.
Although the paper deals primarily with the use of multivariate
statistical relations to estimate cobalt endowment, some consideration
also is given to the use of these relations within a potential supply
system that is designed to estimate the amount of metal that is
discoverable and producible for specific economic circumstances.

INTRODUCTION

Scope and Perspective

The objectives of this paper are twofold:

* Identify parallels in concepts and practices of appraising
  land mineral resources with appraising marine mineral
  resources.

* Examine in some detail procedures for estimating mineral
  endowment for a hypothetical case: a region having bathymetric information but only partial sample information on
  the ferromanganese crust, contained metals, and specific
  geology.

P. G. Teleki et al. (eds.), Marine Minerals, 433–466.
© 1987 by D. Reidel Publishing Company.

   This paper is not intended to be a statement about the genesis and occurrence characteristics of ferromanganese nodules and crusts. To the contrary, the author acknowledges his naïveté on these matters and focuses instead on the use of the available meager information to estimate mineral endowment and potential mineral supply.

   Although economics, geology, and technology of production differ markedly for land versus marine deposits, many principles for mineral resources appraisal apply equally well to both kinds of resources; however, differences in the kinds of geological information may lead to some differences in endowment estimation methodologies.

Communication Needs

   One lesson from the last decade of experience is that mineral resources terms are used variously to refer to different things. Based upon the persistence of nonuniform usage, it appears likely that nonuniformity of terminology will remain. Thus, if clarity of communication is to be achieved, every author should define his terms precisely. The necessity of this increases sharply when dealing with quantitative descriptions of mineral resources. For any estimate to be useful, the estimator first must have an unambiguous view of that thing being estimated. Second, the estimator must convey to the user precisely what has been estimated. It is in the spirit of these needs that the definitions of the following section are presented. Mathematical symbols and relations are defined within the body of the paper. Even so, they also are summarized in Table I for easy reference.

Table I.   Definitions of Mathematical Symbols and Notation

B = universal set of NM metal occurrences within the crust of the earth or region to be appraised.

K_i = set of all characteristics that define the ith metal occurrence.

Z_i = a subset of K_i consisting of those NCM $\leq$ NC characteristics that are to be used in partitioning set B into sets $\bar{D}$ and D: $B = D \cup \bar{D}$.

D = a subset of B that consists of the metal occurrences that comprise metal endowment of the region.

m = metal endowment, which is the sum of metal contained in all of the NMM (NMM $\leq$ NM) metal occurrences in set D.

E = set of NE economic factors, e.g., product (metal) price, that affect the economic value of a metal occurrence.

V^D = a set of net present values formed from D by placing in V^D the net present value of the ith occurrence, given Z_i, $i = 1,2,\ldots,$NMM and E.

R = a set of NR (NR $\leq$ NMM) metal occurrences formed from D by placing in R each element of D for which the corresponding element in V^D is greater than or equal to zero. In other words, set R contains all metal occurrences that could be produced profitably, given economic circumstances of E:

$$v_i \geq 0 \text{ for all } r_i \; \varepsilon \; R$$
$$v_i \; \varepsilon \; \bar{R}, \text{ otherwise.}$$

u   = quantity of metal resources, given economic circumstances of E. Specifically, u is the sum of metal in all occurrences belonging to set R.

R' = a set of metal occurrences formed from D, given economic circumstances E'.

R^d = a set of metal occurrences formed from R by selecting those elements from R that also could be discovered economically.

p   = potential supply (stock), which is the sum of metal in all metal occurrences belonging to set R^d.

s_t = supply (flow) during year t.

n   = number of deposits.

N   = random variable for number of deposits.

t   = deposit tonnage (mineralized material).

T   = random variable for deposit tonnage.

$\bar{q}$ = deposit average grade.

$Q,\bar{Q}$ = random variables for grade and average grade, respectively.

δ = intradeposit grade variance.

Δ = random variable for intradeposit grade variance.

H,h = random variable and a specific value, respectively, for depth to deposit.

μ,δ = parameters of distribution of block grades within a deposit, given that the logarithm of block grades is normally distributed: $\ln Q \backsim N(\mu,\delta)$

$\ln\bar{q}$ = $\mu +\frac{\delta}{2}$ is a relationship between arithmetic average grade and the parameters of the underlying lognormal distribution.

$t_{q'}$ = tonnage of ore in blocks having grades above cutoff grade, q'.

$\bar{q}_{q'}$ = average grade of all ore in blocks having grades above q'.

G   = a set or vector of geological variables.

$\eta(n;q',G)$ = probability function for number of deposits; conditional upon cutoff grade (q') and geology (G).

$\tau(t;q',G)$, $j(\bar{q};q',G)$ = probability density functions for deposit tonnage (mineralized material) and average grade, respectively, given q' and G.

$\ell(\delta;q',t)$ = probability density function for intradeposit grade variance, given q' and t.

X   = a set of eight measurements on the location of the cell and its bathymetric features: $X = \{x_1, \ldots, x_8\}$.

X_4 = $\{x_4, \ldots, x_8\}$ = bathymetric features only.

Y   = a set of nine measurements of geological features.

M   = a set of seven endowment features.

M^* = a set of seven endowment features which are orthogonal, i.e., uncorrelated. M^* is a transformation of M.

$\hat{M},\hat{M}^*$ = estimates of M and M^*, respectively.

Y^* = a set of w measurements of geological features which are orthogonal, $w \leq 9$.

$\hat{Y},\hat{Y}^*$ = estimates of Y and Y^*, respectively.

$\tilde{Y}_1,\tilde{Y}_1^*$ = random variables for the first elements of Y and Y^*, respectively.[†]

$\tilde{M}_1,\tilde{M}_1^*$ = random variables for the first elements of M and M^*, respectively.[†]

$E[\tilde{Y}_1|x_2,x_3,x_4]$ = the expected (mean) value of random variable Y_1, given states x_2,x_3, and x_4.

$t^{(\alpha)}$ = the value of a standardized random variable that is distributed
according to the Student "t" distribution having α degrees of
freedom.

---

†By convention, mathematicians use upper case to refer to a random
value of the variable and lower case to refer to a specific value.
For example $P(Y_1 = y_1)$ represents the probability that the random
variable Y_1 will take on the value of y_1.

MINERAL RESOURCES CONCEPTS AND TERMS

Consider, for convenience, at a given point in time, a single
metallic element in a single region of the earth, so that notationally
we may ignore time, metal varieties, and places. Our experience has
shown that the ultimate deposit of this metal in that region of the
earth consists of many smaller deposits occurring in varied geologic
environments and possessing various characteristics of grade, size,
shape, mineralogy, depth, host rock, and other attributes.

Suppose that there are NM of these occurrences and that they
constitute set B:

$$B = \{r_1, \ldots, r_{NM}\} \tag{1}$$

Let us represent our knowledge about the ith member of set B by a set,
K_i, of NC characteristics:

$$K_i = \{k_{i1}, \ldots, k_{i,NC}\} \tag{2}$$

The set of NC characteristics includes all properties of the NM metal
occurrences.

Suppose that a subset of NCM characteristics, NCM $\leq$ NC, is
identified for the purpose of partitioning set B into two subsets, D
and $\bar{D}$: $B = D \cup \bar{D}$. This subset of characteristics is $Z_i = \{k_{i,1}, \ldots, k_{i,NCM}\}$,
$Z_i \subset K_i$. Consider Z_i' to be a set of cutoff (minimum acceptable) values
for the NCM characteristics. Then,

$$\begin{aligned} Z_i > Z' &\rightarrow r_i \,\varepsilon\, D;^† \\ r_i &\,\varepsilon\, \bar{D}, \text{ otherwise.} \end{aligned} \tag{3}$$

This general notation can represent undesirable characteristics as well
as those that are desirable, provided that each characteristic is
appropriately defined. Consider, for example, that NCM = 3 and that
the three characteristics are grade, tonnage, and depth. Then, define
k_{i1} = grade, k_{i2} = tonnage, and $k_{i3} = \dfrac{1}{\text{depth}}$. Suppose that minimum
acceptable grade and tonnage were 0.10% and 10 tons of mineralized
material, respectively, and that maximum depth were 2000 ft. Then,
$k_1' = 0.10$; $k_2' = 10$; $k_3' = 0.0005$; and $Z' = \{0.10, 10, 0.0005\}$. The

---

†The symbol $\rightarrow$ means implies: $X \rightarrow Y$ is interpreted as X implies Y.

quantity of metal in all elements of D is defined as metal endowment, m:

$$m = \sum_{D} \gamma(r_i)^{\dagger} \tag{4}$$

Suppose that a function, f, is known, which for specified economic and technological conditions, describes the present value, v_i, for each $r_i \in D$:

$$v_i = f(k_{i,1}, k_{i,2}, \ldots, k_{i,NCM}; e_1, e_2, \ldots, e_{NE}), \tag{5}$$

or in vector notation, $v_i = f(Z_i, E)$

where Z_i is the set of NCM characteristics of the ith deposit, as
           previously described;
       E   is the set of NE economic factors:

$$E = \{e_1, e_2, \ldots, e_{NE}\} \tag{6}$$

The set E includes operating costs, capital costs, prices, rate of return, etc. Naturally some of these factors reflect the states of technologies.

Suppose that values of the NE economic conditions for currently feasible and near-feasible technology are specified. While these values must reflect currently feasible and near-feasible technology, some of them, such as product price, need not be those that currently prevail. Given set D and the function f, the present value for each $r_i \in D$ can be computed, giving rise to set V^D:

$$V^D = \{v_1, v_2, \ldots, v_{NMM}\} \tag{7}$$

where   NMM are the number of metallizations in D;
        NMM $\leq$ NM.

By selecting all v_i in V^D that are greater than or equal to zero, the set of NR metallizations that constitute R is formed from D:

$$R = \{r_1, r_2, \ldots, r_{NR}\} \tag{8}$$

where $r_i \in R \rightarrow v_i \geq 0$
        $r_i \in \bar{R}$, otherwise
        NR $\leq$ NMM $\leq$ NM.

Then, quantity of resources, u, given E, is defined as follows:

$$u = \sum_{R} \gamma(r_i) \tag{9}$$

$\dagger$The function γ maps the characteristics of the ith deposit, r_i, into m_i, which is quantity of metal in deposit r_i. $\bar{D}$ means "not D". $D \cup \bar{D}$ means union sets of D and $\bar{D}$. $r_i \in \bar{D}$ means r_i belongs to set $\bar{D}$.

Thus, given the perspective employed here, $R \subset D \subset B$; and
$u \leq m \leq$ resource base.

Suppose that a set $V^{R'}$ is formed by specifying fully the currently
prevailing economic conditions and technology.[†] Then a different set,
R', can be formed from R by selecting all $r_i \epsilon R'$ for which $v_i \geq 0$ for
the current status of economics and technology:

$$R' = \{r_1, r_2, \ldots, r_{NER}\} \tag{10}$$

where $R = R' \cup \bar{R}'$
        $r_i \epsilon R' \rightarrow v_i \geq 0$
        $r_i \epsilon \bar{R}'$, otherwise
        $NER \leq NR \leq NMM \leq NM$.

Let us designate u' as the quantity of metal in R' and u'' as the
quantity of metal in $\bar{R}'$. Then

$$u' = \sum_{R'} \gamma(r_i) \tag{11}$$

Similarly,

$$u'' = \sum_{\bar{R}'} \gamma(r_i) \tag{12}$$

The quantity u' is economic resources, and u'' is subeconomic
resources. Thus, $R' \subset R \subset D \subset B$; $u' \leq u \leq m \leq$ resource base; and
$u = u' + u''$.

Potential supply, p, can be formed directly from R. Consider set
R, which contains those metal occurrences which would be economic to
produce for specified economic conditions and currently feasible or
near-feasible technology, if the occurrences were known. Suppose that
R were partitioned to R^d and $\bar{R}^d$ such that R^d contains those metal
occurrences of R that would be discovered by the optimum exploration
effort, EX^*:

$$R = R^d \cup \bar{R}^d \tag{13}$$

where $r_i \epsilon R^d \rightarrow v_i - c_i \geq 0$ and discovery
        $r_i \epsilon \bar{R}^d$, otherwise.

Then, c_i = the share of EX^* for the ith metal occurrence that was
discovered by EX^*

and,

$$p = \sum_{R^d} \gamma(r_i). \tag{14}$$

[†]Here, it is understood that E', which is associated with R', is less
favorable than E, which is associated with R, making $R' \subset R$.

Thus, $R^d \subset R \subset D \subset B$ and $p \leq u \leq m \leq$ resource base.

The concept of optimum level of EX rests upon the requirement that all deposits contributing to potential supply must be of such a quality that their exploitation ·covers production costs and the costs of discovering them. Allocating to each deposit discovered its share of the exploration effort, EX, gives a net present value (net of exploration and production costs). Naturally, increasing EX to a higher level than EX* discovers more deposits, but since EX is charged against only those deposits discovered, increasing EX beyond EX* loses more economic and discoverable resources than can be gained. This occurs, because at any progression in the optimizing path of exploration the deposits which remain to be discovered require a greater expenditure per unit of resources than those already discovered, and because when exploration and exploitation interact with endowment in a "one-shot" or "single contract" kind of optimization -- and the sequential timing of incremental exploration expenditures is suppressed -- there are no "sunk" exploration costs. Therefore, diminishing returns to exploration, on account of the greater difficulty of discovery of progressively larger fractions of the endowment, means that exploration costs, allocated to the deposits discovered, increase, while additional deposits are discovered by a greater expenditure. Since the set R^d consists of only those occurrences that would be discovered by EX and would be economic to produce, when all exploration and exploitation costs are considered, then there is an optimum level of EX, which is EX*.

If we now relax the assumption of unlimited markets, but invoke the assumptions that a single, large firm would seek to maintain prices and that there is no technological change across time, then exploration would be spread across time as warranted by demand and depletion. Thus, at any moment, EX_t would be less than or equal to EX*; consequently, the set of known and exploited deposits, $\tilde{R}^d_t$, would be some subset of R^d:

$$R^d = \tilde{R}^d_t \cup \bar{\tilde{R}}^d_t \tag{15}$$

Of course, the sum of metal in deposits of $\tilde{R}^d_t$ must be less than or equal to the sum of metal in deposits belonging to R^d.

$$\sum_{\tilde{R}^d_t} \gamma(r_i) \leq \sum_{R^d} \gamma(r_i) \tag{16}$$

Let us designate this sum as a stock measure of supply, s_t:

$$s_t = \sum_{\tilde{R}^d_t} \gamma(r_i) \tag{17}$$

Then, at any moment in our simplified and hypothetical world, supply is less than or equal to potential supply:

$$s_t \leq p \tag{18}$$

Obviously,

$$\text{limit } (s_t) = p \tag{19}$$
$$t \to \infty$$

In terms of an individual area and a single metallic ore type hypothe-
sized here, the more intensely the area has been explored the more
closely s_t will approach p.   The relationships of the stock terms
from resource base to supply can be summarized as follows:

$$s_t \le p \le u \le m \le \text{ resource base } \to R_t^d \subset R^d \subset R \subset D \subset B. \tag{20}$$

GENERAL INFORMATION NEEDS AND APPRAISAL PROCEDURES -- LESSONS AND EXTENSIONS FROM EXPERIENCE WITH LAND MINERAL RESOURCES

Geological Estimation of Endowment -- A Necessity

Because of the lack of an economic history of exploration for and
production of marine minerals, potential supply estimation must be
based upon indirect, geological inference about mineral endowment.
Such a methodology takes as an input the probabilistic description of
mineral endowment (see Figure 1) and performs the economic evaluation
of this endowment as a means of estimating potential supply.   Clearly,
the foundation of potential supply estimation is geological inference
about the quantity and quality of the mineral endowment.   Generally,
inference about mineral endowment can be made by several methods,
including multivariate statistical models, quantitative geologic
analysis combined with subjective probability, and computer expert
systems.   However, for the circumstance cited as reference for this
paper, cobalt in ferromanganese crusts, multivariate statistical models
appear to be particularly well suited to the task of estimating cobalt
endowment in this mode of occurrence when data are very limited.

Establishing Endowment Cutoff Criteria

Irrespective of whether estimation is by subjective (qualitative)
consideration of the geological information or by strict quantitative
analysis, it is necessary to identify the elements comprising the
endowment stock by specifying endowment cutoff criteria.   For example,
cutoff criteria for crusts could include the following:

> mode of occurrence, e.g., cobalt in ferromanganese crusts
> minimum crust thickness
> minimum concentration (grade) of the element in crusts
> minimum density of crusts, i.e., percent of a unit of
>   sea floor or seamount area covered by crusts
> maximum water depth

Establishing these criteria facilitates estimation by removing from
consideration the numerous, widespread very small occurrences, focusing estimation on occurrences which are more compatible with geological
experience and with available data on known occurrences.

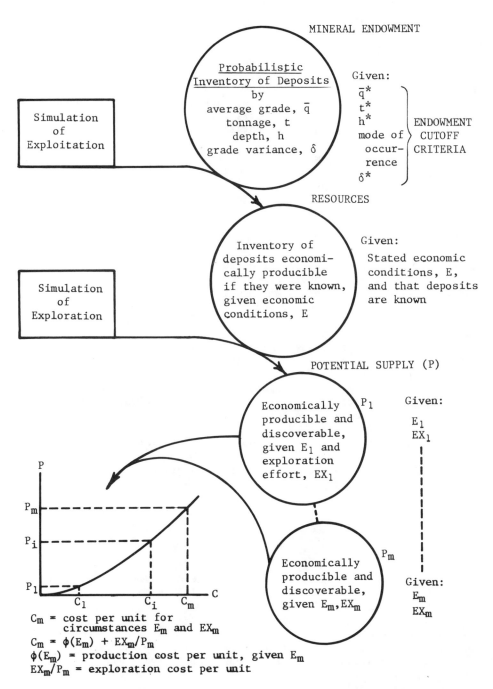

Figure 1.   Schematic representation of a potential supply system.
            (After Harris, 1978)

Essential Descriptors of Mineral Endowment

Land Mineral Deposits
    Identification of descriptors which are essential is not
independent of the objective of the endowment appraisal. Here, it is
assumed that the objective is to support the estimation of potential or
dynamic supply. Given that objective, there are, for a given type of a
deposit (mode of occurrence), at least five fundamental and essential
descriptors for the region to be appraised:

n	number of deposits in the region
t	tonnage of mineralized material per deposit
$\bar{q}$	average grade of mineralized material per deposit
δ	intradeposit grade variance
h	depth to deposit

    The selection of n, t, and $\bar{q}$ needs no explanation, because together
they describe the total metal endowment. Descriptors δ and h are
needed to model exploration and production and to perform economic
analysis. Clearly, for specified technology, factor costs, and product
price, depth to deposit (h) is an important determinant of potential
supply. The greatest effect of depth is on exploration, where it
influences effectiveness, i.e., probability, of discovery and discovery
costs. Depth also affects mining costs, particularly of the small
deposits. Intradeposit grade variance describes the variability of
grade within the deposit; therefore, it is especially important in providing a means for modeling the mine optimization decision, i.e., the
selection of cutoff grade, throughput, life, capital investment, and
operating costs. Without this descriptor, a deposit must be considered
to be of homogeneous grade, $\bar{q}$. For example, suppose a deposit of size
t, average grade $\bar{q}$, and intradeposit grade variance δ is expected to
occur. Having an estimate of δ allows description of the distribution
of grades within the deposit.
    Given t, $\bar{q}$, δ, and that Q is lognormally distributed,

$$\mu = \ln(\bar{q}) - \frac{\delta}{2}^{\dagger} \tag{21}$$

and

$$\ln Q \backsim N(\mu, \delta). \tag{22}$$

Therefore, given cutoff grade q´, the tonnage of material having grades
larger than q´, $t_{q'}$, and the average grade of this tonnage, $\bar{q}_{q'}$, can be
routinely computed (see Harris, 1984a for derivations):

----

†When $\ln Q \backsim N(\mu, \delta)$, the mean grade, $\bar{q}$, is related to the parameters
 μ and δ, as follows (see Harris, 1984a for derivation):

$$\bar{q} = e^{\mu + \delta/2}$$

$$t_{q'} = t \cdot \int_{lnq'}^{\infty} \frac{1}{\sqrt{2\pi\delta}} e^{-\frac{1}{2}(lnq - \mu)^2/\delta} \, dlnq \qquad (23)$$

$$\bar{q}_{q'} = e^{\mu+\delta/2} \left[\frac{\int_{lnq'}^{\infty} \frac{1}{\sqrt{2\pi\delta}} e^{-\frac{1}{2}(lnq - \mu - \delta)^2/\delta} \, dlnq}{\int_{lnq'}^{\infty} \frac{1}{\sqrt{2\pi\delta}} e^{-\frac{1}{2}(lnq - \mu)^2/\delta} \, dlnq} \right] \qquad (24)$$

These relations make it possible to search for that cutoff grade that maximizes net present value for the deposit.

Although these descriptors clearly are essential for potential supply analysis, many previous studies have not included δ, the intra-deposit grade variance. One reason for this neglect is the limited amount of data available on δ. Data required to estimate δ are grades of ore blocks for each deposit; such data usually are not available to individuals other than personnel of the mining firm. In the absence of such data, attempts have been made to estimate δ by indirect means for certain types of deposits (Agterberg, 1982; Charles River Associates, 1978; Harris, 1984b).

Typically, one or more of these descriptors is not known with certainty for the region to be appraised, the extreme case being a frontier region for which there is no direct information about any of the descriptors. For this circumstance, the descriptors are random variables, the states of which are appropriately described by probability distributions. Estimation of mineral endowment is the estimation of the probability distributions for the endowment variables. Such estimation must consider the presence of dependencies among variables.

Most appraisals have used only three variables: n, t, and $\bar{q}$. Apart from the crustal abundance model for uranium, constructed by Harris, et al. (1981), n, the number of deposits, has been considered to be statistically independent of tonnage and grade per deposit (t,q). However, several studies have allowed for dependency of t with q (Harris, 1973; Harris, 1984b; Singer and Ovenshine, 1979).

Marine Mineral Deposits

Endowment for many types of mineral deposits on land is well represented by a number-of-deposits distribution, because the deposits comprising endowment occur naturally as discrete grade or mineralogic anomalies within a lithologic fabric, within which like mineralization is absent or is extremely rare. Additionally, these mineral deposits seldom occur in dimensions too large to be considered as an economic unit. These conditions together make a number-of-deposits distribution both well suited to endowment characterization and useful support for the modeling of exploration and production in a potential supply system.

Crust and nodule deposits may not be as well represented by a number-of-deposits distribution as are many of the mineral deposits on land. Perhaps, these deposits can be considered as discrete phenomena if viewed from a distant and broad perspective. Accordingly, a single deposit would cover a very large area. Consequently, even if such a representation were appropriate, it would be of limited usefulness if thickness and grade vary in systematic ways within this large deposit. When the deposit is widespread geographically and relatively continuous, as are some ferromanganese crusts, economic analysis requires that a single very large geologic deposit be considered as the coalescence of many smaller anomalies of thickness and concentration. In other words, the number of identified deposits within this large geologic deposit is a function of the cutoff values for grade and thickness. Similarly, both tonnage and grade per deposit are functions of these cutoff values. A circumstance similar to this was encountered in the sandstone uranium deposits of the San Juan Basin of New Mexico (Harris, et al., 1981; Harris and Chavez, 1984). What earlier had been viewed as many small, discrete deposits was later found to be simply many grade anomalies in one or two very large geologic deposits. Figure 2 is a schematic illustration of how the number of deposits and the distributions of size and grade per deposit vary with cutoff grade. In concept, such an occurrence is represented by making the probabilities for n, t, q, and δ conditional upon cutoff grade, or equivalently treating cutoff grade (q´) and geology (G) as parameters of the probability functions:

$$\eta(n;q´,G)$$
$$\tau(t;q´,G)$$
$$j(\bar{q};q´,G)$$
$$\ell(\delta;q´,t)$$

Here, N(number of deposits), T(deposit tonnage), and $\bar{Q}$ (deposit average grade) are random variables whose states $(n,t,\bar{q})$ are conditional upon geology (G) and cutoff grade (q´). And Δ (intradeposit grade variance) is a random variable whose state (δ) is conditional upon deposit size and cutoff grade.

APPRAISING COBALT ENDOWMENT IN FERROMANGANESE CRUSTS--A HYPOTHETICAL CASE

Motivation

Parameterizing the probability distributions for endowment variables on cutoff grade may be conceptually useful to endowment assessment, but it is not very informative about how this may be done in practice. Having not had the time, resources, or data to conduct an actual endowment estimation for ferromanganese crusts, the author has no actual case study that elaborates upon these generalities regarding methodology. Consequently, a hypothetical situation is used to explore some particulars of estimation.

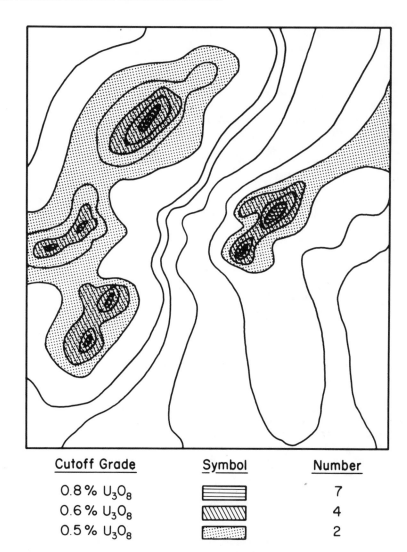

Cutoff Grade	Symbol	Number
0.8 % U_3O_8		7
0.6 % U_3O_8		4
0.5 % U_3O_8		2

Scale: 1" = 10,000'

Figure 2. Schematic diagram showing both number of uranium deposits and size of deposit to be functions of cutoff grade. (After Harris and Chavez, 1984)

The Setting

Consider the following hypothetical circumstances:

The general outline of a region permissive to the
formation of ferromanganese crusts has been determined.

The region (study area) to be appraised has been subdivided by an equi-spaced grid into 50 subregions, here
referred to as cells. Figure 3 depicts the study area
and those cells for which endowment information is
available.

Observations have been made from a bathymetric map for
all cells (square subdivisions of the region--see Figure 3)
on locational parameters (x_1, x_2, x_3) and on five bathymetric
variables $(x_4, x_5, x_6, x_7, x_8)$ -- see variable set X of Table II
for variable identification.

For a small number of cells, sampling activities, e.g.,
dredge and bottom photography, have yielded observations on
nine specific geologic conditions $(y_1, \ldots, y_9)$ and endowment features $(m_1, \ldots, m_7)$, e.g., crust occurrence, crust
thickness, crust density.

For some cells, observations have been made on both X and
Y, but not on M.

The observations identified in Table II are not presented as those
that should be made. Rather they are hastily contrived after a quick
perusal of a limited amount of literature. About all that is offered
in their defense is that they are sufficiently realistic to serve as a
vehicle for exploring resource estimation methodology--see Craig, et al.
1982 and Clark, et al. 1984.

Perspective on Estimation Methodology

It is useful at this point to reflect on the objective and the
overall approach in general before considering the details of
methodology. The fundamental objective is to describe the endowment of
the region in cobalt and associated metals. Given that cell area is
constant, an estimate of cobalt endowment in a sampled cell is simply

$$Co = m_1 \cdot (m_6 * 10^{-6}) \cdot m_2 \cdot m_7 \cdot \Gamma/W \tag{25}$$

where Γ = area in cm^2
         W = grams/metric ton
         Co = metric tons of cobalt.

Endowment of the cell in copper and manganese would be computed
similarly, using m_5 and m_3, respectively. Of course, only a fraction

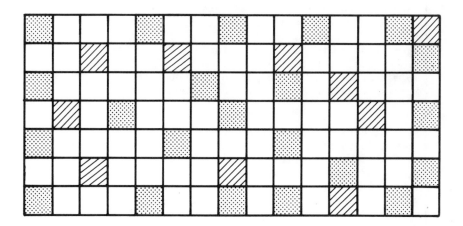

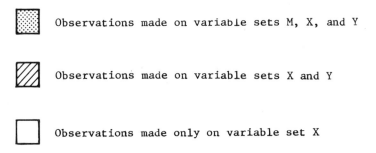

Observations made on variable sets M, X, and Y

Observations made on variable sets X and Y

Observations made only on variable set X

Figure 3. Grid network of cells for the hypothetical appraisal area, showing cells having sample information.

Table II.   Identity of Observations

$$
X \begin{cases}
x_1 = \text{cell number} \\
x_2 = \text{N-S coordinate of cell center} \\
x_3 = \text{E-W coordinate of cell center} \\
x_4 = \text{water depth to seafloor} \\
x_5 = \text{water depth to seamount top} \\
x_6 = \text{water depth to seamount slope} \\
x_7 = \% \text{ of cell area occupied by seamount top} \\
x_8 = \% \text{ of cell area occupied by seamount slope}
\end{cases}
$$

$$
Y \begin{cases}
y_1 = \text{direction of prevailing current} \\
y_2 = 1 \text{ if hydrothermal activity is present, 0 otherwise} \\
y_3 = 1 \text{ if sedimentation site, 0 otherwise} \\
y_4 = 1 \text{ if substrate is fresh basalt, 0 otherwise} \\
y_5 = 1 \text{ if substrate is weathered basalt, 0 otherwise} \\
y_6 = 1 \text{ if substrate is tuff, 0 otherwise} \\
y_7 = 1 \text{ if substrate is pumice, 0 otherwise} \\
y_8 = 1 \text{ if substrate is coral, 0 otherwise} \\
y_9 = 1 \text{ if substrate is volcanic conglomerate, 0 otherwise}
\end{cases}
$$

$$
M \begin{cases}
m_1 = \text{thickness of crust (cm)} \\
m_2 = \text{dry weight density of crust (g/cm}^3\text{)} \\
m_3 = \text{PPM manganese} \\
m_4 = \text{PPM iron} \\
m_5 = \text{PPM copper} \\
m_6 = \text{PPM cobalt} \\
m_7 = \text{fraction of sample area comprised by crust}
\end{cases}
$$

of the total cells was sampled; therefore, estimating the endowment of
a metal, say cobalt, for the entire region requires that the endowments
of cells not sampled be estimated. While this may be done in various
ways, *the preferred approach to estimation would be one which utilizes
all relevant information, and provides a probabilistic description of
the endowments for all cells.* An immediately appealing approach is to
develop, on the sampled cells, a relationship between a metal endowment
variable, say m_1, and the X and Y variables of Table II. The value of
m_1 for a cell sampled for X and Y but not for M could be estimated by
evaluating the relationship (determined on the sampled cells) on the X
and Y. For cells not sampled for Y, the states of the Y variables first
would have to be estimated; then the statistical relationship of X and
Y to m_1 would be evaluated on the X and estimated Y of each cell.

The approach in principle has merit; however, to the extent that
correlations exist among the variables within each set of variables (X,
Y, and M), it must be implemented by appropriate statistical methods.
For example, m_1, thickness of crust, may be correlated to concentration
of cobalt, m_6, and the fraction of the sample area comprised by crust,
m_7. Furthermore, concentrations of the various elements (Mn, Fe, Co,
Cu) may be correlated. For such circumstances, it would be inappro-
priate to estimate each of the M variables separately, because such
estimation ignores correlations and may not be consistent. Instead,
the set of seven variables in M should be factored to a set of seven
orthogonal variables (M^*). This could be accomplished by submitting
the data on these variables for all sampled cells to factor analysis,
in which communalities are set at 1.0 and as many factors are extracted
as there are variables.[†] Such analysis would yield seven factor
equations:

$$m_1^* = a_{11}z_1 + \ldots + a_{17}z_7$$
$$\vdots$$
$$m_7^* = a_{71}z_1 + \ldots + a_{77}z_7 \tag{26}$$

where a_{ij} are coefficients determined by factor analysis of
             data on M

$$z_i = \frac{m_i - \mu_i}{\sigma_i} = \text{standardized score} \tag{27}$$

[†]Principal factor extraction with all communalities set to 1.0 is also
referred to as principal components analysis. Imposing the condition
that the number of factors equals the number of variables runs counter
to one common use of factor analysis, which is to reduce the dimensions
of the problem. Even so, for the application considered in this paper
equality is necessary to permit the later inverse transformation from
factor scores to original measurements. Since the writing of this
paper, it was pointed out to the author by Dr. Donald Myers of the
Department of Statistics, University of Arizona, who reviewed the
paper, that if correspondence analysis is used in place of factor
analysis, equality of factor and variable space dimensions is not
necessary to assure an inverse transformation.

μ_i, σ_i would be replaced by their estimates:

$$\hat{\mu}_i = \bar{m}_i; \left(\frac{\sqrt{n}}{\sqrt{n-1}} \right) s_i = \hat{\sigma}_i,$$

> where $\bar{m}_i$ and s_i are sample mean and standard deviation, respectively, for the ith endowment variable.

In matrix notation,

$$M^* = A \cdot Z. \tag{28}$$

Estimation of Endowment When X_4 and Y Have Been Observed

Given the factor equations and data on M, the seven orginal observations $(m_1, \ldots, m_7)$ for each of the sampled cells can be replaced by seven factor scores, $m_1^*, \ldots, m_7^*$, each of which can be considered to be statistically independent. Therefore, using the data from the sampled cells for X_4 and Y, regression analysis could be employed to yield a statistical relationship between each of the M^* variables and the X_4 and Y variables:

$$\hat{m}_1^* = \hat{\beta}_{10} + \hat{\beta}_{14}x_4 + \ldots + \hat{\beta}_{18}x_8 + \hat{\gamma}_{11}y_1 + \ldots + \hat{\gamma}_{19}y_9$$
$$\vdots$$
$$\hat{m}_7^* = \hat{\beta}_{70} + \hat{\beta}_{74}x_4 + \ldots + \hat{\beta}_{78}x_8 + \hat{\gamma}_{71}y_1 + \ldots + \hat{\gamma}_{79}y_9 \tag{29}$$

Alternatively, in matrix form

$$\hat{M}^* = \left[\hat{\boldsymbol{\beta}} \vdots \hat{\boldsymbol{\gamma}} \right] \cdot \left[\begin{matrix} X_4 \\ \cdots \\ Y \end{matrix} \right] \tag{30}$$

where $\hat{m}_i^*$ = an estimate of m_i^*, given X_4 and Y, $i = 1, \ldots, 7$
$$X_4 = \left[\begin{matrix} X_4 \\ \vdots \\ X_8 \end{matrix} \right].$$

Of course, some of these coefficients may be zero. Stepwise regression could be used to estimate the best equation for each m_i^* variable, which includes determining which coefficients for each equation should be set to zero.

The completion of the foregoing analysis would provide seven equations, one for each m_i^*. Given that X_4 and Y had been observed, an m_i^* would be estimated by substituting X_4 and Y into the ith equation. For such a circumstance, seven estimates would be produced for that cell: $\hat{m}_1^*, \ldots, \hat{m}_7^*$. These estimates would have little direct value for the estimation of cobalt endowment, as such estimation requires estimates for m_1, m_6, m_2, and m_7. However, estimates for these quantities can be obtained from the estimates of m_i^*, $i = 1, \ldots, 7$:

Given $\hat{M}^*$ and equation (28),

$$\hat{Z} = A^{-1} \hat{M}^* \tag{31}$$

and

$$\hat{m}_i = \hat{z}_i \hat{\sigma}_i + \hat{\mu}_i \tag{32}$$

where z_i is the ith element of Z, i.e., the standardized score for the ith endowment variable, m_i.

Subsequently, these estimates of the values of M could be combined to estimate amount of cobalt endowment of the cell according to equation (25).

Estimation of Endowment When Y Has Not Been Observed

Consider the hypothetical appraisal area and circumstances illustrated in Figure 3 and Table II. There are some cells for which M was not observed and both X and Y were observed.[†] Since information on X and Y extends beyond the cells sampled for M, one approach to estimating M for those cells not sampled is to employ the spatial distribution of observed Y and the relationship of observed Y to X to estimate Y on the cells for which Y was not observed; then, M^* is estimated for a cell not sampled for M or Y by evaluating equation (30) on its X and estimated Y.

In practice, implementation of this procedure is complicated by the need to estimate Y in a way that takes into account the correlations of the variables of Y. Furthermore, this estimation must also consider the relations between the variables of X and Y. Such estimation requires two major steps. The first one is analogous to the estimation of M^*. This step is estimation of Y^* on the sampled cells by subjecting the original data on Y to factor analysis, and computing for each sampled cell w factor scores, y_i^*, $i = 1, 2, \ldots, w$; $w \leq 9$:[††]

$$Y^* = D \cdot V \tag{33}$$

where

$$V = \begin{bmatrix} v_1 \\ \vdots \\ v_9 \end{bmatrix}, \quad v_i = \frac{y_i - \mu_{y_i}}{\sigma_{y_i}} \tag{34}$$

[†]Since X consists of locational and bathymetric descriptions, it is considered observable on all cells.

[††]Here, the number of factors need not be equal to the number of Y variables. Conventional methods of factor analysis can be used to reduce the factor space, as long as orthogonality of factors is preserved.

and   D = the matrix of coefficients for the factor equations,
       determined by factor analysis of Y:

$$y_i^* = d_{i1}v_1 + \ldots + d_{i9}v_9; \quad i = 1, \ldots, w \tag{35}$$

The original observations on Y for each sampled cell then could be
replaced by the set of factor scores, Y^*. Using data from sampled
cells only, an endowment relationship would be estimated by regression
analysis:

$$\hat{M}^* = \left[\hat{\beta} \vdots \hat{\delta}\right] \cdot \left[\begin{array}{c} X_4 \\ \cdots \\ Y^* \end{array}\right] \tag{36}$$

Equation (36), which is the basic relationship for the estimation
of M^*, is a function of X_4 and Y^*. Before this relationship could be
used to estimate M^* for cells for which Y is not observable, Y^* would
have to be estimated. This estimation would be based upon the relationship of Y^* to X_4, x_2, and x_3 determined on those cells for which Y is
observable. Specifically, w (w $\leq$ 9) regression equations are estimated,
one for each element of Y^*:

$$\hat{y}_1^* = \hat{\lambda}_{10} + \hat{\lambda}_{14}x_4 + \ldots + \hat{\lambda}_{18}x_8 + \hat{c}_{11}\phi_1(x_2,x_3) + \ldots + \hat{c}_{1m}\phi_m(x_2,x_3)$$

$$\cdot \quad \cdot \quad \cdot \quad \cdot \quad \cdot \quad \cdot \quad \cdot \quad \cdot \quad \cdot \quad \cdot \quad \cdot \quad \cdot \quad \cdot \quad \cdot \quad \cdot \quad \cdot$$

$$\hat{y}_w^* = \hat{\lambda}_{w0} + \hat{\lambda}_{w4}x_4 + \ldots + \hat{\lambda}_{w8}x_8 + \hat{c}_{w1}\phi_1(x_2,x_3) + \ldots + \hat{c}_{wm}\phi_m(x_2,x_3)$$

$$\tag{37}$$

Alternatively, in matrix notation

$$\hat{Y}^* = \left[\hat{\lambda} \vdots \hat{c}\right] \cdot \left[\begin{array}{c} X_4 \\ \cdots \\ \phi \end{array}\right] \tag{38}$$

where

$$X_4 = \left[\begin{array}{c} x_4 \\ \vdots \\ x_8 \end{array}\right], \quad \phi = \left[\begin{array}{c} \phi_1(x_2,x_3) \\ \cdots \\ \phi_m(x_2,x_3) \end{array}\right]^\dagger \tag{39}$$

$\phi_j(x_2,x_3)^\dagger$ is the jth function of the coordinates of location
(x_2 and x_3). The $\phi_j(x_2,x_3)$, $j = 1, \ldots, m$, express systematic trends
in the y_i^* values--across the sampled cells--that are not explained by
x_4, x_5, x_6, x_7 and x_8. They represent missing geological variables, i.e.,
geological variables needed to explain the distribution of y_i^* on the
sampled cells. Y^* for each nonsampled cell could be estimated by
evaluating equation (38) on values of X_4 and ϕ for each cell. In this
hypothetical case, X_4 and ϕ, which are bathymetric measurements and
functions of locational descriptors, respectively, would be known for
all cells. Given $\hat{Y}^*$'s for the cells not sampled, M^* could be estimated
for each of these cells by:

---

†This notation as used here represents interpolation relations in
general, including but not limited to polynomials in x_2 and x_3.

$$\hat{M}^* = \left[\hat{\beta} \vdots \hat{\delta}\right] \cdot \left[\begin{matrix} X_4 \\ \cdots \\ \hat{Y}^* \end{matrix}\right]$$ (40)

Then, given the estimated M^*, $\hat{M}$ could be computed from the factor equation (28):

$$\hat{Z} = A^{-1} \hat{M}^*$$ (41)

where

$$\hat{Z} = \left[\begin{matrix} \hat{z}_1 \\ \vdots \\ \hat{z}_7 \end{matrix}\right], \quad \text{and}$$

$$\hat{z}_i = \frac{\hat{m}_i - \hat{\mu}_{m_i}}{\hat{\sigma}_{m_i}}$$

Therefore,

$$\hat{m}_i = \hat{z}_i \cdot \hat{\sigma}_{m_i} + \hat{\mu}_{m_i} , \quad i = 1, 2, \ldots, 7$$

Finally, using M for sampled cells and $\hat{M}$ for nonsampled cells, cobalt endowment would be estimated for all cells by equation (25). Since this estimation procedure would produce estimates of the endowment variable set, M, for each cell, endowment of other elements associated with cobalt, say manganese, could be estimated readily by replacing m_6 with m_3 in the basic equation (25).

Recapitulation and Linkage

The foregoing estimation procedures seem to be far removed from the simple concept of mineral endowment being described by five variables $(h, n, t, \bar{q}, \delta)$ appropriately parameterized on geology, cutoff grade, or tonnage. However, the foregoing multivariate estimation procedure can be used as a means to estimate these probability relations. In order to demonstrate this linkage, endowment cutoff criteria must be established. In this hypothetical case, the basic reference scheme is a network of cells, each having equal area. Consequently, the endowment criterion of minimum tonnage could be replaced without significant loss of completeness by a minimum thickness of crust, m_1'. Additionally, let m_6' and m_7' be minimum concentration of cobalt and minimum extent of crust, respectively. For this demonstration, assume that water depth throughout the region is less than the maximum depth of interest, therefore, water depth in this case would not limit endowment.

Recall the basic relationship of the endowment variables to X_4 and Y when both X_4 and Y have been observed:

$$\hat{M}^* = \left[\hat{\beta} \vdots \hat{\gamma}\right] \cdot \left[\begin{matrix} X_4 \\ \cdots \\ Y \end{matrix}\right]$$

Select one equation from this system, that for m_1^*:

$$\hat{m}_1^* = \hat{\beta}_{10} + \hat{\beta}_{14}x_4 + \ldots + \hat{\beta}_{18}x_8 + \hat{\gamma}_{11}y_1 + \ldots + \hat{\gamma}_{19}y_9. \qquad (42)$$

The constant $(\hat{\beta}_{10})$ and coefficients $\hat{\beta}_{14}, \ldots, \hat{\beta}_{18}$ and $\hat{\gamma}_{11}, \ldots, \hat{\gamma}_{19}$ would have been estimated by regression analysis; thus, by substituting the values for X_4 and Y into this equation, a value for m_1^* would be estimated. This estimate $(\hat{m}_1^*)$ is referred to as the expected (mean) value of the conditional random variable $\tilde{M}_1^*$, given X_4 and Y:

$$\hat{m}_1^* = E[\tilde{M}_1^* \mid X_4, Y]$$

Let's define error in the estimate of m_1^* as the difference between the actual value, m_1^* and its conditional expectation:

$$m_1^* - E[\tilde{M}_1^* \mid X_4, Y] = m_1^* - \hat{m}_1^* = e_{m_1}$$

Thus,

$$m_1^* = \hat{m}_1^* + e_{m_1}$$

Suppose that e_{m_1} is normally distributed, having a mean of zero and a standard deviation of $\sigma_{e_{m_1}}$. Generally, $\sigma_{e_{m_1}}$ is not known, and we must use an estimate provided by regression analysis, $\hat{\sigma}_{e_{m_1}}$. Given these circumstances, e_{m_1} is the product of $\hat{\sigma}_{e_{m_1}}$ and a $t^{(\alpha)}$ distributed random variate--random variable from a standardized Student's "t" distribution having α degrees of freedom:

$$e_{m_1} = \hat{\sigma}_{e_{m_1}} \cdot t_1^{(\alpha)}$$

Therefore,

$$m_1^* = \hat{m}_1^* + \hat{\sigma}_{e_{m_1}} \cdot t_1^{(\alpha)} \qquad (43)$$

Equivalently,

$$m_1^* = \hat{\beta}_{10} + \hat{\beta}_{14}x_4 + \ldots + \hat{\beta}_{18}x_8 + \hat{\gamma}_{11}y_1 + \ldots + \hat{\gamma}_{19}y_9 + \hat{\sigma}_{e_{m_1}} \cdot t_1^{(\alpha)} \quad (44)$$

where α = degree of freedom = number of observations in sample - the number of parameters in the estimated equation.

Thus, by repeated random sampling for $t_1^{(\alpha)}$, a population of values for m_1^*, conditional upon X_4 and Y could be created.
    Consider the implications of the foregoing:
      - Regression analysis of data on m_1^*, X_4, and Y would
         yield estimates of the unknown parameters
         $(\beta_{10}, \beta_{14}, \ldots, \beta_{18}; \gamma_{11}, \ldots, \gamma_{19})$ of equation (42).
      - Thus, the expected value of $\tilde{M}_1^*$ for a cell for which m_1 would
         not have been observed could be estimated, given its
         observed values for X_4 and Y.
      - Furthermore, since regression analysis would provide an

estimate of the standard error of $\hat{m}_1^*$, it would be possible to estimate e_{m_1}, given X_4 and Y.

- Thus, equation (44) implies an entire population of values for m_1^*, one for each value of $t_1^{(\alpha)}$, where $t_1^{(\alpha)}$ is a random variable distributed according to the Student t-distribution for α degrees of freedom.
- An individual of this population could be obtained by Monte Carlo methods, i.e., drawing randomly from the $t^{(\alpha)}$-distribution, multiplying the drawn value by $\hat{\sigma}_{e_{m_1}}$ and adding the product to $\hat{m}_1^*$, as indicated in equations (43) and (44).

Returning to the equation for m_1, a value for m_1 for each cell having observed X_4 and Y would be generated by evaluating the equation on the observations and by adding an error term created by the sampling procedure just demonstrated. This same procedure would be employed to create values for m_2^*, ..., m_7^*. Finally, $m_1, m_2,$ ..., m_7 for each cell would be computed from the factor equation (4).

Consider now, those cells for which Y has not been observed. Estimation of M for these cells is compounded by first having to estimate Y. Recall the equation system for Y^*:

$$\hat{Y}^* = \left[\hat{\lambda} \,\vdots\, \hat{c}\right] \cdot \left[\begin{array}{c} X_4 \\ \hline \phi \end{array}\right]$$

Let us take the first equation of this system and add its error of estimation to define y_1^*:

$$y_1^* = \hat{\lambda}_{10} + \hat{\lambda}_{14}x_4 + \ldots + \hat{\lambda}_{18}x_8 + \hat{c}_{11}\phi_1(x_2,x_3) + \ldots + \hat{c}_{1m}\phi_m(x_2,x_3) + e_{y_1}$$

(45)

Regression analysis of data on y_1, X_4 and ϕ would have produced estimates of the constant and coefficients in this equation. Therefore, the expected value of Y_1^* for a cell with unknown y_1 would be estimated by evaluating this equation on the observed X_4 and ϕ. A value for y_1^* is created by adding an error term to the expected value:

$$y_1^* = E[Y_1^*|\hat{X}_4,x_2,x_3] + e_{y_1} = E[Y_1^*|\hat{X}_4,x_2,x_3] + \hat{\sigma}_{y_1} \cdot t_1^{(\alpha)}$$

(46)

This procedure is identical to that previously described for m_1^*. Similar estimation would be made for y_2^*, ..., y_w^*. Given the created Y^*, it could be substituted into the basic estimation relationship for M^*, as if Y^* were known, equation (36). In expanded form equation (36) is as follows:

$$m_1^* = \beta_{10} + \beta_{11}x_4 + \ldots + \beta_{18}x_8 + \delta_{11}y_1^* + \ldots + \delta_{1w}y_w^* + \sigma_{e_{m_1}} \cdot t_1^{(\alpha)}$$

$$\cdots \cdots \cdots \cdots \cdots \cdots \cdots \cdots \cdots \cdots$$

$$m_7^* = \beta_{70} + \beta_{71}x_4 + \ldots + \beta_{78}x_8 + \delta_{71}y_1^* + \ldots + \delta_{7w}y_w^* + \sigma_{e_{m_7}} \cdot t_7^{(\alpha)}$$

(47)

In this form, it can be seen that by evaluating these equations on

X_4 and the generated Y^* and by sampling for seven random $t^{(\alpha)}$-variates, seven values from the 7-dimensional population of M^* could be created, and by using the basic endowment relationship (25) and these values, a value from the population of cobalt endowment could be created.

Suppose that this procedure were performed for each cell of the study area, giving for each cell Co, m_1, m_6, m_2, and m_7. One or more of these values or a combination, e.g., cobalt endowment, for each cell could be plotted on a map (see Figure 4). Cells without endowment would be those for which m_1, m_6, or m_7 would not exceed endowment cut-offs. From this map the number of deposits for each of three cutoff grades q´, q´´, q´´´ (see Table III) could be determined (Figures 5, 6, and 7). Table III also shows tonnage of metal in each deposit and the total endowment across deposits for each cutoff grade. An important result is that the estimated endowment variables contain information that allows computation of deposit average grade and intradeposit grade variance for each deposit. Thus, these measures also could be made available for each cutoff grade, thereby providing all of the desired endowment descriptors.

Table III. Number of Deposits and Sizes of Deposit for Three Cutoff Grades

	q´	q´´	q´´´
Number of Deposits	2	5	2
Amount of Metal in Each Deposit			
1	778,000	405,000	290,000
2	252,000	195,000	90,000
3		55,000	
4		60,000	
5		20,000	
Total Metal Endowment	1,030,000	735,000	380,000

Suppose that this entire process were repeated, starting with the generation of M^* and M for cells with observed Y, proceeding to the generation of Y^* for cells not sampled for Y, and culminating with the generation of M^* and M for cells with generated Y^*. Upon completion of this process, another set of maps showing the distribution of endowment and another table like Table III could be generated. The new maps and the numerical values in the new table would be different from those just generated because of the different random numbers and the different error terms that they generate. Imagine repeating this process thousands of times and classifying the results into histograms for number of deposits, tonnage of ore per deposit, average grade per deposit, and intradeposit grade variance. There would be a different set of these

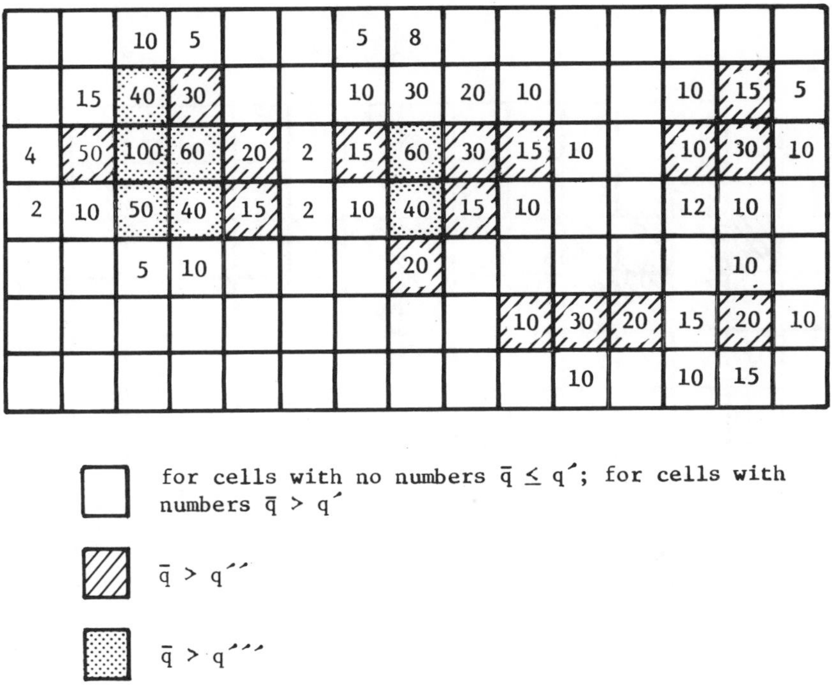

		10	5			5	8							
	15	40	30			10	30	20	10			10	15	5
4	50	100	60	20	2	15	60	30	15	10		10	30	10
2	10	50	40	15	2	10	40	15	10			12	10	
		5	10				20						10	
								10	30	20	15	20	10	
									10		10	15		

☐ for cells with no numbers $\bar{q} \leq q'$; for cells with numbers $\bar{q} > q'$

▨ $\bar{q} > q''$

▦ $\bar{q} > q'''$

Figure 4. Distribution of cobalt endowment (×1000 tons); $q' < q'' < q'''$

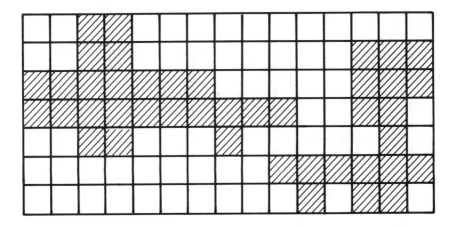

Figure 5. Presence of 2 deposits, given q'; $q' < q'' < q'''$

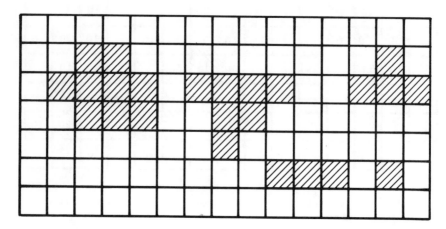

Figure 6. Presence of 5 deposits, given q''; $q' < q'' < q'''$

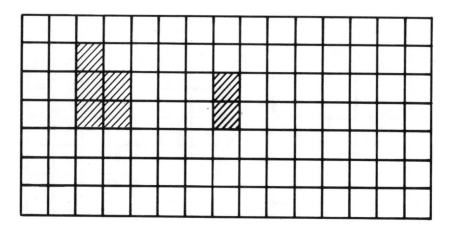

Figure 7. Presence of 2 deposits, given q'''; $q' < q'' < q'''$

for each cutoff grade, fulfilling, at least in part, the desired
objective of describing the mineral endowment of the region by the
basic probability distributions.

ECONOMIC ANALYSIS AND POTENTIAL SUPPLY

A request for a so-called "mineral resource appraisal" usually
reflects a desire to have estimates of potential supply of a mineral
commodity for various levels of product price, given specific states of
other economic or policy variables, or a desire for an estimate of
dynamic supply (supply across time), given estimated demand functions
and states of technology. Very seldom is the request made for an estimate of mineral endowment alone; rather an estimate of mineral endowment may be requested as a means to estimating potential mineral supply,
as indicated by Figure 1. This figure shows that, besides mineral
endowment, the estimation of potential supply must consider exploration
costs and efficiency, production costs, and price. The system would
simulate exploration and mine development and perform the costing and
cash flow analysis necessary for an economic description of potential
supply.

Little has been said in this paper about the modeling of exploration, development, and mining in a potential supply system for marine
minerals. Uncertainties about performance criteria and costs for
these technologies must be great simply because our collective experience is so limited. It must be true that whether economic analysis of
marine mineral deposits is to promote resource development or to
support the estimation of potential and dynamic supply, such analysis
is of limited value without a comprehensive description of uncertainty.
In fact, if uncertainty is great, as in the case of the economic
viability of some marine mineral deposits, economic analysis that does
not describe and consider this uncertainty may lead to incorrect
results. To the extent that there is *unresolved risk*, i.e., risk above
that which is commensurate with the cost of capital used in the economic
analysis to discount to present value terms all projected cash flows,
the economic analysis is biased, ascribing greater economic viability
than really exists.

In concept, cost of capital--the discount rate--is directly
related to risk. An important component of risk is uncertainty about
the capital value of a project. In the case of marine minerals
deposits, uncertainty about capital value reflects uncertainty about a
host of things, including size of deposit, deposit grade, grade variation, mineralogy and processing, environmental costs, and future
markets. Where is the discussion leading? Precisely to this:
  - Economic analysis requires, where possible, the
      parameterization of performance of activities and/or
      costs on variables of the supply system, e.g., water
      depth, deposit size, etc.
  - Equally as important as parameterization of performance
      and/or costs of activities is the description of
      their error of estimation, thereby making it possible

to model likely variations in performance and costs,
which in turn leads to a more comprehensive description of uncertainty and risk.
- Economic analysis is not complete without (1) the
estimation of uncertainty and risk, and (2) consideration of the economic cost of this risk, i.e., linking
cost of capital, which affects capital value, to
risk.[†]

Such a system could be supported by the basic probability distributions for n, t, $\bar{q}$, δ, and h, the estimation of which was described in
this paper. However, given the information on spatial variation that
is carried by the multivariate equation systems, a more comprehensive
and accurate analysis of potential supply could be made by designing
the system to use the equations directly, as contrasted to using them
to generate the probability relations for t, q, δ, h, and n, as was
described in the foregoing. The capability of modeling jointly the
spatial variations of the bathymetric, geologic, and endowment variables would facilitate a more realistic modeling of sampling and other
exploration activities. Likewise, such representation would provide
for a more realistic modeling of development and mining, e.g., the
optimizing decisions that determine scale of operation, investment,
operating costs, and mine life.

Picture a system so designed, which means sampling for states of
bathymetry, geology, and endowment variables, and then simulating the
activities of exploration, development and mining upon these states.
Suppose for simplicity that the cells of the region are so small and
numerous that it is reasonable to delineate deposits by continuous iso-
contours. Figures 8 and 9 depict schematically the use of the multivariate endowment equations and Monte Carlo to simulate the endowment,
exploration, and development as a means to describing potential supply.
Consider the first overall iteration through the system. Figure 8
shows the grade contours of the mineralization created in the first
iteration, using all geologic and bathymetric information and Monte
Carlo methods, as described above. Figure 9 shows endowment of the
first iteration, given the endowment cutoff grade q´. Figure 10 shows
the results of simulating the exploration of the endowment when both
costs and efficiency are considered, namely that only two of the three
deposits that occur are discovered. Finally, Figure 11 shows that when
development and mining costs are simulated, given price and factor
costs, profit maximization defines ore bodies $\ddot{t}_1$, $\ddot{q}_1$, and $\ddot{t}_2$, $\ddot{q}_2$, where
$\ddot{t}_1 < t_1$; $\ddot{t}_2 < t_2$ and $\ddot{q}_1 > \bar{q}_1$; $\ddot{q}_2 > \bar{q}_2$. Potential supply for the first
iteration is $\ddot{t}_1 \cdot \ddot{q}_1 + \ddot{t}_2 \cdot \ddot{q}_2$. Figures 12 – 15 depict schematically
the second overall iteration. Notice that the grade contours of
Figure 12 differ somewhat from those of Figure 8. As geologic and
bathymetric data are the same in both iterations, these differences
reflect random influences introduced through Monte Carlo sampling of

---

[†]To the author's knowledge, the only potential supply system yet
designed with all of these features is for roll-type uranium deposits
(Harris, et al. 1981).

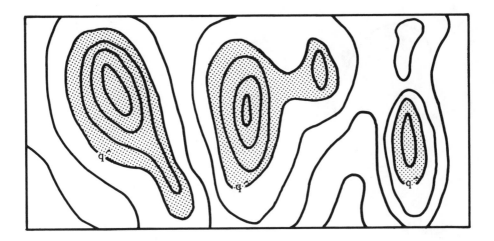

Figure 8.   Contours of grade of endowment simulated by sampling
            multivariate endowment. Stipled areas are deposits,
            given cutoff grade, q´.
            First Iteration

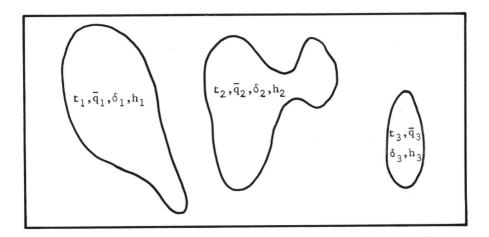

Figure 9.   Map and description of endowment, given q´.
            First Iteration

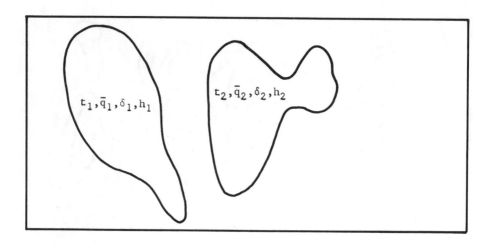

Figure 10.   Discovered deposits, given q´ and economics and efficiency
of exploration.
First Iteration

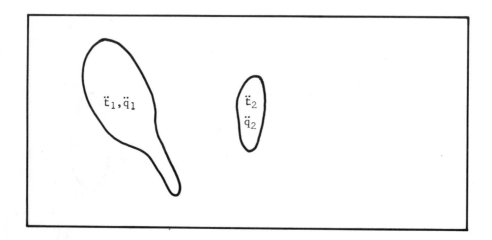

Figure 11.   Discovered and producible resources, given q´, exploration
economics and efficiency, and development and mining costs.
First Iteration

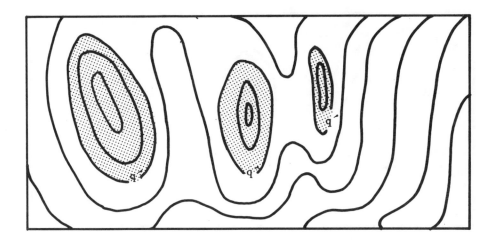

Figure 12.   Contours of grade of endowment simulated by sampling
multivariate endowment equations.
Second iteration

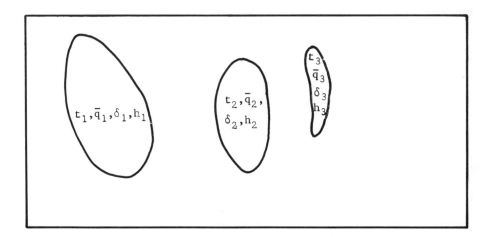

Figure 13.   Endowment, given cutoff grade q´.
Second Iteration

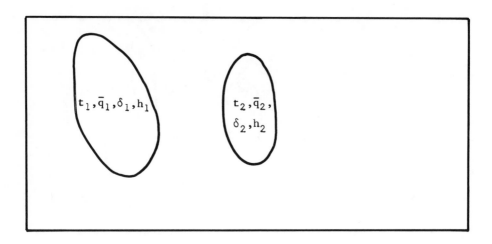

Figure 14.   Discovered deposits.
Second Iteration

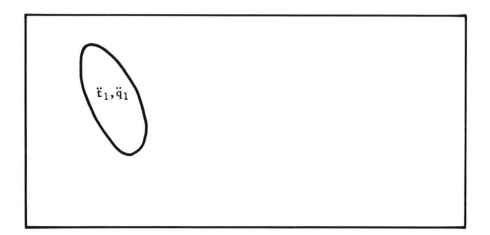

Figure 15.   Discovered, producible ore.
Second Iteration

the equations. The second iteration yields only one ore body, as depicted in Figure 15; potential supply for the second iteration consists of $\ddot{t}_1 \cdot \ddot{q}_1$, where these are the quantities from Figure 15. Selected measures, e.g., potential supply, would be recorded in a file and the entire process repeated for hundreds, or thousands, of iterations. Classification of these files would yield histograms of potential supply, conditional upon the economic parameters. Fitting probability distributions to these histograms provides the desired objective, a set of conditional probability distributions for potential supply. Economic parameters could subsequently be changed and the entire process repeated, thereby showing the effect of economics and policy variables on potential supply, when potential supply is described probabilistically, representing errors in the estimation of endowment, errors in cost estimation, and the uncertainty of exploration outcomes.

CONCLUDING REMARKS

An important parallel of mineral resources appraisal for a marine region with appraisal for a land region is the need for a potential supply system to support appraisal activities. This need reflects the desire of users of resource appraisals -- policy analysts and decision makers -- for potential supply estimates for many different economic circumstances and for measures of uncertainty about the estimates. Only through the use of a system of computer programs can the geologic, economic, and technologic dimensions of potential supply be integrated so as to describe potential supply in both economic and probability terms.

A second parallel is that construction and use of a potential supply system allows separation of geologic analysis from economic and engineering analyses, provided that the geologic analysis describes the estimated endowment by those descriptors, e.g., average grade, size, depth of water, and intradeposit grade variance, that are important determinants of exploration cost and efficiency and of production costs.

Methods for the estimation of mineral endowment descriptors must accommodate some modes of occurrence and kinds of information that differ considerably from those of land mineral deposits. Some marine mineral deposits tend to have great lateral extent, very small thickness, and considerable lateral variation in endowment descriptors. The presence of correlations among descriptors requires that estimation of descriptors be made by means that consider these correlations. Moreover, the values of some descriptors may be related to bathymetric, geologic, and geophysical measurements, and some measurements of each of these kinds of information also may be correlated. When this is the case, estimation of endowment descriptors should also consider these relations and correlations. Simply stated, statistical estimation of mineral endowment of a marine region often will require a holistic resolution of diverse data that are interrelated. Multivariate statistical techniques of regression, factor, and correspondence analyses should prove useful in such estimation.

REFERENCES

Agterberg, F. P., 1982, Regional mineral appraisal: an analytical
    approach, In: Newcomb, R. T. (ed.), Future resources; West
    Virginia University Press, Morgantown, West Virginia, pp. 37-51.

Charles River Associates, 1978, Mineral endowment, resources, and
    discoverable resources, In: The economics and geology of mineral
    supply: an integrated framework for long-run policy analysis; CRA
    Rept. no. 327, Boston, pp. 2-1 -- 2C-7.

Clark, A., Johnson, C., and Chinn, P., 1984, Assessment of cobalt-rich
    manganese crusts in the Hawaiian, Johnston, and Palmyra Island's
    exclusive economic zones; Natural Resources Forum, v. 8, no. 2,
    pp. 163-174.

Craig, J. D., Andrews, J. E., and Meylan, M. A., 1982, Ferromanganese
    deposits in the Hawaiian Archipelago; Marine Geology, v. 45,
    pp. 127-157.

Harris, D. P., 1984a, Mineral resources appraisal - mineral endowment,
    resources, and potential supply: concepts, methods, and cases;
    Oxford Geological Sciences Series, Clarendon Press, Oxford, 445 p.

Harris, D. P., 1984b, Estimation of mineral resources and potential
    supply by geostatistical crustal abundance models--some recent
    investigations; Unpub. manuscript, presented to 27th Int. Geol.
    Cong., Moscow, U.S.S.R., Aug. 4-14, 1984.

Harris, D. P., 1978, Undiscovered uranium resources and potential
    supply, In: Workshop on concepts of uranium resources and
    producibility, National Research Council, National Academy of
    Sciences, pp. 51-81.

Harris, D. P., 1973, A subjective probability appraisal of metal endow-
    ment of Northern Sonora, Mexico; Economic Geology, v. 68, no. 2,
    pp. 222-242.

Harris, D. P. and Chavez, M. L., 1984, Modeling dynamic supply of
    uranium--an experiment in the integration of economics, geology,
    and engineering; Proc. 18th Int. Sym. Appl. of Computers and Mathe-
    matics in the Mineral Industry, Inst. Mining and Metallurgy,
    London, pp. 817-892.

Harris, D. P., Ortiz-Vértiz, S. R., Chavez, M. L., and Agbolosoo, E. K.,
    1981, Systems and economics for the estimation of uranium potential
    supply; Res. Rept. to U.S. Dept. of Energy, Grand Junction Office,
    Grand Junction, Colorado, 609 p.

Singer, D. A. and Ovenshine, A. T., 1979, Assessment of metallic
    resources in Alaska; American Scientist, v. 67, no. 5, pp. 582-589.

ESTIMATION OF THE PROBABILITY OF OCCURRENCE OF POLYMETALLIC
MASSIVE SULFIDE DEPOSITS ON THE OCEAN FLOOR

F.P. Agterberg and J.M. Franklin
Geological Survey of Canada
601 Booth Street
Ottawa K1A 0E8
Canada

ABSTRACT.   Probabilistic methods for mineral resource evaluation
previously developed for land-based polymetallic massive sulfide
deposits in the Abitibi Volcanic Belt of the Canadian Shield are
applied to occurrences of hydrothermal vents on the East Pacific Rise
near 21°N.   It is shown that volcanic and tectonic features on the
ocean floor can be quantified for cells belonging to a grid and used
to estimate probabilities of occurrence of the vents.   The leverage
and influence of multivariate cell observations on the estimated
probabilities are also studied.

INTRODUCTION

Methods of image analysis and multivariate statistical analysis
have been developed for estimating the probability of occurrence of
land-based polymetallic massive sulfide deposits (Agterberg, 1984).
These methods can also be applied to ocean-floor deposits to test more
rigorously the relationship between the occurrence of hydrothermal
vents and polymetallic sulfide deposits with respect to specific
volcanic forms, such as sheet flows, lobate flows and pillow lavas;
relative ages of volcanic units, presence of tectonic features
including fissures, and height above the seafloor of the volcanic
flows as a function of water depth.   This paper contains the results
of a pilot study recently performed on occurrences of hydrothermal
vents along the central axis of the East Pacific Rise at 21°N.   The
volcanic, tectonic and hydrothermal processes of this study area were
studied by Ballard et al. (1981) and facts established by them were
taken as the starting point for this pilot study.   Regression analysis
(cf. Harris, 1984; Tukey, 1984) was employed.   Recently methods of
studying leverage and influence of multivariate observations on
regression results have become available (Welsh, 1982; Gray and
Ling, 1984).   The usefulness of these new methods is explored in this
paper.

467

P. G. Teleki et al. (eds.), Marine Minerals, 467–483.

QUANTIFICATION OF MAP PATTERNS AND CORRELATION ANALYSIS

A coded version of maps published by Ballard et al. (1981) is presented in Figure 1, the úpper part of which shows Zone 1 of Ballard et al. consisting of relatively young lava flows along the central axis of the East Pacific Rise. These authors introduced a relative age scale for lava flows ranging from 1.0 for flows with no sediment cover to 2.5 for areas dominantly covered with sediments, with isolated lava forms protruding through the cover. In their Zone 1, a distinction could be made between pillow flows and sheet flows belonging to three age groups (1.0-1.4; 1.4-1.7; and 1.7-2.0). The resulting 6 lithostratigraphic units are shown by symbols for subcells in the lower part of Figure 1. Each square subcell measures approximately 85 metres on a side. A rock unit code was assigned only if one of the 6 rock units occurred at the centre of a subcell. The hydrothermal vents in the study area were allocated to subcells using the same grid (see upper part of Figure 1).

Variables x_1 to x_6 were defined for the 6 rock units, and x_7 for other, older rocks. Values for x_1 to x_6 were determined by counting subcell codes for rock units in 51 square cells consisting of (4 x 4)

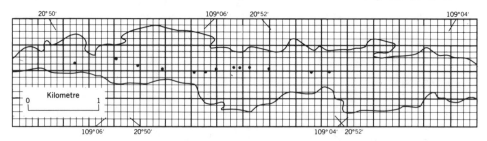

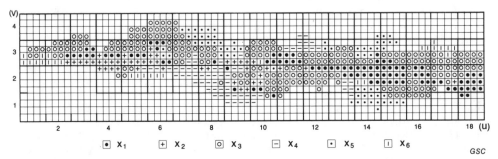

Figure 1. Coded version of area on East Pacific Rise at 21°N studied by Ballard et al. (1981). Upper part shows outline of young volcanic flows and occurrences of hydrothermal vents. Lower diagram shows codes for variables x_1-x_6 in subcells; x_1, x_3 and x_5 represent pillow flows, x_2, x_4, x_6 are sheet flows; age groups are: 1.0 - 1.4 (x_1, x_2), 1.4 - 1.7 (x_3, x_4), and 1.7 - 2.0 (x_5, x_6). There are 16 subcells in a 'small' cell (s) and 4 small cells (64 subcells) in a 'large' cell (1) (see Figure 2).

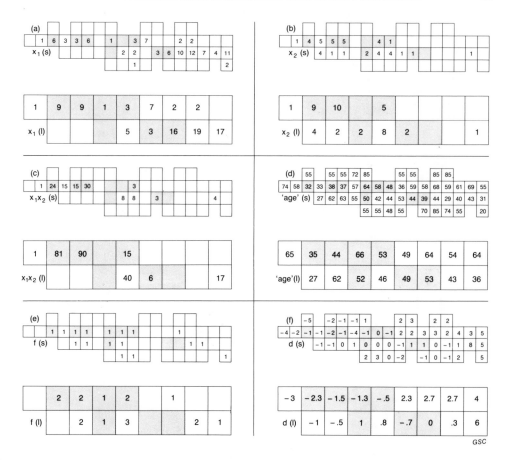

Figure 2. Examples of variables coded for small (s) and large (ℓ)
          cells. Shaded cells contain at least one hydrothermal vent;
          i.e. y = 1. For further explanation, see text.

Table I. Product–moment correlation coefficients between variables for small cells (upper triangle) and large cells (lower triangle). Values in absolute value greater than 0.395 are shown in bold print. y – presence of vents; x₁ to x₆ – rock units; x₇ – absence of x₁ to x₆; a – relative age; f – absence of fissures in x₁, x₂; d – water depth; u, v – map co-ordinates.

	y	x₁	x₂	x₃	x₄	x₅	x₆	x₇	a	f	d	u	v
y	1.00	0.19	**0.46**	0.12	0.13	−0.07	−0.14	**−0.41**	−0.25	**0.41**	−0.21	−0.22	0.07
x₁	0.04	1.00	0.12	0.05	−0.13	−0.09	−0.10	**−0.52**	**−0.61**	0.20	0.11	0.24	−0.12
x₂	0.36	0.02	1.00	0.08	−0.02	0.07	−0.11	−0.37	**−0.45**	0.43	−0.27	**−0.44**	0.02
x₃	0.19	0.19	0.20	1.00	−0.26	−0.19	0.07	**−0.61**	−0.16	−0.16	−0.05	0.01	0.21
x₄	0.12	−0.16	0.06	**−0.40**	1.00	−0.19	−0.12	−0.09	−0.06	−0.06	0.09	−0.15	−0.24
x₅	0.36	0.16	−0.29	0.19	−0.07	1.00	−0.13	−0.10	**0.51**	−0.01	−0.06	0.10	0.09
x₆	−0.34	−0.33	−0.27	−0.19	−0.21	−0.20	1.00	−0.11	0.25	−0.07	0.14	−0.19	0.07
x₇	**−0.40**	**−0.56**	−0.26	**−0.63**	−0.15	**−0.49**	0.34	1.00	0.36	−0.28	0.10	0.06	−0.03
a	−0.04	**−0.45**	**−0.51**	0.08	0.05	0.39	**0.57**	−0.02	1.00	−0.34	−0.07	0.03	0.26
f	0.12	0.21	**−0.64**	0.22	0.20	−0.01	−0.28	**−0.47**	−0.18	1.00	−0.34	−0.13	−0.15
d	**−0.47**	0.19	**−0.44**	−0.06	0.07	−0.15	−0.16	0.13	0.01	−0.19	1.00	**0.74**	−0.12
u	−0.27	**0.41**	**−0.53**	0.17	−0.11	0.10	−0.20	−0.11	0.10	−0.20	**0.84**	1.00	−0.27
v	0.07	−0.30	0.06	0.38	−0.32	0.18	0.19	−0.06	0.39	−0.12	−0.09	−0.10	1.00

adjoining subcells. Each of the square cells in Figure 1 contains at least one subcell with a rock unit code for x_1-x_6. Figures 2a and 2b illustrate cell values for variables x_1 and x_2 representing youngest (age: 1.0-1.4) pillow flows and sheet flows, respectively. Similar values were determined for 17 larger cells. Each large cell (ℓ) consists of (2 x 2) small cells (s) as illustrated in Figure 2, or (8 x 8) of the original subcells.

Cell values of two variables x_i and x_j (i = 1,...,7; j = 1,...,7; i ≠ j) can be cross-multiplied to form product variables $x_i x_j$ such as $x_1 x_2$ in Figure 2c. The average value of a product variable Ave ($x_i x_j$) is related to the covariance Cov (x_i, x_j) of the variables x_i and x_j from which it was formed by

$$\text{Ave } (x_i x_j) = \text{Cov } (x_i, x_j) + \text{Ave } (x_i) \cdot \text{Ave } (x_j) \tag{1}$$

Three other variables, for relative cell age, absence of fissures in youngest flows and water depth, respectively, were coded for the small cells (see Figures 2d, 2e, and 2f). 'Age' (a) satisfies:

$$a = \sum_{i=1}^{6} a_i x_i \Big/ \sum_{i=1}^{6} x_i \tag{2}$$

where $a_1 = a_2 = (1.0 + 1.4)/2 = 1.2$; $a_3 = a_4 = (1.4 + 1.7)/2 = 1.55$; and $a_5 = a_6 = (1.7 + 2.0)/2 = 1.85$. This type of average cell age is shown as $100(a-1)$ in Figure 2d. The variable f in Figure 2e assumes value 1 if fissures in youngest flows (x_1 and x_2) in a small cell are absent; otherwise f = 0. Ballard et al. (1981, Fig. 5, p. 5) published a topographic map of the study area with contours in corrected meters, with a 10 m contour interval. The values of d in Figure 2f represent depth of contours closest to cell centers; 2600 m was subtracted from these cell depths, and the differences were divided by 10. Consequently, a 0-value in Figure 2f represents a depth of approximately 2600 m. The depth increases in the northeasterly direction (= to the right in Figure 2f).

Clearly, the digitized results shown in Figures 1 and 2 are approximations only and could be refined. However, a sufficiently large number of characteristic features of the patterns has been captured to allow, by geomathematical analysis, reproduction of a number of facts established by Ballard et al. (1981) as illustrated in Tables I and II. The upper triangle in Table I contains product-moment correlation coefficients for the variables coded for the small cells; the lower triangle contains similar results for the large cells. The variable y denotes presence (y = 1) or absence (y = 0) of hydrothermal vents in a cell; u and v are geographic co-ordinates as in Figure 1. It should be kept in mind that, because of spatial auto- and cross-correlation effects, the product-moment correlation coefficient of two variables with cell values on a grid, in a statistical sense, does not have the same meaning as the correlation coefficient of two correlated random variables. For example, the

Table II. Correlation coefficients for three product variables.

	y	x_1	x_2	x_3	x_4	x_5	x_6
$x_1 x_2$	0.47	0.30	0.77	0.09	-0.13	-0.06	-0.12
$x_2 x_5$	0.41	0.03	0.47	-0.10	-0.06	0.30	-0.18
d x_2	-0.39	-0.12	-0.64	-0.05	0.11	0.03	0.02

Table III. Squared multiple correlation coefficients (R^2) for various regression experiments.

	R^2 (small cells)	R^2 (large cells)
x_2 (youngest sheet flows)	0.2135	0.1274
a ('age')	0.0641	0.0016
f (fissures absent)	0.1693	0.0152
d (depth)	0.0429	0.2168
u	0.0485	0.0751
v	0.0042	0.0050
u, v	0.0485	0.0770
$x_1 - x_6$ (6 rock units)	0.3136	0.3952
$x_1 - x_6$ (logistic model)	0.3530	0.6229
$x_1 - x_6$, 'age'	0.3136	0.3986
$x_1 - x_6$, f	0.3434	0.4882
$x_1 - x_6$, depth	0.3258	0.4739
$x_1 - x_6$, u, v	0.3234	0.4102
$x_1 x_2$	0.2242	0.0737
$x_1 - x_7$, $x_1 x_2 - x_6 x_7$	0.6609	1.0000
8 'best' variables	0.5516	0.7497
ditto (logistic model)	0.5974	1.0000
ditto (log transformation)	0.3156	0.5306
ditto (0-1 transformation)	0.3628	0.1881

Table IV. Threshold values F_c used in stepwise regression experiments. Last column shows cumulative probability corresponding to F_c value. For further explanation, see text.

F_c	k	R^2	P_f
0.00	27	0.6609	0.00
0.05	24	0.6605	0.17
0.50	11	0.5698	0.52
1.00	8	0.5516	0.68
2.00	4	0.3861	0.83
3.00	3	0.3571	0.92

commonly used significance test to establish whether or not a
correlation coefficient is greater or less than 0 cannot be applied
without modifications to account for the spatial cross-correlation
effects. However, by comparing the elements of Table I with one
another, and by visually clustering them, the following pattern
emerges.

Occurrence of hydrothermal vents is positively correlated with
occurrence of youngest sheet flows (x_2) and absence of fissures in
youngest flows (f). It is negatively correlated with older flows
situated outside Zone 1 of Ballard et al., with depth (d), and with
distance along the positive u-axis in Figure 1. The largest values in
Table I (0.74 for small cells; 0.84 for large cells) indicate that
depth increases in the northeasterly direction. Correlation
coefficients, which in absolute value exceed 0.40, are printed in bold
type in Table I. The variable x_2 for youngest sheet flows has a
relatively large number of bold values. Comparison of its correlation
coefficients with those of x_1 indicates that the ratio of youngest
pillow to sheet flows increases in the northeasterly direction. A
similar relationship applies to the other pairs (x_3, x_4) and (x_5, x_6)
but to a lesser extent. Product variables $x_i x_j$ can be relatively
strongly correlated with one another and with the original variables
(see Table II). They tend to confirm the pattern obtained by the
visual clustering of correlation coefficients in Table I.

It may be concluded that most of the factual relationships
established by Ballard et al. (1981, p. 6-8) are present in the coded
data base described herein and can be extracted by geomathematical
analysis. An example of a relation not coded is process 5 (d) of
Ballard et al. which states that the hottest exiting temperature in
the hydrothermal field increases to the southwest. The remainder of
this paper deals with multivariate methods of estimating the
probability of occurrence of hydrothermal vents (y) from the other
variables shown in quantified form in Figures 1 and 2.

PROBABILITY OF OCCURRENCE OF HYDROTHERMAL VENTS

Early work on estimating the probability of occurrence of mineral
deposits (polymetallic massive sulfide deposits and also vein-type
gold deposits) in the Abitibi Volcanic Belt on the Canadian Shield
included use of the following three methods:

(1) Multiple regression (Agterberg and Cabilio, 1969);
(2) Stepwise regression using product variables (Agterberg et al.,
    1972);
(3) Logistic regression (Agterberg, 1974a).

Computer experiments performed on the coded data of Figures 1
and 2 using these 3 methods are summarized in Table III. In each
experiment, the presence or absence of hydrothermal vents was the
dependent variable with $y = 1$ (vent present) or $y = 0$ (vent absent).
For (1) and (2) above, the Statistical Package for the Social Sciences
(SPSS) subprogram REGRESSION (Nie et al., 1975) and its more recent

implementation in SPSS-6000 were used.   For (3), Chung's (1978)
FORTRAN program was employed. The estimated probabilities for a
number of these experiments are shown in Figures 3 and 4.

The squared multiple correlation coefficient R^2 is shown in
Table III for regression runs performed on the 51 small as well as on
the 17 large cells. Values of R^2 for the logistic model were obtained
by computing $R^2 = SSR/SST$ as in the linear regression model, with SSR
and SST representing sums of squared deviations from the mean of
calculated and original values, respectively. If there is only one
variable, R^2 represents the square of the correlation coefficient
(cf. Tables I and II). Simultaneous use of all 6 rock types increases
R^2. Adding f (fissures absent) or d (depth) produces further
increases. Multiple regression can yield small negative probabilities
or values slightly greater than one as shown in Figure 3. Such
possibilities are not allowed in logistic regression which also
generally gives higher R^2 because of increased flexibility of the
model. The estimated probabilities are exactly equal to the observed
values of the dependent variable (1 or 0) when $R^2 = 1.0000$ as in
2 experiments for large cells in Table III. In the first of these two
runs, the 7 original variables and all 21 possible product variables
(by pairs) were considered for use as independent variables. The
variable x_7 is a linear combination of x_1 to x_6 and was automatically
excluded. The remaining 27 variables gave $R^2 = 0.6609$ for small cells
and the probabilities shown in Figure 3f. For the 17 large cells,
only the first 16 independent variables entered were used providing
100 percent of fit. For this run, as well as for several other runs
on the large cells, the number of observations was too small in
comparison with the number of variables to produce meaningful results.

Because of the large gain in R^2 for small cells after addition of
the product variables to x_1-x_6, stepwise regression was used to refine
this result by excluding redundant variables. A threshold value F_c
can be set to test additional variables for statistical significance
and to evaluate variables already in the regression equation.
Table IV shows total number of independent variables retained (k), and
corresponding R^2 for different threshold values F_c. Individual steps
are shown for F = 0.05 in Table V. The number in the first column
gives the number of independent variables included at a step which
consists of either adding or deleting one variable. After inclusion
of approximately 8 variables, R^2 begins to increase linearly,
suggesting that the contributions of additional variables are not
significant.

As shown in Table IV, a slightly higher R^2 (=0.5516) for
8 variables was obtained by setting $F_c = 1.00$, in comparison with
$R^2 = 0.539$ for the first 8 variables in Table V. The 8 'best'
variables for $F_c = 1.00$ and their coefficients are given in Table VI.
The probabilities for this solution are shown in Figure 3d. All
patterns of probabilities in Figure 3 explain the pattern of observed
occurrences rather well. However, the occurrences of the two
hydrothermal vents in the cells with u = 12 and 13 (cf. Figure 1) are
not explained in the small cell patterns of Figures 3a, 3b and 3c.
The patterns of Figures 3d, 3e and 3f are similar to one another.

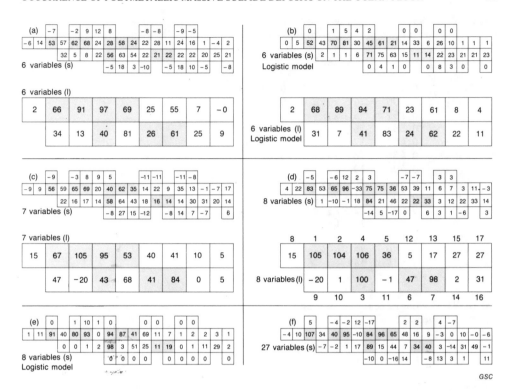

GSC

Figure 3. Probabilities (in percent) estimated by linear regression on (a) 6 rock units, (c) 6 rock units plus variable f for absence of fissures in youngest flows, (d) 8 'best' variables (see text), and (f) x_1 to x_6 and all product variables for x_1 to x_7. Logistic regression results are shown in (b) and (e) for same variables as in (a) and (d), respectively. Shaded cells contain at least one hydrothermal vent. Large cells are numbered in (d).

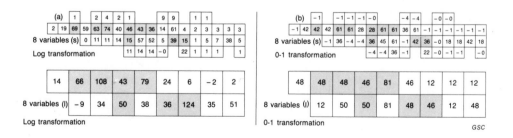

GSC

Figure 4. Probabilities for 8 variables (see Figure 3d) after (a) logarithmic and (b) 0 - 1 transformation.

Table V. Variables added (or deleted)
during stepwise regression for $F_c = 0.05$.

Variable		R^2	Variable		R^2
1	x_1x_2	0.224	13	x_3x_5	0.578
2	x_2x_5	0.300	14	x_1x_5	0.589
3	x_7	0.357	15	x_1x_3	0.597
4	x_4	0.386	16	x_1	0.606
5	x_4x_7	0.407	17	x_1x_7	0.610
6	x_3x_4	0.438	18	x_6x_7	0.611
7	x_4x_5	0.498	19	x_5	0.617
8	x_1x_4	0.528	20	x_5x_7	0.639
9	x_3	0.539	21	x_2x_4	0.642
8	(x_7)	0.539	22	x_2x_6	0.645
9	x_3x_7	0.558	21	(x_1)	0.645
10	x_7	0.564	22	x_5x_6	0.651
11	x_2x_3	0.570	23	x_2x_7	0.652
12	x_1x_6	0.574	24	x_2	0.661

Table VI. Coefficients B of
8 'best' variables. Variance
ratio $\hat{F}$ and corresponding
cumulative probability P_f apply to
deletion of single variables.

Variable	B	$\hat{F}$	P_f
x_3	−0.53	0.05	0.17
x_4	8.31	0.41	0.46
x_1x_2	1.20	7.56	0.97
x_1x_4	−0.50	0.91	0.63
x_3x_4	−0.61	1.95	0.80
x_3x_7	0.09	1.37	0.72
x_4x_5	1.91	11.13	0.99
x_4x_7	−0.06	0.04	0.14
constant	−25.25	0.75	0.59

They suggest that an additional hydrothermal vent could have occurred in small cells with u = 16 or u = 17. In two additional experiments (see Figure 4), cell values greater than zero for the 8 'best' variables were logarithmically transformed and replaced by ones, respectively. This confirmed the validity of the pattern of Figure 3d.

In the evaluation of patterns of probabilities obtained by linear regression, it should be kept in mind that the sum of all probabilities is equal to the number of cells that contain one or more hydrothermal vents. For the 51 small cells, this sum is equal to 9; whereas for the large cells it is equal to 7. The pattern for large cells in Figure 3d was derived from the same 8 variables as were used for the small cells in Figure 3d. Its cells are numbered 1 to 17 beginning with the numbers 1 to 7 for the cells containing one or more hydrothermal vents. The probabilities for the 17 large cells in Figure 3d will be analyzed in more detail in the next section.

LEVERAGE AND INFLUENCE OF MULTIVARIATE OBSERVATIONS

During the past 8 years, various leverage and influence measures have been developed in linear regression analysis (Welsh, 1982; Vinod and Ullah, 1982; Hocking, 1983). Use is made of the hat matrix

$$H = X(X^tX)^{-1}X^t \qquad (3)$$

so called because $\hat{Y} = HY$ where $\hat{Y}$ and Y are column vectors consisting of the n estimated and observed values of the dependent variable, respectively. H is an (n x n) symmetrical matrix derived from the (n x (p + 1)) X matrix of which the first column consisting of n ones is followed by p columns with n values, one for each of the p independent variables. The rows X_i of X each consist of (p + 1) values. The element h_{ij} in H can be interpreted as the amount of leverage exerted by Y_j on $\hat{Y}_i$ (i = 1,...,n; j = 1,...,n). The diagonal elements h_{ii} satisfy $0 \le h_{ii} \le 1$. Leverage indicates potential influence because the observations Y_j have not been used to construct H.

Table VII shows the hat matrix for the 17 large cells in Figure 3d. Agterberg (1974b) showed that the i-th row (or column) of H can be used to measure similarity of the j-th cell with the i-th cell. If probabilities are estimated with $Y_j \in (0,1)$, $\hat{Y}_i$ is simply the sum of the elements in the ith column of H for which $Y_j = 1$. For example, the estimated probability (in percent) for cell 16 in Figure 3d is: 6 + 10 - 1 - 1 + 7 + 9 + 1 = 31 using the first 7 elements of column 16 in Table VII.

Recently, Gray and Ling (1984) have proposed to find influential subsets of cases by application of cluster analysis to the modified hat matrix H*. In their nomenclature, a case consists of a row X_i of the X matrix plus an observation Y_i, and

$$H^* = Z(Z^tZ)^{-1}Z^t \qquad\qquad (4)$$

where Z is the X matrix with the Y vector appended. Each element of H^* satisfies $h_{ij}^* = h_{ij} + e_i e_j/SSE$ where $e_i = Y_i - \hat{Y}_i$, $e_j = Y_j - \hat{Y}_j$, and $SSE = \Sigma\, e_i^2$. The element h_{ij}^* in H^* can be interpreted as a measure of the amount of influence exerted by Y_j on $\hat{Y}_i$. The modified hat matrix for the example is shown in Table VIII. Gray and Ling (1984) rearranged the order of the columns (and rows) in the modified hat matrix to better describe subsets of influential cases. We suggest that this method can be applied to the hat matrix as well, in order to find subsets of cells with relatively high leverage.

Inspection of Table VII shows that the cells are basically of three types (A, B and C). Type A cells (5, 8, 13 and 16) have diagonal elements h_{ii} which are less than 20 (percent) and moderately large values in a relatively large number of other positions. Type B cells (1, 2, 6, 9, 10, 12 14, 15, and 17) constitute the largest group. They have diagonal elements between 20 and 95, and one or more non-diagonal elements greater than 20 (marked by asterisks in Table VII). Type C cells (3, 4, 7 and 11) possess diagonal values between 95 and 100, with relatively small non-diagonal elements.

For the construction of probability maps, we are primarily interested in cells of types A and B. Type C cells can be regarded as unique events because they are not similar to any other cell in the study area. Type A cells are useful for extrapolation for exactly the opposite reason. Type B cells can be further subdivided into subsets by clustering, as proposed by Gray and Ling (1984). Table IX shows the type B cells of Table VII after reordering. Clearly, there are 3 separate subsets each consisting of 3 cells which jointly exert high leverage. The same type of analysis was applied to the modified hat matrix of Table VIII with nearly identical results (see Table X). The main differences between Tables VII and VIII are for cells 5 and 6 of which the diagonal elements are greater in Table 8. All 7 cells with one or more hydrothermal vents are influential (Class A) or belong to influential subsets (Class B) in Table VIII. The values of elements adjoining diagonal elements in Tables VII and VIII can be used as distances between similar cells to construct dendrograms (see Figure 5). The two classifications (T1 and T2) based on hat matrix and modified hat matrix, respectively, are listed in Table XI.

Contributions made by each of the eight variables to the probabilities (in percent) of the large cells in Figure 3d are shown in Table XI. These contributions were obtained by multiplying the cell values of the 8 variables by their coefficients (see Table VI) and adding the constant term (= −25, also see Table VI). The relative significance of a variable is expressed by the F-value and its corresponding cumulative probability in Table VI. The regression results as shown in Tables VI and XI could be obtained routinely on the computer during the past 20 years. The methods for studying leverage and influence applied in this section have been developed more recently. They provide useful new information on which other cells in a study area contribute to relatively high probabilities in some cells, and should be made part of future regression analysis.

Table VII. Hat matrix for 17 large cells of Figure 3d. Elements greater than 9.5 are shown in bold print; elements greater than 19.5 have asterisks.

	1	2	3	4	5	6	7	8	9	10	11	12	13	14	15	16	17
1	**70***	**23***	-1	-2	**14**	-0	2	0	3	9	2	-5	-1	**-33**	9	6	6
2	**23***	**78***	1	1	1	3	-1	-4	-7	-5	-1	2	-4	**27***	**-13**	**10**	**-10**
3	-1	1	**99***	-1	-1	2	2	1	-4	8	1	-3	-1	-0	-0	-1	0
4	-2	1	-1	**99***	7	1	-1	2	-5	6	1	-3	-1	-0	-1	-1	-1
5	**14**	1	-1	7	9	7	-1	8	9	**10**	0	4	8	-1	**10**	7	9
6	-0	3	2	1	7	**38***	-3	**12**	**-23**	**-16**	-3	6	**15**	8	**22***	9	**23***
7	2	-1	2	-1	-1	-3	**97***	1	1	**-12**	-2	5	1	0	-1	1	0
8	0	-4	1	2	8	**12**	1	**12**	**12**	5	1	-3	**12**	**11**	**13**	9	**13**
9	3	-7	-4	-5	9	**-23**	1	**12**	**55***	**32***	6	**-14**	8	**14**	-1	**10**	-1
10	9	-5	8	6	**10**	**-16**	**-12**	5	**32***	**42***	**-11**	**23***	1	2	-0	5	-2
11	2	-1	1	1	0	-3	-2	1	6	**-11**	**98***	5	1	0	0	1	-0
12	-5	2	-3	-3	4	6	5	-3	**-14**	**23***	5	**90***	**13**	-1	-1	-3	-0
13	-1	-4	-1	-1	8	**15**	1	**12**	8	1	1	**13**	**14**	**11**	**14**	9	**14**
14	**-33**	**27***	-0	-0	-1	8	0	**11**	**14**	2	0	-1	**11**	**51***	-3	9	**14**
15	9	**-13**	-0	-1	**10**	**22***	-1	**13**	-1	-0	0	-1	**14**	-3	**22***	8	**20***
16	6	**10**	-1	-1	7	9	1	9	**10**	5	1	-3	9	9	8	**12**	9
17	6	**-10**	0	-1	9	**23***	0	**13**	-1	-2	-0	-0	**14**	**14**	**20***	9	**19**

F. P. AGTERBERG AND J. M. FRANKLIN

Table VIII. Modified hat matrix corresponding to Table VII.

	1	2	3	4	5	6	7	8	9	10	11	12	13	14	15	16	17
1	**70***	**23***	-1	-2	11	-3	2	1	2	9	2	-5	0	-33	10	7	7
2	**23***	**78***	1	1	-2	1	-1	-3	-8	-5	-1	2	-4	**27***	-12	11	-9
3	-1	1	**99***	-1	-1	-1	2	-1	-4	8	1	-3	-0	-0	0	-1	0
4	-2	1	-1	**99***	3	-2	1	-1	-6	6	1	-3	-0	0	1	-1	1
5	11	-2	-1	3	**49***	**40***	-4	-2	**21***	10	1	0	-3	-2	-7	-12	-7
6	-3	1	-1	-2	**40***	**65***	-4	4	-12	-16	-2	3	6	8	8	-7	9
7	2	-1	2	1	-4	-4	**97***	1	8	-12	-2	5	1	0	-0	1	-1
8	1	-3	-1	-1	-2	4	1	**14**	9	6	6	-2	15	11	17	14	17
9	2	-8	-4	-6	**21***	-12	8	9	**59***	**32***	6	-15	5	14	-6	4	-7
10	9	-5	8	6	**10**	-16	-12	6	**32***	**42***	-11	**23***	3	2	-0	5	1
11	2	-1	1	1	1	-2	-2	6	6	-11	**98***	4	1	0	-0	1	-0
12	-5	2	-3	-3	0	3	5	-2	-15	**23***	4	**90***	-1	-1	1	-1	1
13	0	-4	-0	-0	-3	6	1	**15**	5	3	1	-1	**16**	**11**	**19**	**14**	**18**
14	-33	**27***	-0	0	-2	8	0	**11**	**14**	2	0	-1	**11**	**51***	-2	**12**	2
15	**10**	-12	0	1	-7	8	-0	**17**	-6	-0	-0	1	**19**	-2	**29***	**16**	**27***
16	7	**11**	-1	-1	-12	-7	1	**14**	4	5	1	-1	**14**	**12**	**16**	**18**	**17**
17	7	-9	0	1	-7	9	-1	**17**	-7	1	-0	1	**18**	2	**27***	**17**	**27***

Table IX. Type B cells from Table VII reordered to show clusters.

	1	2	14	6	15	17	9	10	12
1	70*	23*	-33	-0	9	6	3	9	-5
2	23*	78*	27*	3	-13	-10	-7	-5	2
14	-33	27*	51*	8	-3	1	14	2	-1
6	-0	3	8	38*	22*	23*	-23	-16	6
15	9	-13	-3	22*	22*	20*	-1	-0	-1
17	6	-10	1	23*	20*	19	-1	-2	-0
9	3	-7	14	-23	-1	-1	55*	32*	-14
10	9	-5	2	-16	-0	-2	32*	42*	23*
12	-5	2	-1	6	-1	-0	-14	23*	90*

Table X. Type B cells from Table VIII reordered to show clusters.

	1	2	14	15	17	6	5	9	10	12
1	70*	23*	-33	10	7	-3	11	2	9	-5
2	23*	78*	27*	-12	-9	1	-2	-8	-5	2
14	-33	27*	51*	-2	2	8	-2	14	2	-1
15	10	-12	-2	29*	27*	8	-7	-6	-0	1
17	7	-9	2	27*	27*	9	-7	-7	1	1
6	-3	1	8	8	9	65*	40*	-12	-16	3
5	11	-2	-2	-7	-7	40*	49*	21*	10	0
9	2	-8	14	-6	-7	-12	21*	59*	32*	-15
10	9	-5	2	-0	1	-16	10	32*	42*	23*
12	-5	2	-1	1	1	3	0	-15	23*	90*

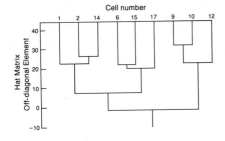 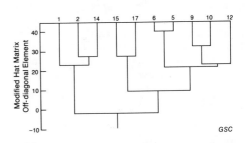

Figure 5. Dendrograms showing groupings of type B cells. Elements adjoining diagonal elements in Tables IX and X were used as 'distances'.

Table XI. Contributions to probabilities P of individual variables in 17 large cells of Figure 3d. Tl and T2 represent types of cells using hat matrix and modified hat matrix for classification respectively.

	X_3	X_4	X_1X_2	X_1X_4	X_3X_4	X_3X_7	X_4X_5	X_4X_7	P	T1	T2
1	-7		97			40			105	B1	B1
2	-15		108			37			104	B1	B1
3	-1	191			-14	4		-53	100	C	C
4	-13	33		-2	-59	38	138	-4	106	C	C
5	-7	8	18	-2	-8	42	11	-2	36	A	B3
6	-12		7			77			47	B2	B3
7	-2	42		-40	-9	7	134	-8	98	C	C
8	-7		1			46			15	A	A
9	-1					5			-20	B3	B3
10	-2	33			-7	14		-12	1	B3	B3
11	-8	133	48	-40	-146	27	31	-19	-1	C	C
12	-5	91		-39	-67	32	42	-23	5	B3	B3
13	-7					50			17	A	A
14	-11					37			2	B1	B1
15	-7					60			27	B2	B2
16	-8		20			43			31	A	A
17	-8					60			27	B2	B2

CONCLUDING REMARKS

Probabilistic resource appraisal of areas containing undiscovered valuable mineral deposits usually takes place in two steps: (1) Estimation of the probability of occurrence of the deposits; and (2) estimation of the amounts of ore and metals present in the deposits. A control group of known deposits of the same type is commonly used during the second step. In this paper, only methods to estimate probabilities of occurrence of ocean-floor deposits were considered. The approach presented in this paper may be useful as a first step in regional resource appraisals. Alternately, automated estimation of probabilities of occurrence from mapped geological features could be used as an aid for discovering new hydrothermal vents and deposits during actual exploration of the ocean floor.

ACKNOWLEDGEMENTS

Thanks are due to S.N. Lew, Geological Survey of Canada, for assistance in computer programming, and to G.F. Bonham-Carter for critical reading of the manuscript.

REFERENCES

Agterberg, F.P., 1974a, Automatic contouring of geological maps to detect target areas for mineral exploration; Jour. Math. Geol., v. 6, pp. 373-395.

Agterberg, F.P., 1974b, Geomathematics; Elsevier, Amsterdam, 596 p.

Agterberg, F.P., 1984, Use of spatial analysis in mineral resource evaluation; Jour. Math. Geol., v. 16, pp. 565-589.

Agterberg, F.P. and Cabilio, P., 1969, Two-stage least-squares method for the relationship between mappable geological variables; Jour. Math. Geol., v. 1, pp. 37-53.

Agterberg, F.P., Chung, C.F., Fabbri, A.G., Kelly, A.M., and Springer, J.S., 1972, Geomathematical evaluation of copper and zinc potential of the Abitibi area, Ontario and Quebec; Geol. Survey of Canada Paper 71-41, 55 p.

Ballard, R.D., Francheteau, J., Juteau, T., Rangan, C, and Normark, W., 1981, East Pacific Rise at 21°N: the volcanic, tectonic, and hydrothermal processes of the central axis; Earth and Planetary Sc. Letters, v. 55, pp. 1-10.

Chung, C.F., 1978, Computer program for the logistic model to estimate the probability of occurrence of discrete events; Geol. Survey of Canada Paper 78-11, 23 p.

Gray, J.B. and Ling, R.F., 1984, K-Clustering as a detection tool for influential subsets in regression; Technometrics, v. 26, pp. 305-318.

Harris, D.P., 1984, Mineral resources appraisal; Oxford University Press, New York, 445 p.

Hocking, R.R., 1983, Developments in linear regression methodology: 1959-1982; Technometrics, v. 25, pp. 219-229.

Nie, N.H., Hull, C.H., Jenkins, J.G., Steinbrenner, K. and Kent, D.H., 1975, SPSS Manual, 2nd edition; McGraw-Hill, New York, 675 p.

Tukey, J.W., 1984, Comments on "Use of spatial analysis in mineral resource evaluation"; Jour. Math. Geol., v. 16, pp. 591-594.

Vinod, H.D. and Ullah, A., 1981, Recent advances in regression methods; Dekker, New York, 361 p.

Welsh, R.E., 1982, Influence functions and regression diagnostics; In: Lanner, R.L., and Siegel, A.F. (eds.), Modern data analysis; Academic press, New York, pp. 149-169.

RESOURCE ASSESSMENTS, GEOLOGIC DEPOSIT MODELS, AND OFFSHORE MINERALS WITH
AN EXAMPLE OF HEAVY-MINERAL SANDS

Emil D. Attanasi, John H. DeYoung, Jr., Eric R. Force, and
   Andrew E. Grosz
U.S. Geological Survey
920 National Center
Reston, Virginia 22092, U.S.A.

ABSTRACT. A resource assessment method for offshore minerals based on
descriptive and grade-tonnage models is proposed. Historical
development and applications of this method are summarized. Based on
this approach, descriptive and quantitative deposit models for strandline titanium placer deposits have been developed. Descriptive
statistics were also computed using the worldwide deposit data set
upon which the grade-tonnage models are based. Certain guidelines and
limitations in applying onshore titanium deposit models to offshore
assessment and exploration must, however, be observed. The descriptive model points out the specific features of strandline titanium
placer deposits which can be of use in selection of areas for exploration; the grade-tonnage models display the expected size distribution
for this type of deposit. Used with an estimate of expected number of
deposits, this information can be applied to quantify probable values
associated with deposits of this type within a given area.

INTRODUCTION

   Nonfuel minerals of the offshore--including continental shelves as
well as the deeper seafloor--are a largely untapped resource. Marine
mineral resources have been the subject of scientific investigations for
many years, but comprehensive resource-assessment methods for evaluating
marine minerals have not been developed for legal, political, and
economic reasons. Except for the work published by Hale and McLaren
(1984) for the Canadian offshore, comprehensive assessments of most
areas are at best still in the beginning stages.
   The procedures used in assessing mineral resources on land have
been the subject of much interest in recent years. This interest
results largely from the use of assessment methods that respond to
requests for information about the value of mineral resources. Such
value estimates are used by government and industrial planners concerned
with land-use problems and commodity-supply forecasts. Resource assessments respond to queries about future supply, the location and size of
operations that will produce minerals, the economic and environmental

P. G. Teleki et al. (eds.), Marine Minerals, 485–513.
© *1987 by D. Reidel Publishing Company.*

effects of these operations, and the implications that resource estimates and related production forecasts have for international trade and national security of sources and supplies of minerals.

In this paper, the application of a resource-assessment method based on descriptive and grade-tonnage models of mineral deposits is proposed for offshore mineral resources. The history of the U.S. Geological Survey's (USGS) involvement in assessing mineral resources on land is described to show the purposes of preparing resource estimates and the evolution of a geologic-deposit-model approach to resource assessment. The construction of a grade-tonnage model that might be applicable to offshore resource estimation is then illustrated with the example of strandline placer deposits containing titanium minerals. Finally, guidelines for and pitfalls of applying this model of an onshore heavy-mineral placer deposit to offshore assessment and exploration are discussed.

RESOURCE ASSESSMENT IN THE UNITED STATES

The history of mineral-resource assessment in the United States extends at least to early explorers who came to the New World with a mandate to search for gold and other precious materials. The idea of a comprehensive mineral-resource picture of the United States was suggested in 1833, when a letter was written to the Secretary of War encouraging preparation of a geologic map of the Union in order to obtain a mineralogical description of every state (Featherstonhaugh, 1833). Featherstonhaugh's ideas focussed on an inventory of identified resources, but in promoting the value of geologic mapping as done in Great Britain, Germany, and France, he saw the value of a systematic survey of earth-science information that would enable trained geologists to look beyond identified resources and make statements about the geologic favorability of areas for undiscovered mineral deposits.

Topographic mapping was one of the major goals of the four Territorial Surveys that the U.S. Congress authorized after the American Civil War. The surveys were directed to explore the western United States, but natural resources, including water, minerals, and timber were included in their studies. In 1879, responding to a recommendation of the National Academy of Sciences, Congress combined the work of these four surveys into the U.S. Geological Survey, which was charged to classify the public lands and to examine the geological structure, mineral resources, and products of the national domain.

One problem that faced government planners in 1879 was to inventory recognized or identified mineral resources on the Nation's land. This was expanded in later years to include estimates of mineral resources that were as yet undiscovered, and also information about resources beneath the Nation's offshore waters. Whether onshore or offshore, the estimation of undiscovered resources required approaches different from the traditional "inventory" methods. The new approach had to be based on a framework of geologic and mineral-deposit information as suggested by Featherstonhaugh, and incorporate ways of organizing and analyzing this information to draw inferences about the relative and absolute possibilities of resource occurrence.

Harris (1984) has grouped methods or models used for resource estimation into five classes: economic resource models, quantity-quality models, geological resource models, geostatistical models, and compound models. The choice of a particular resource-estimation method depends on several factors including the existence of or possibility of obtaining required geologic information, the purpose for which the estimate is needed, and the opinions of the estimators and users of the resource assessments about economic versus physical causes of resource scarcity. This last point is exemplified by the long-standing debate between those who predict mineral scarcity on the basis of how long it will take to "exhaust" a current stock of identified resources and those who think that market forces on demand and on technology can provide an adequate supply of mineral resources in spite of lower ore grades and deeper deposits. Specifically, Vogely (1983) has questioned whether efforts spent on mineral-resource assessments are worthwhile because they do not adhere to consistent terminology and because incomplete data about undiscovered resources are often misinterpreted.

If the planner whose problems include marine mineral resources believes that "results of resource assessment methods are so imprecise that a policy based on them may be worse than a policy that recognizes the complete ignorance of potential reserves" (Vogely, 1983), then the choice of an assessment method for offshore minerals becomes a moot question. Because it is proposed here that a resource-assessment method that incorporates geologic and mineral-deposit information be adapted to offshore minerals, it is appropriate to continue with a description of how such a resource-assessment method came to be used by the USGS.

Resource terminology

The terms used to describe different economic classes and levels of geologic knowledge about mineral resources tend to confuse both resource analysts and users of resource estimates. The "McKelvey box" classification is used in this paper (Fig. 1), but it should be noted that equivalent terms have been agreed upon by other groups. The definitions agreed upon by a United Nations expert group (Fig. 2) has rough equivalence to the terms of the McKelvey box and has been successfully used by several international mineral study groups (DeYoung et al., 1984). The problem of non-uniformity in resource terminology is discussed in the papers by Archer and Harris (this volume).

Deposit-type resource assessments

One of the first comprehensive regional resource assessments in the United States was a study by the USGS of the region around Boulder Dam (Arizona-California-Nevada-Utah). The purpose of this study was to estimate the availability of mineral resources that could serve developing industries that, in turn, would consume power from the dam (Hewett et al., 1936). This report is essentially a descriptive listing of identified deposits organized by mineral commodity and area. Nolan (1950), one of the authors of the 1936 report, took a retrospective look at the Boulder Dam study and made some inferences about undiscovered

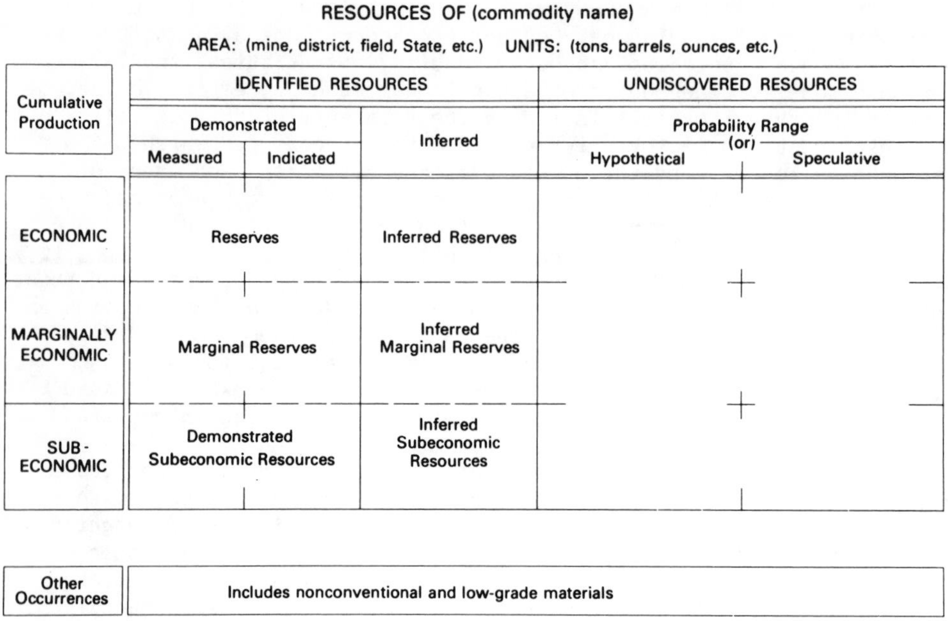

RESOURCES OF (commodity name)
AREA: (mine, district, field, State, etc.) UNITS: (tons, barrels, ounces, etc.)
Cumulative Production
IDENTIFIED RESOURCES
UNDISCOVERED RESOURCES
Demonstrated
Measured
Indicated
Inferred
Probability Range
(or)
Hypothetical
Speculative
ECONOMIC
Reserves
Inferred Reserves
MARGINALLY ECONOMIC
Marginal Reserves
Inferred Marginal Reserves
SUB-ECONOMIC
Demonstrated Subeconomic Resources
Inferred Subeconomic Resources
Other Occurrences
Includes nonconventional and low-grade materials

Figure 1. Major elements of USBM/USGS mineral-resource classification scheme (from U.S. Bureau of Mines and U.S. Geological Survey, 1980).

R*
IN SITU RESOURCES—
Quantities of economic interest
for the next few decades
R1
KNOWN DEPOSITS—
Reliable estimates
R2
EXTENSIONS OF KNOWN AND
NEWLY DISCOVERED DEPOSITS—
Preliminary estimates
R3
UNDISCOVERED DEPOSITS—
Tentative estimates
R1E
Economically exploitable
R1M
Marginally economic
R1S
Subeconomic
R2E
Economically exploitable
R2S
Subeconomic

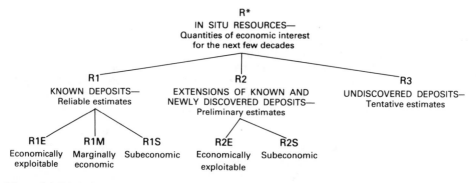

*The capital "R" denotes resources *in situ*; a lower case "r" expresses the corresponding *recoverable* resources for each category and subcategory. Thus, r1E is the recoverable equivalent of R1E.

Figure 2. United Nations resource categories (modified from Schanz, 1980).

resources. He noted that the number of mining districts in a given geologic province in each State was proportional to the area of that province in each State, and concluded that, for very large areas, there is roughly the same amount of ore material introduced per unit area.

Assessment of undiscovered mineral resources in proposed "wilderness areas" is a part of the responsibility given to the USGS and the U.S. Bureau of Mines by the Wilderness Act of 1964. In a publication that summarizes nearly 800 areas studied during 20 years of minerals investigations in the areas, maps show areas that were judged to have "substantiated" or "probable" mineral-resource potential (Marsh et al., 1984). Subjective opinion based on geologic, geochemical, and geophysical information is the basis for these assessments, which in many cases aggregate different mineral commodities and mineral-deposit types.

Grade-tonnage models and deposit-type assessment in Alaska

In 1974, the USGS started a program of mineral-resource-assessment studies of 1:250,000-scale quadrangles in Alaska in order to provide information for decisions about the future disposition of land in that State. A new type of resource-assessment method was used in these studies. This method was based on explicit statement of several physical characteristics of the mineral deposits--the characteristics being observed for the identified resources and expected for the undiscovered resources. Such a method recognizes that the effects of future changes in economics and technology should be able to be judged from a resource assessment that attempts to separate physical factors from their technologic and economic implications. However, resource estimates should be conditioned on earth-science theory; technology for exploration, mining, processing, and mineral use; and economics of producing and selling the mineral raw materials (Singer, 1975).

The first quadrangle studied under this Alaskan Mineral Resource Assessment Program (AMRAP) was the Nabesna quadrangle in eastern Alaska. The resource-assessment map (Richter, Singer, and Cox, 1975) was the result of a three-step process shown in Figure 3:

1) Descriptions of mineral-deposit types (deposit models) were prepared and statistical data on the tonnages and grades of well-explored examples of these deposit types were collected for use as a quantitative measure of possible occurrences;

2) tracts of land that are geologically permissive for the occurrence of the types of mineral deposits specified in step 1 were delineated; and

3) the number of undiscovered deposits of specified types were estimated. This is a subjective estimate that is based on known deposits and the possibility of undiscovered deposits as inferred from knowledge of similar geologic settings elsewhere.

This method of assessment produced a listing of deposit types

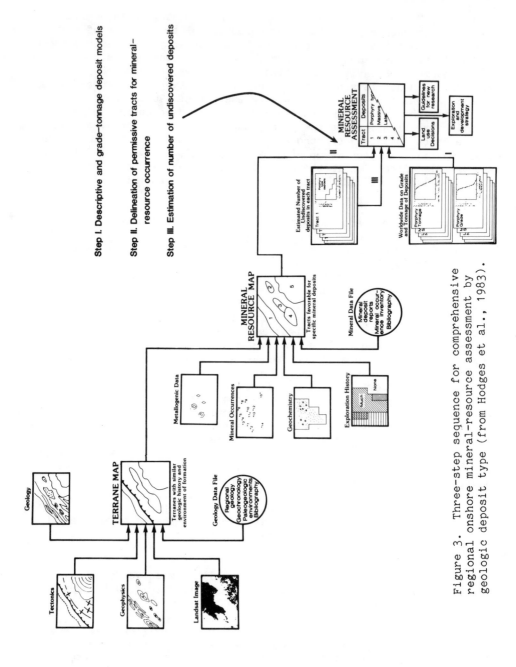

Figure 3. Three-step sequence for comprehensive regional onshore mineral-resource assessment by geologic deposit type (from Hodges et al., 1983).

containing identified and estimated undiscovered resources in a tract, the geologic controls on these mineral resources, production and resource information concerning the tract's identified resources, a statement about the status of geologic information and exploration history in the tract, and a mineral-resource summary including a probability distribution of estimated number of deposits and references to the appropriate grade-tonnage models (Singer and Ovenshine, 1979). A distinction should be made between what deposits are expected to occur and what deposits are expected to be discovered in some future time period. The latter would require information about exploration techniques and discovery probabilities that are explicitly omitted in this assessment method.

This type of resource assessment can, with the use of Monte Carlo simulation, produce frequency distributions of metal (or mineral) contained in deposits that are expected to occur in the area studied. A deposit-model-based assessment and simulation of contained metal curves was applied to wilderness areas in the western United States (Drew et al., 1986). This study was limited to 14 deposit types and 11 metals contained in them.

Evolution of the grade-tonnage data base

When the first AMRAP quadrangle, Nabesna, was studied in 1975, only two grade-tonnage models were available for the assessment. In 1978, data had been collected to construct 11 statistical grade-tonnage models for mineral deposits for use in the 1:1,000,000-scale USGS assessment of Alaskan public lands (examples are shown in Table I). Grade-tonnage models were also used in the assessment of quadrangles in the conterminous United States, where graphical presentations of the grade and tonnage frequency curves were presented (Singer et al., 1983). Subsequent applications of this assessment method by the USGS in the United States and in other countries have resulted in 65 descriptive models of mineral-deposit types (Cox, 1983a, b) and in 37 associated numerical grade and tonnage models (Singer and Mosier, 1983a, b). The Geological Survey of Canada also has recently published a set of models of geologic deposits (Eckstrand, 1984).

Available grade-tonnage models that may be applicable to offshore mineral resources

The collection of tonnage and grade statistics for offshore deposits is a formidable task because so little is known about the size, physical characteristics, or even location of marine mineral deposits. However, some of the grade-tonnage models constructed for deposits on land may provide useful analogs for offshore resource assessments. The models may also be used as a "base case" by changing calculated parameter estimates in the summary statistics to agree better with what is known about the resources being studied. For example, in the assessment of resources in porphyry-type copper deposits in the Nabesna, Alaska, quadrangle; Richter et al. (1975) used tonnage data from a model based on information about similar-type deposits in western Canada. However,

Table I. Four of the eleven grade-tonnage deposit models used in the 1:1,000.000-scale assessment of public interest lands in Alaska. Source: Hudson and DeYoung, 1978, p. 54. [Related data occur on line from column to column; all data in metric units; NS, not significant]

Deposit type	Tonnage & grade variables	Number of deposits used in developing model	Correlation coeff. (listed variable with variable on line with it in column 2)	90% of deposits	50% of deposits	10% of deposits
Porphyry	Ore tonnage (10^6 tons)	41		20	100	430
Copper	Average copper grade (% Cu)	41	with ore tonnage = -0.07 NS	0.1	0.3	0.55
	Average molybdenum grade (% Mo)	41		0.0	0.008	0.031
Podiform Chromite	Tonnage of Cr_2O_3 (tons)	268		15	200	2.700
Nickel	Tonnage of ore (10^6 tons)	48		0.23	1.20	5.90
Sulfide	Average nickel grade (% Ni)	48	with ore tonnage = -0.03 NS	0.32	0.61	1.20
	Average copper grade (% Cu)	48	with ore tonnage = 0.03 NS	0.18	0.47	1.20
			with nickel grade = 0.04 NS			
Vein Gold	Contained gold (tons)	43		0.29	3.30	38.00

grade values for the analogue Canadian deposits were reduced because of the lower grades observed in the few identified porphyry systems in the Alaskan study area. Such modifications could also be made when onshore deposit models are used as analogs for offshore resources.

The list below suggests some models that are being prepared for a comprehensive publication of USGS deposit models and that may be applicable to several types of offshore resources (D. A. Singer, written communication, June 1985). Models do not include all mineral commodities that are of economic interest. Noteworthy among the omitted commodities are construction materials such as sand and gravel.

        Heavy-mineral placers
            placer gold (some platinum-group minerals)
            placer platinum-group-minerals (some gold)
            strandline placer titanium
            alluvial placer tin

        Metalliferous oxides
            volcanogenic manganese
            sedimentary manganese

        Metalliferous sulfides
            Cyprus massive sulfide
            Besshi massive sulfide
            kuroko massive sulfide
            Blackbird cobalt-copper

        Phosphate
            phosphate, upwelling type
            phosphate, warm-current type

The following section describes the construction of one of these grade-tonnage models--for placer titanium mineral deposits.

ONSHORE BEACH-COMPLEX TITANIUM PLACER DEPOSITS--EXAMPLE OF MODEL CONSTRUCTION

Descriptive model

    The descriptive and grade-tonnage models discussed here apply to placer deposits of detrital heavy titanium minerals and associated co-products and by-products which occur in beach-complex or strandline deposits, broadly defined so as to include beach, aeolian dune, inlet, and washover-fan deposits. This type of placer deposit is important as it produces a large portion of the world's titanium minerals, zircon, and other minerals.

    Most of the deposits included in the model were formed within the last million years or so (some are still forming) during stands of sea level that are the same elevation or above that of the present day. The major strandline deposits considered are located in Australia, the eastern United States, Africa, and the Indian subcontinent. Many of

these deposits still have geomorphic expressions one would expect from
old beaches, and have neither been buried nor removed by subaerial
erosion. Only a few of the deposits are indurated. The model also
includes older deposits which were buried by other sediments and some
which were lithified or indurated.

The deposits consist essentially of well-sorted, rounded sand,
generally fine- or medium-grained. Sedimentary structures and
lithologic sequences within these sands are typical of beach, dune,
inlet, and/or washover-fan environments.

The economic minerals concentrated in these deposits are ilmenite
(and its alteration or weathering products), rutile, zircon, some
aluminosilicates, monazite, and locally a few others (see Table II).
Other heavy minerals such as magnetite, amphibole, and pyroxene are
commonly only minor constituents, and where present are currently viewed
by the industry as detracting from the value of a deposit. Thus, this
placer deposit type is usually quite distinct in mineralogy from those
in which chromite, gold, etc., are the important economic minerals.

The mineral assemblage of this deposit type results from a
coincidence of favorable factors which may not be readily apparent in
the deposit itself. The two most important are a source terrane (most
commonly of high-grade metamorphic rocks) that contains the proper
minerals, and a conduit from source to depositional site which not only
supplies the minerals but weathers them, thereby upgrading the economic
component of the assemblage. Further discussion is beyond the scope of
this paper but is treated in some detail by Force (1976). A last
factor, vigorous hydraulic sorting at the depositional site, may be
observed from the deposit itself.

Geophysical exploration for the deposits can be successfully
accomplished with gamma aeroradiometry (preferably spectral) and with
induced polarization (latter by Wynn, Grosz, and Foscz, 1985).

In the grade-tonnage model described in this report, great care has
been taken to list different geologic entities separately to give a true
picture of deposit size. This approach requires separating two or more
such entities, which may be reported together because they are mined
together. This disaggregation has not been available in previous
listings of titanium mineral resources.

Empirical analysis

Data set and deposit characteristics

Data on grades and tonnages of ore[1] of strandline titanium placer
deposits were gathered from sources that included the U.S. Bureau of
Mines Minerals Availability System, U.S. Government (Geological Survey

---

[1] The term "ore" is frequently used in this report to refer to tonnages
of heavy-mineral-bearing sand that may not meet the miner's
profitability criteria usually associated with that term. Thus "ore
tonnage" is the sand body, as distinguished from the tonnages of
minerals contained in the sand.

Table II.   Descriptive deposit model for strandline placer titanium
            Source: deposit model draft authored by E. R. Force (1985)
                     for U.S. Geological Survey deposit-model compilation

Description: Ilmenite and other heavy minerals concentrated by beach
processes and enriched by weathering (Force, 1976).

GEOLOGIC ENVIRONMENT

Rock Types: Well-sorted medium- to fine-grained sand in dune, beach,
and inlet deposits commonly overlying shallow marine deposits.
Age Range: Commonly Miocene to Recent but may be any age.
Depositional Environment: Stable coastal region receiving sediment from
deeply weathered metamorphic terranes of sillimanite or higher grade.
Tectonic Setting(s): Margin of craton. Crustal stability during
deposition and preservation of deposits.

DEPOSIT DESCRIPTION

Mineralogy: Altered (low-iron) ilmenite ± rutile ± zircon. Trace of
monazite. Magnetite, pyroxene, and amphibole are rare or absent.
Quartz greatly exceeds feldspar.

    Economic minerals concentrated in this type of deposit:
      Primary products
            ilmenite - $FeTiO_3$ (46-65% TiO_2)
            leucoxene - alteration product of ilmenite (60-75% TiO_2)
            rutile - TiO_2 (about 95% TiO_2)
      Co-products or by-products
            zircon - $ZrSiO_4$ (about 67% ZrO_2)
            monazite - (Ce, La, Nd, Th) PO_4 (56-70% rare-earth oxides)
            xenotime - YPO_4
            staurolite - $(Fe, Mg, Zn)_2Al_9 (Si, Al)_4O_{22}(OH)_2$
            sillimanite - Al_2SiO_5
            kyanite - Al_2SiO_5

Texture/Structure: Elongate "shoestring" ore bodies parallel to
coastal dunes and beaches.
Ore Controls: High-grade metamorphic source; stable coastline with
efficient sorting and winnowing; weathering of beach deposits.
Weathering: Leaching of iron from ilmenite and destruction of labile
heavy minerals results in residual enrichment of deposits.
Geochemical and Geophysical Signature: High titanium, zirconium,
rare-earth element, thorium, and uranium. Gamma radiometric anomalies
resulting from monazite content. Electrically induced polarization
anomalies from ilmenite.

Examples: Green Cove Springs, Florida   Richards Bay, South Africa
           Trail Ridge, Florida   Eneabba, Australia
           Lakehurst, New Jersey

and Bureau of Mines) files, publications, and industry sources. The
final data set includes 62 deposits or occurrences around the world.
There were only a few large deposits around the world for which no data
were available.

The deposit file includes currently operating and closed mines, as
well as prospects or occurrences that have not been mined to date. For
each deposit, an estimate of the original ore in place was computed
using estimates of remaining resource tonnage and tonnage of past
production, where applicable. Either rutile, ilmenite, leucoxene, or a
combination of these three minerals was present in all deposits where at
least one of the three minerals would be the primary product of mining.
Where available, grades of zircon and monazite were also collected.

Data from separate mining operations on a single ore body were
aggregated to maintain the geologic integrity of the deposit record.
Data were also aggregated into one deposit in a few cases where several
small ore bodies occurred within a relatively small area and where these
ore bodies appeared to be geologically related. The grades of contained
TiO_2 in rutile, ilmenite, and leucoxene were aggregated to calculate a
total TiO_2 grade for each deposit. This provided a "least common
denominator" for comparison among deposits, where several
titanium-mineral products with arbitrary definitions were typically
reported in the original deposit data. Another reason for proceeding in
this way was that although one might consider a weighted grade of
TiO_2-heavy minerals based on market value, market prices change daily.
Note that TiO_2 figures used here are not equivalent to such figures
obtained by bulk chemical analysis of the host sand, a procedure that
includes TiO_2 in all mineral forms in the sample, including those of no
economic value.

The observed deposit size distribution (by ore tonnage) is highly
skewed as is the distribution of deposit TiO_2 grades. Chi-square tests
show that the hypotheses that the logarithms of tonnage and total TiO_2
grade were normally distributed could not be rejected. This suggests
that the lognormal distribution provides a reasonable approximation to
the observed distributions of these variables.

Table III shows the proportion of cumulative ore contained in
various size classes of deposits along with the associated sample
statistics. In both cases, the mean is greater than the median (and
mode). This indicates a positive skew to the distribution. The table
shows that the largest deposit alone accounts for 37 percent of the ore
and 19 percent of the contained TiO_2 tonnage in all deposits. The
largest three deposits (5 percent of all deposits) account for 67
percent of the ore and 26 percent of the contained TiO_2. Two of the 38
deposits in the smallest size class together account for 13 percent of
the contained TiO_2 tonnage in all deposits. The other 36 deposits in
this class account for less than 14 percent of the total TiO_2.
Comparison of the percentage of total ore and percentage of total TiO_2
contained in each size class indicates that tonnage and grade might vary
independently.

When the deposits are ordered by their size in terms of contained
TiO_2, the distribution shows characteristics similar to Table III. The
largest three deposits (in terms of contained TiO_2) account for 45

Table III. Deposit size distribution by ore tonnage contained in
            onshore strandline deposits[1]

Classes (10^6 t)	No. of deposits in class	Cumulative no. of deposits	% of total ore in class	Cum. % of total ore	% of total TiO_2 in class	Cum. % of total TiO_2
2250-2400	1	1	37.3	37.3	19.3	19.3
2100-2250	0	1	0.0	37.3	0.0	19.3
1950-2100	0	1	0.0	37.3	0.0	19.3
1800-1950	0	1	0.0	37.3	0.0	19.3
1650-1800	1	2	19.9	57.2	2.6	21.9
1500-1650	0	2	0.0	57.2	0.0	21.9
1350-1500	0	2	0.0	57.2	0.0	21.9
1200-1350	1	3	9.7	66.9	4.6	26.5
1050-1200	0	3	0.0	66.9	0.0	26.5
900-1050	1	4	6.5	73.4	1.4	27.9
750-900	2	6	9.1	82.5	4.6	32.5
600-750	2	8	7.0	89.5	2.8	35.3
450-600	2	10	4.4	93.9	14.6	49.9
300-450	2	12	1.9	95.8	5.0	54.9
150-300	12	24	2.9	98.7	18.8	73.8
0-150	38	62	1.3	100.0	26.3	100.0

[1] Sample statistics are: mean - tonnage 251×10^6 t, TiO_2 grade 3.26%;
   median - tonnage 102×10^6 t, TiO_2 grade 1.98%; standard deviation -
   tonnage 426×10^6 t, TiO_2 grade 4.36%.

percent of total TiO$_2$ tonnage and the largest 11 deposits account for 72
percent of the total TiO$_2$. The smallest 43 deposits (or nearly 70
percent of the deposits) account for less than 15 percent of total
contained TiO$_2$.

The linear correlation coefficient between the logarithm of tonnage
and the logarithm of the TiO$_2$ grade variable is -0.35. For the raw
data, the correlation coefficient is -0.16. Only the correlation based
on the logarithms is statistically significant at the 1 percent level.
Significance tests of raw data correlations are technically not
appropriate because these data are not normally distributed. The
correlations between the log transforms of ore tonnage and grades for
rutile, ilmenite, and leucoxene are smaller and none are statistically
significant.

Figure 4 shows a scatter plot of the logarithm of the TiO$_2$ grade
versus the logarithm of ore tonnage. The figure appears to show little
correlation between grade and tonnage. The correlation that does exist
is due to the limitations of the data set. Usually, such grade and
tonnage plots exhibit a degree of economic truncation. There are only a
few observations in the lower left hand part of the scatter plot.
Typically, very small and low-grade occurrences are not reported because
they are of little economic interest. The smaller occurrences shown in
Figure 4 were identified principally as a result of academic
investigations or because of their association with desirable
radioactive raw materials. The relationships between deposit tonnage
and individual mineral grades in the empirical grade-tonnage models are
considered in more detail in the next section.

Empirical grade-tonnage models

Figures 5 through 11 present the empirical grade and tonnage model
based on the strandline placer titanium deposit data set. Table IV
provides basic sample statistics including the median and first and
third quartile values for each variable that is considered. Figure 5
shows the deposit ore-tonnage values compared to the associated
proportion of deposits with tonnages of at least that value. The figure
shows that 75 percent of the deposits contain at least 36 million metric
tons of ore. Alternatively, only 25 percent of the deposits contain
more than 219 million metric tons. The correlations between ore tonnage
and mineral grades for the titanium minerals (rutile, ilmenite, and
leucoxene), zircon, and monazite are negative but very weak and none
are significant at the 1 percent level. This is also true for the
correlation between the logarithm of tonnage and the logarithm of grades.
This relation between grade and tonnage is usual for many mineral-
deposit types.

Figure 6 shows the deposit TiO$_2$ grade (from rutile, ilmenite, and
leucoxene) compared to the associated proportion of deposits having
grades of at least that value. The median value or 50 percent fractile
was 1.98 percent. Total TiO$_2$ grades for 75 percent of the deposits are
at least 0.84 percent and 25 percent of the deposit TiO$_2$ grades are
higher than 3.50 percent. As expected, a positive and statistically
significant correlation exists between total TiO$_2$ grade and each of the

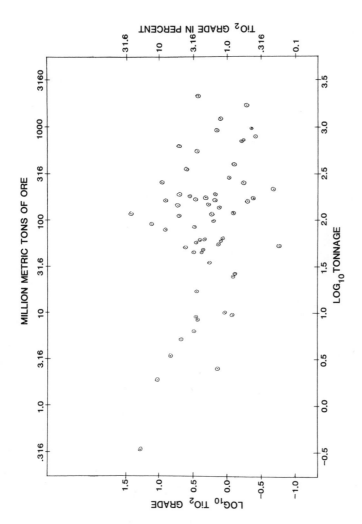

Figure 4. Tonnage of ore versus deposit TiO$_2$ grades for 62 strandline placer titanium mineral deposits.

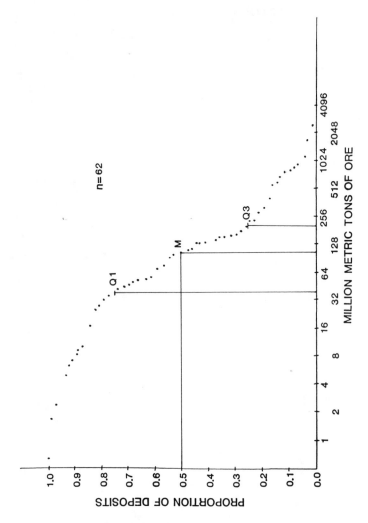

Figure 5. Deposit tonnages of 62 strandline placer titanium deposits
compared to the associated proportion of total deposits with at least
that tonnage value. Symbols marked with Q1 and Q3 denote first (lower)
quartile and third quartile and M is the sample median. See Table III
for other sample statistics.

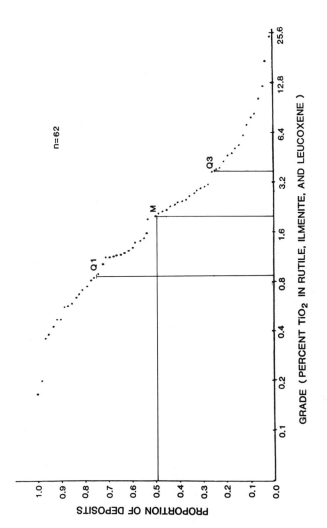

Figure 6. TiO$_2$ grades of 62 strandline placer titanium deposits compared to the associated proportion of total deposits with at least that grade. Symbols marked with Q1 and Q3 denote the first (lower) quartile and third quartile and M is the sample median. See Table IV for other sample statistics.

Table IV.--Statistics of strandline placer titanium deposits (grades are
percent oxide in ore body, not percent oxide in mineral)

Variable	n	Mean	Standard deviation	Median	First quartile (Q1)	Third quartile (Q3)
Ore tonnage (10^6 metric tons)	62	251	426	102	36	219
Grades						
Total TiO_2 (%)	62	3.25	4.36	1.98	0.84	3.50
Rutile (% TiO_2)	50	.332	.339	.206	.107	.423
Ilmenite (% TiO_2)	62	2.76	3.95	1.39	0.68	3.17
Leucoxene (% TiO_2)	25	.552	.923	.270	.079	.480
Zircon (% ZrO_2)	53	.398	.368	.280	.125	.530
Monazite (% rare-earth oxides)	29	.181	.477	.020	.010	.120

rutile, ilmenite, and leucoxene grade levels. A statistically
significant correlation also exists between the logarithmic transforms
of total TiO_2 and zircon grade ($r = 0.57$) and monazite grade ($r = 0.66$).
    In Figures 7, 8, and 9 the grades for rutile, ilmenite, and
leucoxene in terms of percent TiO_2 are plotted versus the proportion of
total strandline placer titanium deposits that have at least those
grades. Chi-square tests of the hypotheses that the logarithms of these
variables are normally distributed could not be rejected. There were 50
deposits in this data set with reported rutile grades and 25 deposits
with reported leucoxene grades. The median rutile grade is 0.206
percent and the first and third quartile values are 0.107 and 0.423
percent, respectively. Only the correlation of the log transforms of
rutile grades and zircon grades ($r = 0.49$) is significant at the 1
percent level.
    Ilmenite grade, in terms of TiO_2 content, is plotted with the
associated proportion of deposits that are at least of that grade in
Figure 7. Ilmenite grades were available for all of the 62 titanium
placer deposits. Grades ranged from 0.04 percent to 24.6 percent and
the median value was 1.39 percent and the first and third quartiles
were 0.68 and 3.17 percent. The log-transformed ilmenite grades were
significantly correlated with leucoxene ($r = 0.75$), zircon ($r = 0.47$),
and monazite ($r = 0.66$) grades.
    Leucoxene grades (only 25 were reported) are shown in figure 9.
It is probable that the leucoxene grade was sometimes combined with the
ilmenite grade because the terms leucoxene and altered ilmenite were
used somewhat interchangably. The correlation between log-transformed
leucoxene and zircon grades is not statistically significant. Simi-

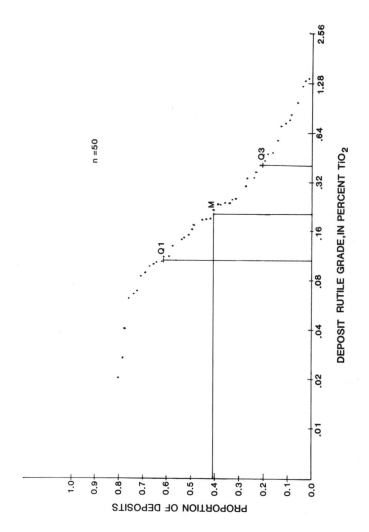

Figure 7. Rutile grades (in percent TiO_2) of 50 strandline placer titanium deposits (where rutile grade is reported) compared to the associated proportion of total strandline titanium deposits with at least that grade. Symbols marked with Q1 and Q3 denote the first (lower) and third quartile and M is the sample median. See Table IV for sample statistics.

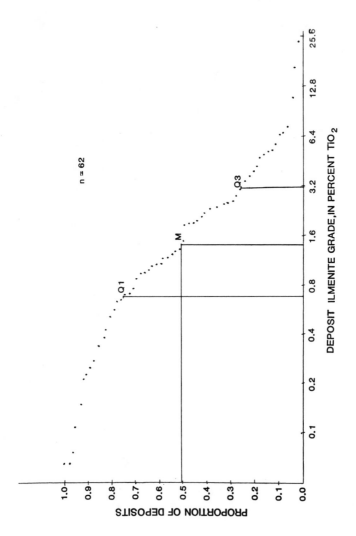

Figure 8. Ilmenite grades (in percent TiO_2) of 62 strandline placer titanium deposits compared to the associated proportion of total deposits with at least that grade. Symbols marked with and Q3 denote the first (lower) quartile and third quartile and M is the sample median. See Table IV for other sample statistics.

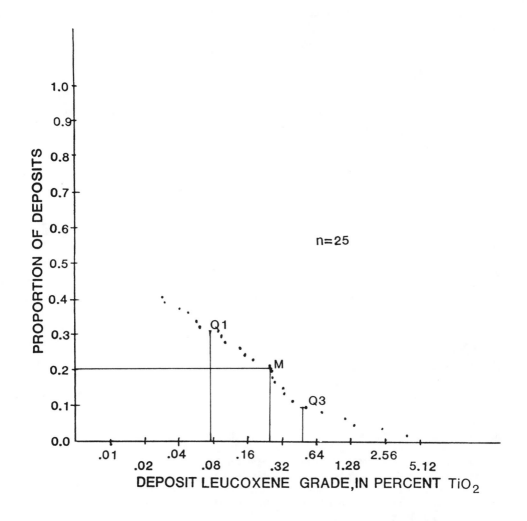

Figure 9. Leucoxene grades (in percent TiO$_2$) of the 25 strandline
placer titanium deposits (where leucoxene is reported) compared to the
associated proportion of total strandline titanium deposits with at
least that grade. Symbols marked Q1 and Q3 denote the first (lower) and
third quartile and M is the sample median. See Table IV for sample
statistics.

larly, the correlation between leucoxene and monazite grades is not significant.

Figures 10 and 11 show grades of by-product zircon (in percent ZrO_2) and monazite, in percent rare-earth oxides, respectively. Zircon grades were reported for 53 deposits and 29 deposits with monazite grades were reported. The log-transformed zircon grades are significantly correlated with (logs of) rutile and ilmenite grades. The log-transformed monazite grades have a significant correlation only with the ilmenite grade logs. Data show that zircon and monazite grades are also not correlated significantly with one another.

Because the foregoing grade-tonnage models are based on a sample of data on identified deposits, caution should be exercised in applying it to the assessment of yet undiscovered resources. The sample is biased in the sense that one would expect a high proportion of even the 'discovered' small low-grade occurrences to go unreported because they were of little commercial interest. However, the model presented here may be regarded as a starting point from which to proceed to further refinement.

APPLICATION OF MODEL TO OFFSHORE AREAS

Offshore deposits - similarities and differences to onshore deposits

Placer deposits of titanium minerals offshore are likely to be in one of three types of sediment deposits which are found onshore. One is well-sorted sands of a beach complex or strandline deposit, formed at a previously lower stand of sea level, and subsequently submerged by marine transgression. For this type of deposit, the descriptive and grade-tonnage model described here (Table II) is well suited, except for possible changes in deposit shape as discussed below. Another type of a well-sorted linear sand body is the shore-tied or isolated sand shoal of Duane et al. (1972). Considerable evidence suggests that these bodies formed in near-shore regions and may be quite different from the deposits used in the strandline model. These shore-tied sand shoals may consist of reworked material from either older strandline sand bodies formed in the same area or from the modern beach, and they may have inherited many sedimentologic characteristics of these sand bodies. A third type is older fluvial deposits of less well-sorted sands and gravels, including buried river channels and former estuaries and deltas. Thus, the grade-tonnage model for strandline titanium-mineral deposits is probably the most important in considering offshore titanium-mineral resources, but does not cover all the possibilities.

An important systematic difference to be expected between onshore and offshore examples of strandline sand bodies is their shapes. A relict offshore deposit by definition has undergone transgression during rising sea level. A large number of studies document the importance of shoreface (and inlet) erosion as transgression takes place, resulting in modification of sand-body shape. Conclusive evidence for transgressed barriers is rare, but those present have had dunes eroded and the upper several meters are completely reworked by wave and current action. In

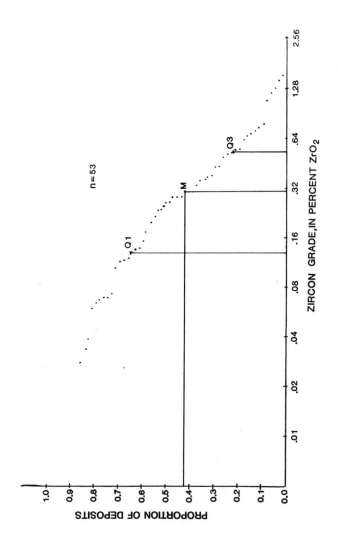

Figure 10. Zircon grades (in percent ZrO_2) of the strandline placer titanium deposits (where zircon is reported) compared to the associated proportion of total strandline titanium deposits with at least that grade. Symbols marked Q1 and Q3 denote first (lower) and third quartile and M is the sample median. See Table IV for sample statistics.

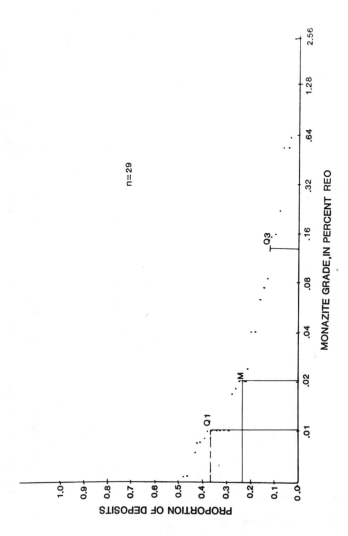

Figure 11. Monazite grades, in percent rare earth oxides (REO), of the
25 strandline placer titanium deposits (where monazite is reported)
compared to the associated proportion of total strandline titanium
deposits with at least that grade. Symbols marked Q1 and Q3 denote the
first (lower) and third quartile and M is the sample median. See Table
IV for sample statistics.

shown that erosion of the entire beach sequence is common during rapid marine transgression, resulting in redistribution of all the material formerly in shoreline deposits.

The strandline type of deposit, if present offshore, might be expected to show a somewhat different mineral assemblage than onshore deposits. Fragmentary evidence available thus far indicates that the offshore deposits contain a less-weathered mineral assemblage.

The geologic knowledge of the continental shelf with respect to heavy minerals is presently limited. Therefore, the continental shelf cannot be precisely subdivided into provinces that have different heavy-mineral potentials. In the future, such divisions will probably become increasingly possible, and the grade-tonnage model (which should by then also be improved by further discrimination of deposit types) should then be reapplied to a zoned continental shelf. The present model should be regarded merely as a first attempt.

Exploration - guidelines and methods

Water depth and the general harsh offshore environment hinder the direct application of onshore exploration techniques to the offshore. Some onshore exploration techniques may be used either at the regional (reconnaissance) level or at the site-specific level and others may be applied in both contexts. Onshore, regional geologic and geomorphologic studies will typically provide evidence about whether conditions are conducive to the formation of titanium-mineral deposits of commercial size and grade. Such regional studies may provide evidence that source rocks, conduits, concentrating mechanisms are present so that deposits can form. Surface geologic investigations, geochemical sampling, and topographic mapping typically provide the basis for these regional studies. From such large-scale surveys, areas may be identified for further examination. Within such areas, high-resolution aeromagnetic data and spectral aeroradioactivity data (Force et al., 1982; Grosz, 1983) might identify or provide clues on anomalies. Additional information on soils and land cover would also aid in the interpretation of these data. Field studies to delineate specific anomalies might proceed from surface geologic studies, gamma-ray spectrometric studies, and induced-polarization studies, to physical sampling by coring or jet drilling and geochemical analysis for the appraisal of identified resources.

Some of these onshore exploration techniques for heavy minerals would require modification for offshore application and others may not be applicable at all. Seawater cover prohibits collection of aeroradio- activity data. The first objective of offshore exploration is the location of large-volume well-sorted deposits of fine-to-coarse sand which is suitable heavy-mineral concentrations. Major capes and high-energy beaches are principal onshore centers of deposition of sand, and these types of areas have high heavy mineral potential. Offshore, drowned analogues may be detected in part by relict physiography. However, the presence of these permissive conditions for favorable hosts or even the presence of nearby onshore deposits is not sufficient to ensure that commercial concentrations of heavy minerals are present in the offshore.

Offshore surveys for heavy minerals rely on the use of high-resolution seismic surveys and side-scan sonar to map seafloor stratigraphy and topography. These will also aid in identifying favorable geomorphologic features such as ancient stream channels or submerged beach-complex deposits. Most of the offshore deposits are expected to be on the continental shelf.

High-resolution seismic profiles should aid in determining the areal and vertical extent of major sand bodies. Spectrometric techniques and induced-polarization techniques are being adapted for offshore conditions. Spectrometer systems must be in contact with the sediments. A newly designed offshore system permits ship-towing of instruments across the seabed at speeds as high as 4 knots (Mining Journal, 1983). Experimental work on an induced-polarization system for offshore application is also in progress (Wynn, Grosz, and Foscz, 1985).

Offshore sediments have been sampled conventionally with bottom grabs, dredges, piston corers, or vibracorers. Vibracores not only provide adequate samples for assessing minerals present, but also aid in deciphering the geology and the geometry in three dimensions of any heavy-mineral concentrations.

CONCLUSION

This example shows both the potential and the limitations of applying mineral-deposit models developed onshore to exploration target areas offshore. The descriptive model points out specific features of strandline sand deposits of titanium minerals which can be used to select offshore areas of higher potential for these deposits, and the grade-tonnage models display their expected size and grade distribu-tions. Coupled with expected-number estimates, this information can be used to quantify probable total values of contained deposits of this type in a given offshore tract. The limitations arise as a result of three factors: 1) strandline deposits will inherently have different shapes and mineralogies because they are offshore, 2) several closely related deposit types are possible offshore, of which only one has been modelled, and 3) several exploration methods used onshore require extensive modification before being used offshore.

ACKNOWLEDGEMENTS

The authors are grateful to Donald A. Singer and S. Jeffress Williams, both of the U.S. Geological Survey, for helpful comments and suggestions on the original manuscript. Assistance provided by Janet B. Chiang in data compilation and drafting is also gratefully acknowledged.

REFERENCES CITED

Archer, A. A., this volume, Sources of confusion: What are marine resources?; pp. 411-422.

Cox, D. P., (ed.), 1983a, U.S. Geological Survey-İNGEOMINAS mineral resource assessment of Colombia; Ore deposit models; U.S. Geol. Surv. Open-File Rept. 83-423, 65 p.

Cox, D. P., (ed.), 1983b, U.S. Geological Survey-INGEOMINAS mineral resource assessment of Colombia; Additional ore deposit models; U.S. Geol. Surv. Open-File Rept. 83-901, 31 p.

DeYoung, J. H., Jr., Sutphin, D. M., and Cannon, W. F., 1984, International Strategic Minerals Inventory Summary Report-- Manganese; U.S. Geol. Surv. Circ. 930-A, 22 p.

Drew, L. J., Bliss, J. D., Bowen, R. W., Bridges, N. J., Cox, D. P., DeYoung, J. H., Jr., Houghton, J. C., Ludington, Steve, Menzie, W. D., Page, N. J, Root, D. H., and Singer, D. A., 1986, Quantitative estimation of undiscovered mineral resources: A case study of U.S. Forest Service wilderness tracts in the Pacific Mountain System; Economic Geology, v. 81, no. 1, pp. 80-88.

Duane, D. B., Field, M. E., Meisburger, E. P., Swift, D. J. P., and Williams, S. J., 1972, Linear shoals on the Atlantic Inner Continental Shelf, Florida to Long Island; In: Swift, D. J. P., Duane, D. B., and Pilkey, O. H., eds., Shelf sediment transport; Process and pattern; Dowden, Hutchinson and Ross, Stroudsburg, Pa., pp. 499-575.

Eckstrand, O. R., (ed.), 1984, Canadian mineral deposit types: A geological synopsis; Geol. Surv. of Canada, Econ. Geol. Rept. 36, 86 p.

Featherstonhaugh, G. W., 1833, Letter dated January 12, 1833, to Lewis Cass, Secretary of War; In: Public documents of the Senate of the United States, 22d Congress, 2nd Session, vol. I, document 35, p. 5-7.

Fischer, A. G., 1961, Stratigraphic record of transgressive seas in the light of sedimentation on the Atlantic Coast of New Jersey; Amer. Assoc. of Petrol. Geol. Bull., v. 45, pp. 1656-1666.

Force, E. R., 1976, Metamorphic source rocks of titanium placer deposits; A geochemical cycle; U.S. Geol. Surv. Prof. Paper 959-B, 16 p.

Force, E. R., Grosz, A. E., Loferski, P. J., and Maybin, A. H., 1982,
    Aeroradioactivity maps in heavy-mineral exploration; Charleston,
    South Carolina area; U.S. Geol. Surv. Prof. Paper 1218, 19 p.

Grosz, A. E., 1983, Application of total-count aeroradiometric maps to
    the exploration for heavy-mineral deposits in the coastal plain of
    Virginia; U.S. Geol. Surv. Prof. Paper 1263, 20 p.

Hale, P. B., and McLaren, P., 1984, A preliminary assessment of
    unconsolidated mineral resources in the Canadian offshore; CIM
    Bull., v. 77, no. 869 (September), pp. 51-61.

Harris, D. P., 1984, Mineral resources appraisal; Mineral endowment,
    resources and potential supply; Concepts, methods, and cases;
    Oxford Univ. Press, Oxford, 445 p.

Harris, D. P., this volume, Thoughts on appraising marine mineral
    resources; pp. 423-456.

Hewett, D. F., Callaghan, Eugene, Moore, B. N., Nolan, T. B.,
    Rubey, W. W., and Schaller, W. T., 1936, Mineral resources of the
    region around Boulder Dam; U.S. Geol. Surv. Bull. 871, 197 p.

Hodges, C. A., Cox, D. P., and Singer, D. A., 1983, International
    minerals resource assessment project; U.S. Geol. Surv. Open-File
    Rept. 83-313, 25 p.

Hudson, Travis, and DeYoung, J. H., Jr., 1978, Map and tables
    describing areas of mineral resource potential, Seward Peninsula,
    Alaska; U.S. Geol. Surv. Open-File Rept. 78-1-C, 62 p., 1 sheet,
    scale 1:1,000,000.

Kraft, J. C., 1971, Sedimentary facies patterns and geologic history
    of a Holocene transgression; Geol. Soc. of Amer. Bull., v. 82,
    pp. 2131-2158.

Marsh, S. P., Kropschot, S. J., and Dickinson, R. G., (eds.), 1984,
    Wilderness mineral potential; Assessment of mineral-resource
    potential in U.S. Forest Service lands studied 1964-1984; U.S.
    Geol. Surv. Prof. Paper 1300, 2 vol., 1183 p.

Mining Journal, 1983, Radiometric techniques for seabed mineral
    surveys; Mining Jour., v. 300, no. 7701 (1 Apr), p. 208.

Nolan, T. B., 1950, The search for new mining districts; Economic
    Geology, v. 45, no. 7, pp. 601-608.

Richter, D. H., Singer, D. A., and Cox, D. P., 1975, Mineral resources
    map of the Nabesna quadrangle, Alaska; U.S. Geol. Surv. Misc. Field
    Studies Map MF-655-K, 1 sheet, scale 1:250,000.

Schanz, J. J., Jr., 1980, The United Nations' endeavor to standardize
    mineral resource classification; Natural Resources Forum, v. 4,
    no. 3, pp. 307-313.

Singer, D. A., 1975, Mineral resource models and the Alaskan Mineral
    Resource Assessment Program; In: Vogely, W. A., (ed.), Mineral
    materials modeling; A state-of-the-art review; Johns Hopkins Press,
    Baltimore, pp. 370-382.

Singer, D. A., and Mosier, D. L., (eds.), 1983a, Mineral deposit
    grade-tonnage models; U.S. Geol. Surv. Open-File Rept. 83-623,
    100 p.

Singer, D. A., and Mosier, D. L., (eds.), 1983b, Mineral deposit
    grade-tonnage models II; U.S. Geol. Surv. Open-File Rept. 83-902,
    101 p.

Singer, D. A., and Ovenshine, A. T., 1979, Assessing metallic resources
    in Alaska; American Scientist, v. 67, no. 5, pp. 582-589.

Singer, D. A., Page, N. J, Smith, J. G., Blakely, R. J., and
    Johnson, M. G., 1983, Mineral resource assessment of the Medford
    1° x 2° quadrangle, Oregon-California; U.S. Geol. Surv. Misc. Field
    Studies Map MF-1383-E, 2 sheets, scale 1:250,000.

Swift, D. J. P., 1968, Coastal erosion and transgressive stratigraphy;
    Jour. of Geology, v. 76, pp. 444-456.

U.S. Bureau of Mines and U.S. Geological Survey, 1980, Principles of a
    resource/reserve classification for minerals; U.S. Geol. Surv.
    Circ. 831, 5 p.

Vogely, W. A., 1983, Estimation of potential mineral reserves, and
    public policy; Earth and Mineral Sciences, v. 52, no. 2, pp. 13-16.

Wynn, J. C., Grosz, A. E., and Foscz, V. M., 1985, Induced
    polarization and magnetic response of titanium-bearing placer
    deposits in the southeastern United States; U.S. Geological Survey
    Open-File Report OF-85-756, 40 p.

ASPECTS OF MARINE PLACER MINERALS: ECONOMIC POTENTIAL OF COASTAL
DEPOSITS IN ITALY, TESTING PROCEDURES AND MARKET CONDITIONS

C. Clerici and A. Frisa Morandini
Centro di Studio per i Problemi Minerari del CNR
Politecnico di Torino
Corso Duca delgi Abruzzi, 24
10129 Torino - Italy

ABSTRACT. Three subjects in the broad field of the utilization of
marine placer minerals are discussed.
    First, investigations of beaches and shallow waters of the Italian
peninsula indicate the existence of heavy mineral placer deposits at
several sites. Results of the research carried out in these areas
during the past few years by the Consiglio Nazionale delle Ricerche
imply that potential resources of commercially useful heavy minerals
exist at these locations. These deposits have not been explored
sufficiently, however, to establish whether they are economically
significant.
    Secondly, such economic evaluation is strongly dependent on the
laboratory-based testing procedures that have to be suitably configured
to approach a commercial separation process.
    Finally, a review of present market conditions and a projection of
future trends for placer minerals indicate that the economically most
significant heavy minerals found in Italian marine placers are rutile,
rare earths and cassiterite.

INTRODUCTION

    In recent years, attention has focused again on marine mineral
resources, although it is commonly believed that in the near future
marine sources of minerals will not have a significant impact in
substituting for traditional terrestrial sources.
    In fact, disregarding the case of the offshore oil industry,
commercial exploitations of marine minerals at present include only
beach and shallow water deposits of sand and gravel and of tin, titanium
and zirconium minerals. Because of a continuing need for certain
critical elements, such as titanium, exploration of the nearshore
waters of the continental shelf and adjacent beaches for placer deposits
continues to be important, at least for a few countries.

P. G. Teleki et al. (eds.), Marine Minerals, 515–532.
© 1987 by D. Reidel Publishing Company.

PLACER DEPOSITS ALONG THE ITALIAN COAST

The search for mineral resources in recent sea-bottom sediments is
currently underway in the Central Mediterranean; the most interesting
indications on the presence of placers that have been found so far along
the Italian coast have been summarized by Brambati (1982). The locations of potentially commercial deposits are shown in Fig. 1, and information about them is elaborated on below.

It is to be noted that, until now, only surface sampling has been
performed on these placer deposits (including submerged placers) and
therefore the preliminary economic evaluations are likely to change as
a consequence of any subsequent deeper exploration.

Percentages of heavy minerals stated below are based on bulk
sediment sample volumes.

Submerged placers east of Elba Island

Deposits of considerable size consisting of detrital heavy minerals, mainly magnetite, hematite and limonite, exist in nearshore waters
along the east coast of the Elba Island (site 1 in Fig. 1).

The content of heavy minerals is between 20% and 50%. The deposit
extends offshore to a distance of 200 to 800 m and to a water depth of
30 m, and is several kilometers long (Lembo et al., 1980). Further
from the coast the heavy mineral content drops abruptly (at a distance
of 2 km, it is only 1%).

Only iron minerals are of interest in this deposit, and indeed
some production has been reported by the same enterprise that exploits
the iron mines on the island.

The placers of Nettuno

A number of placer deposits are scattered along the coast of
Lazio; the most important area, with respect to the quantities, grades
and variety of heavy minerals, is situated southeast of Nettuno, shown
as site 2 in Fig. 1 (Occella and Mancini, 1967).

The deposit spans a distance of about 10 km along the coast. Both
the beach and the recent dunes are mineralized. The width of the
deposit is variable, averaging 150 m, and the thickness is in the 0.5 to
3 m range. The minerals are derived from the nearby volcanic tuffs by
erosion. Heavy minerals have been concentrated afterwards on the shore
by waves, currents, and wind.

Grades change markedly from one point to the other; the following
percentages have been determined for beach sand samples: titaniferous
magnetite 1 to 13%; ilmenite and hematite 0.5 to 1.3%; rutile 0.05 to
0.4%; zircon 0.2 to 6.8%; monazite 0.02 to 0.9%; thorite 0.01 to 0.25%;
garnet 4.5 to 17%; and pyroxenes 27 to 55%. The richest areas are very
narrow; thus, the average grade of the deposit should approach the
lowest values quoted above. Uranium (as much as 100 ppm), tin and
mercury (as high as 20 ppm), and gold (approaching 2 ppm), determined

by chemical methods, have also been noted in a few samples.

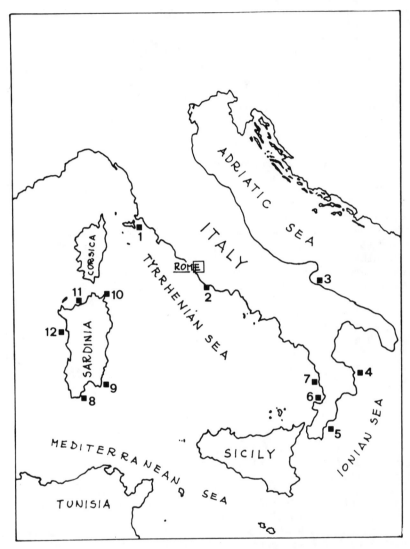

Figure 1. Areas with promising indications for marine placers along the Italian coast; 1: Elba Island (Fe); 2: Nettuno (Fe, Ti, Zr); 3: Gulf of Manfredonia (Fe, Ti, Zr); 4: Punta Alice (Fe, Ti, Zr); 5: Bovalino Marina (Fe, Ti, Zr); 6: Gulf of S. Eufemia (Fe, Ti, Zr, rare earths); 7: Fuscaldo (Ti); 8: Area between Torre di Cala d'Ostia and Capo Spartivento (Fe, Ti, Zr, Sn); 9: Area between Porto Corallo and Capo Carbonara (Fe, Ti, Zr, Sn); 10: Islands of Maddalena and Caprera (rare earths); 11: Castel Sardo (Fe, Ti, Zr, Sn); 12: Torre Argentina (Fe, Ti, Zr).

Several attempts to exploit the deposit have been made in the past, unsuccessfully. The main problems are:
- the most abundant useful mineral (titaniferous magnetite with 8-10% TiO_2) is unsuitable for conventional steelmaking;
- beneficiation is difficult on account of the variety of heavy minerals (in particular the high content of heavy Fe-Mg silicates interferes with the separation of rutile, zircon and monazite whose concentrates do not attain market specifications);
- the size of the deposit is not large enough (2-3 million tons according to the most optimistic estimate) to justify the required sophisticated processing for beneficiation and metallurgy.

The placers of the Gulf of Manfredonia

According to Clerici et al. (1979) the southern coast of the Gulf of Manfredonia contains black sand deposits for a distance of about 40 km (site 3 in Fig. 1). Their source is the Ofanto River alluvial materials. The width of the area containing heavy minerals, that is the present shoreline, is, on the average, 15 to 20 m, and the total volume contained has been estimated to approach 1 million m^3. Several inland dunes are also rich in heavy minerals and so are the submerged sediments to a water depth of 5 m.

The heavy mineral content of the samples falls in the following ranges: titaniferous magnetite 2.5 to 7.3%; ilmenite 0.6 to 2.2%; rutile: traces; zircon 0.01 to 0.05%; garnet 0.2 to 1.1%; and pyroxenes 23 to 42%. Useful minerals are magnetite and ilmenite, however, the magnetite is too titaniferous for use in conventional steelmaking and ilmenite content is too low to justify commercial exploitation for titanium.

Placers of Calabria

Four areas have been found on the Calabria coast that could justify further research for commercial deposits of heavy minerals (Clerici et al., 1978, Clerici, 1983).

At two of these areas (sites 4 and 5 in Fig. 1) significant concentrations of titaniferous magnetite, ilmenite and zircon have been detected. The richest samples are from submerged placers (to a water depth of 50 m) that have shown a content of titaniferous magnetite and ilmenite of 2 to 4% and a content of zircon of 0.4 to 0.5%.

At site 7 in Fig. 1, concentrations of ilmenite, rutile and garnet have been found both onshore and in nearshore sediments. In the beach deposits samples contain as much as 15% ilmenite, 1.5% rutile and 60% garnet.

More detailed research has been carried out on the beach and submarine placers of the Gulf of S. Eufemia (site 6 in Fig. 1). A deposit, spanning 4 km, has been found on the beach in the middle of the gulf, where the width of the deposit is 50 m on the average. Concentrations determined from the collected samples are: magnetite 0.1 to 0.3%;

ilmenite: 1 to 3%; rutile: 0.1 to 0.3%; zircon: 0.01 to 0.03 %;
monazite: 0.01 to 0.03%; sillimanite 0.1 to 0.3%; garnet 5 to 15%;
and pyroxenes: 3 to 5%. Only surface samples have been collected from
offshore in this area, with inconclusive results so far in finding high
concentrations of heavy minerals.

Commercial exploitation of this beach deposit does not appear
promising at present, because of the small volume of useful minerals;
however, deeper sampling of both the onshore and offshore sediments and
the exploration of larger areas could favorably alter the prospects.
Beneficiation does not appear to be a simple task. Preliminary tests
have shown that, whereas it is easy to obtain high-grade concentrates
of both magnetite and ilmenite, the separation of rutile, zircon and
monazite is very difficult because of the variety of heavy minerals
present. To obtain concentrates of rutile, zircon and monazite attaining market specifications, very sophisticated processing schemes will
probably be required.

As pointed out before, the size of the deposit will be the governing factor in future exploitation.

Placers of Sardinia

Along the coast of Sardinia many areas have been found where placer
deposits of possible commercial significance exist. Sites 8 to 12 in
Fig. 1 show where the most promising discoveries have been made. At
two of them studies by the Centro Studi Geominerari e Mineralurgici del
Consiglio Nazionale delle Ricerche (1980) have been carried out to
sufficient detail to allow a preliminary economic evaluation. These
are described below.

From Torre di Cala d'Ostia to Capo Spartivento.

The heavy mineral deposits in this area (site 8 in Fig. 1) extend
over a 10 km distance of the shore and the submerged part of the deposits reach a water depth of 30 m (1-2 km offshore). Constituent minerals
are: magnetite (often titaniferous), ilmenite, hematite, leucoxene,
zircon, rutile, anatase, cassiterite. The following percentages have
been determined for sand samples: FeO: 3 to 5%; TiO_2: 0.5 to 1.5%;
ZrO_2: 0.1 to 0.15%; Sn: 100 to 400 ppm. Titanium is present in the
form of rutile, ilmenite and titaniferous magnetite. The last of these
has a vanadium content as high as 0.9%. Tin occurs in cassiterite.

From Porto Corallo to Capo Carbonara.

In this area (site 9, Fig. 1) the placers on the continental shelf,
spanning a length of approximately 30 km along the coast and a few
kilometers in width, occur between the shore and the 60-m bathymetric
contour. Constituent minerals are the same as in the previously
described area, with concentrations of: FeO: 5 to 11%; TiO_2: 1 to
5%; ZrO_2: 0.1 to 0.2%; Sn 100 to 200 ppm. Titanium is contained partly
in rutile and ilmenite, partly in titaniferous magnetite. Tin occurs

in the form of cassiterite.

It can be inferred that some of the placer deposits of Sardinia are assuredly interesting, in particular owing to their cassiterite content that compares favorably with the commercially exploited Southeast-Asian placers and on account of the size of the deposits. Moreover, the water depth is usually less than 50 m, and that is not excessive for a commercial exploitation.

CURRENT TECHNOLOGY AND ITS INFLUENCE ON COMMERCIAL EVALUATION PROCEDURES FOR PLACERS

Many factors have an influence upon the commercial evaluation of a placer deposit. Broadly speaking they can be classified into three groups:

a) geological and mineralogical factors: mineralogical composition (valuable and gangue minerals), grades, grain size distribution, size of the deposits, etc.;

b) economic factors: market value and trend for potentially recoverable minerals, geographic location of the deposit, labor, power and water availability, port facilities, etc.;

c) factors affecting mining and processing: mining conditions (thickness of the deposit, water depth, etc.), suitability of current technology to recover marketable concentrates (presence of weathered minerals and of grain-coating materials, interference between the different mineralogical components during beneficiation, etc.).

In general the first group of factors is a task of a geologist, the second is that of an economist and the third that of a mining engineer. The issue is that coordinated action of three specialists is required to obtain a reliable evaluation. In the following the connections between a) and c) are examined.

The broad range of commercial or potentially commercial placer deposits can be subdivided into three classes:

1) high-grade deposits containing large reserves, plus a wealth of easy-to-reach customers; examples are silica sand deposits, gravel and sand deposits, garnet deposits and other comparatively low-priced commodities;

2) high-grade deposits and large reserves alone can suffice, provided that capital investment is large; a typical example is the industry exploiting ironsands;

3) even low grade deposits can be economically enticing, provided that reserves are large and commercial concentrates are economically obtainable with current beneficiation technologies; the most typical placer minerals which belong to this class are cassiterite, rutile, zircon, ilmenite, monazite, W, Cb and Ta minerals and precious metals (usually a by-product).

As most placer exploitation projects belong to the third class, only this case will be considered here. Most of the minerals mentioned are ubiquitous in occurrence, and plainly, the commercial evaluation of a placer deposit cannot be accomplished by means of just routine mineralogical and chemical analysis performed on a multitude of samples, because the subject of the investigation cannot merely be a statement on the abundance of any particular mineral in the sediment. Rather, it has to be a reliable estimate of the amounts of those minerals that can actually be recovered, at a specified purity, from one ton of "run-of-mine" ore.

Consequently, the conventional mineralogical laboratory gives way to what can be termed a small, batch or semi-continuous processing plant. This is not simply a "pilot plant," being still the general-purpose type and very flexible, but a plant where the main processing technologies of the commercial plants can, in principle, be duplicated, and the testing procedures tend to approach a commercial separation process.

In the following, two examples of testing procedures are given. The first scheme (Fig. 2, after Brambati et al., 1986) is suitable for small samples (1 to 5 kilograms); it has been adopted by an Australian mining company to determine the content of useful minerals that can actually be recovered in each sand sample collected during prospecting campaigns. According to this scheme the coarsest fraction, usually barren, is screened away and grain-coating materials are removed by means of a laboratory attrition cell. Further, slimes are removed by fine mesh screen (a procedure that does not exactly duplicate the actual commercial process in which hydraulic classifiers are used).

As the valuable minerals are heavier than the gangue minerals, pre-concentration is usually achieved by hydrogravimetric methods; Reichert cones and Humphreys spirals are usually adopted in large commercial plants, but a shaking table can reproduce the action of these devices and is more suitable to the needs of a testing facility, because of its comparatively low productivity. In commercial plants, several stages of preconcentration are needed to obtain a clean separation of the light minerals; in the testing facility this cleaning step is duplicated by means of heavy-liquid testing. The commercial beneficiation process is not reproduced exactly, but the laboratory is forced to accept the compromise on account of the comparatively small amounts of sample material.

Magnetic separation is a conventional process in commercial plants and in testing facilities too, used to separate magnetic minerals (magnetite, ilmenite, hematite) from non-magnetic heavy minerals (rutile, zircon and monazite). The composition of the non-magnetic minerals is determined by counting under the microscope.

An example of testing procedure suitable to treat large samples (up to 1-2 tons) is shown in Fig. 3 (after Clerici, 1983). Such a scheme can be used to process representative bulk samples of a placer deposit with the following aims:
- evaluation of overlap effects between the different mineral components in the separation process;
- detection of minute amounts of non-removable contaminants;
- anticipation of the need of further refining steps;
- evaluation of yields and recoveries of the intended processing plant;
- production of sizeable amounts of concentrates that are required in order to perform, besides the chemical and mineralogical analyses,

technological tests to ascertain their suitability for the intended
industrial uses.

This testing procedure is based upon the joint utilization of hydrogravimetric preconcentration, high-tension separation, low-intensity
and high-intensity magnetic separation. In this manner, concentrates
of the different useful minerals can be obtained.

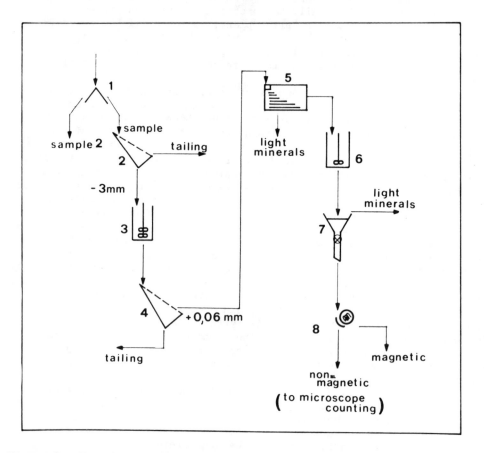

Figure 2. Testing procedure on small placer samples: 1 sampler; 2,4
sieves; 3 attrition cell; 5 shaking table; 6 acid leaching; 7 heavy-
liquid separation; 8 magnetic separator (after Brambati et al., in
press).

Much as magnetic separation, high-tension separation can also be
duplicated in the laboratory by means of small-scale laboratory units.
Admittedly, large-scale units usually perform better, moreover in the
laboratory a large amount of rough concentrate is seldom available and
the typical multiple treatment procedures of commercial plants cannot
be duplicated on the comparatively small sample tested.

A representative bulk sample from a beach deposit of the Gulf of S. Eufemia (site 6 in Fig. 1) has been tested according to the flow chart of Fig. 3.

As already pointed out in the previous section, marketable concentrates of magnetite and ilmenite have been obtained. On account of the high content of heavy silicates, which interferes with the separation of rutile, zircon and monazite, it was impossible to obtain, by high-tension methods, a concentrate of conductive minerals containing only zircon and monazite. Thus, there is a need for additional refining steps in order to obtain marketable concentrates of rutile, zircon and monazite.

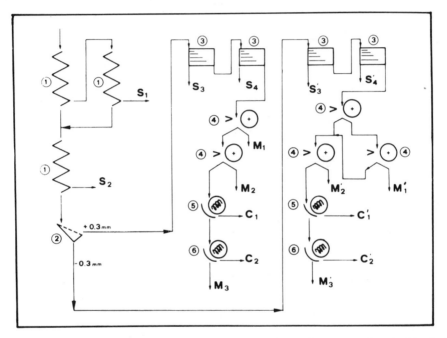

Figure 3. Flow chart for laboratory processing of a representative bulk sample from a beach deposit of the Gulf of S. Eufemia: 1, Humphreys spiral; 2, vibrating screen; 3, shaking table; 4, high-tension separator; 5, low-intensity magnetic separator; 6, high-intensity magnetic separator; S, tailings; C_1 and C_1' magnetite concentrates; C_2 and C_2', ilmenite concentrates; M_1, M_1', M_2, M_2', middlings containing zircon, mon-azite, garnet and pyroxenes; M_3 and M_3', middlings containing rutile and unidentified opaque minerals (after Clerici, 1983).

WORLD PRODUCTION AND USE OF PLACER MINERALS

Cassiterite

Eighty percent of cassiterite, the only commercial source of tin, is produced from detrital deposits such as in southeast Asia, leaving only 20% to be produced from rock deposits. World production of tin in 1982 is shown in Table I.

Reserves of tin are abundant; Thailand, Malaysia, Indonesia and Bolivia account for 50% of world reserves, China for 25%. According to the International Tin Council (1984), tin consumption in 1983 falls into four categories (Fig. 4).

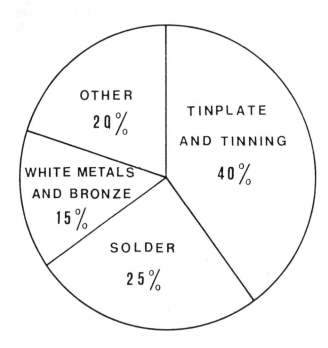

Figure 4. World consumption of tin by end use in 1983 (after Int. Tin Council, 1984).

The market for tin has lately been characterized by a considerable surplus of supply over demand, notwithstanding the International Tin Council's measures of increasing the Buffer Stock, and the corollary export restrictions in the six member countries. Indeed, tin concentrate production has declined, attaining the lowest level since 1969 in 1983; even so, tin prices declined too, from $18.5/kg in 1980 to $12.5/kg in 1984. The decline in the demand for tin is to be credited to the economic recession as well as to the adoption of technological changes in the consuming industry. A new international organization, the Association of Tin Producing Countries (including the six I.T.C. members plus Bolivia) was founded in 1983, aimed mainly at maintaining and increasing tin consumption and promoting products containing tin.

Titanium minerals

Two minerals, rutile and ilmenite, and two artificial intermediate
products, titanium slag and synthetic rutile obtained from low-grade
ilmenites and titaniferous magnetite, are the commercial titanium
sources. World production in 1981, by country, is shown in Table II
(Lynd and Lefond, 1983).
Ilmenite reserves are much more abundant than rutile reserves.
Referring to the widely used "life index," although the limits of this
concept are well known (Zwartendyk, 1974), ilmenite reserves at the
present production rate could last 100 to 200 years while rutile reserves could last only 30 years.
The present structure of the titanium industry is shown in Fig. 5
(Hardwick, 1981). Two trends are apparent. First, besides the two
conventional titanium-dioxide production processes (sulfate and chloride
routes), a new chloride process is expanding that can exploit lower
grade, cheaper raw materials (the DuPont process). Secondly, the
artificial intermediate products such as titaniferous slag and synthetic
rutile, are gaining acceptance and are more in demand now than unprocessed ilmenites.
As for the market, a fast expansion up to 1974 has been followed by
an overproduction crisis caused in part by a world-wide recession, and
in part by the start-up of several large production units planned during
the previous years. Prices have been affected by this general trend in
production capacity and demand. Prices climbed steadily, reaching
a maximum in 1974, when ilmenite was quoted at 35 and rutile at 320
Australian dollars per ton, then declined. Since 1983, however, they
are slowly rising again. The following prices were in effect in 1984
(f.o.b. producers, bulk):
- Australian ilmenite (min. 54% TiO_2): 33 to 43 A$/ton
- Sorel slag (74% TiO_2): 160 US$/ton
- Australian rutile (min. 95% TiO_2): 310 to 420 A$/ton
- United States synthetic rutile: 350 US$/ton.

Table I. World production of tin concentrate in 1982 by country (tons
of constituent metal) (after the Int. Tin Council, 1984).

Country	Production (t)
Australia	12615
Bolivia	26773
Brazil	8279
China	22000
Indonesia	33800
Malaysia	52342
Nigeria	1708
Thailand	26207
United Kingdom	4175
U.S.S.R.	16000
Zaire	2174
TOTAL	206073

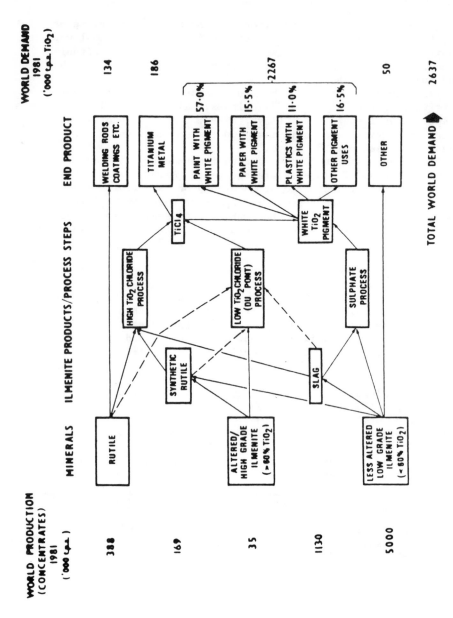

Figure 5.   Flow chart of the titanium industry (after Hardwick, 1981).

Table II. World production of titanium concentrates in 1981 by country
(after Lynd and Lefond, 1983)

Concentrate type and country	Production (10^3 t)	Concentrate type and country	Production (10^3 t)
ILMENITE		RUTILE	
Australia	1337	Australia	239
Brazil	15	India	9
China	136	Sierra Leone	51
Finland	161	South Africa	50
India	189	Sri Lanka	14
Malaysia	172	U.S.S.R.	9
Norway	658	United States	16
Portugal	1	TOTAL	386
Sri Lanka	80		
U.S.S.R.	426	RUTILE, SYNTHETIC	
United States	462		
TOTAL	3637	Australia	56
		India	12
TITANIFEROUS SLAG		Japan	46
		United States	55
Canada	759	TOTAL	169
South Africa	370		
TOTAL	1129		

Zircon

World production of zircon comes entirely from placers deposits,
as a by-product or a co-product of titanium minerals and other heavy
minerals. Table III shows world production data in 1982, by country
(Coope, 1983).

Reserves of zircon are large, more than 50 million tons, mainly in
Australia and the United States. Zircon consumption data, by end use
and country, are shown in Fig. 6 (Hardwick, 1981).

The zircon market suffered an overproduction crisis, even larger
than titanium minerals: prices dropped from A$300/ton in 1975 to A$60/
ton in 1978. Subsequently, prices started to recover slowly; in 1984
foundry-grade zircon was quoted at A$115/ton, premium grade at A$145/
ton. Exceptionally high prices in 1975, however, were the result of
fear of supply cuts. Zircon prices cannot rise freely, because of the
great number of competitors in one of the most important uses (foundry),
and because it is a "must sell at any price" product, being constantly
associated, in placers with other minerals whose cumulative value
represents the larger fraction of the earnings.

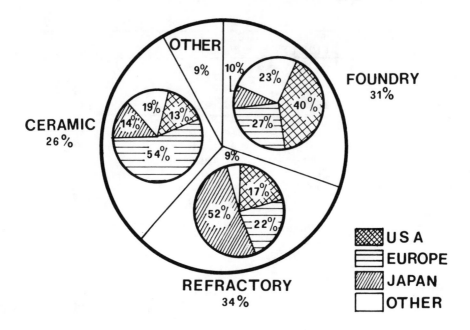

Figure 6.   Zircon consumption by industry and region in 1981 (after Hardwick, 1981).

Table III.   World production of zircon concentrates in 1982 by country (after Coope, 1983)

Country	Production (10^3t)
Australia	462
Brazil	6
China	10
India	15
South Africa	105
USSR	75
United States	75
Other	8
TOTAL	756

Table IV. World production of rare-earth concentrates in 1982,
           by country (after Griffiths, 1984)

Concentrate type and country	Production (t)
MONAZITE	
Australia	9562
Brazil	2100
India	4200
United States	1660
Other (Malaysia, Sri Lanka, Tailand)	1048
TOTAL	18570
BASTNAESITE	
United States	29160
China	12000
TOTAL	41160

Rare-earth minerals

    The main source of rare earths are the minerals monazite and
bastnaesite; monazite is produced from placer deposits as a by-product
or co-product with rutile, zircon, ilmenite and, seldom, cassiterite.
Bastnaesite is produced from primary deposits.
    World production of rare-earth concentrates in 1982, by country,
is shown in Table IV (Griffiths, 1984). Reserves of rare-earth minerals
are plentiful. Rare-earth minerals consumption by end use is shown in
Fig. 7 (Robjohns, 1984). Prices of monazite concentrates (min. 55%
rare-earth oxides) began to climb in 1977 from A$150/ton to the present
price of A$500/ton.

Titaniferous magnetite for steelmaking

    The largest reserves of titaniferous magnetite are in New Zealand
(about 1000 million tons, most of them in beach deposits). In New
Zealand, commercial exploitation of sand deposits is a young industry
because until 15 years ago the titaniferous magnetite was considered to
be unsuitable for steelmaking, and the TiO_2 content was too low to be
used in titanium-dioxide production. A tailored metallurgical process
had to be developed to give value to these deposits.
    Present production in New Zealand approaches 3.5 million tons
annually. The main uses are domestic steelmaking (titaniferous magnetite and non-coking domestic coal are used in a direct reduction process)

and exports to Japan, where the mineral is added (in amounts less than 5%) to the conventional sintered iron ore, in order to increase the life of the refractory lining in the steelmaking furnaces (Brambati et al., 1986).

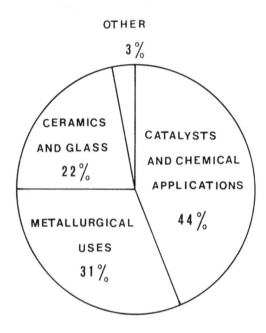

Figure 7.   World consumption of rare earths by end use in 1983 (after Robjohns, 1984).

CONCLUSIONS

Results of research carried out along the Italian coast over the past few years by the Consiglio Nazionale delle Ricerche indicate that potential resources of commercially useful heavy minerals exist in several areas. Only surface sampling has been performed until now and therefore deeper exploration is required to establish whether these resources are economically significant. Nonetheless, the concentrations of useful heavy minerals determined at some of these locations compare favorably with commercially exploited placers in other countries.

The review of market conditions and trends indicates that among the heavy minerals found along the Italian coast the economically most important are rutile, rare earths and cassiterite (in this connection it is to be noted that, in spite of the present surplus of tin supply over requirements, tin prices are still high). Moreover, for all these minerals Italy is entirely dependent on a few world producers; the developing of local resources, if economically exploitable, would therefore be of great importance.

In evaluation procedures of Italian placer mineral resources laboratory testing for processing is essential. In most cases beneficiation does not appear to be a simple task on account of the variety of heavy minerals present, both commercially useful (magnetite, titaniferous magnetite, hematite, ilmenite, rutile, zircon, monazite, cassiterite) and useless or nearly so (pyroxenes, amphiboles, garnet, sillimanite). Sophisticated separation flow charts will be needed to obtain marketable concentrates.

Finally, it is to be noted that the studies concerning processing may be of general interest, even if the Italian placer mineral resources will prove to be economically insignificant. In fact, results of these studies can be applied to the beneficiation of similar placer deposits in other countries.

REFERENCES

Brambati, A., 1982, Some aspects of the mineral resources of the Mediterranean Sea; Lo Spettatore Internazionale, v. 17, pp. 307-317.

Brambati, A., Bernabini, M., Clerici, C., Fontanive, F. and Ulzega, A., 1986, Attivita estrattive in giacimenti di sabbie metallifere (Placers) nel SE Asiatico, Australia e Nouva Zelanda: Rapporto della missione di studio, In: P.F. Oceanografica e Fondi Marini, Sottoprogetto Risorse Minerarie - Rapporto tecnico finale: Consiglio Nazionale delle Ricerche, Rome.

Centro Studi Geominerari e Mineralurgici del Consiglio Nazionale delle Ricerche, Cagliari, 1980, Risultati delle campionature della piattaforma continentale della Sardegna. Prime indicazioni sulle ricerche svolte; Atti Convegno Scientifico Nazionale sui Placers Marini, Consiglio Nazionale delle Ricerche, Trieste, pp. 27-59.

Clerici, C., Frisa Morandini, A., Mancini, R., Rossi, S. and Zasso, G., 1978, Risultati della campionatura di depositi marini e di spiaggia effettuata lungo le coste ionica e tirrenica della Calabria per la ricerca di sabbie metallifere; Memoria del Centro di Studio per i Problemi Minerari del Consiglio Nazionale delle Ricerche, Torino, 81 p.

Clerici, C., Frisa Morandini, A., Giorgetti, F. and Mancini, R., 1979, Prospezione dei depositi marini e di spiaggia del Golfo di Manfredonia per la ricerca di sabbie metallifere; Boll. Associazione Mineraria Subalpina, v. 16, pp. 231-252.

Clerici, C., 1983, Placers della piattaforma calabra; Boll. Associazione Mineraria Subalpina, v. 20, pp. 25-44.

Coope, B.M., 1983, Zircon in good shape after a turbulent decade; Industrial Minerals, no. 195, pp. 19-33.

Griffiths, J., 1984, Rare-earths attracting increasing attention;
    Industrial Minerals, no. 199, pp. 19-37.

Hardwick, R.T., 1981, Prospects for the 1980's; Proc. First Mineral
    Sands Symposium "Minerals and the State," Mineral Sands Prod.
    Assoc., Sydney, pp. 2-6.

Int. Tin Council, 1984, Tin, In: Mining Annual Review; The Mining
    Journal Ltd., London, pp. 41-43.

Lembo, P., Maino, A. and Salvati, L., 1980, Notizie sulle concentrazioni
    di minerali pesanti riscontrate negli strati superficiali dei
    fondali marini antistanti le aree minerarie dell'Isola d'Elba; Atti
    Convegno Scientifico Nazionale sui Placers Marini, Consiglio
    Nazionale delle Ricerche, Trieste, pp. 99-108.

Lynd, L.E. and Lefond, S.J., 1983, Titanium Minerals, In: Lefond, S.J.
    (ed.), Industrial Minerals and Rocks; Am. Inst. Mining, Metallurgi-
    cal and Petroleum Engineers, New York, pp. 1303-1362.

Occella, E. and Mancini, R., 1967, Natura e possibilita di concentra-
    zione di minerali pesanti contenuti nelle sabbie marine lungo la
    linea di costa nella zona di Nettuno; Boll. Associazione Mineraria
    Subalpina, v. 4, pp. 599-631.

Robjohns, N., 1984, Rare Earths, In: Mining Annual Review; The Mining
    Journal Ltd., London, pp. 88-89.

Zwartendyk, J.A., 1974, The Life Index of Mineral Reserves - A statis-
    tical mirage; Canadian Inst. Mining and Metallurgy, Bull. v. 67,
    pp. 67-70.

GEOSTATISTICAL PROBLEMS IN MARINE PLACER EXPLORATION

Robert M. Owen
Department of Atmospheric & Oceanic Science
The University of Michigan
Ann Arbor, Michigan 48109-2143, USA

ABSTRACT. Current developments in in-situ geochemical analysis systems undoubtedly will result in significant cost-reductions in marine placer exploration programs. These systems cannot be used to full advantage, however, until more sophisticated mathematical techniques, to process and interpret the geochemical data they generate, are also developed and employed. In particular, methods are needed to facilitate the identification of specific geochemical anomalies associated with particular types of marine placers, to discern the relationship between geochemical signals used in exploration and the geochemistry of the source rocks from which the placer minerals are derived, and to improve the understanding of sediment dispersal patterns in nearshore areas. Simple correlation analysis of geochemical data has been used to gain insight into the formation of both gold and platinum placers. For platinum placers, it has been shown that the results of simple correlation analysis can be refined by employing a comparitive principal-components analysis in order to obtain a better understanding of the geochemical basis for any observed statistical correlations. Recent research results suggest that newly developed geostatistical techniques, such as end-member composition factor analysis and modelling with linear programming methods, may also be useful in solving the types of geostatistical problems encountered in marine placer exploration.

INTRODUCTION

Three features of marine placers make them a particularly attractive resource. First, they include many minerals which are generally considered to be "strategic" or "critical," such as gold, platinum-group metals, tin, titanium, chromium, tungsten, and rare earth elements. Secondly, they occur within the 200-nautical-mile Exclusive Economic Zone (EEZ) declared by several nations. Hence, from the legal viewpoint, access to these deposits is either direct or can be achieved through international bilateral agreements--a legal situation far less complex than for many other categories of marine minerals, particularly those in deep-ocean, international waters. Finally, marine placers can be mined using existing technology, usually some form of dredging.

The exploitation of marine placers is constrained primarily by the cost of exploration. Compared to other types of marine mineral deposits, economic-grade placers tend to be relatively small and highly localized. Unless they can be located rapidly and economically, there is a considerable

P. G. Teleki et al. (eds.), Marine Minerals, 533–540.

risk that exploration costs will offset potential profits. Herein lies the crux of the main problem associated with developing this resource: at present there is a notable lack of rapid, accurate, and inexpensive methods for use in marine placer exploration.

The major focus of research aimed at solving this problem is on the development of in-situ geochemical analysis systems, which can be towed behind an exploration vessel and linked by means of a coaxial cable to a computer-assisted data base on shipboard. Prototypes of these kinds of devices, which are capable of generating substantial amounts of real-time geochemical data, already have been successfully tested in the field (Tixeront et al., 1978; Noakes and Harding, 1982). However, the most time-consuming stage in the exploration program involves data reduction and interpretation. In order to take full advantage of the data acquisition capabilities of these devices, one needs to determine how they should be programmed; i.e., what kinds of geochemical information should the device measure, and how should these data be processed to facilitate decision-making about the rest of the survey while the vessel is underway.

The object of the present study is to examine the potential for adapting existing, and/or developing new, geostatistical methods to address the specific needs of marine placer exploration programs.

GEOSTATISTICAL PROBLEMS IN MARINE PLACER EXPLORATION

The identification of anomalous samples within a sample group, or of anomalous values within a set of measurements, is the most basic statistical problem in exploration geochemistry. Unfortunately, sampling techniques which are valid from the standpoint of pure statistical theory often require a number of samples and sample tests that are prohibitively expensive, time consuming, or both. These difficulties are the primary reason why applied statistics has become one of the fastest growing fields of mathematics. Applied statistical techniques are designed specifically to deal with data reduction and interpretation problems, where the inherent differences between physical and mathematical reality are most pronounced. For this reason, exploration geochemists have been quick to adopt and to adapt new developments in areas of applied statistics such as anomaly identification, multivariate analysis, and statistical modeling.

Geostatistical problems or research needs associated with marine placer exploration include:
1.   The identification of specific geochemical anomalies associated with particular types of marine placers.
2.   The development/application of the appropriate mathematical techniques for separating a true geochemical "marine placer signal" from the background "normal marine sediment signal."
3.   Understanding the relationship between geochemical signals used in exploration and the geochemistry and lithology of the source rocks from which the placer minerals are, or have been, derived.
4.   Adapting existing methods of modeling sediment dispersal patterns in the littoral zone to the specific problem of understanding placer formation in relation to seafloor topography and the extant wave and current regimes.

GEOSTATISTICAL APPROACHES

Pathfinder Elements

    The most common geostatistical method employed in marine placer
exploration involves a simple correlation or regression analysis of bulk
geochemical data, obtained from a chemical analysis of samples collected
during the exploration survey. These methods seek to identify statistically
significant relationships between individual geochemical parameters, which
are then thought to be indicative of meaningful geochemical relationships
between the same parameters. For example, investigations conducted in areas
of known marine placer occurrences have used simple correlation analysis in
an attempt to identify "pathfinder elements" for various types of placer
deposits. The use of pathfinder elements is a common tactic in geochemical
exploration programs, when it is known that the actual target of the search
is a mineral or element which is difficult to find because it has a limited
areal distribution, is relatively immobile in a geochemical sense, and/or
presents analytical problems. Marine placers are prime examples of target
deposits which display these characteristics. Pathfinder elements typically
consist of some combination of chemical elements (often transition metals)
which are consistently associated with a particular type of mineral deposit,
and which occur in characteristic concentrations, numerical ratios, or
anomalous patterns in the general vicinity of a placer deposit.
    To date possible pathfinder element groups have been reported only for
marine gold and platinum placers, and are based on only one or two studies.
Moore and Welkie (1976) reported a strong statistical relationship among Zn,
Cu, Ni, Co, V, and Fe in marine sediments collected near Bluff, Alaska (an
area of known gold placers), and among Mn, Zn, Cu, Ni, and V in sediments
from Kuskokwim Bay, Alaska (an area of known platinum placers). Marine gold
placers have also been investigated by Kogan (1978), who suggests a
pathfinder element group which includes the elements Zn, Cu, Ni, Co, V, Cr,
Sn, and Pb. (Kogan's work was done in a bay along the Arctic coast of the
Soviet Union, but he did not report the specific study area. It was
probably somewhere near South Primor'ye in the Laptev Sea—an area known for
gold placers (cf. Burns, 1979).) It is noteworthy that the studies of Moore
and Welkie (1976) and Kogan (1978) involved samples from different geologic
and geographic regions, yet many of the same elements (V, Fe, Co, Ni, Cu,
Zn) were found to be positively correlated in the case of gold placers.
This at least suggests the possibility that the observed statistical
relationship among these elements reflects a fundamental geochemical
association. If so, these elements could serve as pathfinders in the search
for gold placers in other regions.
    In each of the above cases, the investigators emphasized that their
observations should be considered empirical, i.e., the observed association
between the pathfinder elements and the target mineral is based on a
statistical relationship rather than on a demonstrated geochemical
relationship. This distinction is important, because strong statistical
correlations generally exist among various chemical elements in marine
sediments, whether or not these sediments are associated with a placer
deposit. In general, whenever a potential set of pathfinder elements is
determined on the basis of a simple correlation analysis, there is a need to
apply more sophisticated mathematical techniques to these data sets, in
order to separate the true "marine placer" signal from the background
"normal marine sediment" signal, and to discern the underlying cause and

effect or generic basis of the observed relationship.

One approach which has been successfully employed in this regard involves the use of comparative principal components analysis. The theoretical development of principal components analysis and its application to geological problems involving multivariate data sets is described in detail by Davis (1973). In essence, this technique (and factor analysis, a closely related method) provides a means of determining underlying patterns or relationships in a multivariate data set, so that the data can be reduced to a small number of groups, or principal components, which account for most of the total variance in the data set. Mathematically, the principal components (or "factors" in a factor analysis) are equivalent to the eigenvectors calculated from a variance–covariance matrix of the data. In comparative principal components analysis, the original data set is initially subdivided into two groups. One group (the "background" group) includes all samples in which the target element is present in concentrations equivalent to geochemical background levels, and the other group (the "enriched" group) includes all samples with target element concentrations significantly greater than geochemical background levels. Separate principal components analyses are calculated for each group, and the results are compared, to determine the nature of the geochemical associations in each group, and whether or not observed differences between the two groups can serve as exploration guides. For example, a comparative principal components analysis (Owen, 1978, 1979) of the same data used by Moore and Welkie (1976) to formulate their pathfinder element group for platinum placers showed that some, but not all, of these elements were generically related to platinum. The key generic relationship was found to be the geochemical association of Mn, V, Ni, Zn, and Pt in ultramafic dunite and diorite lithic fragments. In contrast, it was found that copper probably was correlated to platinum only because particulate copper–organic complexes occurred in some of the sediment samples which contained the lithic fragments. Clearly, there is a need to make this same kind of distinction in other cases where the existing pathfinder element groups are based solely on simple statistical correlations.

Development of Sediment Dispersal Models

The dispersal of placer minerals from source areas to deposition sites is a complex process. We know, for example, that marine placers definitely are not simply lag deposits (Moore, 1979). For this reason, there is a need to adapt existing methods of modelling sediment dispersal patterns in the littoral and nearshore zone to the specific problem of understanding placer formation and identifying the source areas of placer minerals. It is important to clarify the type of modelling effort which is being discussed here. To many geologists, the term "sediment dispersal model" implies an attempt to quantify some aspect of the relationship between the fluid dynamics of transport processes and the hydraulic characteristics of suspended or sediment particulate matter. To a placer mineral explorationist, however, the problem of understanding sediment dispersal patterns implies something far more pragmatic; in effect, he needs to solve a sediment mixing problem.

Marine sediments in coastal areas are usually mixtures of sedimentary materials from different source areas. At some point in the exploration program, usually following a reconnaissance survey, the explorationist will have in hand a data set which includes various geochemical, mineralogical, and textural measurements on sediment samples collected from the depositional basin of interest. If some of these data are promising (e.g., based on an analysis of pathfinder elements), it is then desirable to obtain a quantitative estimate of how much of the sediment at each sampling point within the depositional basin is derived from each source area. Finally, the information concerning the relative amount of source area material in each sample can be plotted on a chart of the area to determine the dispersal pattern of material from each source area.

A quantitative approach to this type of problem involves a three-step process. First, it is necessary to determine the number of source areas which are represented in the data set, as well as their characteristic chemical compositions. This step was sometimes troublesome to take in past applications because it was often necessary to make an a-priori postulation about the number of source areas, but having only limited knowledge of the regional geology and local oceanographic conditions of these areas. Leinen and Pisias (1984) recently have developed a modified version of factor analysis which can be used to overcome this difficulty. The key mathematical modification involves the use of a non-orthogonal rotation of end-member vectors toward the mean vector to bring each of the end-members into positive vector space. In more practical terms, this method is an objective means for identifying both the number of source areas and their chemical compositions.

The second step involves screening the various parameters which represent the chemical composition of each source area in order to select those parameters which can best serve as sediment tracers for each source area. Methods for making these selections have previously been described (Owen, 1975, 1980). Briefly, it involves subjecting the end-member composition data to an analysis of variance, which serves to identify those parameters which are "most distinguishing" (in a statistical sense) for each source area. Those parameters which meet this criterion are subsequently screened by eliminating any which are unlikely to remain chemically or physically conservative (e.g., because of size fractionation) during sediment transport. The remaining parameters represent the best possible set of quantitative "tags" or sediment tracers for the sedimentary material from each source area.

The third and final step in the model development requires a determination of the relative amount of source area material which is present in each sediment sample. Here it is reasonable to assume that the sediments in any part of the depositional basin are composed of some linear combination of the sedimentary material from each source area. This implies that the parameters which characterize each source area also can be represented as linear combinations. For example, if we assume that the element Zn is a usable tracer for each of three source areas which are supplying sediments to a depositional site, then the total concentration of Zn in any sample collected from the site must be equal to a linear combination of the Zn contributed from each of the source areas. This situation can be represented in equation form as

$$Zn_t = x_1 Zn_A + x_2 Zn_B + x_3 Zn_C$$

where Zn_A, Zn_B, and Zn_C = the respective Zn concentrations of each of the source areas;

Zn_T = the measured Zn concentration in the sample; and

x_1, x_2, and x_3 = unknowns representing the relative amount of material from each source area.

Similar equations containing the same unknowns can be written for every other parameter which has been identified as a sediment tracer. Provided that the number of parameters, which are suitable as sediment tracers, is greater than the number of source areas, the final system of equations for each sample will contain more equations than unknowns (i.e., it will be overdetermined) and can easily be solved. (This requirement generally proves no difficulties in practice because the number of source areas seldom exceeds three, and it is almost always possible to identify more than three parameters for use as sediment tracers.)

Mathematical solutions to overdetermined systems of linear equations of the type described above are most often obtained by calculating a solution by least squares, multiple linear regression (e.g., Edwards, 1979). In sediment dispersal problems, however, the linear regression approach is limited because it can sometimes yield solutions which are mathematically correct, but which are inconsistent with the physical constraints of the problem. For example, a linear regression analysis of an overdetermined system of equations might determine that the best mathematical "solution" is one in which one of the unknowns has a negative value. This is equivalent to saying that one of the source areas has contributed a negative amount of sedimentary material to some point in the depositional basin. In other words, linear-regression solutions can be quite feasible from a mathematical standpoint, but nevertheless may represent physically impossible conditions. Because of this problem, geochemists recently have turned to linear programming methods (e.g., Chvatal, 1980) to solve such overdetermined systems of equations. A linear programming approach is favored, because it allows the incorporation of certain physical constraints into the mathematical calculations, and it results in an optimum solution for which the values of the unknowns are not negative. Recent applications of this technique to sediment dispersal problems in the South Pacific Ocean (Dymond, 1981) and Lake Michigan (Ruhlin and Owen, 1983) have demonstrated that it is a very viable method to use in cases of this kind. For example, the initial data set in the Ruhlin and Owen (1983) study included bulk composition analyses for 15 chemical elements in each of 210 surficial sediment samples collected from Lake Michigan. The number and composition of geochemically significant end-members in these sediments were determined by first performing a Q-mode factor analysis (Klovan and Imbrie, 1971) on the data set and then applying the vector rotation procedure of Leinen and Pisias (1984) to estimate true end-member compositions. These end-members included a coarse-grained detrital fraction, a fine-grained aluminosilicate fraction, a carbonate fraction, and a ferromanganese oxide fraction. The linear programming model was then used to determine the relative amount of each end-member in each of the samples. Contour maps prepared from this information were shown to reveal significant sedimentary processes, such as coastal erosion and the dispersal of particulate fluvial inputs.

Two significant advantages should be realized by developing a sediment dispersal model for marine placers along the lines described above:
1. The model can be applied in any coastal setting, and it should be particularly useful in high latitude mineralized regions (such as the Bering Sea coastline) where the multimodal character of the sediments precludes the determination of sediment dispersal patterns based on variations in textural parameters only.
2. The only data set required is that which is conventionally obtained during the course of a geochemical reconnaissance survey.

CONCLUSIONS

A common finding from marine mineral exploration programs in recent years, regardless of the specific target mineral, is that the standard exploration techniques used in the search for terrestrial resources often are inadequate for prospecting in the marine environment. In effect, a knowledge gap exists, which can only be closed by modifying present exploration techniques and developing new ones. Recent advances in the development of geostatistical methods certainly represent one means of closing this gap. Beyond the pragmatic goal of improving exploration efficiency, these methods also represent an important tool for viewing the occurrence of these deposits in a broader framework. For example, Moore (1979) has observed: "We do not yet sufficiently understand the origin of marine placers ... (hence) ... our searches are still based on accident, empiricism, analogy, or some combination thereof." The formation of any mineral deposit implies a combination of physical and/or chemical processes which, by virtue of the rarity of these deposits, represents an anomalous situation within the spectrum of geological processes. Our past experience has clearly demonstrated that the study of such aberrations can often provide a different perspective and an insight into these geological processes which could not otherwise be obtained. Marine placer deposits, for example, apparently cannot form if mechanical energy levels are too low, nor could they be preserved during a transgression if mechanical energy levels were too high. Similarly, the composition of marine placers and associated sediments is a product of geochemical interactions, as yet poorly understood, which occurred during formation and subsequent halymrolysis and diagenesis in a marine system. To the extent that applied mathematical techniques can reveal the generic basis for what are now only statistical relationships associated with marine placers, they provide us with the opportunity to assess the range and temporal and spatial variations of both geochemical and physical processes which have occurred on modern and ancient shelves and coastlines.

ACKNOWLEDGEMENTS

Research on geostatistical methods of The University of Michigan is supported by grants from the Sea Grant Program of the National Oceanic and Atmospheric Administration and the National Science Foundation.

REFERENCES

Burns, V.M., 1979, Marine placers, In: Burns, R.G. (ed.), Marine Minerals; Mineral Soc. Am., v. 6, pp. 347-380.

Chvatal, V., 1980, Linear programming; W.H. Freeman, San Francisco, 478 p.

Davis, J.C., 1973, Statistics and data analysis in geology; J. Wiley and Sons, New York, 550 p.

Dymond, J., 1981, Geochemistry of Nazca Plate surface sediments: an
    evaluation of hydrothermal, biogenic, detrital, and hydrogenous
    sources; Geol. Soc. Am., Mem. 154, pp. 133-174.

Edwards, A.L., 1979, Multiple regression and the analysis of variance and
    covariance; W.H. Freeman, San Francisco, 221 p.

Klovan, J.E. and J. Imbrie, 1971, An algorithim and FORTRAN IV program for
    large-scale Q-mode factor analysis and calculation of factor scores;
    Jour. Math. Geol., v.3, pp. 61-76.

Kogan, B.S., 1978, Method of geochemical prospecting for coastal marine
    placers; Internat. Geol. Rev., v. 20, pp. 1309-1318.

Leinen, M. and N. Pisias, 1984, An objective technique for determining end-
    member compositions and for partitioning sediments according to their
    sources; Geochem. Cosmochim. Acta, v. 48, pp. 47-62.

Moore, J.R., 1979, Exploration problems and sites for new discoveries;
    Proc., Int. Sem. Offshore Mineral Resources, Bur. Rech. Geol. et Min.,
    Orleans, France, Doc. no. 7-1979, pp. 131-163.

Moore, J.R. and C.J. Welkie, 1979, Metal-bearing sediments of economic
    interest, coastal Bering Sea; Proc. Symp. on Sedimentation, Alaska
    Geol. Soc., pp. K-1 to K-17.

Noakes, J.E. and L.J. Harding, 1982, Nuclear techniques for seafloor mineral
    exploration; Proc., Oceanology International, v. 1, paper OI82 1.3.

Owen, R.M., 1975, Sources and deposition of sediments in Chagvan Bay,
    Alaska; unpub. Ph.D. thesis, Univ. of Wisconsin-Madison, 201 p.

Owen, R.M., 1978, Geochemistry of platinum-enriched sediments: applications
    to mineral exploration; Marine Mining, v. 1, pp. 85-102.

Owen, R.M., 1979, Geochemistry of platinum-enriched sediments of the coastal
    Bering Sea; Proc. 7th Internat. Geochem. Explor. Symp., pp. 347-356.

Owen, R.M., 1980, Quantitative geochemical models of sediment dispersal
    patterns in mineralized nearshore areas; Marine Mining, v. 2, pp. 231-
    249.

Ruhlin, D.E. and R.M. Owen, 1983, A model of sediment dispersal patterns in
    Lake Michigan, Geol. Soc. Am. Ann. Mtg., (abs), v. 15, p. 675.

Tixeront, M., F. LeLann, R. Horn, and G. Scolari, 1978, Ilmenite prospection
    on the continental shelf of Senegal: methods and results; Marine
    Mining, v. 1, pp. 171-188.

GEOSTATISTICAL RESERVE MODELING AND MINING SIMULATION OF THE ATLANTIS II DEEP'S METALLIFEROUS SEDIMENTS

Peter Diehl
PREUSSAG
Arndtstr. 1
D-3000 Hannover 1
Federal Republic of Germany

ABSTRACT.   A detailed geostatistical block model of the Atlantis II metalliferous sediments has been computed in order to guide planning for future pilot mining. Assay values from 3900 samples taken at 631 core locations were the principal source of data for the study. Specific problems encountered were the large number of incomplete core sections and the highly stratified nature of the deposit, the combination of which required using a two-dimensional approach with the sedimentary thicknesses of and accumulation products in each layer as input variables to the model. For mining simulation the original two-dimensional model was transformed into a three-dimensional block matrix and tested for various mining options. Two examples are given for selective mining; one method allows a selection of the sediments only by area, the second technique has the additional option to control the penetration depth of the suction head. The choice of the mining method affects mud volumes, metal recovery, and metal grades.

INTRODUCTION

The exploitation of novel types of deposits, such as the metalliferous sediments of the Red Sea, presents unique problems to offshore mining technology, processing, and economic assessment. A hostile environment at a water depth of more than 2000m under a pool of hot, saturated brines is a challenge for the design of suitable mining equipment. Other serious problems are the high water content of the sediments (up to 95%), their abrasive characteristics and their abnormal thixotropic behavior (Lück and Nawab, 1981). Consequently, investment even for a modest pilot mining operation may well exceed those required to place a typical intermediate-sized conventional base metal mine into production. In this situation, computer modeling and simulation techniques are relatively fast and inexpensive tools to assess the potential mineral resources of the Atlantis II Deep under various economic and technical scenarios and to reduce the number of possible options for test mining. This paper describes an example of computer modeling and simulation for future pilot mining. Two problems

P. G. Teleki et al. (eds.), Marine Minerals, 541–558.
© 1987 by D. Reidel Publishing Company.

542   P. DIEHL

are examined: The development of a numerical three-dimensional block
model which takes account of the sedimentary structures in the deposit
and the influence of selective mining on recoverable reserves.

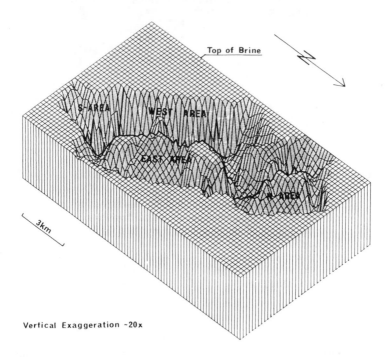

Figure 1. Isometric plot of the Atlantis II Deep's submarine
topography. Shaded areas are below the seawater-brine contact. The
delineation of subareas of this study are informal and approximate.

GEOLOGICAL SETTING

     The Atlantis II Deep is economically the most significant of a
number of submarine metalliferous mud deposit sites discovered along
the axial trough of the Red Sea. The Deep has a complex seafloor
morphology with steep slopes and isolated buttes under a pool of brines
whose temperature exceeds 60° C (Fig. 1). Sedimentary layers as thick
as 20 m contain the economically interesting mineralized zones composed
mainly of iron, zinc, copper, and silver. Minor quantities of gold and
cobalt are also present and have been recovered from concentrates
produced during a pre-pilot mining operation conducted in 1979 (Mustafa
and Amann, 1980). The sediments overlie a bedrock of basaltic pillow
lavas. The deposit covers an area of about 60 km^2 at a water depth of
2050 m to 2200 m. The origin of the sediments has been interpreted as
"hydrothermal in statu nascendi" (Bäcker and Schoell, 1972). The
current location of the sources of the metal-bearing brines is believed

to be in the south area (Fig. 1), where the brines have been observed to have the highest temperatures. It is assumed that the salt content of the brines is related to adjacent evaporitic formations, while the metals may be derived from the basaltic bedrock. More detailed descriptions of the geology of the Atlantis II Deep may be found in Bäcker and Schoell, 1972, and Bischoff, 1969.

Variations in the sedimentary environment during the Deep's 15,000 years of history resulted in a complex stratigraphic sequence with rapid changes in the chemical and mineralogical composition of individual layers, both vertically and laterally. Below is a generalized stratigraphic sequence of five geological units that may be distinguished by appearance and chemical composition (Fig. 2):

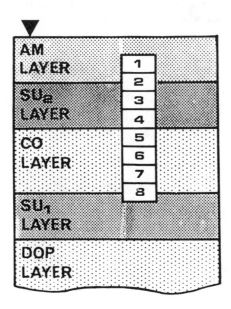

Figure 2. Generalized stratigraphic sequence of the metalliferous sediments (except SW-area). The stratigraphic units are:

AM   - Amorphous-Silicate zone, semi-liquid, low or intermediate content of base metals.
SU_2 - Upper Sulfidic zone, relatively high zinc, copper, and silver grades.
CO   - Central Oxidic zone, thick, low-grade or barren intercalation with often high content of solid material.
SU_1 - Lower Sulfidic zone, second metal-rich layer with zinc, copper, and silver mineralization.
DOP   - Detrital-Oxidic-Pyrite zone, relatively low-grade with high content of solid material.

In the SW-area the sedimentary sequence is different:
SAM - Sulfidic-Amorphous-Silicatic zone, enriched in base metals.
OAN - Oxidic-Anhydritic zone, low base-metal content.
SOAN - Sulfidic-Oxidic-Anhydritic zone, relatively high zinc,
       copper, silver content.
DOP - as above

Dark shading indicates high-grade mineralization. The typical position
of a core section with eight 1 m-long subsamples (see numbers) is
indicated. Only SU_2-layer and CO-layer are completely penetrated in
this example while the AM-layer and SU_1-layer are partly sampled due to
toploss and insufficient penetration depth of the corer.
Two histograms of Zn grade for SU_2-layer and CO-layer illustrate the
different statistical distribution of this metal within two adjacent
layers (Fig. 3). Variations **within** individual geological units are low.

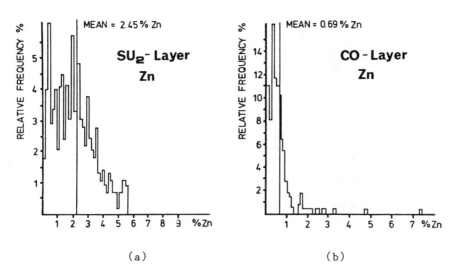

(a)   (b)

Figure 3. Histograms of Zn grade for Su_2-layer (left) and CO-layer
(1 m-samples from E-Area).

THE DATABASE

     The available information at the time of this study consisted of
the coordinates of the sediment cores, the assays for nine chemical and
three physical grades of core samples, the geological interpretation of
the core sections, and the bathymetric depth measured by a 30-kHz echo
sounder on profiles at 200 m to 400 m spacing.

Six exploration campaigns between 1968 and 1979 provided 631 piston and
kasten core samples taken at an irregular grid. The sample length was

1 m for the bulk of the sediment cores, but varied considerably for a
few of the samples taken during the early exploration stage. For these
older core samples 1 m-long composite samples have been calculated by a
weighted average method that takes the length of the original sample
and its solid content into account. A total of 3906 1 m-long samples
(original samples and composites) is the principal source of data for
this study. The routine analysis on each sample included density
(g/cm^3), water content (%), dry saltfree material (%), iron (%)
sulfur (%), zinc (%), copper (%), silver (ppm), calcium oxide (%),
manganese (%), carbon dioxide (%), and silicon dioxide (%). All
chemical grades refer to dry saltfree material. Flotation tests have
shown that all of these parameters must be considered when designing
the dressing plant. A specific problem in deep-sea core sampling is the
difficulty in controlling the penetration depth of the corer. Most of
the recovered core material represents only a portion of the
sedimentary sequence. This is the result of insufficient depth
penetration of the corer and/or no core recovery from the uppermost
(semi-liquid) layers, a so-called toploss. A typical position of a core
section within the sedimentary column is given in Fig. 2.

Table I shows the number of complete penetrations achieved in each of
the five sedimentary layers (SW-area exluded). Obviously, the
information for the uppermost and lowermost layers is considerably
reduced by toploss or insufficient penetration by the corer.

Table I. Number of complete penetrations

AM	SU_2	CO	SU_1	DOP	TOTAL
No. 155	380	256	20	76	887

THE NUMERICAL BLOCK MODEL

The main purpose of block modeling the Atlantis II Deep was to
estimate the mineable ore reserves under various economic and technical
constraints. One of its requirements was that the model should be
three-dimensional and as detailed as possible with full use of the
available input data. The layered structure of the deposit were to be
fully represented and the distinct chemical and mineralogical characteristics of each geological units not "smeared" by an unqualified
interpolation method. Local features such as differences in sedimentation between various basins and ridges were also to be reflected by the
block model. It was thought that in a global sense, the block estimates
should be unbiased, i.e. systematically neither too high nor too low,
and the model ought to provide estimates for all twelve assay values,
thickness of layers, and water depth.

A geostatistical approach was selected for estimation of a block model with these characteristics. The theory of geostatistical ore reserve estimation was developed, during the last 30 years, mainly by French workers and is now well established in the mining industry worldwide. Its superiority over conventional methods is mainly based on the fact, that in the first step of a geostatistical study one quantifies the spatial characteristics (fluctuations, continuity, trends,...) of a mineralization (variography, histography) and then builds an optimum algorithm for interpolation between sample points (kriging technique), which in turn is custom-tailored for a specific type of mineralization and sampling pattern. The method is described in detail by David (1977), Journel and Huijgbregts (1978) and Clark (1979).

Three-Dimensional or Two-Dimensional Approach?

A crucial point when modeling a layered deposit is the choice of a reference system for the location of the sample values, particularly the decision, whether to work in three- or two- dimensional space. In sedimentary deposits continuity of mineralization is usually parallel to the strata, which means not necessarily parallel to the (horizontal) axes of the original coordinate system. In former geostatistical studies of the Atlantis II Deep resources (Guney and Marhoun, 1984, Preussag report, 1980) the reference system for the vertical position of a sample was its distance below the seafloor. In variogram computations samples are grouped according to their vertical and horizontal distances. As the thickness of individual geological units varies considerably, adjacent samples taken at the same depth may represent strata with completely different characteristics. In this approach, the chance to group samples from different zones increases with sample depth and horizontal distance of the samples. Typically these variograms show relatively low overall variations (plateaus, sills) and a reasonable zone of influence for the uppermost layers but become more and more "noisy" with increasing depth (Fig. 4). Formerly, this was interpreted as a decrease of continuity of the mineralization downwards, possibly the result of tectonic movements and sedimentary slumping. In reality, it is a pure artifact caused by an unsuitable coordinate system.

As an alternate solution a "natural" coordinate system was proposed for sedimentary deposits, where samples are grouped according to their stratigraphic position instead of their vertical depth (Dagbert, et al., 1984). As a consequence, for each block of the model the proportion of all layers which form the block must be digitized, which entails a detailed interpretation of the geology. Such an approach is not feasible at the present stage of knowledge about the deposit at the Atlantis II Deep. Instead, classical two-dimensional geostatistics is preferred for this study.

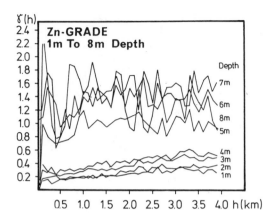

Figure 4. Horizontal variograms for zinc grade in 1 m-long samples grouped according to depth (omnidirectional absolute variograms). The sill values (plateaus) and fluctuations of the variograms increase from 1 m to 8 m depth; apparently indicating that spatial variations of zinc grade increase with depth below subsurface.

Creation of a two-dimensional database

A classical two-dimensional reserve estimation requires, for each core location, the mineral accumulation products (thickness times grade) of all stratigraphic layers. Accumulation products are preferred to the straight arithmetic mean of the subsamples of a core in order to establish a database with constant statistical "support", i.e. to counterbalance the different sample length of a layer ("regularization", Journel and Huijbregts, 1978, p. 77). In the case of this study chemical grades (iron, zinc, copper,...) refer to dry salt-free material. Obviously the "support" of a core section varies as the content of solids varies. For this reason, three component accumulation products have been computed for all chemical grades (thickness x solid content x grade). Although all grades were measured in weight-percent, density was not considered as a weighting factor for two reasons: (1) the poor accuracy of density measurements and (2) low overall variations of mud density relative to other parameters.

A serious problem in this approach is the poor recovery of the bottom and top layers due to insufficient penetration depth and toploss respectively. Considering only samples from complete core sections would have resulted in a loss of about 35% of the information from 1 m-samples. Information on samples from incomplete intersects could only be used by estimating the thickness of the respective layer at this

point. In a first attempt, the Co-Kriging technique was used for
estimating thicknesses (Journel and Huijbregts, 1978, p. 324).
This method was developed to improve the estimation of poorly-sampled
variables (thicknesses of the bottom and top layers in this study) by
considering a possible spatial correlation between this variable and
other better-sampled variables (the thicknesses of the SU_2 and CO
layers, for instance). Unfortunately, no such correlation was observed
between the thicknesses of layers at points with complete intersects.
Instead, the following iterative procedure was applied in which
Ordinary Kriging (Journel and Huijgbregts, 1978, p. 306) is combined
with geological reasoning:

1. Define a stratigraphic indicator for each sample according to
    the geologist's interpretation of core sections.

2. Indentify all complete intersects in a stratigraphic layer.

3. Compute variograms for thicknesses at complete intersects.

4. Apply Ordinary Kriging to the thickness of layers at core locations
    with incomplete intersects.

5. Compute the top and bottom elevation of layers by "stacking"
    observed or estimated thicknesses.

6. Compare the results with the geologist's description of the
    core sections.

7. Correct obvious inconsistencies between the computed
    stratigraphic position of a sample and its chemical
    characteristics.

8. Reiterate this sequence

     After several iterations, a consistent model with four thickness
values at each core location was derived. In computing the accumulation
products, only those intersects were considered with at least 50% of
the (estimated) thickness of the layer sampled. Finally, statistical
parameters of incomplete intersects were investigated for the presence
of any possible bias that could have been introduced by considering
only samples from the bottom or the top of a layer. No significant
differences were found between parameters from accumulation products
derived from complete and incomplete intersects.

Homogeneous units of mineralization

     A prerequisite in geostatistical theory is the existence of
(stationary) subzones with homogeneous mineralization. More precisely,
it is assumed that the variations of grades are about the same within
the area under study and that grades follow no trend. By sorting the

database according to the relative position within geological units
much of the overall variances of grades have already been removed.
However, there are still significant horizontal variations caused by
variations in sedimentary conditions from basin to basin. After
numerous test runs, four subareas were defined to have fairly
homogeneous mineralization.

For the creation of a block model classical geostatistical methods
have been applied. Most of the computer programs used were modified
versions of source code provided by GEOSTAT SYSTEMS, Montreal.
Following are a few highlights of the results.

Variography

Experimental variograms of all parameters have been computed in
four directions along the main axes of the Deep. Because of a
pronounced proportional effect and partly skewed distributions of the
parameters, the use of relative variograms was preferred (Fig. 5).
These variograms typically show a "nugget effect" of 10% - 30% of the
sill value, a short range structure with about a 800 m wide influence
zone, and a second structure of 2000 m to 5000 m range. With the
exception of CaO and Fe, no anisotropies were detected. In most cases,
double spherical or exponential schemes have been used as variogram
models. Other than for the 1 m-long samples (Fig. 3a) overall variances
(sills) are largest for the SU_2-layer and CO-layer.

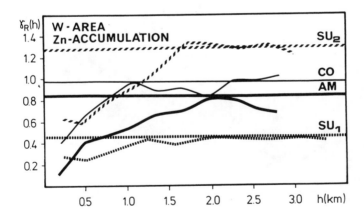

Figure 5. Variograms of Zn-accumulation for the upper four sedimentary
layers, i.e. samples are grouped according to stratigraphic position
rather than to depth. The sill values (horizontal lines) indicate, that
pertubations of Zn-accumulation do not increase with depth below
subsurface.

Kriging

The block size of 200 m by 200 m (Fig.6) chosen for kriging is
relatively small compared to the average sample spacing of about 500 m.
The small grid was selected for a better approximation of the
complicated contours of the deposit. Resource estimates for each of the
four subareas were calculated separately, although with some overlap,
allowed for the data neighbourhood, because gradual transitions of the
mineralization were known to exist among the four areas. The final two-
dimensional "kriged" block matrix comprised 1559 200 m x 200 m blocks,
each block with estimates for the thicknesses and 11 accumulations in
the four layers, thus, a total of 74,832 estimates. Because of the
small block size, relative "kriging" errors were rather high ranging
from 25% to 40%. From the two-dimensional model a three-dimensional
block matrix containing 200 m x 200 m x 1 m slices, was derived by computing, for each slice, the grades as a (volume) weighted average from
the respective geological layers which are part of the slice. The
derived pile of slices allowed average grades to be computed;
subsequently water depth values were added using available bathymetric
data. The final block model comprised about 12,500 slices with 180,000
estimates. The result were the principal input to the mining simulation
described in the following.

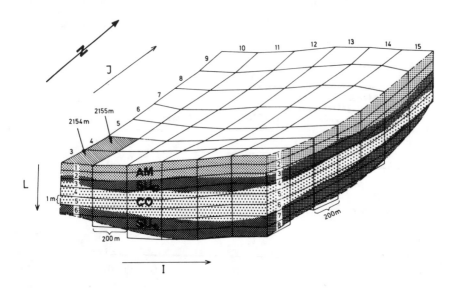

Figure 6.   Isometric drawing of a portion of the block model with
slices of 200 m by 200 m by 1 m. Numbers are block indices along the
axes I, J, L of the matrix. Waterdepth is shown for the two shaded
blocks on the left. Locations of sedimentary layers are approximated.

Quality of the block model

The performance of a mining simulation is greatly influenced by the local accuracy and (conditional) unbiasedness of the numerical model (Journel and Huijbregts, 1978). Here, a satisfactory local accuracy cannot be expected from estimates of individual slices on account of the high "kriging" error. Reasonable accuracy, however, may be obtained when major areas (mining panels) - for example nine or more slices - are averaged for simulation. Those areas would represent about one month of production of a pilot mining operation. Hence, the model is suitable for intermediate and long-term, but not for short-term planning. In addition the model is not suited for the simulation of highly selective mining, as Ordinary Kriging is generally not a conditionally unbiased estimator in the presence of skewed grade distributions (Fig. 3). In the vertical direction, problems with local accuracy are less severe, as the model fully honors the layered structure of the deposit.

MINING SIMULATION

The proposed scheme for pilot mining includes lowering a pipe string, which carries a suction head with cage cutters, into the muds, fluidizing and possibly diluting the sediments in-situ and then pumping the slurry to the surface (Fig. 7). This can be achieved by a submerged pump situated just on top of the seawater-brine interface and which is part of the pipe string. Onboard the vessel, the muds will be pumped from holding tanks to a flotation plant, where the metals will be recovered in a bulk sulfide concentrate. The concentrate will be shipped to a metallurgical plant on-shore, the tailings being pumped back to sea at depth. Within this general concept further options are possible regarding the type of equipment operated and the strategies for mining and processing.

Below are examples for the simulation of three different scenarios together with a set of technical and economic constraints. In all cases it is assumed that the maximum solid content of one liter of slurry is 120g. Muds with a higher solid content have to be diluted with seawater prior to being pumped to the surface to avoid excessive energy losses in the pipe string. The dilution factor V, is defined as:

$$V = S \ / \ S_{max}; \quad \text{if} \quad S > S_{max} \quad\quad\quad (1)$$

$$V = 1 \quad\quad ; \quad \text{if} \quad S \leq S_{max}$$

where S (in weight-percent) is the estimated (saltfree) solid content in a block. S_{max} is the respective percentage of solids of a slurry of 120g solids/liter given the estimated density (g/cm^3) of the mud of the block in-situ.

$$S_{max} = 12 \ / \ \varrho \ \ (\%) \quad\quad\quad (2)$$

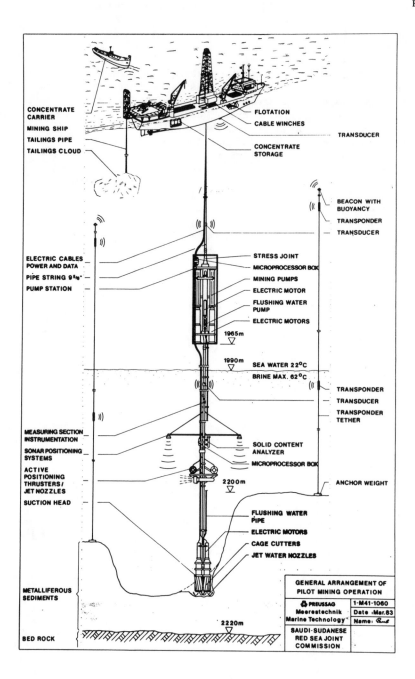

Figure 7. General arrangement of the mining system for a pilot operation.

As a general rule, no dilution is required for the uppermost AM and SU_2 layers, but the CO and SU_1 layers require, in most cases, some dilution with seawater. It should be stressed that the solid content is as important for the economic assessment of the sediments as their metal grade in-situ. This is a direct consequence of the high cost for pumping, which consumes about 25 kWh/m^3 of raised material. The amount of slurry that has to be pumped and processed to produce about one cubicmeter of zinc sulfide is illustrated in Fig. 8.

A simple selection method is mining only areas where the average grade of all layers surpasses a cut-off grade. More precisely, it is necessary to discriminate between 200 m by 200 m squares by a (mining) cut-off grade ZnE_c expressed in zinc-equivalent grade ZnE (%):

$$ZnE\ (\%)\ =\ 1Zn\ (\%)\ +2Cu\ (\%)\ +\ 400Ag\ (\%) \qquad (3)$$

The coefficients reflect the relation of the market prices for the three metals at time of the study. It is assumed that, during mining, the cut-off grade will be applied directly to the estimates of the present block model with no further sampling prior to selection. This type of selective mining is assumed in scenario 1.

Here, the suction head is lowered into the sediments digging a crater with steep walls, thus, mixing high grade and low grade material. If the estimated average grade ZnE of the pile of slices surpasses the cut-off grade ZnE_c a 200 m by 200 m square is mined and processed, otherwise it is rejected.

In scenario 2 and 3 it is assumed as a further option that the mining equipment allows to seperately extract slices of 1 m thickness. Then a further discrimination between ore-grade material and waste material may be achieved by use of an on-stream analyzer onboard the mining vessel and applying a concentrator cut-off grade ZnE_m. Flotation of the sediments is rather difficult, because particle sizes are generally extremely small. Below a concentrator cut-off grade ZnE_m (Scenario 2: ZnE_m = 1.7%; scenario 3: ZnE_m = 2.0%), the solid content is considered as too low-grade for beneficiation.

With the option to control mining depth, the economic value of a slice of 1 m thickness of the deposit depends not only on its metal content, but also on the amount of overburden that has to be removed, just as in conventional open-pit mining. The cost of the latter may be dealt with as a reduction of metal content. The in-situ metal content $q_{ZnE,i}$ of a slice i of 1 m thickness is:

$$q_{ZnE,i}(kg/m^2)\ =\ ZnE_i\ x\ S_i\ x\ 0.1 \qquad for\ ZnE_i\ \geq\ ZnE_m \qquad (4)$$

$$q_{ZnE,i}\ =\ 0 \qquad for\ ZnE_i\ <\ ZnE_m$$

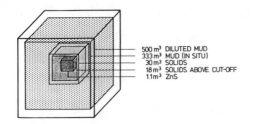

Figure 8. Illustration of the amount of mud volumes handeled in a pilot mining operation with a 500 m^3/h extractive capacity.

Note that ZnE_i is zero for slices below the concentrator cut-off grade as this material will be pumped back to sea at depth. Similarly, the extraction cost C ($ per m^3 diluted material) may also be expressed in terms of a metal content q_e in a slice of 1 m thickness:

$$q_e \ (kg/m^2) \ = \ (V \ . \ C) \ / \ (P \ x \ 0.001) \tag{5}$$

with P the market price of zinc (700 $ US/t at the time of the study). Because of the high initial cohesiveness of the muds, mining long walls may be near vertical. Therefore it is not necessary to introduce slope constraints into the simulation model. The reduced metal content $q_{r,i}$ of a slice of 1 m thickness depends only on the cost of pumping the slurry, the market price of the metal and the in-situ metal content:

$$q_{r,i} \ (kg/m^2) = q_{ZnE,i} - q_e \tag{6}$$

Extraction cost per m^3 is $ 0.4 in scenario 2 and $ 0.6 in scenario 3. For low-grade material, i.e. for most of the CO-layer, calculated reduced metal-content values are negative numbers.

The cumulative reduced metal content $Q_{r,n}$ of a stack of n slices is:

$$Q_{r,n} = \sum_{i=1}^{n} q_{r,i} \tag{7}$$

Mining is at an optimum for a given depth of j slices when $Q_{r,j}$ reaches a maximum:

$$\max_{j} \ \left\{ Q_{r,j} \right\} \tag{8}$$

Consequently for the optimum mining depth corresponding to j slices it is simple to compute averages of all parameters and volumes (in-situ and diluted). In this simplistic approach it is assumed that high-grade ore is not diluted by waste material at the time of extraction.

Figure 9a.   Schematic profile with thicknesses of geological layers, ZnE grades, and cumulative metal content (percentage of total metal value as a function of depth).

Figure 9b.   Same profile but with apparent thicknesses (after dilution), reduced ZnE grade and reduced metal content.

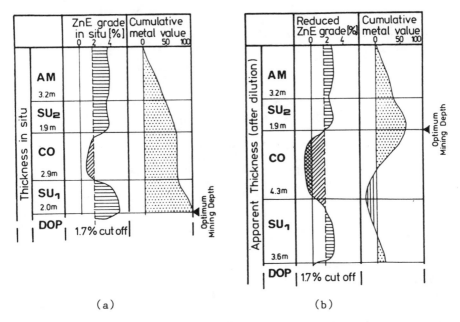

(a)   (b)

      Fig. 9a and 9b show the same typical stratigraphic section. In Fig. 9a, thicknesses, ZnE grades and cumulative metal value, refer to in-situ material. The cumulative metal content for a specific depth of j slices is expressed as a percentage of the total metal content of all slices. A concentrator cut-off grade of 1.7% ZnE is indicated. As no dilution and pumping cost are considered the cumulative metal content is at a maximum at the bottom of the SU_1 layer.

      Fig. 9b shows the same profile but with apparent thicknesses (in-situ thickness times dilution factor), reduced equivalent grades and reduced cumulative metal content. Reduced grades are derived from $q_{r,i}$ assuming an average density of 1.3 (g/cm^3) of the mud in-situ for all slices. Reduced grades are negative for the low-grade CO layer where the metal content is too low to cover the extraction cost which excludes the flotation charges. Now the cumulative metal content curve $Q_{r,j}$ is at a maximum at the bottom of the SU_2 layer and then drops within the "overburden" CO layer to negative values. The otherwise payable SU_1 layer proves uneconomic under these condition as more than its metal value is consumed to cover the extraction cost of the CO layer. Consequently the optimum mining depth in this example is at the bottom of the SU_2-layer. Relations between mining cut-off grade and a)

gross mud volume (diluted); metal recovery ZnE; and c) ZnE grade
recovered for the three scenarios, are shown in Figs. 10,11,12.

As a general rule the deposits of Atlantis II Deep are most
sensitive to mining cut-offs in the range of 3% to 6% ZnE. In this
range differences are greatest between scenario 1 (horizontal selection
only) and scenarios 2 and 3 (vertical optimization and subsequent
horizontal selection). If we focus on a mining cut-off ZnE_C of 4%,
mining options 2 and 3 would increase metal recovery by a factor of
about 1.85 at a grade that is slightly higher than in option 1.

CONCLUSIONS

The classical two-dimensional approach is a suitable method to
build a numerical model of the deposit of Atlantis II Deep for mining
simulation. Though the local accuracy of the model is relatively low
because of wide sample spacing and the necessity to convert the 2D
model into a 3D block matrix, it is suitable for developing equipment
and strategies of a pilot mining test. One of the results is that
mineable reserves could be significantly increased for intermediate
cut-off grades if mining equipment is developed that allows a slice-
-by-slice extraction of the metalliferous sediments.

ACKNOWLEDGMENTS

The work described was financed by the Saudi Sudanese Red Sea
Commission, Jeddah, whose permission to publish part of the results is
greatly appreciated. The author wishes to thank his colleagues at
PREUSSAG, notably Rosemarie Scharringhausen and Dr. Wolfgang Ehrismann
for their help with and criticism of an early draft of this paper.

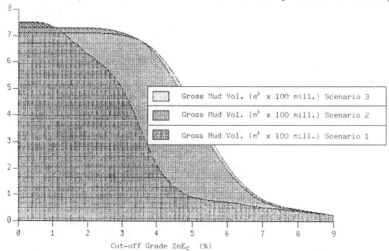

Figure 10. Relationship between ZnE_C mining cutoff grade and gross mud
          volume for three scenarios.

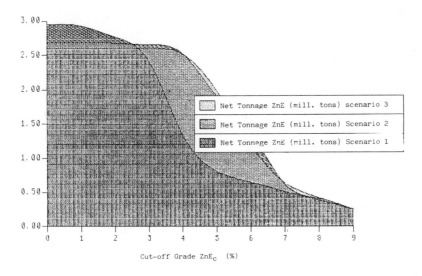

Figure 11. Relationship between ZnE_C mining cutoff grade and metal tonnage recovered for three scenarios.

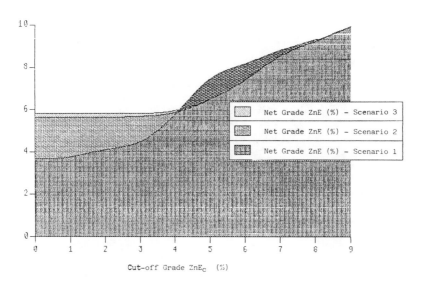

Figure 12. Relationship between ZnE_C mining cutoff grade and recovered grade ZnE for three scenarios.

REFERENCES

Bäcker, H., and Schoell, M., 1972, New deeps with brines and
    metalliferous sediments in the Red Sea; Nature Physical Science,
    v. 240, 103, pp. 153 - 158.

Bischoff, J. L., 1969, Red Sea geothermal brine deposits:
    Their mineralogy, chemistry, and genesis, In: Degens, E.T. and
    Ross, D. A. (eds.), Hot brines and recent heavy metal deposits
    in the Red Sea; Springer, New York, pp. 368 - 401.

Clark, Isobel, 1979, Practical geostatistics; Applied Science Publ.,
    London, 129 p..

Dagbert, M., David, M., Crozel, D., Desbarats, A., 1984,
    Computing Variograms in folded stratacontrolled deposits;
    In: Verly, G., David, M., Journel, A.G., Marechal, A.,
    Geostatistics for natural resources characterization, Part 1;
    NATA ASI series, A. Reidel Publ., pp. 71 - 89.

David, Michel, 1977, Geostatistical ore reserve estimation; Dev. in
    Geomathematics, v.2, Elsevier, Amsterdam, 364 p..

Guney, U., and Marhoun, M.A., 1984, Atlantis-II-Deep's metal reserves
    and their evaluation; Proc. 16th Offshore Technol. Conf., Houston,
    Texas, Reprint No. 4780, 12 p..

Journel, A. G. and Huijbregts, Ch.J., 1978, Mining geostatistics;
    Academic Press, London, 600 p..

Lück, K. and Nawab, Z., 1981, Metalliferous sediments in the
    Atlantis-II-Deep, recovery and selected special problems;
    Proc. Inter Ocean, Düsseldorf, Reprint 402/01, pp. 103 - 115.

Mustafa, Z. and Amann, H., 1980, The Red Sea pre-pilot mining test,
    1979; Proc. 12th Offshore Technol. Conf., Houston, Texas,
    Reprint No. 3874, 13 p..

Preussag Meerestechnik, 1980, Atlantis-II-Deep deposit evaluation
    on the basis of data collected during exploration cruises,
    1969 - 1979; Internal tech. rep. 23, 36 p..

AN INVESTIGATION OF THE APPLICABILITY OF TREND SURFACE ANALYSIS TO MARINE EXPLORATION GEOCHEMISTRY

S.A. Moorby
R.J. Howarth*
P.A. Smith**
and D.S. Cronan

Applied Geochemistry Research Group, Dept. of Geology,
Imperial College, London, SW7 2BP, UK.

*ISS (TESA), BP International Ltd.,
Britannic House, Moor Lane, London, EC2Y 9BU, UK.

**Enterprise Oil, Griffin House, Strand, London, WC2, UK.

ABSTRACT. A modified trend surface analysis technique is presented and used to attempt to distinguish between potentially mineralized and un-mineralized offshore areas on the continental shelf of Greece. Using data from two areas, each receiving sediments of different provenance, the method successfully distinguished the two, by producing anomalies which were stable against perturbations of the data in the potentially mineralized area. This effect is particularly noticeable when only samples anomalous for more than one element are considered. In this study, the most satisfactory results were obtained by treating residuals falling in the top 20th percentile as anomalous. In future applications, this upper limit may need to be adjusted as necessary. The overall success of the technique suggests that it may have consider-able potential for future offshore mineral exploration studies because it can successfully distinguish between true and false anomalies.

INTRODUCTION

Interest in nearshore minerals has increased in recent years, and this interest is likely to be enhanced by the recent declaration of 200 nautical mile exclusive economic zones by many coastal states. Several studies have been undertaken throughout the world in search for conti-nental shelf hard mineral deposits (e.g. Tixeront et al., 1978; Jones and Davies, 1979; Siddique et al., 1979; Yim, 1979; Beiersdorf et al., 1980; von Stackelberg, 1982; Meyer, 1983), almost all of which have adopted a basic geological and/or geophysical approach. Statistical approaches to the search for offshore hard minerals have generally not been used so widely. However, Owen (1978, 1980) has shown the

559

P. G. Teleki et al. (eds.), Marine Minerals, 559–575.
© 1987 by D. Reidel Publishing Company.

usefulness of principal components, and vector and regression analysis
in the search for offshore Pt-placers, and Burger et al (1980) used
moving weighted averages, kriging (Clark, 1979) and trend surface
analysis to investigate potential resources of deep-sea manganese
nodules.

Multivariate statistical techniques are being used with increasing
frequency on land in the search for mineral deposits, and the success of
these techniques in locating geochemical anomalies related to mineralization has established them as a powerful tool in mineral exploration.
Successful recognition of anomalies is the main aim of exploration
geochemistry on land and at sea, and in both environments a major
problem is to distinguish between real and 'false' anomalies caused,
for example, by scavenging of the metals of interest by Mn-oxides.
This paper examines this problem, using a modification of trend-surface
analysis to identify geochemical anomalies on the continental shelf of
Greece and to assess their significance in relation to potential
exploration targets.

Trend surface analysis was widely applied to geological problems
in the 1960's and the early 1970's, but, with the realization that it
has several shortcomings (Howarth, 1983), it has been used less
frequently during the last decade. Marine sediments, however, are a
particularly suitable subject for trend surface analysis because the
variations in their composition in a given area tend to be relatively
gradual.

GEOLOGY OF THE SELECTED AREAS

Two areas in the Aegean Sea were chosen for the purpose of this study.
The first area is offshore of an area of extensive Pb-Zn mineralization
at Cape Sounion on the Attic Peninsula (Fig.1). Offshore bedrock
mineralization in this area is likely because the trend of the onshore
deposits continues seawards. This area was chosen because dispersed
Pb-Zn-rich terrigenous material could create offshore geochemical anomalies and their presence could then be used to test the proposed method
of locating samples of anomalous composition. An area lying offshore
of the unmineralized island of Naxos in the Cyclades group was chosen
for control purposes (Fig.2). Both areas are similar with respect to
the number of samples available, evenness of grid spacing and areal
extent. A brief summary of the geochemistry and sedimentology of these
deposits is given below.

Sounion

The rocks comprising the Sounion peninsula are predominantly marbles,
phyllites and mica-schists of the Attic-Cyclades massif (Marinos and
Petrascheck, 1956). Hydrothermal mineralization, comprised largely of
a mixed sulphide assemblage of argentiferous-galena, sphalerite and
pyrite, trends generally north-south (Fig.1). The area has been mined
for Ag, Pb and Zn since ancient times.

The bathymetry and the sample distribution are shown in Fig.1. A

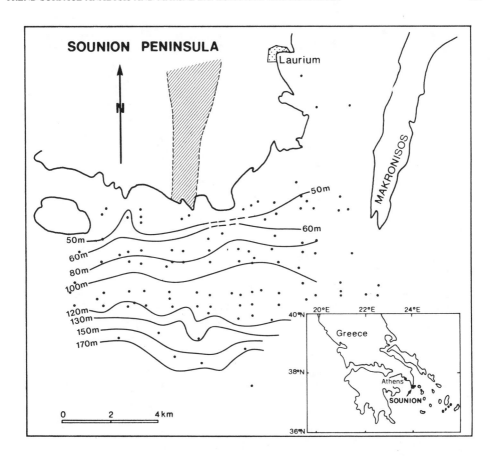

Figure 1. Map showing sample sites and bathymetry in the Sounion area.
The location of the sample area is shown in the inset by a
shaded rectangle. The shaded area in the main part of the
figure indicates the approximate limits of Pb-Zn mineralization.

gently sloping "inner-shelf" exists shorewards of the 55 m-isobath, and
a gently sloping terrace exists at about 100-120 m, which may represent
the former coastline during the lowered sea-level stand of the last
glaciation. The seafloor between 50 and 100 m and deeper than 120 m
is more steeply sloping and irregular in form than the two areas
discussed. Sediments in the eastern and northeastern parts of the area
are predominantly gravelly sands, giving way to sandy and muddy sediments to the west. Muddy sands are particularly prominent on the
terrace at 100-120 m water depth. Coarser sediment occurs on the
steeper slopes seawards and, in particular, landwards of this feature.

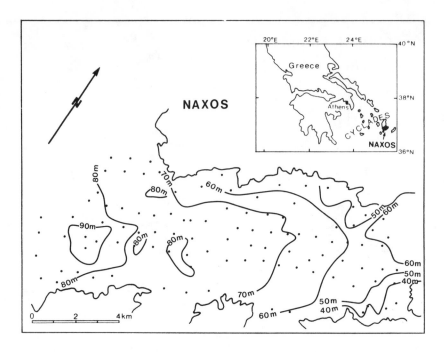

Figure 2.   Map showing sample location and bathymetry in the Naxos
            area. The location of the sample area is shown in the inset
            by a shaded rectangle.

Naxos

The island of Naxos is composed largely of marbles but in contrast to
the Sounion area, these are not known to be mineralized (Zachos and
Maratos, 1965). The bathymetry offshore is rather different to that at
Sounion (Fig.2). The seafloor is nowhere deeper than about 90 m, and
shoals very gently to the north-east, reaching a minimum depth of less
than 50 m. Much of the area sampled is a fairly flat sea floor, from
60 m to 80 m in depth. The sediments in this area are predominantly
muddy sands, with a significant gravel component in the north-east
sector. At both Sounion and Naxos, the sediments are carbonate-rich
(50%–80% $CaCO_3$, Smith, 1979).

ANALYTICAL METHODS

A total of 101 surface sediment samples from Sounion and 101 from Naxos
were collected at the sites shown in Figs. 1 and 2. The <2 mm size-
fraction of each was dried, crushed, dissolved by mixed hydrofluoric,
nitric and perchloric acids and analyzed by atomic absorption spectrophotometry for Ca, Mg, Al, Fe, Mn, Pb and Zn. Analytical precision,

defined on the basis of duplicate samples in terms of maximum percentage variation from the mean concentration at the 95% confidence level was $\pm$10% or better, except for samples low in Fe ($\pm$20%) and for samples very low in Pb and Zn ($\pm$30% and $\pm$20% respectively). Levels of Pb and Zn were found, on average, to be lower in Naxos sediments than in Sounion sediments, hence the analytical precision for these elements was, on average, slightly poorer for Naxos sediments than for Sounion sediments.

REGRESSION ANALYSIS

In order to best define the trends of background variation for the two elements of primary interest, Pb and Zn, the concentrations of those elements were first regressed on Ca, Mg, Fe, Al and Mn, using a linear model. From what is known of their geochemistry, it is these elements which will be most likely to influence variation in Pb and Zn. Levels

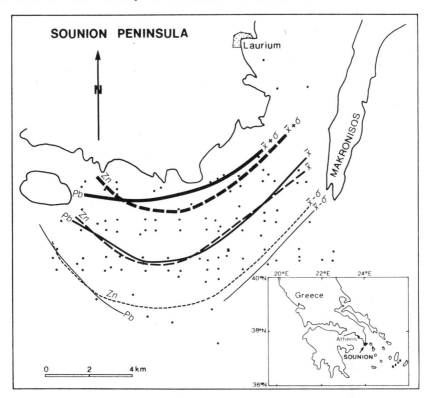

Figure 3. Trend surfaces for residual Pb and Zn values at Sounion. Contours for each surface are drawn at $\bar{x}+\sigma$, $\bar{x}$ and $\bar{x}-\sigma$, where $\bar{x}$ is the mean and σ the standard deviation. Solid lines are contours for Pb, dashed lines for Zn.

of Mn, for example, may provide a guide to the amount of Mn-oxide
present which, because of its marked scavenging properties, is likely
to enrich Pb and Zn in the sediments. Fe and Al provide a guide to
the amount of terrigenous material in the sediment, which could be
potentially Pb- and Zn-bearing, while Ca and Mg reflect the carbonate
fraction of the sediment.

    The residual Pb and Zn values from this stage of the analysis
(i.e. that proportion of Pb and Zn remaining unaccounted for by the
chemical model) were then regressed on a combination of the geographic
coordinates (x, y) and their cross-products (x^2, xy and y^2). The
prediction equation so obtained is a quadratic 'trend surface'. Strong
positive residuals from this surface should indicate anomalous concentrations of Pb and Zn remaining unaccounted for by both broad compositional and spatial trends in the sediments.

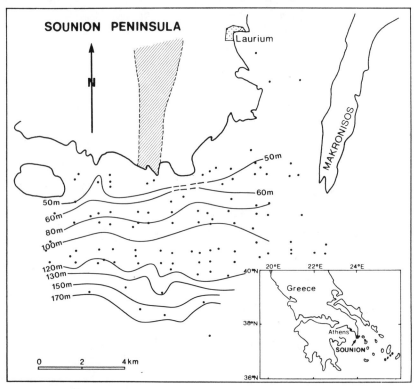

Figure 4. Trend surfaces for residual Pb and Zn values at Naxos.
           Contour intervals same as Fig. 3.

    The procedure used in practice was to write a 'macro' file of
commands for the MINITAB statistical package (Ryan, Joyner and Ryan,
1985) which could be used to repeatedly execute the following
operations sequentially for each data set:

1. Log-transformation of raw data for those elements with a high positive skew to their distributions.
2. Standardisation of all elements (to minimise the risk of truncation or round-off numerical errors during subsequent calculations).
3. Linear regression of the dependant elements (Pb and Zn) on the prediction set of 'major' elements (Ca, Mg, Al, Fe and Mn).
4. Obtain the residuals from 3.
5. Quadratic regression of the residuals (4) on the prediction set x, y, x^2, xy, y^2.
6. Obtain the residuals from 5 in standardized form.
7. Plot the quadratic trend surface (5) and residuals (6) on the basic geographic coordinates.
8. List the calculated fitted values and residuals from 3-6.

Care is required to ensure numerical accuracy by avoiding the risk of incurring truncation and round-off errors during numerical calculation (2, above). It is also necessary to recognise that the presence of extreme outliers in either the dependant variables, the prediction set, or both, may adversely bias the fitting process so that the true background relationships are not adequately represented by the regression equation. The high sensitivity of trend surface geometry to values of isolated sampling points at the margins of, or outside, areas of good spatial sampling control has long been recognised (Doveton and Parsley, 1970). Such 'influential' points can be readily identified using modern regression diagnostics such as leverage (Cook and Weisberg, 1982; Vinod and Ullah, 1981; Velleman and Welsh, 1981; Howarth, 1984), and attention is drawn to the presence of such samples in the standard MINITAB regression output.

For this reason, a preliminary statistical analysis of the data sets from both areas was carried out. One sample with abnormally high Mn was excluded from the Naxos data set and one sample with exceptionally high Fe was omitted from the Sounion data on the basis of cumulative probability plots. The three samples lying south-east of Laurium, almost certainly derived from a different provenance to that of the rest of the Sounion data, were also eliminated as they were shown to be highly influential data points. The chemical composition of these three Laurium samples also suggested the possibility that, unlike the rest of the Sounion data, they had been subject to urban or industrial contamination.

Results

The resultant quadratic trend surfaces for residual Pb and Zn in the Sounion sediments are shown in Fig. 3, contoured at the mean, and one standard deviation above and below the mean (on the basis of the MINITAB TPLOT command). As primary interest is focussed on the residuals from the regression, such a plot is perfectly adequate to assess the similarity of surface shape between the Pb and Zn background trends in the areas of good sampling control. It is apparent that at Sounion the two surfaces are almost coincident and slope away offshore.

This pattern reflects the dispersion away from the coast of Pb-Zn-rich
material derived from erosion of the many spoil heaps on land, and from
the mineralized rocks themselves(Smith, 1979). In contrast, the trend
surface for residual Pb at Naxos is almost exactly inverse to that of
Zn (Fig.4).

The main purpose in fitting the trend surfaces to these data sets,
however, is to be able to identify samples which are anomalous with
respect to their predicted "background" values. For the purposes of
this study, we define anomalous samples to be those in which the
positive standardized residuals from the quadratic trend surface fall
above the 80th percentile for the ordered residuals, i.e. fall into the
uppermost 20 percent of the data. The coincidence of such Pb and Zn
anomalies is 58% at Sounion, but only 15% at Naxos. Figures 5 and 6
show the spatial distribution of these anomalous samples.

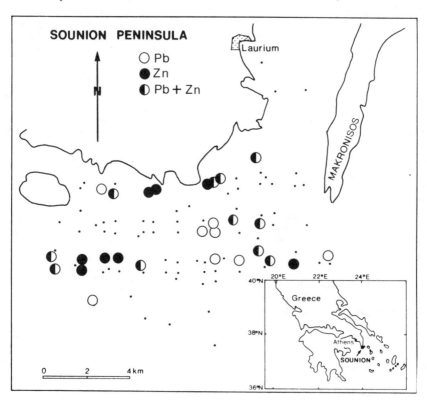

Figure 5.   Distribution of samples above the 80th percentile of the
            standardized residuals resulting from trend surface analysis
            of the Sounion data.

In order to check whether these anomalies could be artefacts of
of analytical variation in the data, a stochastic simulation technique
has been used. Assume that the maximum likely analytical error (E) at

a given analyte concentration (C) is $\pm$100 percent at the detection
limit, and approaches a minimum value asymptotically at high values
of C. The model may then be stated as $E = K_1 + K_2/C$. K_1 and K_2 may
be determined from experimental laboratory data, and hence E will be
defined for any value of C. Ten independent sets of simulated data
were then generated for both the Sounion and Naxos areas, by perturbing
each original concentration value by an amount proportional to the
original concentration to obtain $C_{new} = C + ER$, where R is drawn from
the standard normal distribution $N(0,1)$.

The effects of the data perturbations are shown by the variation
in shape of the quadratic trend surfaces for the ten sets of simulated
data for Pb and Zn at Sounion and Naxos in Figs. 7 and 8. It is evident
that the trend surfaces for both Pb and Zn at Sounion are very stable,
while those at Naxos are not, showing in some cases a complete reversal
of the original trend.

The behaviour of the "anomalous" samples after introduction of
these random errors is summarized in Fig. 9. This figure shows the
frequency with which samples, originally defined as being anomalous by
falling over the 80th and 90th percentiles of the ordered residual
values, remain anomalous in 1,2, ..., 10 of the simulation trials.

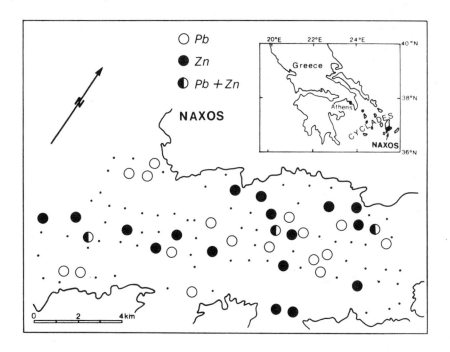

Figure 6. Distribution of samples above the 80th percentile of the
standardized residuals resulting from trend-surface analysis
of the Naxos data.

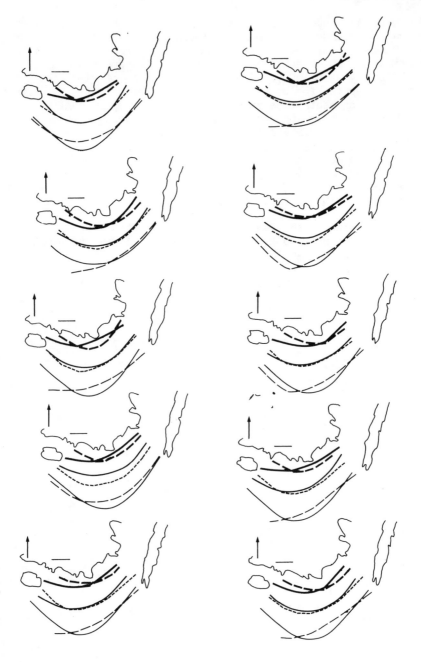

Figure 7.   Variations in trend surfaces for Pb and Zn at Sounion
            produced by introduction of random error in ten different
            trials.

Samples originally anomalous for both Pb and Zn (see Figs. 5 and 6)
show a very different behaviour in the two areas. At Sounion 8 of the
11 samples remained anomalous for both Pb and Zn in all 10 simulation
trials; and the remaining 3 samples were retained in 10 trials for one
element and 6 or more of the 10 trials for the other. At Naxos,
however, the three coincident Pb-Zn anomalies are far less stable, Pb
remaining anomalous in the three samples in 6, 6 and 5 of the 10 trials,
and Zn remaining anomalous in 9, 4 and 3 of the 10 trials respectively.

   The distribution of the stable anomalies at Sounion can readily
be explained. Those close to the coast are likely to include amounts
of Pb-Zn contaminated beach sands or similar Pb-Zn-rich material reaching these sites. The anomalies to the south-east may reflect dispersion
of Pb-Zn-rich material southwards from Laurium harbor since strong
southerly currents occur between the mainland and Makronisos Island,
and sediments close to Laurium harbor are known to contain up to
1000 ppm or more of Pb and Zn (Smith, 1979). Other anomalous samples
occur in an east-west belt across the southern part of the area. These
coincide with the terrace at 100-120 m and may represent reworked old
beach deposits enriched in Pb and Zn. There appears to be no evidence
of bedrock mineralization seaward of the land outcrops that would
cause local enrichment, by submarine weathering of outcropping bedrock.
It is possible, however, that the sampling grid was too coarse to
detect anomalies produced in this way, since they are likely to be of
very local extent. The pattern of anomalies at Naxos is puzzling but
on account of the general instability of the trend surfaces (Fig.8) and
lack of robustness of the residuals, these are not regarded as geologically significant.

DISCUSSION

One of the main aims of exploration geochemistry is recognition of
significant anomalies. To assist this aim, a statistical technique is
needed that is capable of revealing trends or anomalies other than those
obvious by simple manual examination of the data. By virtue of this
criterion, the method must be able to produce statistically reliable
trend surfaces explicable by known geological factors and, most importantly, it must be able to distinguish adequately between real and false
anomalies.

   The pattern of Pb and Zn distribution in surface sediments at
Sounion shows clearly the distribution and dispersion of Pb-Zn-rich
terrigenous material. In circumstances of this kind, the trend surfaces
for Pb and Zn, following removal of the influences of major element
variations, would be expected to show this trend even more distinctly
and should be highly resistant to the presence of random errors in the
data set. This seems to be the case. In contrast, in an area where
there is no known onshore mineralization, such as offshore of Naxos,
a spatially random pattern of trace element residuals would be expected
to occur, especially when the effects of major element control on trace
element variability are eliminated. Such a pattern would also be
expected to be less resistant to the introduction of random changes in

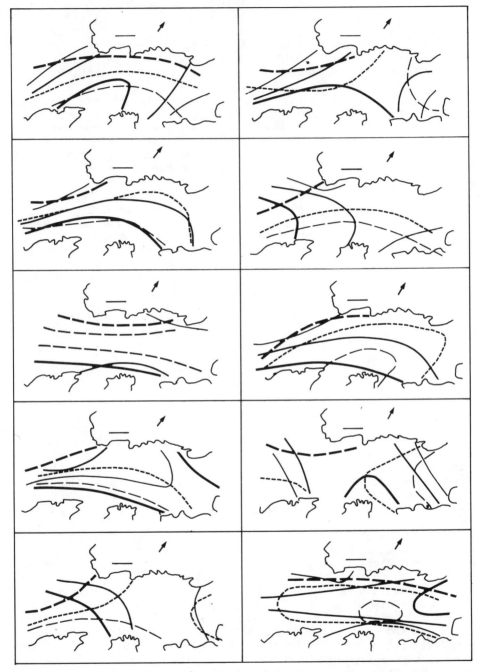

Figure 8.   Variation in trend surfaces for Pb and Zn at Naxos produced
by introduction of random error in ten different trials.

trace metal values. The large fluctuations observed in the simulated trend surfaces for Pb and Zn in Naxos sediments confirm this postulate. The original trend surfaces produced for Naxos (Fig.4) are, therefore, not reliable. The main geological conclusion which can be drawn from this is that, at Naxos, residual Pb and Zn do indeed behave independently of one another.

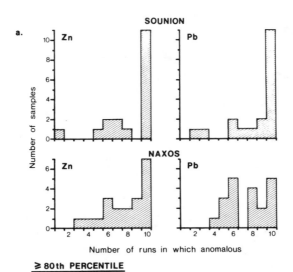

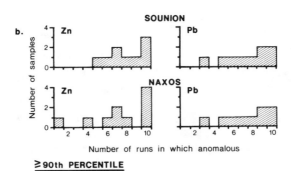

Figure 9. Reproducibility of Pb and Zn anomalies at Sounion and Naxos in terms of the number of times an anomalous sample in the original data set remains anomalous during 10 simulation trials in which the data values are randomly perturbed.

A greater variation in the behaviour of the trend surface, upon the introduction of random errors, need not in itself lead to a conclusion about the lower reliability of anomalies, insofar that one cause of this variation is the low value of the residuals to which the surface is fitted. This is shown quite clearly by the behaviour of the most highly anomalous samples in each area (those residuals in the top 10th percentile). These anomalies are just as stable, overall, at Naxos as at Sounion (Figs.9a and 9b). One may conclude, therefore, that although little reliance can be placed on the original trend surfaces at Naxos, the largest residuals from that surface could be reliable. However, important information is lost by considering only the largest anomalies. For example, no samples in either area retain coincident Pb and Zn anomalies above the top 90th percentile in all 10 trials. More information can be obtained by considering either the less highly anomalous

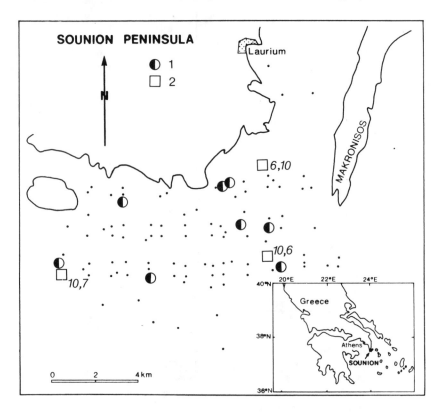

Figure 10. Regional distribution of Pb-Zn anomalies at Sounion after
          checking by simulation. 1 = Pb-Zn anomalies remaining above
          the 80th percentile in all 10 trials. 2 = Pb-Zn anomalies
          remaining in fewer than 10 trials, the exact number of occur-
          rences is indicated (the first figure is for Pb, the second
          for Zn).

samples or those which remain highly anomalous fewer than 10 times in the simulation trials. It can be seen from Figs. 9 and 10 that the first of these alternatives is preferable. By lowering the threshold for anomalies to include those values above the top 80th percentile, 8 out of 11 samples originally anomalous for Pb and Zn, remain so at Sounion (Fig.10). These anomalies must, therefore, be considered as real. There are no stable Pb-Zn anomalies at Naxos and even the single element anomalies show much less stability.

TABLE 1. Summary of the effects of reducing the Pb and Zn values of anomalous samples in the Sounion and Naxos data by their respective maximum calculated analytical error.

		percentile defining anomaly level	no. of anomalous samples in original data	no. of samples remaining anomalous following shift	% agreement
SOUNION	Pb	≥ 80	19	16	84
		≥ 90	9	6	67
	Zn	≥ 80	19	12	63
		≥ 90	9	9	100
NAXOS	Pb	≥ 80	20	9	45
		≥ 90	10	6	60
	Zn	≥ 80	20	9	45
		≥ 90	10	6	60

As levels of Pb and Zn are lower on the average in Naxos samples than in those from Sounion, the maximum likely analytical error will be correspondingly higher. The random addition of analytical error in the simulation trials will therefore lead to proportionately more variation in the data set for Naxos compared to that for Sounion. It could, therefore, be argued that for this reason alone, the trend surfaces at Naxos will be less stable than those at Sounion and anomalies on the original Naxos surfaces are unlikely to be as reliable. In order to test this, the MINITAB macro was used in a further simulation trial, this time lowering Pb and Zn values in all the originally anomalous samples by the maximum likely analytical error, instead of adding or subtracting a percentage of it at random. The results (Table 1) show that the most highly anomalous samples (above the 90th percentile) are relatively robust against such variations in the data even in the Naxos samples but those above the 80th percentile are less stable in the Naxos than the Sounion data. Hence the overall lower Pb and Zn values in the Naxos sediments are not likely to compromise the conclusions drawn above.

ACKNOWLEDGEMENTS

This work was carried out under a contract from the Commission for the European Communities which we acknowledge with gratitude.

REFERENCES

Beiersdorf, H., Kudrass, H-R. and von Stackelberg, U., 1980,   Placer deposits of ilmenite and zircon on the Zambesi shelf; Geol. Jahrb. v.36, pp.5-85.

Burger, H., Ehrismann, W and Stala, W., 1980, Aspects of the statistical analysis of marine ore deposits; Mineralium Deposita, v.15, pp.335-350.

Clarke, I., 1979, Practical geostatistics; Applied Science Publishers, London, 129pp.

Cook, R.D. and Weisberg, S., 1982, Residuals and influence in regression; Chapman and Hall, London, 230pp.

Doveton, J.H. and Parsley, A.J., 1970, Experimental distortions induced by inadequate data-point distributions; Trans. Inst. Min. Metall., London, Sect.B, v.79, pp.197-207.

Howarth, R.J., 1983, Mapping; In: Handbook of Exploration Geochemistry; Series Editor J.G. Govett; vol.2, 'Statistics and data analysis in geochemical prospecting', R.J. Howarth (ed.), pp.111-206, Elsevier, Amsterdam.

Howarth, R.J., 1984, Statistical applications in geochemical prospecting; A survey of recent developments; Jour. Geochem. Explor, v.21, pp.41-61.

Jones, H.A. and Davies, P.J., 1979, Preliminary studies of offshore placer deposits, eastern Australia; Marine Geology, v.30, pp.243-268.

Marinos, G. and Petrascheck, W.E., 1956, Laurium; Inst. for Geology and Subsurface Research, Athens, v.4,

Meyer, K., 1983, Titanium and zircon placer prospection off Pulmoddai, Sri Lanka; Marine Mining, v.4, pp.139-166.

Owen, R.M., 1978, Geochemistry of Pt-enriched sediments : applications to mineral exploration; Marine Mining, v.1, pp.259-282.

Owen, R.M., 1980, Quantitative geochemical models of sediment dispersal patterns in mineralised nearshore environments; Marine Mining, v.2, pp.231-249.

Ryan, T.A., Joiner, B.L. and Ryan, B.F., 1985, The MINITAB Student Handbook; Duxbury Press, North Scituate, Mass., 2nd edn., 500p.

Siddique, H.N., Rajamanickam, G.V. and Almeida, F., 1979, Offshore ilmenite placers of Ratnagiri, Konkan Coast, Maharashtra, India; Marine Mining, v.2, pp.91-110.

Smith, P.A., 1979, Geochemical investigations of recent sediments from the Aegean Sea; Unpubl. Ph.D. thesis, Univ. London, 320pp.

Tixeront, M., Le Lann, F., Horn, R. and Scolari, G., 1978, Ilmenite prospection of the Continental shelf of Senegal : methods and results; Marine Mining, v.1, pp.171-188.

Velleman, P.F. and Welsch, R.E., 1981, Efficient computing of regression diagnostics; Amer. Statist. v.34, pp.234-242.

Vinod, H.D. and Ullah, A., 1981, Recent advances in regression methods; Marcel Dekker, New York, 361pp.

von Stackelberg, U. (ed), 1982, Heavy mineral exploration of the East Australian shelf, SONNE cruise SO-15, 1980; Geol. Jahrb., v.56, pp.1-215.

Yim, W. W-S., 1979, Geochemical exploration for tin placers in St. Ives Bay, Cornwall; Marine Mining, v.2, pp.59-78.

Zachos, K. and Maratos, G., 1965, Carte métallogenique de la Grèce; Inst. for Geology and Subsurface Research, Athens, 1 sheet, (scale 1:1 million).

LIST OF PARTICIPANTS AT THE
NATO ADVANCED RESEARCH WORKSHOP ON
MARINE MINERALS: RESOURCE ASSESSMENT STRATEGIES

F.P. Agterberg
Geological Survey of Canada
601 Booth Street
Ottawa, Ontario K1A OE8
Canada

Alan A. Archer
4 Brook Rise
Chigwed, Essex IG7 6AP
United Kingdom

Dennis A. Ardus
Marine Earth Sciences Research
    Programme
British Geological Survey
Murchison House
West Mains Road
Edinburgh EH9 3LA
United Kingdom

Harald Bäcker
PREUSSAG AG
Marine Technology Department
Arndstrasse 1
D-3000 Hannover 3,
Federal Republic of Germany

Jean Paul Barusseau
Laboratoire de Recherches
    de Sédimentologie Marine
Université de Perpignan
Avenue de Villeneuve
66025 Perpignan
France

William C. Burnett
Department of Oceanography
Florida State University
Tallahassee, Florida   32306
USA

David S. Cronan
Marine Mineral Resources
    Programme
Applied Geochemistry
    Research Group
Royal School of Mines
Imperial College
Prince Consort Road
London SW7 2BP
United Kingdom

John R. Delaney
School of Oceanography
University of Washington
Seattle, Washington 98195
USA

John H. DeYoung
Office of Mineral Resources
U.S. Geological Survey
920 National Center
Reston, Virginia   22092
USA

Peter Diehl
PREUSSAG AG
Arndstrasse 1
D-3000 Hannover 3
Federal Republic of Germany

M.R. Dobson
Department of Geology
University of Wales
Aberystwyth, Dyfed SY23 3DB
United Kingdom

David B. Duane
National Sea Grant College
   Program
National Oceanic and
   Atmospheric Administration
6010 Executive Boulevard
Rockville, Maryland 20852
USA

John M. Edmond
Department of Earth,
   Atmospheric, and Planetary
      Sciences
Massachusetts Institute of
   Technology
Cambridge, Massachusetts 02139
USA

Mustafa Ergün
Institute of Marine Sciences
   and Technology
Dokuz Eylül University
P.K. 478 Konak
Izmir, Turkey

T.J.G. Francis
Institute of Oceanographic
   Sciences
Wormley,
Godalming, Surrey GU8 5UB
United Kingdom

DeVerle P. Harris
Department of Mining and
   Geological Engineering
University of Arizona
Tucson, Arizona 85721
USA

Benjamin W. Haynes
U.S. Bureau of Mines
Avondale Research Center
4900 LaSalle Road
Avondale, Maryland 20782
USA

J.J.H.C. Houbolt
NATO Scientific Affairs
   Division
B-1110 Bruxelles
Belgium

Randolph A. Koski
Branch of Pacific Marine
   Geology
U.S. Geological Survey MS 999
Menlo Park, California 94025
USA

H.R. Kudrass
Bundesanstalt für Geowissen-
   schaften und Rohstoffe
Postfach 51 01 53
D-3000 Hannover 51
Federal Republic of Germany

Helmar Kunzendorf
Risø National Laboratory
P.O. Box 49
DK-4000 Roskilde
Denmark

Augusto Mangini
Institut für Umweltphysik
Universität Heidelberg
Im Neuheimer Feld 366
D-69 Heidelberg
Federal Republic of Germany

Jose H. Monteiro
Servicos Geologicos de Portugal
R. Academia das Ciencias, 19-2
1294 Lisbon
Portugal

Angela Frisa Morandini
Consiglio Nazionale delle
   Ricerche, Centro di Studio
   per i Problemi Minerari
Politecnico di Torino
Corso Duca delgi Abruzzi, 24
10129 Turin
Italy

S.A. Moorby
Applied Geochemistry Research
   Group, Department of Geology
Royal School of Mines
Imperial College
Prince Consort Road
London, SW7 2BP
United Kingdom

J. Robert Moore
Department of Marine Studies
University of Texas
P.O. Box 7999 University
    Station
Austin, Texas   78712
USA

Elisabeth Oudin
Direction de la Technologie
Départment Minéralogie-
    Géochimie-Analyses
Bureau de Recherches
    Géologiques et Minières
B.P. 6009
45060 Orleans
France

Robert M. Owen
Department of Atmospheric
    and Oceanic Science
University of Michigan
Ann Arbor, Michigan   48109
USA

Constantine Perissoratis
Institute of Geology and
    Mineral Exploration
70 Messoghion St.
Athens 608
Greece

Ulrich von Rad
Bundesanstalt für Geowissen-
    schaften und Rohstoffe
Postfach 50 01 53
D-3100 Hannover 51
Federal Republic of Germany

Stanley R. Riggs
Department of Geology
East Carolina University
Greenville, North Carolina
    27834   USA

Steven D. Scott
Department of Geology
University of Toronto
Toronto, Ontario M5S 1S1
Canada

William D. Siapno
Deepsea Ventures, Inc.
Gloucester Point, Virginia
    23062   USA

Richard Sinding-Larsen
Geological Survey of Norway
Leiv Eirikssons vei 39
P.O. Box 3006
N-7001 Trondheim
Norway

F.N. Spiess
Marine Physical Laboratory,
Scripps Institution of
    Oceanography, University of
    California at San Diego
La Jolla, California   92093
USA

Ulrich von Stackelberg
Bundesanstalt für Geowissen-
    schaften und Rohstoffe,
Postfach 51 01 53
D-3000 Hannover 51
Federal Republic of Germany

Paul G. Teleki
Office of Energy and Marine
    Geology,
U.S. Geological Survey
915 National Center
Reston, Virginia   22092
USA

R. Whittington
Department of Geology
University of Wales
Aberystwyth, Dyfed SY23 30B
United Kingdom